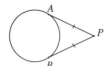 If $\overline{PA}$ and $\overline{PB}$ are tangents to a circle at A and B, then $PA = PB$.

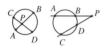 If two chords (or secants) $\overline{AB}$ and $\overline{CD}$ of a circle meet at P, then $AP \cdot BP = CP \cdot DP$.

$m\angle AOB = r$
(O center)

$m\angle ABC = {}^1/_2 r$

$m\angle APB = {}^1/_2(r + s)$

$m\angle APB = {}^1/_2(r - s)$

HIGH SCHOOL TRIGONOMETRY

The Six Trigonometric Ratios

Pythagorean Relation

$a^2 + b^2 = c^2$

$\sin A = \text{opposite/hypotenuse} = a/c$ $\csc A = c/a$
$\cos A = \text{adjacent/hypotenuse} = b/c$ $\sec A = c/b$
$\tan A = \text{opposite/adjacent} = a/b$ $\cot A = b/a$

Basic Identities ($A, B \leq 180$)

$\sin^2 A + \cos^2 A = 1$

$1 + \tan^2 A = \sec^2 A$

$\sin (A + B) = \sin A \cos B + \cos A \sin B$

$\cos (A + B) = \cos A \cos B - \sin A \sin B$

$\tan (A + B) = (\tan A + \tan B)/(1 - \tan A \tan B)$

$\sin 2A = 2 \sin$ $\cos A$

$\cos 2A = \cos^2 A - \sin^2 A$ $\cos {}^1/_2 A = {}^1/_2\sqrt{1 + \cos A}$

$\qquad = 1 - 2 \sin^2 A$

$\qquad = 2 \cos^2 A - 1$ $\tan {}^1/_2 A = \sqrt{\dfrac{1 - \cos A}{1 + \cos A}}$

$\tan 2A = (2 \tan A)/(1 - \tan^2 A)$

Miscellaneous Formulas

AREA $\triangle ABC = {}^1/_2 bh$

AREA $\triangle ABC = \sqrt{s(s - a)(s - b)(s - c)}$

where $s = {}^1/_2(a + b + c)$

AREA (PARALLELOGRAM) $ABCD = bh$

AREA (CIRCLE): $K = \pi r^2$

CIRCUMFERENCE (CIRCLE): $C = 2\pi r$

(θ in radians)

AREA (SECTOR): $K = r^2\theta$ ARC LENGTH: $s = r\theta$

STANDARD CONSTRUCTIONS IN GEOMETRY

MIDPOINT
OF SEGMENT
($AP = BP = AQ = BQ$)

PERPENDICULAR
BISECTOR
OF SEGMENT

BISECTOR
OF ANGLE

PERPENDICULAR
TO LINE AT GIVEN
INTERNAL POINT
($AP = AQ, PR = QR > AP$)

COLLEGE GEOMETRY

A DISCOVERY APPROACH

COLLEGE GEOMETRY

A DISCOVERY APPROACH

DAVID C. KAY

University of North Carolina at Asheville

HarperCollins*CollegePublishers*

Sponsoring Editor: George Duda
Design Administrator: Jess Schaal
Text Design and Production: Monotype Composition Company
Cover Design: Roseanne Lufrano/Andrea Eisenman
Cover Illustration: Andrea Eisenman
Picture Research: Judy Ladendorf
Production Administrator: Randee Wire
Compositor: Monotype Composition Company
Printer and Binder: R.R. Donnelley & Sons Company
Cover Printer: R.R. Donnelley & Sons Company

For permission to use copyrighted material, grateful acknowledgment is made to the following copyright holders: p. 21 Milt and Joan Mann/Cameraman Intl.; p. 74 From *Survey of Geometry*, Volume 1 by Howard Eves, p. 390. Reprinted by permission of the author; p. 98 Courtesy of the University Library, Gottingen, Germany; p. 133 Chicago Tribune Company, all rights reserved, used with permission. Photo: Charles Osgood; p. 141 Courtesy of the University Library, Gottingen, Germany; p. 153 Bohdan Hrynewych/Stock Boston; p. 188 Archiv fur Kunst und Geschichte; p. 214 Steve Monti/Bruce Coleman, Inc.; p. 221 Michael George/Bruce Coleman, Inc.; p. 245 Adam Woolfitt/Woolfin Camp & Associates; p. 362 Courtesy of the University Library, Gottingen, Germany

pp. 121, 151, 181, 223, 228, 236, 237, 322, 323, 330, 331: From *The University of Chicago School Mathematics Project*. Reprinted by permission of ScottForesman.

Library of Congress Cataloging-in-Publication Data

Kay, David C.,
     College geometry: a discovery approach / David C. Kay.
          p.   cm.
     Includes index.
     ISBN 0–06–500006–4
     1. Geometry.   I. Title.
     QA445.K383   1993
     516--dc20

                                                                    93–32085
                                                                    CIP

95 96 9 8 7 6 5 4

Dedicated to my grandchildren
(may their generation preserve our geometric heritage)

Andrew	*Josh*	*Justin*
Kristin	*Melissa*	*Michael*
Ryan	*Stephen*	*Tyler*

INTERDEPENDENCE OF TOPICS

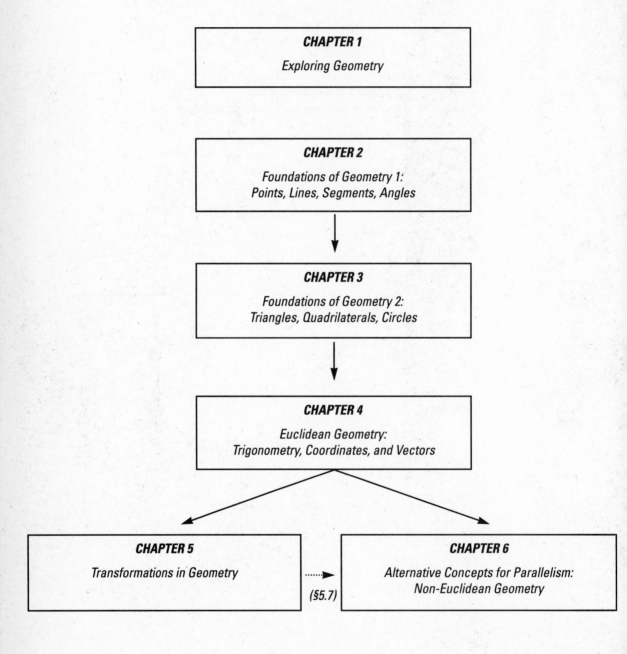

CHAPTER 1

Exploring Geometry

CHAPTER 2

Foundations of Geometry 1:
Points, Lines, Segments, Angles

CHAPTER 3

Foundations of Geometry 2:
Triangles, Quadrilaterals, Circles

CHAPTER 4

Euclidean Geometry:
Trigonometry, Coordinates, and Vectors

CHAPTER 5

Transformations in Geometry

(§5.7)

CHAPTER 6

Alternative Concepts for Parallelism:
Non-Euclidean Geometry

CONTENTS

APPENDIXES

SPECIAL TOPICS

An Introduction to Higher Dimensional Geometry
An Introduction to Projective Geometry
An Introduction to Convexity Theory

Any of these topics will be available upon request to adopters of *College Geometry* by contacting your local HarperCollins College Publishers representative.

*Optional

*P*REFACE

This book was written for an introductory, college level course in geometry for mathematics majors or students in mathematics education seeking teacher certification in secondary school mathematics. The latter purpose is fully recognized throughout the text, with the development following traditional lines, and numerous problems and examples coming from current secondary school textbooks.

GUIDING PHILOSOPHY

The following three underlying principles directed the design of this book:

1. The many aspects of geometry should be as visual as possible, and students should be inspired to develop a questioning, curious nature about geometry.
2. Students should be encouraged to discover many principles and phenomena of geometry for themselves.
3. The gentle art of proof-writing, as a *tool* for answering questions in geometry and discovering what is actually true, should be effectively conveyed to students.

In short, the focus of this book is not only to provide an introduction to geometry, but to help the instructor achieve a high level of student participation in and outside the classroom. The writing style is itself intended to maintain student interest, and the presentation of geometry is designed to promote student participation, interest, satisfaction, and fulfillment. The main objective in this whole endeavor is to help students master the subject of geometry at this introductory level.

DISTINGUISHING FEATURES

The book has special units for self-discovery, frequent examples for proof-writing involving the concepts being presented, and graded problem sets for each section.

MOMENTS FOR DISCOVERY. Appearing in most sections of the text, these units are specifically directed towards student participation and are designed for the average student. Many of them are purely experimental in nature, directing the student to conduct a drawing experiment or investigate an example, always involving a surprising revelation or significant phenomenon of geometry. Others involve abstract reasoning, guiding the student in a step-by-step procedure to make warranted conclusions regarding certain geometric phenomena. These units can be used in a variety of ways by the instructor: As a device to arouse likely classroom discussion, as team projects, or just for outside reading assignments on which to report.

TOPICAL AND GRADED PROBLEM SETS.   The problem sets have also been designed to foster self-discovery, and at the same time to allow every student to have successful problem-solving experiences. The wide variety and large quantity of problems also serves to stimulate greater interest and participation. Occasionally, problems which correspond to major named theorems in geometry, or problems using a different approach than that used in the text, appear. These are presented with a topical heading to catch the reader's attention. Longer problems thought to be appropriate for undergraduate research, either partly or totally unsolved, or suitable for an expository treatment, are so labelled.

To make it easier for the instructor, each problem set is graded according to difficulty.

Group A: Largely of a computational nature, or involve short proofs.

Group B: Involve a higher level of sophistication, requiring formal multiple-step proof-writing and some original thinking.

Group C: Require yet a higher level of original thinking, and includes longer projects and topics for undergraduate research.

Short answers and problem solutions for most of the odd-numbered problems appear in Appendix B.

VISUAL IMPACT.   The text is heavily illustrated, with over three times the ordinary number of figures. In presenting a concept, often several stages of a line of development pursued in the text appear in the illustrations. Frequently, the steps of a proof are illustrated separately, and each major case of a proof always appears with a separate figure.

GEOMETRIC CONTENT.   Although most texts at this level concentrate on either axiomatics (at the exclusion of transformation theory and modern applications), or transformation theory (at the exclusion of axiomatics), this text contains a careful axiomatic treatment beginning with Chapter 2 and continuing through Chapter 3, then branches out into coordinates and vectors in Euclidean geometry (Chapter 4), transformations and the group concept (Chapter 5), and finally, returning again to the axiomatic foundation, where an introduction to Lobachevskian geometry and the Poincaré model appears (Chapter 6).

COMPUTER FRIENDLY.   Although independent of computer software, the text is compatible with a parallel use of the latest in computer technology to make the subject more lively in the minds of students. Occasional comments about the usefulness of computers and the correct context for this technology occur. There are a few exercises in the problem sections explicitly designed for investigation on the computer.

CROSS-REFERENCING.   For the further convenience of the instructor (and student), the cross-referencing of problems and text material, backward and forward, appears. If a problem is used in any significant way in some future development, it is starred, with the forward reference stated, and problems involving a previous concept or problem include the appropriate information and location in the text.

PEDAGOGICAL FEATURES

This text was written primarily for the student, although it contains much material which the average instructor should find interesting, and even exciting when it is used to engage the minds of students.

HISTORICAL NOTES.   The book includes numerous biographical anecdotes to show students the human side of mathematics. Rather than present a complete chronological account of events as would be found in a more formal treatment of the history of mathematics, special or unusual features concerning the lives of geometers and their contributions were singled out.

OUR GEOMETRIC WORLD.   Excerpts that relate geometric concepts to the physical world in which we live are liberally sprinkled throughout.

CHAPTER OVERVIEWS AND SUMMARIES.   Each chapter contains an overview (where we intend to go) and a summary (where we have been). The overviews briefly state what we intend to study, and why, often with helpful reminders of what one should look out for or to avoid. The summaries at the end of each chapter discuss the major results covered in that chapter in broad outline, but detailed enough to provide pertinent information. At the very end, there is a short True-False test for quick, student self-evaluation on general mastery of the concepts.

LOCATION OF PREVIOUS RESULTS.   The problem of quickly locating previous results has been eliminated by placing the statements of all major lemmas, theorems and corollaries, by section number, in Appendix C. To find the idea or theorem you are looking for, a quick glance is often all it takes.

CLASS TESTED

Classes which field-tested a preliminary version of the manuscript include the students of James Spencer, of The University of South Carolina at Spartanburg, the students of David Dezern, of The University of North Carolina at Asheville, and the students of James Tattersall, Providence College. Comments from students were carefully considered in making revisions or deletions of material found confusing or ambiguous. Because of this feedback, the book has been greatly improved.

PREREQUISITES

It is not necessary for a student to have a fresh memory of the content of high school geometry in order to succeed. (A cursory review appears in the inside front cover for the convenience of students.) However, it is recommended that at least one year of calculus be required of students entering a course using this text. This will guarantee a fairly recent encounter with trigonometry which is assumed here to be common ground for all students. Informal references are made to limits and continuity (mostly confined to Group C problems). We do not assume an ability to use abstract reasoning, but sooner or later students will be expected to make some progress in this area, and a foundations course could be considered as another prerequisite in colleges where this is a viable option.

COURSE FLEXIBILITY

The text contains 29 sections, each designed for one or two 50-minute classroom sessions, in addition to 15 optional sections (starred) which the instructor can use for enrichment. After Chapters 1–4 are covered, Chapter 5, Chapter 6, or combination of topics from both can be covered independently (one brief reference involving the Poincaré model in Chapter 6 which uses a previous discussion in Chapter 5 is the only exception). It is recommended that students at least read Chapter 1 on their own; if class time is devoted to this chapter, then it should be kept to a minimum. The intention was to have Chapter 1 whet the student's appetite for geometry before a more serious axiomatic treatment takes place, but one should not devote a major part of the semester or term to do so. Some topics introduced in Chapter 1 are dealt with later, at a more appropriate stage of development.

AXIOMATIC GEOMETRY.   A course emphasizing axiomatics might include Chapters 1-4 and Chapter 6 (24 sections, not counting optional sections).

MODERN GEOMETRY.   A course stressing primarily the techniques of geometry could be based on Chapter 1–5, with emphasis on Chapter 5 and concentrating on solving the problems occurring there (25 sections). A shortened version of Chapter 2 may be feasible here, consisting of Sections 2.5, 2.7, and 2.8. Doing this would allow the addition of several optional sections.

SURVEY COURSE.   A course intending to give students a broader view of geometry at this level can be created using Chapter 1, the short version of Chapter 2 (Sections 2.5, 2.7, 2.8), Chapter 3 (omitting sections 3.3, 3.8), Chapter 4 (omitting Sections 4.1, 4.2, 4.3), Chapter 5, Chapter 6, and content from one of the available topics (mentioned below).

SPECIAL TOPICS

Since opinions vary about the specific topics which should be included in an introductory course in geometry, we make available to anyone adopting this text one or more of the following additional topics:

- *An Introduction to Higher Dimensional Geometry*

- *An Introduction to Projective Geometry*

- *An Introduction to Convexity Theory*

Each of these is about the same length as the previous chapters, and includes the same features (units for self-discovery, historical notes, graded problem sets). Solutions to selected problems for each topic appear at the end of that topic. These may be obtained by contacting your local HarperCollins College Publishers representative or by writing to the College Mathematics Division, 1900 East Lake

Avenue, Glenview, IL, 60025

SUPPLEMENTS

Complete worked-out solutions to all discovery units and all problems in the text (except undergraduate research projects) is available in the *Instructor's Solutions and Resource Manual.* A brief running commentary of portions of the text, answering all open-ended questions raised, is included. A software package, *GeoExplorer* (available for IBM and Macintosh), enables students to draw, measure, modify, and transform geometric shapes on the screen. Both supplements are available to adopters only by contacting your local representative or by writing to the address previously mentioned.

ACKNOWLEDGMENTS

I would like to acknowledge those who helped make this project possible, and express my deep appreciation.

REVIEWERS.   Those who reviewed the manuscript at various stages of writing were: L. Ray Carry, University of Texas, Austin; Kenneth Evans, Southwest Texas State University; William Fitzgerald, Michigan State University; Karen J. Graham, University of New Hampshire; Paul Green, University of Maryland; Mark Hughes, Florida State University; Jerry Johnson, Western Washington University; Mark Kon, Boston University; R. Padmanabhan, University of Manitoba.

EDITORIAL STAFF.   The author deeply appreciates the expert editorial staff at Harper Collins—true professionals, who are responsible for whatever success this venture may have. In particular, I would like to single out one of those professionals, who provided just the right amount of encouragement when most needed, who was my strongest advocate for the project throughout its development: My "personal editor" and good friend, George Duda. I also appreciate the fine work done by other individuals on this project: Shirley Cicero and Carol Zombo at HarperCollins, and John Budz, Julie Klavens, and Margaret Tatro at Monotype Composition.

We welcome comments and criticism from the readers and users of this book. Feedback from varying classroom situations would be most important for making improvements that could lead to a more effective text in the future.

David C. Kay
University of North Carolina at Asheville

To the Student

You may be wondering what to expect from a college level geometry course. Indeed, you might be thinking "If I had a hard time with high school geometry, how will I be able to learn something of value here?" Or you might wonder if a college course in geometry is totally different from high school geometry, with a far-ranging subject matter. These are natural questions for you to ask, just as you are about to embark on a new journey. We are going to try to give you some honest answers here.

Many of you are obliged to take a college course in geometry because it is required for teacher certification. If you are in that group, this course is designed especially for you. This course will prepare you for the time when you will face your own class in geometry. However, our aim is not to show you explicitly how to teach geometry, but rather to emphasize the concepts and inherent beauty of geometry so you will be inclined to teach geometry as an exciting subject yourself—one which your own students will be able to enjoy, just as you have, rather than as a subject to endure.

We also believe in the old saw "teachers should know more than they teach." Mathematics teachers should always be in a position of strength, able to field any unpredictable question which a student may ask. An unexpected question from a student can often catch a teacher off guard, and the better prepared you are as a teacher, the better you will be able to respond to those students in a helpful and effective manner.

You may be taking this course as a major elective and not as a prospective teacher. This book is also written for you, because the presentation is introductory in nature, and the variety of subjects covered, particularly in the problems, will show you some very unusual aspects of geometry. We think you too will come to enjoy the subject of geometry, and may even get excited at times when you find that you can really "do geometry".

The MOMENTS FOR DISCOVERY are designed for everybody. It is strongly recommended that you work through each of these units as you come to them in your reading of the text. In each unit, you will be carefully led through a logical sequence of steps, sometimes consisting of just a few simple calculations, drawing experiments, or elementary logic, based on what you already know. When you get to the end of the unit the chances are very good that you will have "discovered" a really neat concept of geometry on your own. As you grow more mature in the subject you will gradually acquire the ability to generalize and discover geometric properties on your own, or at least make good guesses about what may be true. These units provide working models to help you make progress towards that goal. As you can see, this course is really very creative, and we are going to encourage your best creative thinking.

At the end of each section a problem section has been carefully crafted for you. These are divided into three groups. Group A contains those problems everyone can do. Well, *almost* everyone—some of you may need a little help from your friends! By and large, these problems are of the pocket calculator, crank-and-grind variety, or one-step proofs mainly involving definitions, and generally do not require much thinking. They are just to test you basic understanding of the material in the section just preceding. Group B consists of a little higher level of abstraction, requiring the writing of short, multiple-step proofs. Some of you may need help from your instructor on these. Then there is Group C, consisting of problems requiring a great deal more thought and involvement on your part. Occasionally, ideas for undergraduate research projects will appear. Here you are on your own. However, even these problems will often include helpful hints and suggestions to help you get started.

Finally, if you wish to test yourself on your basic understanding, a short True-False test appears at the end of each chapter, beginning with Chapter 2. Answers to these appear with the problem solutions in Appendix B.

In summary, we have attempted to present a good introduction to geometry within the bounds of a one-semester, undergraduate course, which not only provides a body of knowledge in mathematics that has a host of practical applications, but reveals a fascinating area of active, on-going research in which it is possible for almost anyone to participate. We hope that you will derive great benefit, personal satisfaction and accomplishment in this very important area of mathematics.

David C. Kay

1

EXPLORING GEOMETRY

OVERVIEW

Geometry may be defined in broad terms as the systematic study of the physical world we live in—the measurement and study of the shapes of common everyday objects, the analysis of artistic patterns and designs, the calculation of distances to far galaxies, or the study of our own solar system and charting the blue sphere on which we live. Taken literally, "geometry" (= "geo-metric") means *to measure the earth.* The earliest studies in geometry were devoted to just such endeavors. As our world advanced, our viewpoint of geometry changed, and the subject now encompasses a great many widely differing branches of mathematics.

One way to begin a study of geometry is to examine one of its elegant theorems. We have chosen an absolute gem—a rare and beautiful theorem of Jacob Steiner discovered in the 1800s, one which involves a variety of interesting, seemingly unrelated topics. Steiner's theorem is to mathematics as a famous poem or essay would be to literature.Our goal is an understanding and appreciation of Steiner's theorem, but these topics and a few other interesting relationships in geometry will be examined on our journey.

We will arrange some of the material so that you can actually make a few geometric discoveries for yourself. You will need to make use of mostly basic trigonometric relations or the geometry of a right triangle. Similar triangles is a staple here.

1.1   DISCOVERY IN GEOMETRY

Many elementary properties in geometry can be discovered if presented in a discovery mode. That is, instead of passively reading about it, one can often discover a geometric property by means of a figure or from a drawing experiment, if led to do so. In this manner, it is possible to participate in what can be a rather exciting venture, the goal of which is a personal observation, or "discovery," of geometric relationships.

Let's look at a few examples.

EXAMPLE 1   The medians of a triangle are concurrent. (The point of concurrency is called the **centroid.**)

(a) Suppose you take two or three triangles at random (as illustrated in Figure 1.1) and carefully construct the three medians (lines joining the vertices to the midpoints of the opposite sides). Subject to experimental error, in each case you can observe that the medians pass through a common point.

Figure 1.1

(b) Another method, also experimental, is based on the physical attributes of the center of mass. Since the medians of a triangle divide it into triangles having equal areas, a cardboard model of a triangle will perfectly balance on a knife edge along any one of its medians (Figure 1.2). If we consider two medians, their intersection is a point of equilibrium for the entire

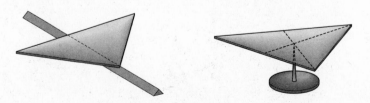

Figure 1.2

triangle—the center of mass. Since this is true of any two medians, and there is only one center of mass, then *all three* medians pass through this center of mass, which is a common point of concurrency for the medians.❑

E X A M P L E   2   The square of the hypotenuse of a right triangle equals the sum of the squares of the two legs.

Start with any right triangle (as shown), and construct three other right triangles congruent to the given one. Next, *by experiment*, you might discover that in moving the triangles about, they may be made to form a square (as shown in Figure 1.3). By calculating the area of the resulting square in two different ways, we might deduce the following:

I.   AREA OF SQUARE FROM STANDARD FORMULA: $c \cdot c = c^2$

II.   AREA OF THE PARTS OF THE SQUARE:

   (a) Small square inside $= (a - b)^2 = a^2 - 2ab + b^2$

   (b) Area of the four triangles $= 4 \cdot \frac{1}{2}ab = 2ab$

   TOTAL: $a^2 - 2ab + b^2 + 2ab = a^2 + b^2$

III. $\therefore c^2 = a^2 + b^2.$❑

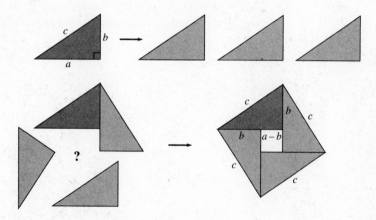

Figure 1.3

E X A M P L E   3   The sum of the measures of the angles of any triangle equals 180.

Start with any $\triangle ABC$, using a paper model. Tear off a portion of the triangle that includes vertex A. Repeat this for vertices B and C (as shown in Figure 1.4). Reassemble the torn pieces so that A, B, and C coincide and the pieces are flush. Take a ruler or straight-edge and place alongside the configuration to check the desired result.❑

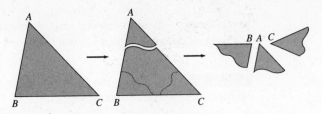

Figure 1.4

EXAMPLE   4   A regular octagon and interior may be dissected into five pieces and reassembled to make a square and interior.

Start with a regular octagon *ABCDEFGH*, each side of length 2 units. Consider the circle of diameter 2 located at the center of the octagon (as shown in Figure 1.5). From the midpoints *L*, *M*, *N*, and *K*, draw the tangents to the circle. Since the interior angles of the quadrilateral thereby formed (circumscribing the circle) are clearly congruent by symmetry, and their sum is 360°, each of these are of measure 90°. Thus a square is formed at the center, and the tangents define a dissection of the octagon and interior into four congruent pentagons (and interiors) and a square of side 2. By working with a paper model, see if you can reassemble the pieces to make a perfect square. If you succeed, verify that a perfect fit really is achieved by your solution, making an analysis of the angles.❏

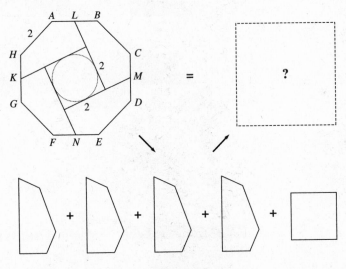

Figure 1.5

Now let's explore two examples involving concepts which you have very likely never seen before. Just for fun, we want *you* to discover the proper conclusion in each case. We therefore set them off as special displays in the first of many discovery units to be found throughout this book.

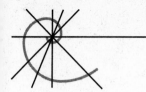

MOMENT FOR DISCOVERY

The Morley Triangle

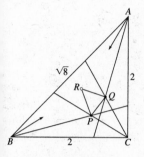

Figure 1.6

1. On a sheet of paper, draw a 45° (isosceles) right $\triangle ABC$ with side lengths proportional to 2, 2, and $\sqrt{8}$, as shown in Figure 1.6. In order to achieve a high degree of accuracy, allow the triangle to fill up most of the page.

2. Construct the **angle trisectors**. Accuracy is important, so use either a protractor or compass/straight-edge construction for a 15° and 30° angle.

3. Suppose the angle trisectors intersect in pairs at the points P, Q, and R. Draw the segments joining those points. By symmetry in the figure, it is clear that $\triangle PQR$ is isosceles ($PR = QR$). Do you observe this in your figure?

4. Carefully measure the segments PQ, QR, and RP with a marked ruler. What do you observe?

5. What conjecture seems reasonable to you?

If you concluded in the preceding discovery unit that $PQ = QR = RP$, or that $\triangle PQR$ is equilateral, your discovery was indeed correct. In fact, this relationship holds true for *any triangle*. This discovery was first made by an English mathematician, Frank Morley (1860–1937), the father of the novelist Christopher Morley, and the equilateral triangle is for this reason known as the **Morley Triangle** of the given triangle. The existence of the Morley Triangle was the result of a more general theory, published by Morley in a paper about 1900. Friends of Morley became fascinated with the result and spread word of it throughout Europe. Its actual direct proof (without involving the more general theory originally presented by Morley) did not appear until 1914. For further historical details, see the informative article by C. Oakley and J. Baker, "The Morley Trisector Theorem," *American Mathematical Monthly*, Vol. 85, No. 9 (1978), pp. 737–745. Most proofs either are in some way indirect, use unusual methods, or employ trigonometric formulas or identities. In Problem 18 we shall indicate a direct elementary proof using only high school geometry (for those who are curious and would like to pursue it). For the specific case of the isosceles right triangle, a short trigonometric proof can be constructed. (See Problem 16.)

Now we turn our attention to another interesting phenomenon of the triangle.

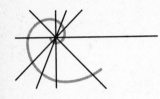

MOMENT FOR DISCOVERY

Triangles and Circles

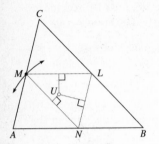

Figure 1.7

1. Let triangle $\triangle ABC$ be given, and construct the midpoints L, M, and N of the three sides. Again, use a large figure for greater accuracy.

2. Locate the intersection U of the perpendicular-bisectors to segments LM and MN, which is the center of a circle passing through L, M, and N (the **circumcircle** of $\triangle LMN$). Draw this circle.

3. Construct the three altitudes AD, BE, and CF to the sides of $\triangle ABC$. Did anything seem to happen?

4. If H is the **orthocenter** of $\triangle ABC$ (point of concurrency of altitudes AD, BE, and CF), locate the midpoints of segments AH, BH, and CH, and label them X, Y, and Z. Did anything seem to happen this time?

5. What conclusion do you reach based on this experiment?

The **Nine-Point Circle** emerged in Europe in the early 1800s. The French mathematician Jean-Victor Poncelet (1788–1867) named it the "Nine-Point Circle," and Karl W. Feuerbach (1800–1834) in Germany (where it is called the **Feuerbach Circle**) made several remarkable discoveries concerning it. These properties are so unique that we shall mention a few of them, without proof. Later, when you know more geometry, we may return to some of these and provide their interesting proofs. (See Section 5.7.)

PROPERTIES OF NINE-POINT CIRCLE AND NINE-POINT CENTER

1. The Nine-Point Circle of $\triangle ABC$ with orthocenter H passes through the midpoints L, M, and N of the three sides, the feet of the altitudes D, E, and F to those sides, and the **Euler points** X, Y, and Z, which are the midpoints of the segments AH, BH, and CH, respectively.

2. The Nine-Point Center U lies on the **Euler Line** of $\triangle ABC$—the line passing through the orthocenter H, the circumcenter O, and the centroid G of a triangle. (See Figure 1.8.)

3. The tangents to the Nine-Point Circle at the midpoints L, M, and N of the sides of the triangle form a triangle that is similar to the **orthic triangle** (the triangle $\triangle DEF$). In fact, the sides of this triangle are parallel to those of $\triangle DEF$. (See Figure 1.9.)

4. **Feuerbach's Theorem** The Nine-Point Circle is tangent to the **incircle** (circle inscribed in a triangle). Furthermore, the other three circles tangent to the extended sides of a triangle (called the **excircles**) are each tangent to the Nine-Point Circle. (See Figure 1.10.)

PROPERTY 2

PROPERTY 3

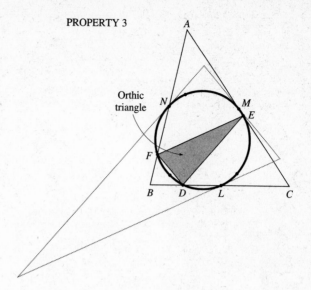

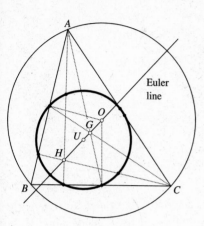

Figure 1.8

Figure 1.9

FEUERBACH'S THEOREM

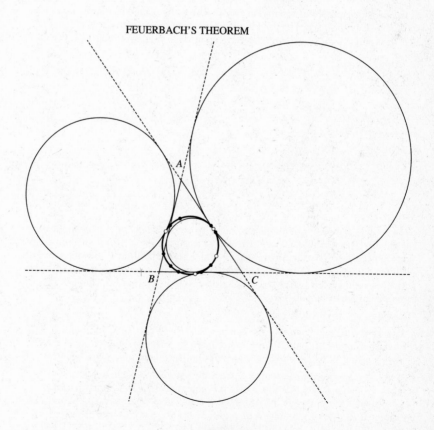

Figure 1.10

PROBLEMS   (§1.1)

_____ **Group A** _____

1. Try the experiment described in Example 3 for a quadrilateral. Construct a quadrilateral of your own, and repeat the experiment indicated here. What seems to take place?

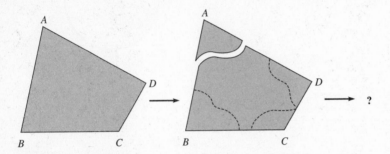

2. A square is constructed externally on a side of a regular octagon forming the following figure. Can this pattern be used to tile a bathroom floor (lopping off the tiles at the edges and corners of the room, as necessary)? What pattern results? Draw a diagram.

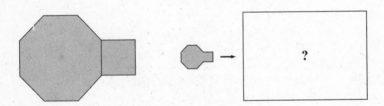

*3. Draw a circle and one of its diameters, *AB*. Choose any point $P \neq A$ or B on that circle and draw segments *AP* and *BP*. Carefully measure angle $\angle APB$ with a protractor. What did you find? Try the experiment several times with different locations for *P* to see if the phenomenon recurs. What theorem in geometry does this experiment support? If the software is available to you, as discussed in Section 1.5, conduct this experiment on the computer.

   *See Corollary to Theorem 1, §4.5.

4. **Conjecture:** The midpoint *M* of the hypotenuse *AB* of right $\triangle ABC$ is equidistant from the vertices *A*, *B*, and *C*. Conduct a drawing experiment either to support or deny this claim. What did you find? Does this property hold for *any* triangle?

5. Perform an experiment in reasoning to see if the theorem illustrated in Problem 3 implies, or is implied by, that illustrated in Problem 4. (**HINT:** Try an argument beginning with the statement, "*If M* is indeed equidistant from *A*, *B*, and *C*, and a circle centered at *M* passes through *A*, *B*, and *C*, *then* two isosceles triangles are formed." What follows from this? See Problem 14.)

6. Draw any circle and locate any point A outside it. Draw any two secants AP and AR, cutting the circle at P, Q, R, and S, as shown. Draw lines PR and QS meeting at U, and lines PS and QR meeting at V. Finally, draw line UV, cutting the circle at B and C. Now draw lines AB and AC. Do you observe anything in particular? Try the experiment two more times, first with point A further from the circle, and then closer to it. (Conduct this experiment by computer if the equipment is available to you.) Write down a conjecture based on this experiment.

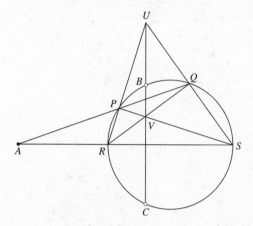

*7. Pick's Theorem The points in the coordinate plane having integer coordinates are called **lattice points**. A **lattice polygon** is an ordinary polygon whose vertices are lattice points. The number of lattice points inside a lattice polygon (and possibly on its boundary) would seem to be related in some way to its actual area K in the plane.

   (a) Try the three experiments illustrated, tallying your results in a table of the sort indicated. The number of points on the sides, or boundary, of the polygon is labelled B and the number strictly interior to it I.

	TRIANGLE	SQUARE	PARALLELOGRAM
K	2	?	?
B	6	?	?
I	0	?	?

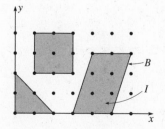
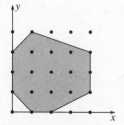

(b) If you believe in a linear relationship, write for the area

$$K = Bx + Iy + z$$

Substitute the data you found into this formula which yields a three-by-three system of equations in x, y, and z, and solve for x, y and z. You have now found a plausible formula for K.

(c) Test this formula with the seven-sided polygon shown, then try some of your own. (This result is *Pick's Theorem*, discovered in 1900 by an English mathematician, George Pick. For a proof, see Coxeter, *Introduction to Geometry*, p. 34.)

*See Problem 20, §2.3.

8. **Dudeney Dissection** An equilateral triangle and interior is dissected according to the specifications shown in the figure below. Make a paper or cardboard model, using a carefully constructed figure, and having the dimensions specified.

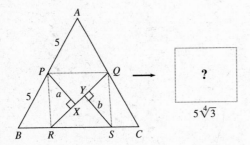

$AB = BC = CA = 10$ cm
$AP = AQ = 5$ cm
$QR^2 = 25\sqrt{3}$ ($QR \approx 6.58$ cm)
$RS = 5$ cm ($BR > SC$)

Find a way to reassemble these four pieces to form a square. (**HINT:** Think of *BPXR* and *CQR* pivoting about points P and Q, and *RYS* about S.) Try to find a proof for the validity of this dissection.

NOTE: This construction is due to H.E. Dudeney (1857-1931) who discovered many ingenious dissections, many of them optimal in terms of the number of pieces required.

9. The first letter of the Kunif alphabet, *ali*, has the form shown, with uniform thickness of one unit and all angles of measure either 60° or 120°. Find, by paper experiment or otherwise, how to assemble six of these letters to form a perfect regular hexagon and interior.

6 copies

10. Based on numerical examples involving rectangles, the following conjecture arises in connection with any parallelogram: If P is any point on the diagonal AC of parallelogram $ABCD$, and if lines are drawn through P parallel to the sides of the parallelogram, the two pairs of opposite smaller parallelograms (shaded) *have equal areas*. Investigate this, either by simple proof, or failing that, by making a paper model of the parts shown here and experimenting with the pieces. Write down any tangible discoveries you make.

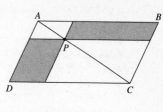

*11. **Pappus's Theorem** Draw any two lines ℓ and m, and locate three points A, B, and C on ℓ and three points A', B', C' on m, arbitrarily and in any order (an example is shown). Intersect the following pairs of "cross-joins": Lines $(AB', A'B)$, $(AC', A'C)$, and $(BC', B'C)$. Determine the respective points of intersection L, M, and N. Do L, M, and N appear to be collinear? Repeat several times, with ℓ not parallel to m. (This theorem was discovered by Pappus of Alexandria around 300 A.D.)

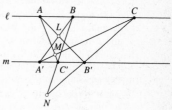

*See Problem 14, §4.8.

*12. **Desargues's Theorem** Draw three lines passing through a common point P, and then draw two triangles $\triangle ABC$ and $\triangle A'B'C'$ with vertices on the three lines (A and A' on ℓ, B and B' on m, and C and C' on n). Now find the points of intersections of the lines $(AB, A'B')$, $(AC, A'C')$, and $(BC, B'C')$ which contain the corresponding sides of the two triangles. This will determine points of intersection L, M, and N. Do L, M, and N appear to be collinear?

*See Problem 12, §2.4 and Example 3, §4.8.

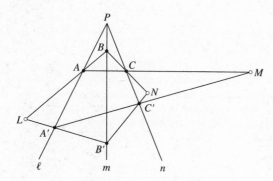

_____ *Group B* _____

13. Find a way to decompose $\triangle ABC$ and its interior into five subtriangles, each similar to the original triangle $\triangle ABC$, if

(a) the given triangle has a right angle

(b) the given triangle is isosceles with vertex angle $120°$.

NOTE: A recent theorem due to Werner Raffke ("Partitions of triangles," *Beitrage Algebra Geometry*, No. 32 (1991), pp. 87–93) shows that the given triangles in (a) and (b) are the *only* ones which will allow such a decomposition.

14. A theorem from geometry states that if a line bisects one side of a triangle and is parallel to a second side, it bisects the third side also. In the figure to the right, *M* is the midpoint of hypotenuse *AB*, and line *MK* is the perpendicular to side *BC*. Then lines *MK* and *AC* are parallel, so line *MK* bisects side *BC* and *KB = KC*. Finish the argument to show that *MC = MB*. (Thus: *The midpoint M of the hypotenuse of right triangle* △*ABC is equidistant from the three vertices, and a circle centered at M passes through A, B, and C.*)

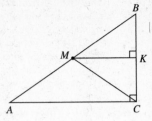

*15. Existence of Nine-Point Circle Observe the following figure, where △*ABC* is a given triangle; *L, M, N* are the midpoints of the sides; *D, E,* and *F* the feet of the altitudes on the sides; *H* the orthocenter (point of concurrency of the altitudes); and *X, Y,* and *Z* the midpoints of segments *AH, BH, CH*, respectively. Observe the quadrilateral *MNYZ* drawn in the figure. Is there any reason why segment *MN* is parallel to segment *BC*? (Use the fact mentioned in Problem 14.) Is *ZY* ∥ *CB*? Why, then, is *MNYZ* a parallelogram? Why is *MNYZ* a rectangle? (Show that line *NY* ∥ line *AD*, which is perpendicular to each of the parallel lines *BC* and *ZY*.) In the same manner, show that quadrilateral *XYLM* is a rectangle, and that *U = U′* = center of a circle passing through six of the nine desired points. To finish the argument, use the result of Problem 14 (on △*DLX*, etc.).

*See Theorem 1, §5.7.

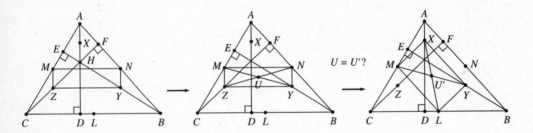

16. The law of sines for △*ACQ* yields *AQ*/sin 30° = *AC*/sin 135°. From this, derive the value *AQ* = $\sqrt{2}$. Show that *AM* = $\sqrt{2}$. Hence *AQ = AM* and △*ARQ* is congruent to △*ARM*. Thus ∠*RQA* is a right angle (complete the details). Now compute the angle measures needed to determine the measure of ∠*PQR*, and show that ∠*PQR* is equiangular, hence equilateral.

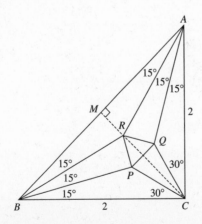

17. **Another Dudeney Dissection** The following sequence of diagrams shows how to
   (a) dissect a parallelogram (and interior) to form a square
   (b) dissect a regular pentagon (and interior) and form a parallelogram.

   Use this to make a paper model demonstrating a six-piece dissection of the regular
   pentagon to a square.

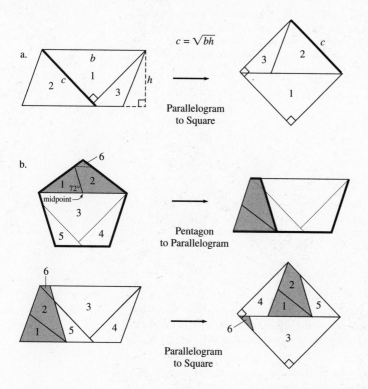

$c = \sqrt{bh}$

Parallelogram
to Square

Pentagon
to Parallelogram

Parallelogram
to Square

———— *Group C* ————

18. **Morley's Theorem** We give here an outline of a direct, elementary proof of the
   existence of the Morley triangle for any triangle. You are supposed to supply all
   missing details. The basic theorems used are

   (a) The angle bisectors of the angles of a triangle are concurrent (the point of con-
        currency is called the **incenter**, which is the center of the incircle of the trian-
        gle).

   (b) The sum of the measures of the angles of any triangle equals 180

   (c) The measure of an exterior angle of a triangle equals the sum of the two oppo-
        site interior angles.

   For convenience, write $A°$ for the value of the angle at A, and similarly for the angles
   at B and C of $\triangle ABC$.

   (1) Let the original trisectors BR and CQ meet at point X. Since P is the point
        where the two bisectors BP and CP meet, then line XP bisects $\angle BXC$. Choose
        points R' and Q' on segments BX and CX, respectively, so that angles $R'PX$ and
        $Q'PX$ each have measure 30°. It follows that $\triangle PQ'R'$ is equilateral and line PX
        is perpendicular to line $Q'R'$. Prove this.

(2) Let the perpendiculars from Q' and R' to sides PR' and PQ' meet lines CP and BP at Z and Y, respectively, and let A' be the point of intersection of lines YQ' and ZR'. Solve for all angles in the figure needed to establish that $\frac{1}{3}A°$ is the measure of the angle at A'. (**HINT:** $m\angle PQ'Y = m\angle Q'PY = m\angle XPY - 30° = m\angle XBP + m\angle BXP - 30° = \frac{1}{3}B° + \frac{1}{2}m\angle BXC - 30° = \frac{1}{3}B° + \frac{1}{2}(180° - \frac{2}{3}B° - \frac{2}{3}C°) - 30°$; simplify. Find $m\angle PR'Z$ in a similar fashion. Thus, the base angles of $\triangle A'Q'R'$ have been determined.)

(3) Show that $m\angle A'BC = B°$ and $m\angle A'CB = C°$, hence point A' coincides with point A. (**HINT:** Locate W on line $A'Q'$ such that $m\angle WBC = B°$. Since R' is the incenter of $\triangle BWY$, line WR' bisects $\angle BWY$. Show that $m\angle R'WQ' = m\angle R'A'Q' = \frac{1}{3}A°$, which is impossible unless point W coincides with point A'.)

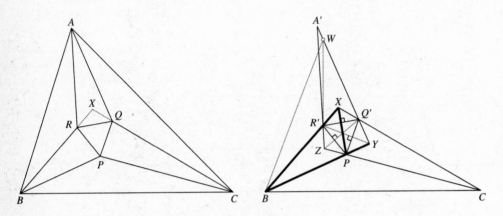

19. Use the following figure to show how to dissect a regular heptagon to form a parallelogram and verify that it is correct. Use the same idea as that used in Problem 17 to obtain a 12-piece dissection of the regular heptagon into a square.

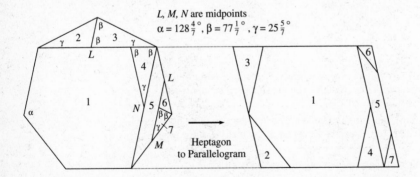

L, M, N are midpoints
$\alpha = 128\frac{4}{7}°, \beta = 77\frac{1}{7}°, \gamma = 25\frac{5}{7}°$

Heptagon to Parallelogram

20. **Project** For some integer n, find an n-piece dissection of a **regular pentagon** into a **regular heptagon**. You will need the result of Problem 19 for this purpose. (**HINT:** One way to accomplish this is to use dissections of previous problems and superimpose one dissection onto another. Superimposed cut lines will create additional pieces in the dissection.)

NOTE: A famous theorem of geometry asserts that any two polygons having equal areas may be dissected into the same number of pairwise congruent triangles. Thus, for any two positive integers m and n (≥ 3), a regular n-gon may be cut into triangular pieces which can be reassembled to form a regular m-gon. This theorem is due to J. Bolyai, one of the discoverers of hyperbolic geometry, discussed in detail later (Section 6.1).

1.2   VARIATIONS ON TWO FAMILIAR GEOMETRIC THEMES

We are going to consider here a few unusual ideas involving two fundamental geometric concepts—the Pythagorean Theorem and the number π.

PYTHAGOREAN-LIKE THEOREMS

There is evidence that the early Egyptians knew of the special Pythagorean relation for the 3-4-5 triangle, as illustrated in Figure 1.11, and that the Chinese were also aware of this special case, perhaps as early as 2,000 B.C. But the school of Greek geometers founded by Pythagoras that flourished in the period 600–500 B.C., whose contributions in other areas of geometry may have been more significant, has nevertheless become inextricably associated with this proposition, known as the **Pythagorean Theorem.**

This relation is one of the most pervasive formulas in geometry and its applications. A variety of unique proofs appeared even in antiquity, and both amateurs and professionals have been deriving new proofs ever since. In his book, *The Pythagorean Proposition*[1], E. S. Loomis presents a collection of over 370 different proofs of the theorem.

EUCLID'S METHOD OF PROOF

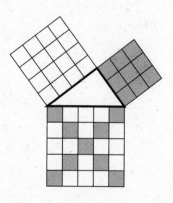

Figure 1.11

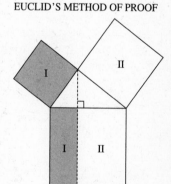

Figure 1.12

The earliest versions of the Pythagorean Theorem were all stated and demonstrated strictly in terms of area. Euclid himself developed the concept of area before

[1]References at the end of text provide exact sources.

introducing the theorem, and both his statement and proof of the theorem involved area rather than ratios and algebra. Euclid demonstrated that the square on the hypotenuse can be split into two rectangles whose areas are separately equal to the areas of the squares on the two legs, as shown in Figure 1.12. (See Problem 18.)

A dynamic pictorial proof (Figure 1.13) was given by H. Baravalle (found in the *Eighteenth Yearbook of the NCTM*). In this sequence of diagrams, it is important to realize that as one proceeds from one step to the next, the total area of the shaded portions remains constant. (This is due to the basic area property of parallelograms with equal bases and altitudes.) It is sometimes refreshing to see a proof unfold so graphically in this manner (only minor details need to be worked out). Another "proof by picture" was invented by the Hindu mathematician Bhaskara (A.D. 1150), which he supposedly drew in the sand, then proclaimed "Behold!" to his students (we already saw the essence of this proof in Example 2, Section 1.1).

PROOF BY PICTURE: Pythagorean Theorem

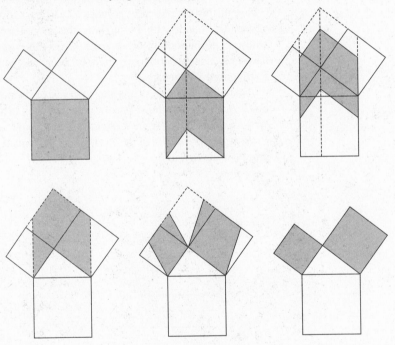

Figure 1.13

A wide variety of well-known generalizations of the Pythagorean Theorem also exists in abundance. One with which the reader is no doubt familiar involves the Law of Cosines (Figure 1.14). If C represents the measure of $\angle ACB$ as well as its vertex, as is customary, then

(1)
$$c^2 = a^2 + b^2 - 2ab \cos C,$$

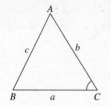

Figure 1.14

which applies to any triangle. In the special case $C = {}^{\pi}\!/_{2} = 90°$, cos $C = 0$ and **(1)** reduces to $c^2 = a^2 + b^2$.

A strictly geometric generalization of the Pythagorean Theorem which you may not know about is very intriguing. It is doubtful that it would ever have been discovered from a strictly algebraic version of the Pythagorean Theorem. Let's experiment with a few calculations.

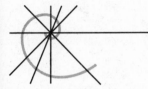

MOMENT FOR DISCOVERY

A Pythagorean-Like Theorem

Instead of *squares* on the sides of a right triangle, consider here *rectangles*, as shown (Figure 1.15). We have a 10-24-26 right triangle △*ABC*, and a 13 × 26 rectangle is constructed on hypotenuse *AB*. Altitude *CD* is extended to *P* so that *PC* = 13, as shown, and lines through *P* parallel to the legs of the triangle are drawn to form rectangles on the other two sides.

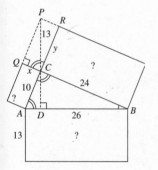

Figure 1.15

1. Observe the equal angles as marked in the figure. Thus, right triangles △*PCQ* and △*PCR* are each similar to △*ABC*. (Do you see why?)

2. Hence, since corresponding sides have proportional lengths,

$$\frac{x}{13} = \frac{10}{26} \quad \text{and} \quad \frac{y}{13} = \frac{24}{26}$$

Verify this.

3. Solve for x and y from Step 2.

4. Calculate the areas of the three rectangles. Did you notice anything unusual? (You should have obtained 288 square units for the rectangle on side *BC*. If you did not find anything unusual, calculate the *sum* of the areas of the rectangles on sides *BC* and *AC*.)

5. Try a similar experiment for a different right triangle (for convenience, choose a Pythagorean Triangle—one whose sides have integer lengths, such as 8, 15, and 17).

6. Write down what the general relationship seems to be.

CALCULATING π WITH A POCKET CALCULATOR

The following formula for π is often found in calculus texts, based on infinite series:

(2)
$$\frac{\pi}{4} = 1 - \frac{1}{3} + \frac{1}{5} - \frac{1}{7} + \frac{1}{9} - \frac{1}{11} \cdots$$

Actually calculating π using this series and a (programmable) calculator would be quite unfeasible due to the very slowly converging nature of the series and accumulating round-off error. Observe, for example,

$$\pi \approx 4(1 - \frac{1}{3}) = \frac{8}{3} = 2.666... \qquad \text{(sum of the first 2 terms)}$$

$$\pi \approx 4(1 - \frac{1}{3} + \frac{1}{5}) = 3.4666... \qquad \text{(sum of the first 3 terms)}$$

$$\pi \approx 4(1 - \frac{1}{3} + \frac{1}{5} - \frac{1}{7}) = 2.8952... \qquad \text{(sum of the first 4 terms)}$$

It would take, as a matter of fact, about 401 terms of **(2)** carried to five or six decimals to guarantee even two-place accuracy!

An altogether different set of calculations for π exists, not based on infinite series, but coming directly from geometry and algebra. Although it involves radical expressions that look rather complicated, they can be easily and efficiently handled with an ordinary pocket calculator. We encourage you to try your hand at some of these calculations as an experiment.

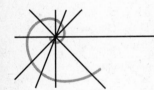

MOMENT FOR DISCOVERY

Calculating π

1. Evaluate these four expressions on your pocket calculator:

$$\frac{2}{\sqrt{2}}, \quad \frac{2}{\sqrt{2+\sqrt{2}}}, \quad \frac{2}{\sqrt{2+\sqrt{2+\sqrt{2}}}}, \quad \frac{2}{\sqrt{2+\sqrt{2+\sqrt{2+\sqrt{2}}}}}$$

   (***HINT:*** You should get 1.004838572 for the last term.)

2. Find the *product* of the above four values you just calculated in Step 1, and then double the result. That is, calculate the following product:

$$2 \cdot \frac{2}{\sqrt{2}} \cdot \frac{2}{\sqrt{2+\sqrt{2}}} \cdot \frac{2}{\sqrt{2+\sqrt{2+\sqrt{2}}}} \cdot \frac{2}{\sqrt{2+\sqrt{2+\sqrt{2+\sqrt{2}}}}}$$

3. Compare your answer in Step 2 with the "exact" value of π (as given by your calculator).

4. If you have a programmable calculator, write a program in BASIC which will enable you to calculate the product of 7, 9, and 11 of the terms in the *infinite product*:

(3)
$$2 \cdot \frac{2}{\sqrt{2}} \cdot \frac{2}{\sqrt{2+\sqrt{2}}} \cdot \frac{2}{\sqrt{2+\sqrt{2+\sqrt{2}}}} \cdots$$

(Note that the 11th factor will have 10 radical signs.)

5. What did you discover?

6. Evaluate the following expression, and compare your answer with that of Step 2.

$$2^4 \cdot \sqrt{2 - \sqrt{2 + \sqrt{2 + \sqrt{2}}}}$$

If you are curious about the strange radical expressions for π in the Discovery Unit, in a moment we will show how they may be derived from elementary geometry. The original idea for these formulas came from Archimedes, a contemporary of Euclid. The first step (which we omit) is to prove from properties of approximating regular polygons that the ratio of the circumference of a circle to its diameter is a constant for all circles. Then, in the manner of Euclid, we *define* π as that constant. Thus, if C is the circumference of a circle having diameter d and radius r, then $C/d = \pi$, or

(4) $C = 2\pi r$

HISTORICAL NOTE

Archimedes (287–212 B.C.), who is regarded as the father of modern physics, was the greatest scientific and mathematical intellect in the world up through modern times (the 18th century). He was born of low state in the ancient Greek city of Syracuse, but because of his ingenious mechanical inventions, he was soon noticed by King Hieron. Tradition has it that the king once asked Archimedes to test his gold crown for its authenticity. As the story goes, a flash of insight struck Archimedes while he was in his bath, and he was so excited that he ran into the streets naked, proclaiming "Eureka! I have found it!" His method was to submerge the crown in water and thereby determine its density, a method known today as **Archimedes' Principle**. Archimedes was proudest of his achievements in pure mathematics, however, such as his development and proof of the volume of a sphere, a problem which had baffled Euclid. During a final siege on Syracuse, the order of the commander Marcelus to spare Archimedes was ignored, and a Roman soldier murdered him while he was purportedly working on a mathematics problem—after Archimedes had admonished the soldier to leave him alone. At his request, Archimedes' tomb bore the inscription of a sphere inscribed in a cylinder, representing his result that the ratio of these two volumes equals $2/3$ — proven entirely by logic.

Archimedes elaborated on Euclid's method of using inscribed and circumscribed polygons having many sides in deducing geometric facts about π. Since $C = 2\pi$ for a circle of radius 1, the value for π lies between one-half the perimeter of a regular inscribed and circumscribed polygon of n sides for all integers $n \geq 3$. If we start with an inscribed square (Figure 1.16) and begin doubling the number of sides to

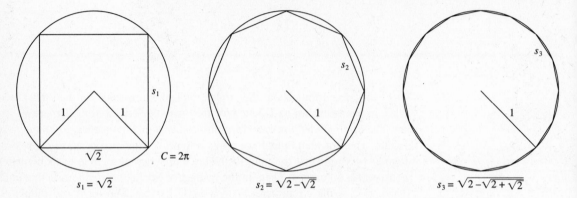

$$s_1 = \sqrt{2} \qquad\qquad s_2 = \sqrt{2 - \sqrt{2}} \qquad\qquad s_3 = \sqrt{2 - \sqrt{2 + \sqrt{2}}}$$

Figure 1.16

obtain a regular octagon, regular 16-sided polygon, and so on, we know the perimeters of these polygons will tend to a limit of twice the value of π. The lengths of the sides of these polygons can be derived strictly from applications of the Pythagorean Theorem. For convenience, let s and t be the sides of two successive regular polygons. Then, from Figure 1.17, where $OA = OB = OC = 1$, $AC = s$, and $AB = t$, we obtain

$$t^2 = x^2 + \left(\frac{s}{2}\right)^2 \qquad (\triangle ABD)$$

$$1 = (1-x)^2 + \left(\frac{s}{2}\right)^2 \qquad (\triangle AOD)$$

$$= (1-x)^2 + t^2 - x^2$$

$$= 1 - 2x + t^2$$

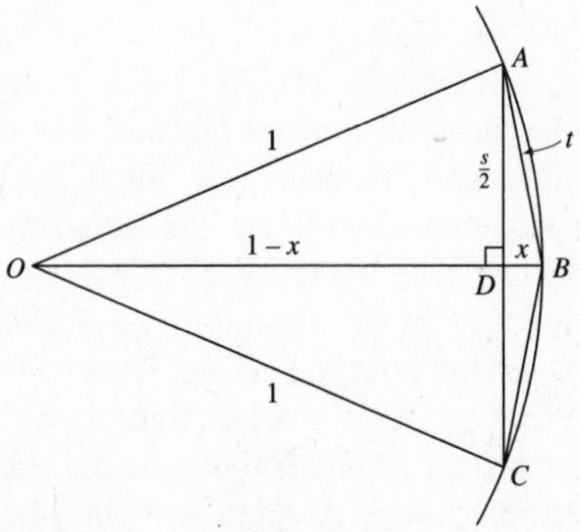

Figure 1.17

Thus, by algebra, $0 = -2x + t^2$ or $x = \frac{1}{2}t^2$. By substitution,

$$1 = \left(1 - \frac{t^2}{2}\right)^2 + \left(\frac{s}{2}\right)^2 \qquad \text{or} \qquad \frac{t^2}{2} = 1 - \sqrt{1 - \frac{s^2}{4}} = 1 - \frac{1}{2}\sqrt{4 - s^2}$$

With $s = s_n$, solve this last equation for $t = s_{n+1}$ and obtain the result

(5)
$$s_{n+1} = \sqrt{2 - \sqrt{4 - s_n^2}}$$

Now s_1 is the side of the inscribed square in Figure 1.16, so

$$s_1 = \sqrt{2}$$

$$s_2 = \sqrt{2 - \sqrt{4 - \sqrt{2}^2}} = \sqrt{2 - \sqrt{2}}$$

$$s_3 = \sqrt{2 - \sqrt{4 - (2 - \sqrt{2})}} = \sqrt{2 - \sqrt{2 + \sqrt{2}}}$$

etc. In general,

$$s_n = \sqrt{2 - \sqrt{2 + \sqrt{2 + \ldots \sqrt{2 + \sqrt{2}}}}} \quad (n \text{ radicals})$$

Since the perimeter of the nth polygon is $p_n = 2^{n+1}s_n$, the sequence $\pi_n \equiv \frac{1}{2}p_n = 2^n s_n$ converges to π.[2] That is,

(6)
$$\pi = \lim_{n \to \infty} \pi_n = \lim_{n \to \infty} 2^n \sqrt{2 - \sqrt{2 + \sqrt{2 + \ldots \sqrt{2 + \sqrt{2}}}}} \quad (n \text{ radicals})$$

Problem 8 will show how to connect this expression for π with the infinite product **(3)** in the previous Discovery Unit, which is a reasonably effective, elementary tool for calculating π on a pocket calculator.

OUR GEOMETRIC WORLD

The Great Pyramid at Gizeh, Egypt, has a square base with sides 755.8 feet in length and rises 481.2 feet. The ratio of the perimeter of the base to its height is very close to 2π. One explanation is that builders used a wheel with a radius of one foot as a convenient measuring device, counting the number of revolutions instead of the actual distance. But some scholars believe this connection with 2π is mere coincidence.

---

[2]Use of $\equiv$ designates a definition or identity in this text.

FURTHER HISTORICAL FACTS ABOUT π

Archimedes's original analysis started with a regular inscribed and circumscribed *hexagon* instead of the square which we used in our construction. This results in a similar expression.

(7) $\pi = \lim_{n\to\infty} \pi'_n \equiv \lim_{n\to\infty} 3 \cdot 2^{n-1} \sqrt{2 - \sqrt{2 + \sqrt{2 + \ldots + \sqrt{2 + \sqrt{3}}}}}$

(*n* radicals)

Archimedes used a 96-sided regular inscribed and circumscribed polygon together with an extensive analysis of square roots to obtain the following famous inequality for the value π:

(8) $$3\frac{10}{71} < \pi < 3\frac{1}{7}$$

or, in decimals,

$$3.140845070... < \pi < 3.142857143...$$

A survey of the historical facts surrounding π reveals the following milestones in its development (further details may be found in H. Eves, *An Introduction to the History of Mathematics*).

A TIME LINE FOR π

2500 B.C. Babylonians and Chinese use the value π = 3.

1650 B.C. Egyptians use the value $\left(\frac{4}{3}\right)^4 = 3.1604...$

240 B.C. Archimedes' inequality **(8)**. His upper bound of $\frac{22}{7} = 3.1428...$ is often adopted as a convenient approximation, even today.

A.D. 480 Chinese mathematician Tsu Ch'ung-chih proposed the value 355/113 = 3.1415929... (See Problem 15.)

1706 John Machin obtained π to 100 decimals using infinite series.

1760 Comte de Buffon devised his famous **needle problem** by which π may be determined by probability theory and tosses of a needle.

1767 J.H. Lambert showed that π is irrational.

1853 William Rutherford of England calculated π to 400 decimal places using infinite series.

1882 F. Lindemann showed that π is **transcendental**. That is, π is not the solution to any polynomial equation having rational coefficients. (Thus, there is no compass/straight-edge construction of π.)

1948 D.F. Ferguson of England and J.W. Wrench of America published the value of π to 808 places using infinite series.

A TIME LINE FOR π (CONTINUED)

1949 The first use of the electronic computer to calculate π was made: The ENIAC at the Army Ballistic Research Laboratories in Maryland calculated π to 2,037 places.

1961 Using an IBM 7090, J. W. Wrench and Daniel Shanks computed π to 100,265 places.

1967 Wrench and Shanks computed π to 500,000 places.

1989 Two Columbia professors, Gregory V. and David V. Chudnovsky, using a Cray 2 and an IBM 3090, calculated π to over a billion decimal places (to be exact, 1,011,196,691 places).[3]

PROBLEMS   (§1.2)

_____ *Group A* _____

1. **Another Discovery Proof of the Pythagorean Theorem** Let right triangle △*ABC* be given, as shown, with the squares constructed on the three sides. Let *P* be the center of the square on the larger leg *BC*, and draw lines ℓ and *m* through *P* parallel and perpendicular to the hypotenuse *AB*. Make a paper model and show how square *ACFG* and the pieces thus formed in square *BCKL* can be made to fit precisely inside square *ABMN*.

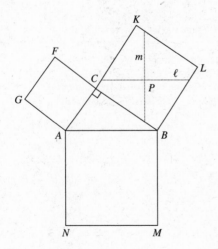

[3]J. Borwein, P. Borwein, and D. Bailey, "Ramanujan, Modular Equations, and Approximations to Pi, or How to Compute One Billion Digits of Pi," *American Mathematical Monthly*, Vol. 96, No. 3 (1989), pp. 201–219.

2. The following diagram, found in the *Arithmetic Classic of the Gnomon and the Circular Paths of Heaven* (the oldest existing Chinese text containing mathematical theory and dating back to 600 B.C.) represents the oldest known proof of the Pythagorean relation. Explain the steps in the proof. Does this proof apply only to the 3-4-5 case exhibited, or may it be generalized?

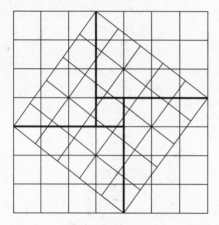

3. (a) Verify the Pythagorean relation for the right triangle △*ABC* shown in the figure at the right.

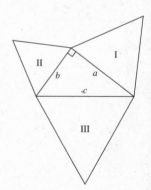

   (b) On each side of this triangle, as shown, a semicircle has been constructed. If the areas of those semicircles are I, II, and III, respectively, what relationship exists between I, II, and III?

   (c) Prove this in general by using standard area formulas.

4. Equilateral triangles constructed on the three sides of a right triangle have areas I, II, and III, respectively. What relationship exists for the numbers I, II, and III? It's alright to guess at this point, but if you can prove it, include that in your solution.

*5. **Experimenting With the Divine Section and Golden Ratio** If a point *B* divides segment *AC* into two other segments *AB* and *BC* such that both the internal and external ratios *AB/BC* and *AC/AB* are equal, this division is called a **divine section**, a term originating with the early Greeks. The common ratio *AB/BC* has come to be known as the **Golden Ratio**, denoted τ. (If *AB* = 1 then τ = *AC*.)

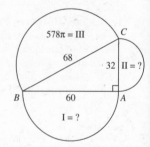

   (a) By experimenting with the location of point *C* on line *AB* using increments of one-tenth of a unit, find which of the positions indicated by hash marks yields the closest approximation to the divine section, and show that 1.6 < τ < 1.7.

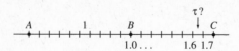

   (b) By letting *x* = *AB/BC*, and from the equation

$$\frac{AB}{BC} = \frac{AC}{AB} = \frac{AB + BC}{AB},$$

show that $x^2 = x + 1$. By finding the roots of this equation, establish the exact value for τ.

*See Problems 13, 14, and 15, §4.4.

6. **The Golden Rectangle** If Rectangle $ABCD$ has sides in the ratio $AB/BC = \tau$, where τ is the Golden Ratio (defined in Problem 5), the rectangle is called a **Golden Rectangle.**

SQUARES IN A GOLDEN RECTANGLE

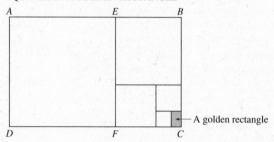

— A golden rectangle

(a) Show that a Golden Rectangle $ABCD$ has the characteristic property that if we cut off square $AEFD$ at one end, the remaining rectangle $EBCF$ will be another Golden Rectangle, and the dissection process of cutting off a square at the end of each succeeding Golden Rectangle continues ad infinitum. (***HINT:*** $\tau^2 = \tau + 1$.)

(b) Show that if $AD = 1$, the squares obtained in the dissection described in (a) have successive areas $1, \tau^{-2}, \tau^{-4}, \tau^{-6}, ...$

(c) Prove that $1 + \tau^{-2} + \tau^{-4} + ... = \tau$

(***HINT:*** First derive the relation $1/\tau = \tau - 1$.)

7. **Euclidean Construction of the Golden Ratio** Start with a unit square $ABCD$ and locate the midpoint M of segment AB. With M as center and MC as radius, swing a circular arc to obtain point E on line AB; finally, complete Rectangle $AEFD$. Show that $AE/AD = \tau$ (hence, $AEFD$ is a Golden Rectangle). (See Problems 5 and 6.)

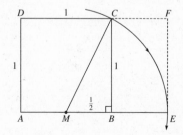

8. By the algebraic process of "rationalizing the denominator," simplify the following radical expressions so there are no radicals in the denominator. Generalize.

(a) $2 \cdot \dfrac{2}{\sqrt{2}} \cdot \dfrac{2}{\sqrt{2 + \sqrt{2}}}$

(b) $2 \cdot \dfrac{2}{\sqrt{2}} \cdot \dfrac{2}{\sqrt{2 + \sqrt{2}}} \cdot \dfrac{2}{\sqrt{2 + \sqrt{2 + \sqrt{2}}}}$

(**HINT:** In (a), multiply numerator and denominator by $\sqrt{2 - \sqrt{2}}$.)

_____ *Group B* _____

9. Let $x_1 = 3$, $x_2 = x_1 + \sin x_1$, $x_3 = x_2 + \sin x_2$, ... Using the radian mode on your calculator, find x_2 and x_3. What did you find? What happens if $x_1 = 2$? Attempt an explanation. (See note by Daniel Shanks, *American Mathematical Monthly*, Vol. 99, No. 3 (1992), p. 263, where it is shown that if P is an approximation to π correct to n places, then $P + \sin P$ is an approximation correct to $3n$ places.)

10. **Euclid's Generalization of the Pythagoras Relation**[4] In $\triangle ABC$, the altitudes are drawn and extended to divide the squares on the sides into the rectangles labelled I, II, III, as shown.

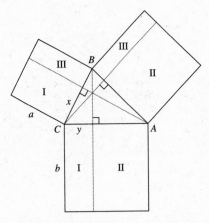

   (a) Verify (prove) this relation as revealed in the figure (equality of certain rectangles is indicated).

   (b) Show that this result has the Pythagorean Theorem as a special case.

   (**HINT:** Use similar right triangles to determine the ratio x/y in terms of a and b, and similarly for b and c, and a and c.)

11. An area proof for the Law of Cosines when $C > 90°$ is exhibited in the figure below

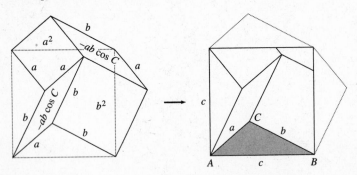

[4]See Heath's *The Thirteen Books of Euclid*, Vol. 2, p. 406 (information provided by Don Chakerian).

(one of the examples given by H. Eves in a discussion of dissection theory, *Survey of Geometry*, Vol. I, p. 268). Analyze the figure and provide a proof.

12. The eminent French mathematician Francois Viete in 1579 discovered the following formula for π.

$$\frac{2}{\pi} = \frac{1}{2} \cdot \sqrt{\frac{1}{2} + \frac{1}{2}\sqrt{\frac{1}{2}}} \cdot \sqrt{\frac{1}{2} + \frac{1}{2}\sqrt{\frac{1}{2} + \frac{1}{2}\sqrt{\frac{1}{2}}}} \cdots$$

Use the infinite product **(3)**, take reciprocals, and verify this formula by elementary algebra.

13. Use the formula **(5)** for s_{n+1} to find the length of the side of the inscribed dodecagon from the hexagon ($s_1 = 1$), and then generalize to obtain Archimedes' expression for π_n:

$$3 \cdot 2^{n-1} \cdot \sqrt{2 - \sqrt{2 + \sqrt{2 + \ldots \sqrt{2 + \sqrt{3}}}}} \quad (n \text{ radicals})$$

--------- *Group C* ---------

14. Observe the two expressions for π introduced previously—as limits of the sequences $\{\pi_n\}$ (in **(6)**) and $\{\pi'_n\}$ (in **(7)**). We could write them together in the form

$$\pi = \lim_{n \to \infty} a \cdot 2^{n-1} \cdot \sqrt{2 - \sqrt{2 + \sqrt{2 + \ldots \sqrt{2 + \sqrt{a}}}}} \quad (n \text{ radicals})$$

where $a = 2$ and 3 (each value resulting in one of the formulas). This leads directly to the conjecture that the above relation holds for all positive values of a. Either verify or disprove the conjecture, by using, perhaps, a very large or very small value of a as a test. If the conjecture is false, try proving that this formula is true *ONLY* for $a = 2$ and 3. (For this last part, see next problem.)

15. To what does the following pseudo-Archimedean sequence converge, if at all, for any positive real $a < 4$?

$$u_n = 2^n \sqrt{2 - \sqrt{2 + \sqrt{2 + \ldots \sqrt{2 + \sqrt{a}}}}} \quad (n \text{ radicals})$$

(**HINT:** The answer may be interpreted geometrically; start with a chord of length $s_1 = \sqrt{4-a}$ in a unit circle, and begin applying Archimedes's algorithm.)

16. Pappus's Extension of the Pythagorean Theorem In $\triangle ABC$, parallelograms I, II, and III are constructed on the three sides AB, BC and CA, respectively, such that if the sides of II and III opposite BC and CA meet at P, the side AQ adjacent to AB of parallelogram I is parallel and equal in length to PC. Then Area I = Area II + Area III. (**HINT:** To prove this, use the fact that parallelograms having congruent bases lying between two fixed parallel lines which contain those bases have equal areas since they have equal bases and altitudes. Thus it follows that Area $AQST =$ Area $ACPQ'$ and Area $BTSR =$ Area $BCPR'$, etc.)

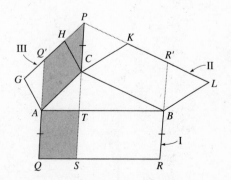

17. Ruler/Compass "Construction" of π   Verify the end result of the following details of a compass/straight-edge construction for a very close approximation for π. Let $CD = CA = 1$, with line AC perpendicular ($\perp$) to line CD. Then

(1) $AB = \dfrac{1}{8}$ (obtainable by successive bisections)

(2) $DE = DM = \dfrac{1}{2}$ (a single bisection)

(3) $EF \perp CD$ (standard construction of perpendicular)

(4) $EG \parallel BF$ (standard construction of parallel lines)

(5) $CD = DH = HK, \ KL = GD$ (segment construction)

(6) Result: CL differs from π by less than 0.0000003. Show this by calculating $CL = GD + 3$ from applications of similar triangles and the Pythagorean Theorem.

Euclidean "Construction" of π

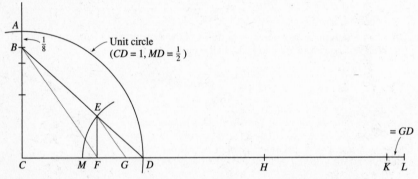

NOTE: This construction was discovered in 1849 by a German mathematician, Jakob de Gelder. It makes use of the Chinese rational approximation of A.D. 480 mentioned previously.

18. In Euclid's proof of the Pythagorean Theorem, he showed that one of the shaded triangles in the accompanying figure was equal in area to one-half the square on the leg (I) and that the other shaded triangle was equal to one-half that of the rectangle below it (I'). He then had to show that the two triangles had equal areas. He accomplished this by showing they were congruent. Show this part only, using standard congruence theory for triangles in geometry (if you have the background).

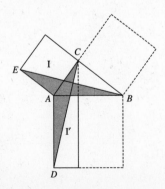

*1.3 PTOLEMY, BRAHMAGUPTA, AND THE QUADRILATERAL

Ancient geometers often used surprisingly clever and sophisticated ideas in deducing from logic the formulas they or their predecessors had verified by observation or experiment, and it is amazing how much they achieved without access to algebra, trigonometry, or the modern theory of limits and calculus—bearing testimony to the power of the deductive method. The basic properties of the quadrilateral will be needed later for our study of Steiner's Theorem in connection with the pedal triangle, so we shall take a look at some of these interesting ideas.

PTOLEMY'S THEOREM

Claudius Ptolemy (c. A.D. 150) was one of the earliest noted astronomers. His work included an extensive sine table, with values worked out by means of ingenious trigonometric identities which he established for just that purpose.

In the course of his work with angles and circles, he discovered an interesting relation between the sides and diagonals of a **cyclic quadrilateral** (a quadrilateral whose vertices lie on a circle). A logical argument for it may be based on the trigonometric relation

(1) $$a = 2r \sin \frac{1}{2} A,$$

where a is the length of a chord of a circle, r is its radius, and A is the degree measure of the circular arc subtended by the chord.

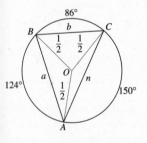

Figure 1.18

EXAMPLE 1 In Figure 1.18, $\triangle ABC$ is inscribed in a circle of radius ½.

(a) Find a, b, and n using **(1)**.

(b) Verify your calculation for n independently by the use of the Law of Cosines for $\triangle ABC$ at angle B.

SOLUTION

(a) By **(1)**, because $r = \frac{1}{2}$, we have

$$a = 2 \cdot \frac{1}{2} \sin \frac{1}{2}(124°) = \sin 62° = 0.88294759... \quad \rightarrow \quad a \approx 0.883$$

$$b = \sin \frac{1}{2}(86°) = \sin 43° = 0.68199836... \quad \rightarrow \quad b \approx 0.682$$

$$n = \sin \frac{1}{2}(124° + 86°) = \sin 105° = 0.96592582... \quad \rightarrow \quad n \approx 0.966$$

(b) The Law of Cosines for the angle at B is
$$n^2 = a^2 + b^2 - 2ab \cos B$$

Since an inscribed angle is measured by one-half its intercepted arc,

$$B = \frac{1}{2}m\angle COA = \frac{1}{2}(150°) = 75°$$

Thus, $n^2 \approx (0.883)^2 + (0.682)^2 - 2(0.883)(0.682)\cos 75° \approx 0.93309$, or $n \approx 0.966$, in agreement with the result in **(a)**. ❏

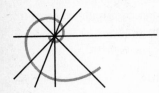

MOMENT FOR DISCOVERY

Cyclic Quadrilaterals

In Figure 1.19, a cyclic quadrilateral $ABCD$ has sides of length a, b, c, and d, and diagonals m and n, subtending arcs whose measures are as exhibited. Assume that the radius of the circle is ½. Thus, by formula **(1)**, $a = \sin \frac{1}{2}A$, $b = \sin \frac{1}{2}B$, etc. Use this to complete the following.

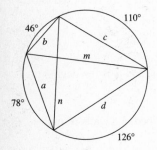

Figure 1.19

1. Find the values for a, b, c, and d, using six-place approximations for accuracy. (**HINT**: You should get $a \approx 0.629320$ and $b \approx 0.390731$, for example.)

2. Compute the products ac and bd.

3. Find m and n using **(1)**, and calculate the product mn.

4. Experiment with sums of various combinations from the products ac, bd, and mn until you find a relationship.

5. What did you discover?

The proof of **(1)** follows from a simple right triangle relation, and should be more or less self-evident from Figure 1.20 (the second case covers an arc greater than a semicircle and uses the identity $\sin(180° - \theta) = \sin \theta$).

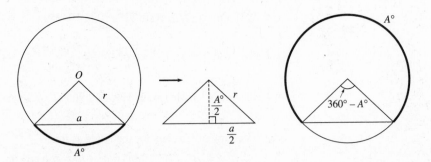

Figure 1.20

$$\frac{\frac{a}{2}}{r} = \sin\frac{A}{2}$$

$$= \sin\left(180° - \frac{A}{2}\right)$$

$$= \sin\frac{1}{2}(360° - A)$$

Ptolemy's relation for a cyclic quadrilateral in general can be easily proven by trigonometry. The Discovery Unit makes it apparent that Ptolemy's Theorem is equivalent to the following trigonometric identity, involving arbitrary positive values for A, B, C, D whose sum equals 360° (Figure 1.21).

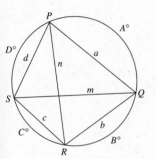

Figure 1.21

(2) $$\sin\frac{A}{2}\,\sin\frac{C}{2} + \sin\frac{B}{2}\,\sin\frac{D}{2} = \sin\frac{A+B}{2}\,\sin\frac{B+C}{2}$$

or, because $D = 360° - (A + B + C)$,

(2′) $$\sin\frac{A}{2}\,\sin\frac{C}{2} + \sin\frac{B}{2}\,\sin\frac{A+B+C}{2} = \sin\frac{A+B}{2}\,\sin\frac{B+C}{2}$$

The latter may be easily obtained from the product-to-sum identity (which Ptolemy knew) from elementary trigonometry.

$$2\sin x \sin y = \cos(x - y) - \cos(x + y)$$

(You are invited to try your hand at providing the details; see Problem 8.) This would prove Ptolemy's Theorem for a circle of radius ½ (as in the Discovery Unit). For a circle having arbitrary radius r, simply multiply both sides of **(2)** by $4r^2$.

PTOLEMY'S THEOREM FOR A CYCLIC QUADRILATERAL (FIGURE 1.21)

$$ac + bd = mn$$

BRAHMAGUPTA'S AMAZING QUADRILATERAL FORMULAS

In view of the difficulty which the quadrilateral has presented to geometers,[5] it is refreshing to find an early penetrating study of its properties in Indian mathematics. Ingenious formulas relating the diagonals of certain quadrilaterals to their sides, in addition to an elegant formula for area, were discovered and proven by Brahmagupta, the most prominent Hindu mathematician of early times. Brahmagupta lived and worked in the astronomical center of Ujjain, located in

---

[5]The ancient Egyptians once proposed the erroneous area formula $K = \dfrac{(a + c)(b + d)}{4}$, where a, b, c, and d are the consecutive sides of a quadrilateral.

central India. In 628, he compiled a work on astronomy of 21 chapters, of which Chapters 12 and 18 deal with mathematics.

You may have already worked with Heron's formula for the area of a triangle in terms of its sides a, b, c, which dates back to 100 B.C.

$$K = \sqrt{(s - a)(s - b)(s - c)}$$

where $s = \frac{1}{2}(a + b + c)$ (the **semiperimeter**). Brahmagupta discovered an analogous formula for the area of certain quadrilaterals in terms of the sides a, b, c, d, which generalizes Heron's formula. For convenience and subsequent study, set

(3) $$E = \sqrt{(s - a)(s - b)(s - c)(s - d)}$$

where $s = \frac{1}{2}(a + b + c + d)$.

EXAMPLE 2   Show that the area of a rectangle with consecutive sides a, b, $c = a$, and $d = b$ is numerically equal to the quantity E in **(3)**.

SOLUTION

We compute s first.

$$s = \frac{1}{2}(a + b + c + d)$$

$$= \frac{1}{2}(2a + 2b)$$

$$= a + b$$

Therefore,

$$s - a = s - c = b$$

$$s - b = s - d = a,$$

and $E = \sqrt{b \cdot a \cdot a \cdot b} = \sqrt{a^2 b^2} = ab$ (the correct value).❑

Now you can discover something for yourself.

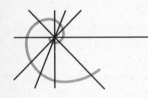

MOMENT FOR DISCOVERY

Quadrilaterals and Semiperimeters

A quadrilateral has consecutive sides 7, 24, 20, and 15, as shown in Figure 1.22 (two angles are marked as right angles). First, verify that $7^2 + 24^2 = 20^2 + 15^2 = 625$. Another quadrilateral (b) has consecutive sides 4, 4, 5, and 7.

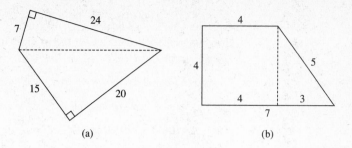

Figure 1.22

1. Using the formula ½*bh* for the area of a triangle in terms of its base and altitude, calculate the area of the quadrilaterals shown in (a) and (b). (Did you get 234 for (a)?)

2. Find *s*, the semiperimeters of both quadrilaterals.

3. Calculate: $s - a, s - b, s - c, s - d$ in each case.

4. Find the product of the four numbers you calculated in Step 3 for the two quadrilaterals.

5. What two values do you then get for the expression

$$E = \sqrt{(s-a)(s-b)(s-c)(s-d)}\,?$$

6. What conclusion did you reach?

7. Can a circle be drawn through the vertices of each of the quadrilaterals in Example 2 and (a) and (b) here?

8. Does a conjecture suggest itself? If so, write it down.

We proceed to prove that the conjecture you might have proposed in the Discovery Unit is indeed valid. In order to prove anything about the quantity E, we need to do some work with it algebraically. First note that

$$s - a = \frac{1}{2}(a + b + c + d) - a = \frac{1}{2}(-a + b + c + d)$$

Similarly,

$$s - b = \frac{1}{2}(a - b + c + d),$$

and so on. Thus,

$$E^2 = (s - a)(s - b)(s - c)(s - d)$$

$$= \frac{1}{2}(-a + b + c + d) \cdot \frac{1}{2}(a - b + c + d) \cdot \frac{1}{2}(a + b - c + d) \cdot \frac{1}{2}(a + b + c - d)$$

or

$$16E^2 = [(c + d) - (a - b)] \cdot [(c + d) + (a - b)] \cdot [(a + b) - (c - d)] \cdot [(a + b) + (c - d)]$$

Now we make repeated use of the product law $(X - Y)(X + Y) = X^2 - Y^2$, first with $X = c + d$ and $Y = a - b$:

$$16E^2 = [(c + d)^2 - (a - b)^2] \cdot [(a + b)^2 - (c - d)^2]$$

$$= [c^2 + 2cd + d^2 - a^2 + 2ab - b^2] \cdot [a^2 + 2ab + b^2 - c^2 + 2cd - d^2]$$

$$= [2(ab + cd) - (a^2 + b^2 - c^2 - d^2)] \cdot [2(ab + cd) + (a^2 + b^2 - c^2 - d^2)]$$

Finally, with $X = 2(ab + cd)$ and $Y = (a^2 + b^2 - c^2 - d^2)$, we obtain

(4) $$16E^2 = 4(ab + cd)^2 - (a^2 + b^2 - c^2 - d^2)^2$$

Now the expression $ab + cd$ can be easily related to the area K of a cyclic quadrilateral $ABCD$ (Figure 1.23), as follows. By a standard area formula,

$$\text{Area } \triangle ABC = \frac{1}{2}ab \sin \theta \quad \text{and} \quad \text{Area } \triangle ACD = \frac{1}{2}cd \sin \phi$$

Because $\phi + \theta = 180°$, we have

$$K = \frac{1}{2}ab \sin \theta + \frac{1}{2}cd \sin \theta$$

or

$$2K = (ab + cd) \sin \theta,$$

$$16K^2 = 4(ab + cd)^2 \sin^2\theta = 4(ab + cd)^2(1 - \cos^2\theta)$$

Finally,

(5) $$16K^2 = 4(ab + cd)^2 - 4(ab + cd)^2\cos^2\theta$$

By the Law of Cosines applied to $\triangle ABC$ and $\triangle ACD$,

$$n^2 = a^2 + b^2 - 2ab \cos \theta$$

$$n^2 = c^2 + d^2 - 2cd \cos \phi = c^2 + d^2 + 2cd \cos \theta$$

Now equate the two right members of the above two equations.

$$a^2 + b^2 - 2ab \cos \theta = c^2 + d^2 + 2cd \cos \theta$$

or

(6) $$a^2 + b^2 - c^2 - d^2 = 2(ab + cd) \cos \theta$$

Thus we see from equations **(4)**, **(5)**, and **(6)** that

$$16K^2 = 4(ab + cd)^2 - 4(ab + cd)^2\cos^2\theta = 16E^2$$

and therefore $E = K = \text{Area } ABCD$.

Brahmagupta's other formulas are included in the following table (refer to Figure 1.24). The proofs will be explored in the problem set which follows.

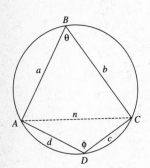

Figure 1.23

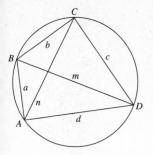

Figure 1.24

$$K = \sqrt{(s-a)(s-b)(s-c)(s-d)}, \qquad s = \tfrac{1}{2}(a+b+c+d)$$

$$m^2 = \frac{(ab+cd)(ac+bd)}{ad+bc} \qquad\qquad n^2 = \frac{(ac+bd)(ad+bc)}{ab+cd}$$

PROBLEMS   (§1.3)

_____ *Group A* _____

1. The lengths of sides and certain line segments are shown in the following figure. Use the relation $AE \cdot EC = BE \cdot ED$ from elementary geometry (established later in Section 4.5) for intersecting chords of a circle and Ptolemy's Theorem to find the lengths of the diagonals.

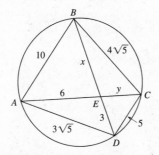

2. Use Brahmagupta's Formulas for m and n to work Problem 1.

3. Find the missing diagonal BD if $AC = 25$ and the sides of the quadrilateral are as indicated. Solve by two methods:

   (a) by Ptolemy's Theorem

   (b) by Brahmagupta's Formula.

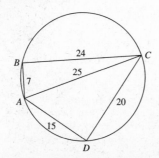

4. **(a)** Find the area of the isosceles trapezoid shown using the traditional formula

$$K = \frac{1}{2}(b_1 + b_2)h$$

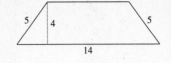

(b) Verify Brahmagupta's Formula for the area of a cyclic quadrilateral using this trapezoid.

5. Suppose that a cyclic quadrilateral has sides $a = b = 10$, $c = 15$, and $d = 21$.

(a) Find its area and the lengths of its diagonals.

(b) Show that $K \neq \frac{1}{2}mn$.

NOTE: This has bearing on whether the diagonals are perpendicular: The area is given by $K = \frac{1}{2}mn$ if and only if they are perpendicular.

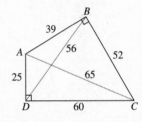

6. Repeat Problem 5(a) for the case $a = 102$, $b = 80$, $c = 136$, and $d = 150$. Compute $\frac{1}{2}mn$ in this case.

7. Verify all of Brahmagupta's Formulas for the cyclic quadrilateral in the accompanying figure.

_____ *Group B* _____

8. Write the formula for the left member of Ptolemy's Theorem, $ac + bd$, in terms of sines of half the intercepted arcs: $2r \cdot \sin \frac{1}{2}A \cdot 2r \cdot \sin \frac{1}{2}C + 2r \cdot \sin \frac{1}{2}B \cdot 2r \sin \frac{1}{2}D = 4r^2(\sin \frac{1}{2}A \sin \frac{1}{2}C + \sin \frac{1}{2}B \sin \frac{1}{2}D)$. Now use the trigonometric identity mentioned previously to find separate expressions for $\sin \frac{1}{2}A \sin \frac{1}{2}C$, $\sin \frac{1}{2}B \sin \frac{1}{2}(A + B + C)$, and $\sin \frac{1}{2}(A + B) \cdot \sin \frac{1}{2}(B + C)$. Complete the proof of Ptolemy's Theorem.

9. **A Nontrigonometric Proof of Ptolemy's Theorem** Given a cyclic quadrilateral *ABCD* as shown in the figure, let *BE* be constructed so that $m\angle ABE = m\angle DBC$. Using the facts

(a) *angles inscribing a common arc of a circle are congruent*

(b) *two triangles are similar if they have two pairs of corresponding angles congruent*

show that $\triangle ABE \sim \triangle DBC$. By similar triangles, the ratio relations indicated may be established (verify these), and by simple algebra, we have $ac + bd = mn$.

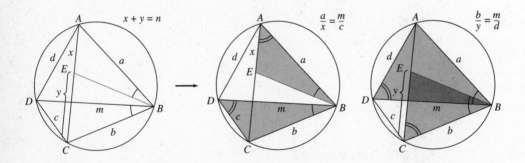

10. **Brahmagupta's Trapezium** A significantly challenging problem is finding a quadrilateral having integer sides, integer diagonals, and integer area. In most of our examples of quadrilaterals so far, this was not the case (review this to your own satisfaction). Brahmagupta's ingenious construction was to start with two **Pythagorean Triples** (a, b, c) and (x, y, z) (that is, $a^2 + b^2 = c^2$ and $x^2 + y^2 = z^2$), then form the cyclic

quadrilateral *QRST* with sides *az*, *cx*, *bz*, and *cy*, as shown in the accompanying figure. This quadrilateral was called a **trapezium.** Using the formulas for cyclic quadrilaterals, find the sides, diagonals, and area of the trapezium generated by (3, 4, 5) and (5, 12, 13), and verify that all these numbers are integers. (Note that the resulting quadrilateral is different from that of Problem 7.)

BRAHMAGUPTA'S TRAPEZIUM

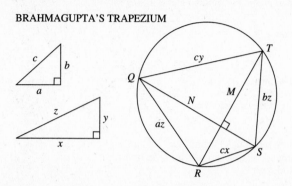

11. From Brahmagupta's Formulas, show by algebra (using $a^2 + b^2 = c^2$ and $x^2 + y^2 = z^2$) that in Brahmagupta's Trapezium (defined in Problem 10), the lengths of the two diagonals are given by

$$M = bx + ay \qquad \text{and} \qquad N = ax + by$$

12. With $A = az$, $B = cx$, $C = bz$, and $D = cy$ as the lengths of sides of the trapezium generated by the Pythagorean Triples (a, b, c) and (x, y, z),

   (a) show that

   $$S - A = \frac{1}{2}(-az + cx + bz + cy),$$

   $$S - B = \frac{1}{2}(az - cx + bz + cy),$$

   etc.

   (b) Use algebra and factoring to reduce the expression

   $$E = \sqrt{(S - A)(S - B)(S - C)(S - D)}$$

to

$$\frac{1}{2}(bx + ay)(ax + by) = \frac{1}{2}MN,$$

where M and N are the lengths of the diagonals. (See Problem 11.)
(***HINT***: Observe that the *first two* factors under the radical may be written

$$\frac{1}{2}(-az + cx + bz + cy)(az - cx + bz + cy) = \frac{1}{2}[(bz + cx)^2 - (az - cy)^2]$$

Square out and simplify, then re-factor (at this point use the relations for a, b, c and x, y, z). Finally, do the same thing to the last two factors.)

_____ *Group C* _____

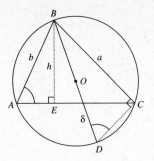

13. In the figure, circle O is the circumscribed circle of $\triangle ABC$. Observe: $m\angle A = m\angle D$ (why?) and $\triangle BCD$ is a right triangle similar to $\triangle BEA$. Show that if $BD = \delta = $ circumdiameter, then $h\delta = ab$.

14. A cyclic quadrilateral has consecutive sides of length a, b, c, and d, respectively, diagonals of length m and n, and circumdiameter δ. Using the result in Problem 13, show that

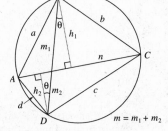

(a) $m\delta \cos\theta = ab + cd$, $n\delta \cos\theta = ad + bc$

(b) $\dfrac{m}{n} = \dfrac{ab + cd}{ad + bc}$

(c) If the diagonals are perpendicular, then

$$mn\delta^2 = (ab + cd)(ad + bc)$$

(Refer to the figure at right.)

15. Use Ptolemy's Theorem to obtain Brahmagupta's Formulas for m^2 and n^2, and also, when the diagonals are perpendicular, a formula for δ^2 in terms of the sides a b, c, and d. (See Problem 14.)

16. Using the result of Problems 11 and 12, prove Ahmd Alhari's Theorem (Hindu mathematician in the time of Brahmagupta): *The diagonals of a trapezium are perpendicular.* (**HINT**: First prove that the diagonals of a cyclic quadrilateral are perpendicular if and only if $K = \frac{1}{2}mn$; see Problem 14.)

17. **Brahmagupta's Theorem** (See Problems 11, 12, 13, and 14.) The culmination of several preceding results is the following.

Theorem: If (a, b, c) and (x, y, z) are two Pythagorean Triples, and if a cyclic quadrilateral (trapezium) be defined having sides of lengths $A = az$, $B = cx$, $C = bz$, and $D = cy$, respectively, then the quadrilateral has integral sides, diagonals, area, and circumdiameter. Moreover, if a multiplication table be set up as shown below, the sides of the quadrilateral are given by alternating the first two elements from the last column and last row; the lengths of the two diagonals are given by summing the elements in the diagonal and off-diagonal positions in the matrix formed by the first two rows and columns; and the remaining element, cz, is the circumdiameter.

	x	y	z
a	ax	ay	az
		$+$	
b	bx	by	bz
c	cx	cy	cz

*18. **Synthetic Proof of Heron's Formula** The proof which Heron himself constructed for the formula bearing his name contains very little algebra. Read and fill in all the missing details of the proof, given below. (See figure which follows, next page.)

Side BC of an arbitrary $\triangle ABC$ is extended to G so that $CG = AE$, the perpendiculars to lines BI and BC at I, the incenter, and C, respectively, meet at H, and segment BH is drawn, with line IH meeting BC at J.

(1) $AE = AF$, $BF = BD$, $CD = CE$, so the perimeter $2s$ of $\triangle ABC$ equals $2AE + 2BD + 2DC = 2AE + 2a$, or $AE = s - a$. Similarly, $BD = s - b$ and $CD = s - c$. Also, $BG = AE + a = s$.

(2) Area $\triangle ABC = ID \cdot BG$ because Area $\triangle ABC$ = sum of areas of triangles $\triangle BIC$, $\triangle CIA$, $\triangle AIB = \frac{1}{2}ra + \frac{1}{2}rb + \frac{1}{2}rc = rs$.

(3) The midpoint O of BH is the center of a circle passing through B, I, C, and H.

(4) $\angle CHB$ is supplementary to $\angle BIC$, hence

$$m\angle CHB = m\angle IBC + m\angle ICB$$

$$= \frac{1}{2}B + \frac{1}{2}C$$

$$= 90 - \frac{1}{2}A$$

$$= m\angle AIE$$

$$\therefore \triangle BCH \sim \triangle AEI$$

(5) $\quad \dfrac{BC}{CG} = \dfrac{BC}{AE} = \dfrac{CH}{EI} = \dfrac{CH}{ID} = \dfrac{CJ}{JD}$

(6) $\quad \dfrac{BC}{CG} + 1 = \dfrac{CJ}{JD} + 1, \quad \text{or} \quad \dfrac{BG}{CG} = \dfrac{CD}{JD}$

(7) $\quad JD \cdot BD = ID^2$ (Use Pythagorean Theorem and $BJ^2 = JD^2 + 2JD \cdot BD + BD^2$.)

(8) $\quad \dfrac{BG^2}{CG \cdot BG} = \dfrac{CD \cdot BD}{JD \cdot BD} = \dfrac{CD \cdot BD}{ID^2}$

(9) $\quad \therefore \ (\text{Area } \triangle ABC)^2 = ID^2 \cdot BG^2 = BG \cdot CG \cdot BD \cdot CD = s(s-a)(s-b)(s-c)$

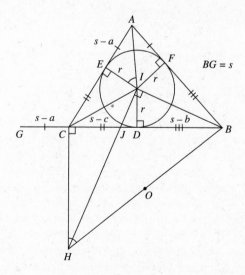

*For yet another proof, see Problem 23, §4.3.

1.4   A GLIMPSE AT MODERN CLASSICAL GEOMETRY

The Renaissance of the 16th and 17th centuries appeared not only in the fine arts, but in mathematics and science as well. Geometric problems arising from both mathematics and physics dominated mathematics research. A proliferation of elegant and intricate new relationships between circles and triangles, special points of collinearity and concurrence of lines associated with triangles—undreamed of by Euclid and his contemporaries—spread throughout the centers of learning in the European continent, and has continued even up through modern times. A prime example of this vast literature is a theorem devised by Jacob Steiner, a prominent 19th century mathematician. This theorem surprisingly ties together several other interesting results in geometry.

NINE-POINT CIRCLE AND THE MORLEY TRIANGLE

Two of these results have already been introduced (Section 1.1): The Nine-Point Circle and the Morley Triangle. For your convenience, we summarize them here. (See Figure 1.25)

NINE-POINT CIRCLE

The midpoints of the sides of a triangle, the feet of the altitudes on the sides, and the midpoints between the orthocenter and vertices lie on a circle.

MORLEY'S THEOREM

Corresponding angle trisectors of any triangle meet at the vertices of an equilateral triangle.

HISTORICAL NOTE

Jacob Steiner was born in 1796 in Germany. Although he did not learn to write until he was fourteen, he made up for it later with his prolific writing in geometry. He became a serious student of mathematics at the age of 18 and soon began his studies at the universities of Heidelberg and Berlin. Steiner was one of the leading contributors to the celebrated *Crelle Journal of Mathematics*—his articles appeared in practically every issue, and everything he submitted for publication was published. The chair of geometry was established for him at Berlin in 1834, a position he held until his death in 1867. Steiner and M.

Chasles, another outstanding 19th century geometer, independently gave a full development of synthetic projective geometry, including a construction of the classical conics by projective methods. All of Steiner's work was synthetic (without the use of coordinates). In spite of this, his arguments in maxima and minima surpassed in power the analytic methods then in vogue. It was once said that Steiner hated analysis as thoroughly as Lagrange disliked geometry. Even today, Steiner's methods and developments remain models for elegance in geometry.

NINE-POINT CIRCLE   MORLEY'S TRIANGLE

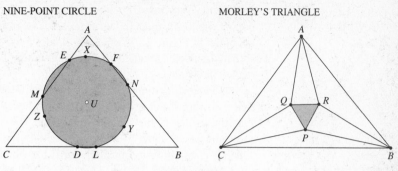

Figure 1.25

PEDAL TRIANGLES AND SIMSON LINE OF A TRIANGLE

Many theorems of classical geometry have to do with concurrence of lines and collinearity of points. Prominent examples are the theorems of Pappus and Desargues. (See Problems 11 and 12 in Section 1.1.) Another is the so-called *Simson Line* of a triangle.

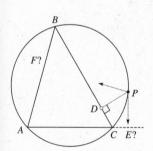

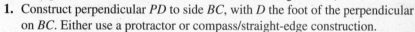

MOMENT FOR DISCOVERY

Pedal Triangles

Draw a large circle on a sheet of paper, and with a ruler draw a triangle inscribed in the circle, as illustrated. (This circle is called the **circumcircle** of the triangle.) Choose any point P on the circle (such as that shown in Figure 1.26).

1. Construct perpendicular PD to side BC, with D the foot of the perpendicular on BC. Either use a protractor or compass/straight-edge construction.

2. Construct perpendicular PE to side AC, with E the foot of the perpendicular on AC. (In this particular figure and for this choice of P, it is necessary to extend side AC, as shown.)

3. Repeat for side AB, constructing point F on segment AB.

4. What do you observe in particular about points D, E, and F?

5. Try this experiment for at least two other distinctly different positions of point P on the circumcircle.

6. Repeat the experiment for two points not lying on the circumcircle.

7. Write down your conclusion, based on this experiment.

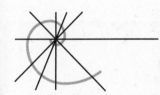

Figure 1.26

We will show how the conclusion you were supposed to reach in the Discovery Unit can be established from Ptolemy's Theorem and a few simple formulas for the

A PEDAL TRIANGLE
OF TRIANGLE $\triangle ABC$

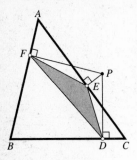

Figure 1.27

triangle derived from trigonometry. In the Discovery Unit, the perpendiculars from point P were constructed. In the general case, this leads to a special triangle $\triangle DEF$, as shown in Figure 1.27, where D, E, and F are the feet of the perpendiculars from P. This triangle is called the **pedal triangle** of $\triangle ABC$ with respect to point P (called its **origin**).

It is interesting to predict what $\triangle DEF$ will look like as one moves point P to different positions. For example, $\triangle DEF$ will be *similar* to the given triangle $\triangle ABC$ if P is the center of the circumcircle. (Can you verify this using simple geometric facts you know about?) Do you think there could be other cases when $\triangle DEF \sim \triangle ABC$? Can you find a position for P for which $\triangle DEF$ is equilateral (even if the given triangle $\triangle ABC$ is not)? Does the pedal triangle assume any shape whatever? A problem like this is a perfect illustration of the kind of experiment best done on a computer, using, for example, the software MacDraw™ or Geometric Supposer™. (This aspect of discovery in geometry will be discussed in more detail in the next section.)

Interesting formulas involving the sides of the pedal triangle lead to an easy proof of the observation you might have made in the previous Discovery Unit. The proof of the theorem will be developed following the informal discussion that follows; the proof of the corollary occurs in Problem 13.

THEOREM 1

Let P be the point of origin for pedal triangle $\triangle DEF$ of $\triangle ABC$, and let $PA = x$, $PB = y$, and $PC = z$. If R is the circumradius of $\triangle ABC$ and the sides of $\triangle ABC$ are a, b, and c, then the sides of the pedal triangle are given by the formulas

$$EF = \frac{ax}{2R}, \qquad DF = \frac{by}{2R}, \qquad \text{and} \qquad DE = \frac{cz}{2R}.$$

COROLLARY

The pedal triangle of $\triangle ABC$ is degenerate if and only if its origin lies on the circumcircle of the given triangle.

The line of collinearity resulting from the corollary is called the **Simson Line** of a triangle after Robert Simson (1687–1768), an anomaly since the line was never discovered by Simson, but rather by William Wallace in 1797. We are going to let you have the pleasure of establishing the corollary from the formulas in the theorem (see Problem 13 at the end of this section). Thus, the corollary is valid, provided we show how to derive the formulas stated in the theorem. The starting point is to recall three basic concepts from elementary (high school) geometry:

RESULT A
Angles inscribed in
a circle which sub-
tend a common arc
(such as angles A
and P in Figure 1.28)
are congruent.

RESULT B
Angles inscribed
in a semicircle
are right angles.
(A corollary to Re-
sult A using the
perpendicular OP.)

RESULT C
The midpoint of
the hypotenuse of
a right triangle
is equidistant
from the three
vertices.

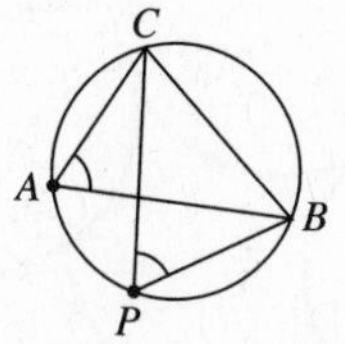

Figure 1.28

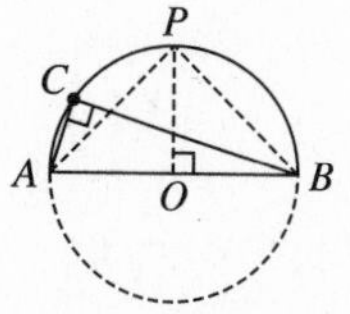

Figure 1.29

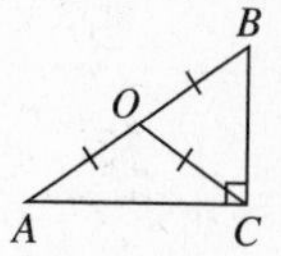

Figure 1.30

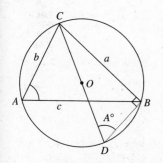

Figure 1.31

Let CD be the diameter of the circumcircle of $\triangle ABC$ (Figure 1.31). Then, from Result A, $m\angle CDB = m\angle A \equiv A$. (Although it is potentially a source of confusion, it is customary to use A, B, and C also to denote the measures of the interior angles of $\triangle ABC$, as in trigonometry.) By Result B, $\triangle BCD$ is a right triangle, with right angle at B. From the trigonometry of the right triangle, $BC/CD = \sin A$. That is, using standard notation,

$$\frac{a}{2R} = \sin A,$$

where R is the radius of the circumcircle of $\triangle ABC$. In exactly the same manner (choosing D at the appropriate locations on the circumcircle), we can prove that

$$\frac{b}{2R} = \sin B \quad \text{and} \quad \frac{c}{2R} = \sin C$$

Now consider pedal triangle $\triangle DEF$ of $\triangle ABC$ (Figure 1.32). The circumcircle of $\triangle AEF$ will pass through point P, with $AP = x$. (By Result C, the midpoint of segment AP is the center of this circumcircle.) Hence, by the relations just preceding (applied to $\triangle AEF$),

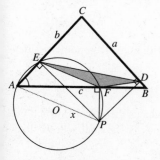

Figure 1.32

$$\frac{EF}{x} = \sin A = \frac{a}{2R} \qquad \rightarrow \qquad EF = \frac{ax}{2R}$$

Similarly,

$$\frac{DF}{y} = \sin B = \frac{b}{2R} \qquad \rightarrow \qquad DF = \frac{by}{2R}$$

and

$$\frac{DE}{z} = \sin C = \frac{c}{2R} \qquad \rightarrow \qquad DE = \frac{cz}{2R}$$

as desired.

**THE ASTROID AND
DELTOID**

You may recall from calculus a curve known as a **hypocycloid**. This is the path of a point on one circle as it rolls inside another circle without slipping (Figure 1.33). Point P will come into contact with the base circle a number of times, and these points are called the **cusps** of the hypocycloid. If the radius of the rolling circle has the appropriate value, the curve will be periodic and repeat after one cycle.

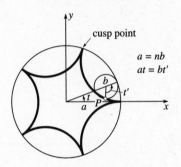

Figure 1.33

The hypocycloid of four cusps, called an **astroid,** has a particularly simple coordinate equation:

(1) $$x^{\frac{2}{3}} + y^{\frac{2}{3}} = a^{\frac{2}{3}}$$

Deriving **(1)** from the parametric form poses a nontrivial challenge to the average calculus student. In general, the hypocycloid of n cusps, where the radius of the base circle circumscribing it is a, is given parametrically by

(2) $$x = \frac{n-1}{n}\, a \cos t + \frac{a}{n}\cos(n-1)t$$

$$y = \frac{n-1}{n}\, a \sin t - \frac{a}{n}\sin(n-1)t, \quad t = \text{any real number}$$

To obtain the astroid, set $n = 4$, which, after working with a few trigonometric identities, yields the familiar form

$$x = a\cos^3 t, \quad y = a\sin^3 t$$

or, upon eliminating t, equation **(1)**. The case $n = 3$ produces the **deltoid**, which will be involved in a very special way with a given triangle, as we shall see. Its equations are

(3) $$x = \frac{2a}{3}\cos t + \frac{a}{3}\cos 2t, \quad y = \frac{2a}{3}\sin t - \frac{a}{3}\sin 2t$$

OUR GEOMETRIC WORLD

One of the components of an automatic transmission is known as a *planetary gear train*. This is an arrangement of three rotating gears within a larger interior gear. A single tooth of one of the inside gears will travel along the path of a hypocycloid. At the instant of shift, an electronic signal causes the gears to lock in place, and the outer gear which is attached to the drive shaft and rear axle begins to rotate at the same speed as the inner shaft attached to the engine, and the vehicle is now in "high gear."

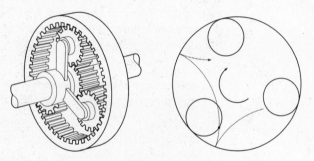

Figure 1.34

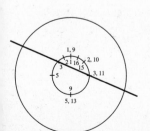

MOMENT FOR DISCOVERY

A Family of Rotating Lines

1. With the center of a sheet of paper as center, draw as large a circle as you can. Concentric with this, draw another circle *one-third* the radius of the large circle.

2. On the smaller of the two circles, mark points at 90-degree intervals. Bisect the four arcs obtained to obtain 45-degree intervals, then bisect these, in turn, to yield a total of 16 arcs of 22.5° each.

3. On the inside of the circle you just divided into 16 arcs, start numbering the 16 division points in the counterclockwise direction (1, 2, ... , 16).

4. On the outside of this same circle, number EVERY OTHER POINT in the CLOCKWISE direction 1, 2, ... , 16, starting at the same place as you did in Step 3. (You will have to go around twice this time, numbering some points twice.)

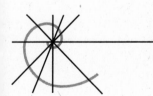

Figure 1.35

5. Start drawing the lines which join points labelled by the SAME NUMBER, (1, 1), (2, 2), ... , (16, 16). Keep going until you reach the very end, drawing a total of 16 lines. (Since the number 1 labels the same point, draw the *tangent* to the circle at that point to get started.)

6. What did you seem to get?

7. *For further reflection*:   What would be the result if you divide the original circle into 32 arcs, or 64 arcs? Try this experiment on the computer.

So far, we have discussed the Nine-Point Circle of a triangle, the Morley Triangle, the deltoid as generated by its tangents, and Simson's Line—seemingly altogether unrelated concepts. It is amazing that they all may be brought together in a beautiful theorem of geometry. We refer you to Figure 1.36 for illustration. We do not attempt a proof, but merely recommend it to you as a singular example of the beauty of geometry.

THEOREM 2 (STEINER'S THEOREM)

As a point revolves about the circumcircle of a triangle, the Simson Line with respect to that point generates the family of tangents to a deltoid, whose center is the Nine-Point Center, whose inner tangent circle is the Nine-Point Circle of the triangle, and whose cusps are located on the three lines through the Nine-Point Center which are perpendicular to the sides of the Morley Triangle.

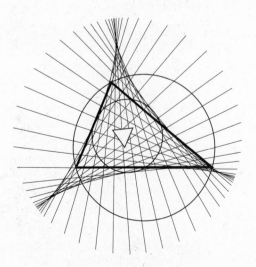

Figure 1.36

PROBLEMS (§1.4)

_____ *Group A* _____

1. Verify to your own satisfaction that the Nine-Point Circle of an equilateral triangle coincides with the incircle (inscribed circle of the triangle), and that it is one-half the size of the circumcircle of the triangle in this case.

2. Using the formulas involving hypocycloids, find the ratios of the radius of the outer circle to the inner circle for a hypocycloid of

   (a) 3 cusps

   (b) 4 cusps

   (c) *m* cusps

   (**HINT:** For **(a)**, we must have $C_1/C_2 = 3$.)

3. Use trigonometry to show that each side of the Morley Triangle of an equilateral triangle (figure to the right) is precisely *q* times as large as that of the given triangle, in which

   $$q = \sin 10° \sec 20°$$

   (**HINT:** The identity $\sin^2\theta = \frac{1}{2}(1 - \cos^2\theta)$ may be useful.)

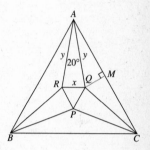

4. Using the idea advanced in the Moment for Discovery involving the tangents to a deltoid, design an experiment which will produce the tangents enveloping an astroid. (**HINT:** Outside numbers should alternate on every *third* point rather than on every other point.)

5. This experiment, appearing in E.H. Lockwood, *Book of Curves*, will show that circles, instead of lines, can be used to generate classical curves. (See figure at right.)

   (1) Draw a circle on a sheet of paper, using a diameter of about ¼ the width of your paper, or slightly less.

   (2) Mark a point *A* on the circle. With another point $P \neq A$ on the circle, draw a circle centered at *P* and passing through *A*.

   (3) Repeat Step **(2)** a number of times, more often the closer *P* is to *A*. (Let *P* vary about the entire circle.) What curve seems to be generated?

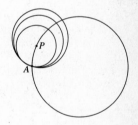

6. Repeat the instructions of Problem 5 to obtain another well-known curve, this time choosing *A* to be some point *outside* the circle. (These experiments are especially adaptable to computer graphics.)

7. Using coordinate geometry and the distance formula, verify the formulas in Theorem 1 for the example shown in the following figure.

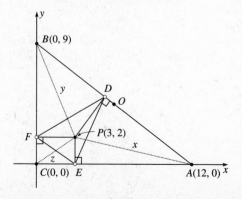

_____ **Group B** _____

8. Use the theorem on pedal triangles to prove:

Corollary The sides of any pedal triangle (with any point P as origin) of an equilateral triangle are proportional to the three distances from P to the vertices.

(This means, for example, that if the distances from P to the vertices of an equilateral triangle form a Pythagorean Triple, we can be certain that the pedal triangle from P is a right triangle.)

9. Prove that the incircle of an equilateral triangle bisects the segments joining its center with the vertices.

10. Verify the classical formula for a median of a triangle, $m^2 = \frac{1}{2}a^2 + \frac{1}{2}b^2 - \frac{1}{4}c^2$, for the specific triangle $\triangle ABC$ in the accompanying figure. (Find m directly, using trigonometry.) If you use the Law of Cosines for $\angle A$ in $\triangle ACD$ and $\triangle ABC$, the proof of the general formula is immediately evident. Try this.

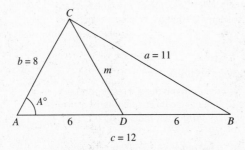

*11. A useful formula in geometry is the **Cevian Formula,** which generalizes the result in Problem 10 and expresses the length of a line segment AD (a **cevian**), where D is a point on side BC, in terms of the lengths of the sides of the triangle and the ratios of BD and DC to BC. Let $BD/BC \equiv BD/a = p$ and $DC/BC \equiv BC/a = q$. (Thus, $p + q = 1$.) The formula is

$$d^2 = pb^2 + qc^2 - pqa^2$$

Complete the details of the following steps used to establish this formula (see figure).

(1) $pa = BD$ and $qa = DC$

(2) $d^2 = (pa)^2 + c^2 - 2(pa)c\cos B$ (Law of Cosines for what triangle?)

(3) $2ac \cos B = a^2 + c^2 - b^2$ (?)

(4) $d^2 = p^2a^2 + c^2 - p(a^2 + c^2 - b^2)$
$= p(pa^2 - a^2) + c^2 - pc^2 + pb^2$ (?)

(5) $\therefore\ d^2 = p(-qa^2) + qc^2 + pb^2$ (?)

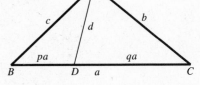

*See Example 5 and Problem 22, §4.3.

*12. Ceva's Theorem In 1678, the Italian mathematician Giovanni Ceva (1647–1736) discovered the geometric property that three cevians of a triangle are concurrent if and only if the ratios into which those cevians divide the sides obey the relation

$$\frac{p_1 p_2 p_3}{q_1 q_2 q_3} = 1$$

(see figure). Here, $p_1 = BD/BC$, $q_1 = DC/BC$, $p_2 = CE/CA$, ... , or $BD = p_1 a$, $DC = p_2 a$, $CE = p_2 b$, ... A simple proof of this makes use of a property of area: If in triangle $\triangle BPC$ cevian PD cuts BC in segments of ratios p_1 and q_1, respectively, then since $\triangle BPD$ and $\triangle DPC$ have the same altitude, $K_1/K_2 = p_1 a/q_1 a = p_1/q_1$. Complete the details in the following proof of Ceva's Formula.

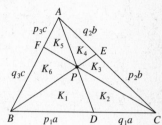

(1) $\dfrac{p_1}{q_1} = \dfrac{K_1}{K_2} = \dfrac{K_1 + K_5 + K_6}{K_2 + K_3 + K_4}$, $\therefore \dfrac{p_1}{q_1} = \dfrac{K_5 + K_6}{K_3 + K_4}$

(2) $\dfrac{p_2}{q_2} = \dfrac{K_1 + K_2}{K_5 + K_6}$, $\dfrac{p_3}{q_3} = \dfrac{K_3 + K_4}{K_1 + K_2}$

(3) $\dfrac{p_1}{q_1} \cdot \dfrac{p_2}{q_2} \cdot \dfrac{p_3}{q_3} = \dfrac{K_5 + K_6}{K_3 + K_4} \cdot ? \cdot ?$ (Complete these details.)

(4) For the converse, assume $\dfrac{p_1 p_2 p_3}{q_1 q_2 q_3} = 1$ and that the first two cevians AD and BE meet at P; if CP meets AB at F', then show that $F' = F$.

(**HINT:** To obtain **(1)**, cross-multiply in the first equation, simplify, and recast in ratio form.)

*See Theorem 1, §4.8.

_____ *Group C* _____

13. Prove the corollary of Theorem 1: The pedal triangle of $\triangle ABC$ is degenerate if and only if the origin P lies on the circumcircle. (**HINT:** See the accompanying figure; use Ptolemy's Theorem in Section 1.3.)

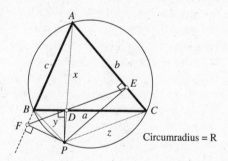

14. UNDERGRADUATE RESEARCH PROJECT Find the polar equation of the envelope of circles in the experiment of Problem 6 using polar coordinates.

NOTE: The subject of envelopes generated by a one-parameter system of curves $F(x, y, t) = 0$ (usually a family of lines) is sometimes found in the older books on differential equations. The main result is the following (under certain hypothesis): The envelope of $F(x, y, t) = 0$ is given by the solution of the system of equations in x and y (where t is eliminated),

$$\frac{\partial F}{\partial t} = 0 \qquad \text{and} \qquad F(x, y, t) = 0$$

You should carefully reference this result as part of the project. (Here's an example: Let

$$F(x, y, t) = \frac{x}{t} + \frac{y}{\frac{1}{t}} - 1 = \frac{x}{t} + ty - 1,$$

the family of lines having x- and y-intercepts at t and $\frac{1}{t}$, respectively. The system

$$\frac{\partial F}{\partial t} = -\frac{-x}{t^2} + y = 0 \qquad \text{and} \qquad F = \frac{x}{t} + ty - 1 = 0$$

reduces to $4xy = 1$, a rectangular hyperbola.)

*1.5 Discovery Via the Computer

There are a variety of software packages available which address the problem of

(a) visualizing by graphics certain properties of geometry, or

(b) verifying those properties by calculation (some software will do both).

We do not intend to provide a complete list of all the packages on the market (which is constantly changing), nor do we intend to evaluate the relative merits of the different systems. It is our purpose here merely to indicate the capabilities of some of them, with an example or two.

One program is called TURBO-BASIC or TURBO-PASCAL, designed for IBM-compatible PCs (available through Borland International, Inc.: Scotts Valley, California). This program includes a graphics mode that is accessed by the command SCREEN. When in this mode, two further commands, LINE and CIRCLE, enable the user to display any line segment specified by its endpoints and any circle specified by its center and radius. Thus, using either a WHILE or FOR/NEXT loop, one can simulate the graph of any function by displaying short segments joining a hundred or more pairs of adjacent points on the graph. Also, one can display lines located in extremely precise positions, as well as triangles, perpendiculars, circles, etc.

While this system is very effective, it often requires the user to write extensive computer programs in either BASIC or PASCAL in order to get the variety needed to work geometry problems. For example, the program given in Example 1 was written in BASIC, and even though a bit complicated, it displays a large number of Simson Lines of a given triangle (user may change the coordinates of the vertices at will) which results in a nice view of the deltoid for a direct verification of Steiner's Theorem (Section 1.4).

E X A M P L E   1   **Steiner's Theorem**

PROGRAM	**EXPLANATION**
INPUT a, b, c	vertices of $\triangle ABC$ are at $A(2a,0)$, $B(2b,0)$, $C(0,2c)$
SCREEN, COLOR,	graphics commands
WINDOW (–25,–14)–(35,30)	parameters chosen to eliminate distortion
LINE $(2 * a, 0) - (2 * b, 0)$	draws side AB of $\triangle ABC$
LINE $(2 * b, 0) - (0, 2 * c)$	draws side BC of $\triangle ABC$
LINE $(0, 2 * c) - (2 * a, 0)$	draws side AC of $\triangle ABC$
PI = 3.1415926	declares constant π
$c_1 = a + b : c_2 = c + ab/c$	coordinates of circumcenter
$r = \text{SQR}(a^2 + b^2 + c^2 + a^2 * b^2/c^2)$	radius of circumcircle
CIRCLE (c_1, c_2), r	draws circumcircle of $\triangle ABC$
FOR $k = 0$ TO 9	loop generating Simson Lines
INPUT	press RETURN to continue
FOR $t = 3 * k$ to $3 * k + 2$	graphs three Simson Lines at a time (for each k)
$p = c_1 + r * \cos(\text{PI} * t/15)$ $q = c_2 + r * \sin(\text{PI}*t/15)$	points on circumcircle occur in increments of 12°
IF abs$(p–a) < 1$ THEN $a = b$	substitutes B for A if P is close to A
$u = a * (a^2+c^2)^{-1}(a * p - c * q + 2 * c^2)$ $v = c * (a^2+c^2)–1(–a * p + c * q + 2 * c^2)$	coordinates of foot of P on segment AC $(= Q(u, v))$
IF abs$(u–p) < .001$ THEN 10 $m = v/(u–p)$	skips if $u = p$ slope of line QR, where $R(u, 0)$ is foot of P on line AB
$x_1 = -30$ $y_1 = m(-30 - p)$	coordinates of a point off-screen on Simson Line QR

$x_2 = 30$ Second point off-screen on
$y_2 = m(30 - p)$ Simson Line QR

IF $abs(y_1) > 30$ OR   Prevents overflow error
      $(y2) > 30$ THEN 20

GOTO 30

20 $y_1 = -30 : y_2 = 30$
  $x_1 = -30/m + p : x_2 = 30/m + p$

30 LINE $(x_1, y_1) - (x_2, y_2)$ Full Simson Line QR appears
                                   on screen

10 NEXT t

NEXT k

END❏

Two recent developments in the area of geometric graphics, exceeding even the capabilities of the excellent GEOMETRIC SUPPOSER™ (available through Geometric Supposer Series: Sunburst Publications, Pleasantville, New York), are the GEOMETER'S SKETCHPAD™ (Key Curriculum Press, Berkeley, California), and CABRI™. The GEOMETER'S SKETCHPAD is a software kit that is specifically designed to enable a user not only to test theories and known theorems visually, but to perform accurate computations for analytical verification.

As an example, let's see how this software would allow us to test the theorem that the altitudes of a triangle are concurrent. We first "draw" a triangle at random on the monitor (by specifying vertices A, B, C using a mouse and then entering the command TRIANGLE (A, B, C). Next, we "draw" two altitudes AD and BE and call for the coordinates of their point of intersection, P. Then we repeat this for altitudes BE and CD, calling for the coordinates of their intersection Q and compare coordinates to verify that $P = Q$. Finally, using the "drag" capabilities (similar to GEOMETRIC SUPPOSER), we can keep all the above procedure while we move one of the vertices of the triangle to any point on the screen and call for the coordinates of P and Q for each new triangle we consider, verifying the geometry principle for as many triangles as we like. Further examples follow.

EXAMPLE 2   **The Sum of the Angles of a Triangle**   Begin with a triangle ABC on the screen. Then command the display of angle measures and angle sum. Using the mouse, drag vertex A to any position, and for each new position observe the sum (displayed automatically without additional commands).

DISPLAYED
ON SCREEN

$$A = 33.4°$$
$$B = 68.1°$$
$$C = 78.5°$$

$$A + B + C = 180.0°\square$$

EXAMPLE 3   Given two circles, and about 20–30 points distributed evenly on the two circles (vertices of regular polygons), find the locus of the midpoint M of the line PP', where P' is the point on the second circle corresponding to P on the first ("locus" is another term for "curve" or "path"). The use of SKETCH-PAD™ shows the following figure appearing on the screen (Figure 1.37). The locus of M is apparently another circle. (This can, indeed, be proven by use of coordinate methods to be found in Chapter 5.)$\square$

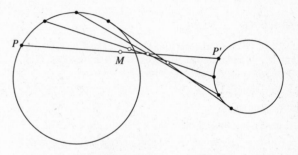

Figure 1.37

NOTE: A more general version of the problem above appeared in the section on Modeling edited by Jonathan Choate in the *UMAP*[6] *Consortium*, Vol. 7 (1992), p. 7.

Another effective piece of software that should be mentioned is EXPLORING GEOMETRY™ (Harper Collins, Pleasantville, Illinois). This software is effective in dealing with images under affine transformations, particularly reflections and rotations, which is the topic of Chapter 5.

It should be remarked that while such capabilities are quite remarkable, the results are only experimental, and they only indicate what *might* be true in *Euclidean* plane geometry. Furthermore, no conjectures about non-Euclidean geometry can be tested directly (without resorting to models). The answers we do obtain and the relations which may be suggested from these experiments do not give us totally dependable results even in Euclidean geometry. Although the accuracy is very great, there is some danger in concluding, for example, that two points are coincident based on a comparison of their (computed) coordinates, or other properties. For that matter, any formula based on computer computations is suspect. In

[6] Undergraduate Mathematics and Its Applications.

fact, because there are only a finite number of pixels on any computer screen, only properties for a *finite geometry* are actually being demonstrated, and it is possible for discrepancies to occur. Any property which we might observe on the computer, therefore, remains unproven, and is merely a conjecture until such time as we can provide a general proof based on logic and abstract reasoning.

The abstract proof is one aspect of mathematics not yet programmable on a computer despite extensive research in the field of artificial intelligence and the ever-increasing capabilities of computer hardware. Thus, the proof or counterexample in mathematics remains at the heart of the discovery of what is true and what is false, and it remains the ultimate tool for verifying new (or old) truths and relationships—the focus of the remainder of this text.

PROBLEMS   (§1.5)

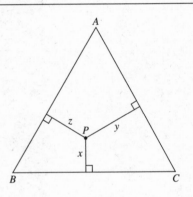

_____ *Group A* _____

1. Let $\triangle ABC$ be an equilateral triangle and make P any interior point. If x, y, and z are the distances from P to the sides, find what location for P makes the sum $S = x + y + z$

   (a) a maximum

   (b) a minimum

   (c) Investigate this problem when P lies outside $\triangle ABC$.

2. Let $\triangle ABC$ be an equilateral triangle, and consider any point P on its incircle. As P varies on the circle, investigate the maximum/minimum properties of

   (a) $x + y + z$

   (b) $x^2 + y^2 + z^2$,

   where $x = PA$, $y = PB$, and $z = PC$.

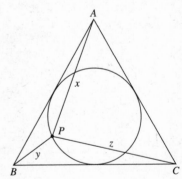

_____ *Group B* _____

3. Find the **third pedal triangle** of a given triangle $\triangle ABC$ from a given point P. (If $\triangle DEF$ is the pedal triangle of $\triangle ABC$ from P, then the *second* pedal triangle of $\triangle ABC$ is the pedal triangle $D'E'F'$ of $\triangle DEF$ from P, and the *third* pedal triangle is the pedal triangle $D''E''F''$ of $\triangle D'E'F'$ from P.) Discover what relationships you can from using any of the software discussed in this section.

   NOTE: The theorem of geometry which this demonstration illustrates is a result due to J. Neuberg, which may be found in Coxeter and Greitzer, *Geometry Revisited*, p. 24.

2

FOUNDATIONS OF GEOMETRY 1: POINTS, LINES, SEGMENTS, ANGLES

OVERVIEW

This book deals extensively with the development of geometry from the axiomatic point of view for three important reasons:

(1) Geometry is easier to understand from simple axioms than from an entire landscape of mathematics, which often includes an overwhelming variety of ideas and methods that are unfamiliar to beginners.

(2) Many high school geometry courses continue to be taught from axioms, and some of you may be teaching a course like this someday.

(3) Without axioms and the traditional development, it is impossible to appreciate or understand one of the great dramas in the history of mathematics, the development of non-Euclidean geometry.

The drawback is that with axioms comes the necessity for proof, and a certain amount of plodding is inevitable. One is not afforded the luxury of prior knowledge. In fact, too much knowledge may even be a handicap; all results, trivial or otherwise, must be derived directly from the axioms. We recognize the difficulty for the student and will therefore provide as much help as possible, with many examples of proofs worked out. At first we will encourage the outline form, or two-column proof (which can actually be applied to any field of mathematics). Then, as we progress in our development, we will gradually move away from this kind of argument toward the freer prose style of proof as found in most mathematics texts, indeed, even as found in Euclid's *Elements*. (Contrary to popular belief, the two-column proof did not originate with Euclid.)

However, it is not our intent to dwell on axiomatics exclusively, so our axiomatic development will be as sparing as possible. We will return to a more comprehensive approach to geometry in Chapter 4 once we understand the foundations, and at that time we will include coordinates and vectors in our study. Transformations and techniques of modern geometry will appear in Chapter 5.

*2.1   AN INTRODUCTION TO AXIOMATICS AND PROOF

Our aim is first to give you an idea of what it means to work in an axiomatic system and how to write simple proofs.

WHY IS IT TRUE?

You may wonder what role discovery plays in proving theorems. This is best illustrated by example.

EXAMPLE 1   Which is larger for positive real x: x or $\sin x$?

SOLUTION

We might approach this first from the pragmatic viewpoint: What can our pocket calculator tell us? Naturally, it can be used to calculate efficiently any rational approximation for $\sin x$. If $x = 0.05$ (radians), then the calculator shows

$$\sin 0.05 = 0.049979169 < 0.05$$

And if $x = 1.5$, we find

$$\sin 1.5 = .997494986 < 1.5$$

Such calculations are evidence of the inequality $\sin x < x$, but they do not by themselves prove the inequality.❑

It is often difficult to find mathematical reasons for various phenomena we may have observed or discovered by experiment. In this particular example, however, there are two sources of information, or tools, with which we have to work.

I.   *Method of Geometry/Trigonometry*   Thinking back to the definition of $\sin t$ for real t, we might remember that the sine and cosine functions are the so-called circular functions: they represent the coordinates of a point on the unit circle. Thus, in Figure 2.1 we see that the question sin $t < t$ actually involves the geometric principle involving the length of the arc of a circle and that of the chord which joins the endpoints of that arc. Also involved is the length of the hypotenuse of a right triangle compared to either leg. Thus,

$$\sin t = PQ < \text{chord } PT < \text{arc } PQ = t,$$

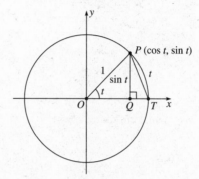

Figure 2.1

and therefore sin $t < t$.

II. *Method of Calculus*   Since we want to examine the inequality sin $x < x$, it makes sense to define the function

$$f(x) = x - \sin x$$

The derivative is

$$f'(x) = 1 - \cos x$$

and since cos $x < 1$ for $x \leq 1.5$, it follows that $f'(x) > 0$. Hence, the function f is increasing on the interval [0, 1], and therefore

$$f(x) > f(0) \text{ for } 0 < x < 1.5$$

That is, since $f(0) = 0$,

$$f(x) > 0, \text{ or } x > \sin x, \text{ for } x > 0$$

EXAMPLE 2   Let's say that you have just discovered that in a 45-degree (isosceles) right triangle $\triangle ABC$ (Figure 2.2), the midpoint of the hypotenuse (M) appears to be equidistant from the three vertices, and that, therefore, a circle centered at M will pass through the vertices of the triangle. (This discovery may have been made from the computer, drawing experiments, or other empirical sources.) We now want to determine if this phenomenon is actually valid. Initial reflection also suggests further questions: If true for a 45-degree right triangle, is it true only for this case, or for all right triangles? Is it true for arbitrary triangles?

SOLUTION

Depending on whether we had actually seen the theorem before, our thoughts might turn to the phenomenon in isosceles triangles where the line from C to the midpoint M (Figure 2.3) of base AB bisects the vertex angle $\angle ACB$. Thus,

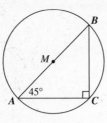

Figure 2.2

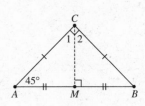

Figure 2.3

$$\angle 1 = \angle 2 = \frac{1}{2}m\angle ACB = \frac{1}{2}(90°) = 45°,$$

making triangles $\triangle MAC$ and $\triangle MBC$ isosceles triangles. Therefore,

$$MA = MC \text{ and } MC = MB$$

It is easy to see that the theorem is not true for triangles in general: we have only to draw an extreme case (to make it really convincing): Say $MC = \frac{1}{2}MA$, as in Figure 2.4. The question for right triangles in general remains.❏

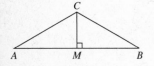

Figure 2.4

As you might guess from the above examples, finding and writing proofs in mathematics is to the mathematician as conducting experiments in a scientific laboratory is to the scientist. Although we might think that an explanation of why a property is valid (its proof) is only to convince other people, equally important is the use of logical reasoning as a *tool* for finding out whether a property is true, long before we try to communicate it to others.

DISCOVERING RELATION-SHIPS FROM AXIOMS

Since we are going to study axioms, let's take a brief look at how we might make discoveries in an axiomatic system. Our example will be one involving points and lines.

EXAMPLE 3   Consider the following three axioms.

Axiom 1: Each line is a set of four points.

Axiom 2: Each point is contained by precisely two lines.

Axiom 3: Two distinct lines which intersect do so in exactly one point.

QUESTION: Do parallel (nonintersecting) lines exist?

SOLUTION

As with most axiomatic systems in geometry, our ideas about axioms can be more clearly understood by representative diagrams. Let's use an ordinary line segment for each "line" mentioned in the axioms, and dots for "points." When three or four dots in the diagram appear on a line segment, it means that those four dots represent points on a line in the axiomatic system. From Axiom 1, we could then represent a line as shown in Figure 2.5.

Assuming points actually exist, we find that lines exist (Axiom 2), and our reasoning then takes us in the following direction indicated by Figure 2.6.

If we continue in this manner indefinitely, we can see that we are generating an example for the axiom system that has an infinite number of points, and many pairs of nonintersecting lines do indeed seem to exist for these axioms. The final figure can be arranged in a slightly different manner, generated in the final stages by computer, as suggested in Figure 2.7.

A FRACTAL GENERATED BY AXIOMS

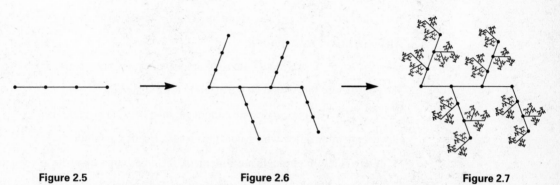

Figure 2.5　　　　　　　**Figure 2.6**　　　　　　　**Figure 2.7**

NOTE: The resulting diagram in Figure 2.7 is a **fractal**, a geometric object which has the property that at any point of the diagram, a window, when enlarged, is an exact copy of the original figure. (See Figure 2.8.)❏

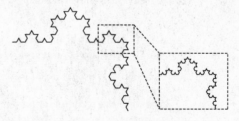

Figure 2.8

IF-THEN STATEMENTS, CONDITIONALS

The most common statement we encounter in geometry (and mathematics in general) is the conditional "if-then" proposition, which may be written symbolically as

$$p \rightarrow q \quad \text{(read: "p implies q," or "if p, then q")}$$

The condition p is called the **hypothesis**, or "given," and q, the **conclusion**—the objective of proof (the "to prove" part).

Given one conditional, several others may be defined. Since they are important, we focus on two such conditionals, the **converse** and **contrapositive**, tabulated for convenience. The prefix ~ stands for the **negation** of a statement: if p is any condition, $\sim p$ means *not p*.

	GIVEN CONDITIONAL	ITS CONVERSE	ITS CONTRAPOSITIVE
Prose Form	if p, then q	if q, then p	if NOT q, then NOT p
Symbolic Form	$p \to q$	$q \to p$	$\sim q \to \sim p$

Each of the above propositions plays its own role in proofs. The contrapositive is logically equivalent to the original conditional, but the converse is not. For example, consider the true statement

Right angles are congruent.

That is, stated as a conditional,

If two angles are right angles, then they are congruent.

The converse of this would be the patently false statement

If two angles are congruent, then they are right angles.

However, a proposition whose converse is also true is the familiar

If two sides of a triangle are congruent, then the angles opposite are congruent.

The converse "If two angles of a triangle are congruent, then the sides opposite are congruent" is equally valid. Thus we have both $p \to q$ and $q \to p$. That is, if p, then q (or p is true ONLY IF q is true), and if q, then p (or p is true IF q is true). Thus,

$$p \to q \quad \text{and} \quad p \leftarrow q \quad \text{means} \quad p \leftrightarrow q$$

often abbreviated in prose form by

p iff q

(read "p **if and only if** q"). Here, we say that p and q are **logically equivalent**, or we can also say that q **characterizes** p. In the above example then, isosceles triangles are *characterized by having congruent base angles.*

DIRECT PROOF

The simplest form of argument is, of course, the **direct proof**, which is the embodiment of the **transitivity law** for implication.

LOGIC RULE 1: TRANSITIVITY

If $p \to q$ and $q \to r$, then $p \to r$.

Note how this is used in a typical direct proof.

GENERAL FORM OF A DIRECT PROOF

Suppose $p \to q$, $q \to r$, and $r \to s$ are valid. Assume p as hypothesis. Then q. Since q, then r. Since r, then s. Therefore, $p \to s$.

The outline form of the above looks like this:

OUTLINE FORM OF A DIRECT PROOF

Previously proven or accepted propositions:

Theorem 1: $p \rightarrow q$

Theorem 2: $q \rightarrow r$

Theorem 3: $r \rightarrow s$

THEOREM: If p, then s.

Proof

Given: p

Prove: s

CONCLUSIONS	JUSTIFICATIONS
(1) p	given
(2) q	*Theorem 1*
(3) r	*Theorem 2*
(4) $\therefore s$	*Theorem 3*

Sometimes a definition is used in a proof in the place of a theorem or axiom. For example, the proof of the proposition that the least distance from a point P to a line ℓ is the perpendicular distance $PQ = a$, where $PQ \perp \ell$ (Figure 2.9) might look like this:

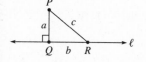

Figure 2.9

CONCLUSIONS	JUSTIFICATIONS
(1) Let P be given, not on line ℓ, and suppose line $PQ \perp \ell$.	given
(2) Let R be any other point on ℓ.	Axiom: A line contains many points
(3) $\triangle PQR$ is a right triangle.	*definition*
(4) If $QR = b$ and $PR = c$, then $a^2 + b^2 = c^2$.	Pythagorean Theorem
(5) $a^2 < c^2$ or $a < c$ $\therefore PQ < PR$	properties of real numbers and inequalities

INDIRECT PROOF

Another kind of mathematical argument was introduced by Euclid himself. It is often described by the phrase *reductio ad absurdum* ("to reduce to an absurdity"). The logical pattern in this kind of argument is to assume that the conclusion you are trying to prove is false, and then show that this leads to a contradiction of the hypothesis. In other words, to show that p implies q, show instead that *not q* implies *not p*. This means we prove instead the *contrapositive* of the proposition instead of the proposition itself.

Two further important rules of logic govern the validity of the indirect proof in mathematics.

LOGIC RULE 2: LAW OF EXCLUDED MIDDLE

For every condition or proposition p, either p is valid, or $\sim p$ is valid.

LOGIC RULE 3: RULE OF CONTRAPOSITIVE

Every conditional proposition is equivalent to its contrapositive.

A passage from Euclid's *Volume 1, Book* I reveals a slightly more involved line of indirect reasoning. Euclid has just proven that if one side of a triangle is greater than a second, then the angle opposite the first side is greater than that opposite the second (Proposition 18). Then we come to the converse, Proposition 19 (as found in Heath's *The Thirteen Books of Euclid's Elements*):

PROPOSITION 19

In any triangle the greater angle is subtended by the greater side.

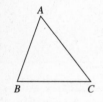

Figure 2.10

Let ABC be a triangle having the angle ABC greater than the angle BCA; I say that the side AC is also greater than the side AB.
For, if not, AC is either equal to AB or less.
Now AC is not equal to AB;
for then the angle ABC would also have been
equal to the angle ACB; [I. 5][1]
but it is not;
      therefore AC is not equal to AB.
Neither is AC less than AB,
for then the angle ABC would also have been less than the angle ACB;
but it is not;
      therefore AC is not less than AB.
And it was proved that it is not equal either.
      Therefore AC is greater than AB.

In this passage, Euclid states two possibilities other than the one he is trying to prove: AC equals AB, or AC is less than AB. Then he proceeds to eliminate two of the three logical possibilities by contradiction, leaving the desired possibility as the only conclusion remaining. This type of argument was referred to by the early Greeks as the *method of exhaustion*, that is, the method of *exhausting all possibilities*. In such an argument, two things are crucial:

(1) You must consider *all* the logical possibilities.

(2) All the logical possibilities must lead to a contradiction *except* the one you are trying to prove.

---

[1]This is Euclid's proposition about isosceles triangles: If two sides of a triangle are equal, the angles opposite are equal.

HISTORICAL NOTE

About all we know concerning the man Euclid is that he was a professor of mathematics around 300 B.C., taught at the University of Alexandria in Greece, and was the author of the celebrated *Elements*. The stories told of Euclid show that he was a serious, uncompromising scholar. When asked by King Ptolemy whether there were an easier way to learn geometry (other than studying the *Elements*), Euclid is supposed to have responded, "There is no royal road to geometry!" Euclid's great contribution to mathematics was the elegant, perfectly logical arrangement of the *Elements*. This monumental work consists of 13 books containing 465 propositions, with detailed

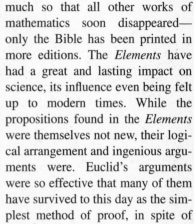

arguments for each, all based on only 10 basic axioms. The work was an immediate success, so much so that all other works of mathematics soon disappeared—only the Bible has been printed in more editions. The *Elements* have had a great and lasting impact on science, its influence even being felt up to modern times. While the propositions found in the *Elements* were themselves not new, their logical arrangement and ingenious arguments were. Euclid's arguments were so effective that many of them have survived to this day as the simplest method of proof, in spite of major revisions of his work to make it more rigorous.

A good symbol to use in proofs when a contradiction has been reached is →×←. This signals the end of one chain of logic that produces a contradiction and the start of another, where the next case is considered.

The mathematical justification for this "process of elimination" is our next, and final, rule for logic.

LOGIC RULE 4: RULE OF ELIMINATION

Suppose that $r, s, t,..., u$ is a complete finite set of logical cases, and that some (or none) of them imply $\sim p$ and all the rest imply q. Then $p \rightarrow q$ is valid.

We present the model for this argument for a typical proof by exhaustion in two-column form.

OUTLINE FORM OF AN INDIRECT PROOF

Previously proven propositions:

Theorem 1: $r \rightarrow \sim p$

Theorem 2: $s \rightarrow \sim p$

Complete set of logical possibilities: r, s, q

OUTLINE FORM OF AN INDIRECT PROOF (CONTINUED)

THEOREM: If p, then q.

Proof

Given: p

Prove: q

CONCLUSIONS	JUSTIFICATIONS
(1) p	given
(2) $r, s,$ or q	all the logical cases
(3) Assume r.	first logical case
(4) ~p →←	*Theorem 1*
(5) Assume s.	second logical case
(6) ~p →←	*Theorem 2*
(7) ∴ q	Rule of Elimination

Because it is used so often in geometry, we state formally the special rule Euclid used in stating the three logical possibilities at the beginning of his proof of Proposition 19.

TRICHOTOMY PROPERTY OF REAL NUMBERS

For any two real numbers a and b, either $a < b$, $a = b$, or $a > b$.

PROBLEMS   (§2.1)

_____ *Group A* _____

1. If p = "John wears his raincoat," q = "it rains," and r = "Mary is happy," translate the following propositions of logic into prose if-then statements:

   (a) $q \to p$ **(b)** $p \to r$ **(c)** $q \to \sim r$ **(d)** $\sim r \to q$

2. Suppose that q is the statement "it rains" and s = "there are no clouds in the sky." Translate the following propositions of logic into prose if-then statements:

   (a) $\sim s \to q$ **(b)** $q \to \sim s$ **(c)** $\sim q \to s$

   (d) Is the statement in **(b)** the same as that in **(a)**?

   (e) Is the statement in **(c)** the same as that in **(a)**?

In each of the following problems (3–6) you are given a statement that is implicitly an "if-then" statement. Treat each of the statements as a proposition to be proven, and write down what is given (the "if" part) and what is to be proven (the "then" part). Also, state the *converse* of each proposition, and decide whether the converse is valid in each case.

3. A right triangle has only one right angle.

4. Every equilateral triangle is equiangular.

5. The medians of a triangle are concurrent lines.

6. *Chinese Proverb*: Success can be achieved only by hard work. (***WARNING:*** To obtain the correct if-then statement, the usage of the word "only" must be carefully considered.)

7. An economist claims that inflation causes high unemployment.

   (a) Write this in the form $p \rightarrow q$, identifying precisely the components p and q.

   (b) Under what circumstances would $p \rightarrow q$ be refuted?

8. Suppose that the following information is at our disposal.

   Definition: a is b iff $r \rightarrow s$

   Axiom 1: $r \rightarrow q$

   Theorem 1: If a is c then $q \rightarrow t$.

   Theorem 2: $t \rightarrow s$

   Fill in the missing reasons in the following two-column proof below.

   Theorem: If a is c then a is b.

   Proof

   Given: a is c

   Prove: a is b (i.e., $r \rightarrow s$)

Conclusions	Justifications
(1) a is c	given
(2) r	assumption for r → s
(3) q	*Axiom 1*
(4) t	(a) _____?
(5) s	(b) _____?
(6) ∴ a is b	(c) _____?

9. Rewrite the following proof in two-column form: John lost his locker key in his home room or on his way home from school. Since he did not stop in his homeroom, he must have lost the key on his way home from school.

_____ *Group B* _____

10. Give an outline proof of $p \rightarrow q$ (Given: p, Prove: q) if the following propositions are valid: ~q → r and either r → ~p or r → q.

11. Suppose it has been established that either $p \rightarrow r$ or $p \rightarrow s$, and

    Theorem 1: $y \rightarrow {\sim}p$ *Theorem 3*: $s \rightarrow y$

    Theorem 2: $r \rightarrow x$ *Theorem 4*: $x \rightarrow q$

    Prove in outline form that $p \rightarrow q$. (***HINT:*** Your first step is to assume p. Then either r or s. Justification: All the logical possibilities.)

_____ *Group C* _____

12. Suppose the following have been established or assumed:

    Axiom 1: $p \rightarrow {\sim}y$

    Axiom 2: ${\sim}q \rightarrow r$

    Theorem 1: $p \rightarrow {\sim}z$

Theorem 2: $x \rightarrow q$ or $x \rightarrow z$

Theorem 3: $r \rightarrow x$ or $r \rightarrow y$

Write a two-column proof of the proposition $p \rightarrow q$.

13. In the axiom system given in Example 3, discover the consequences of the following additional assumption: Every pair of lines intersects at exactly one point.

14. Research the usage of the term "exhaustion" in mathematics, or "method of exhaustion." Eudoxus (370 B.C.) was one of the first to have used the term.

*2.2   THE ROLE OF EXAMPLES AND MODELS

Without examples, very little new mathematics would ever be discovered. Examples literally point the way for modern mathematical research, showing the possibilities of what may be true, and sometimes what is not. A model is like an example, but it is usually more elaborate and for the purpose of simulating a system of axioms. (The currently popular area known as "mathematical modeling" involves an entirely different purpose for models; in this field, models are constructed in some mathematical system to simulate real-world situations which occur outside of mathematics.)

EXAMPLES AND FALSE PROOFS IN MATHEMATICS

As often happens, an argument we use to prove a theorem may have unsuspected flaws, and sometimes a concrete example will shed light on the error.

EXAMPLE 1   Consider the following theorem and the reasoning used to prove it.

THEOREM: The sum of the measures of the angles of a parallelogram equals 360°.

Proof: Consider Figure 2.11, where the parallelogram is divided into two triangles, each of whose angle measures sum to 180°. Hence, in the original parallelogram,

$$m\angle A + m\angle B + m\angle C + m\angle D = m\angle 1 + m\angle 2 + m\angle B + m\angle 3 + m\angle 4 + m\angle D$$

$$= (m\angle 1 + m\angle 3 + m\angle B) + (m\angle 2 + m\angle D + m\angle 4)$$

$$= 180° + 180° = 360°$$

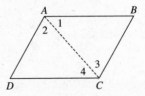

Figure 2.11

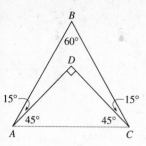

Figure 2.12

It may seem difficult at first to find fault with this argument. The thought might even occur to us to try our luck at generalizing it. Since we did not actually use the fact that the figure was a parallelogram, maybe we could prove this for any quadrilateral. But consider the example illustrated in Figure 2.12. Here we have a quadrilateral $ABCD$ with angles at A, B, and C having measures 15°, 60°, and 15°, respectively, and with a right angle at D. Since a right angle has measure 90°, we obtain the sum

$$m\angle A + m\angle B + m\angle C + m\angle D = 15° + 60° + 15° + 90° = 180°!$$

Hence, something is wrong with the argument, since it should also apply to the quadrilateral in Figure 2.12.❏

NOTE: It may seem that a different measure for the angle at D should be used, namely, 270° instead of 90°, since the *interior* of the quadrilateral is on the opposite "side" of $\angle D$. But this is contrary to the essential property of an angle (*defined only by its sides*), which can have only one unique angle measure.

The next example establishes a blatantly false theorem, and serves as a warning about the danger of misusing diagrams to aid in proof-writing.

EXAMPLE 2   Consider the following result, with an apparently rigorous argument to prove it. (See Figure 2.13.)

Theorem (!): Any obtuse angle $\angle ABC$ is a right angle.

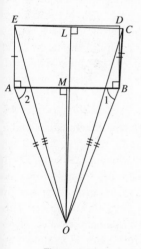

Figure 2.13

Proof: Construct a rectangle $ABDE$, as shown, so that $AE = BD = BC$, and complete the quadrilateral $ABCE$. Then construct the perpendicular bisectors of segments EC and AB at L and M, respectively. Since lines AB and EC are not parallel, their perpendiculars at M and L will not be parallel and must meet at some point O. Construct the remaining line segments OA, OE, OC, and OB. Now $OA = OB$ and $OE = OC$, since O is equidistant from A and B, and from E and C. Then by SSS $\triangle OBC$ is congruent to $\triangle OAE$, and

$$m\angle OBC = m\angle OAE$$

But these two angles are made up of their parts, $\angle 1$ and $\angle ABC$, and $\angle 2$ and $\angle BAE$. Hence

$$m\angle ABC + m\angle 1 = 90° + m\angle 2$$

But since $\triangle ABO$ is isosceles, $m\angle 1 = m\angle 2$. Then the above equation reduces to

$$m\angle ABC = 90°$$

Therefore, $\angle ABC$ is a right angle.❏

Can you find the fatal flaw with this reasoning? Problem 9 at the end of this section will provide you with some ideas.

AXIOMS, AXIOMATIC SYSTEMS

An axiomatic system always contains statements which are assumed without proof—the **axioms**. These axioms are chosen

(a) for their convenience and efficiency

(b) for their consistency

and, in some cases (but not always)

(c) for their plausibility

UNDEFINED TERMS

Every axiom must, of necessity, contain some terms that have been purposely left without definitions—the **undefined terms**. To attempt a precise definition for every significant term that might appear in a particular development would result in failure. For, as in a dictionary, we may define A in terms of B and C, then B and C in terms of D, E, and F, and so on. But inevitably, we will find that at some point the term W will have been defined in terms of one defined earlier, A, B, C, ... , resulting in a circular chain.

In geometry, the most common undefined terms are "point" and "line." We may think of a point as a dot on a sheet of paper, computer screen, or chalkboard, but in reality a dot has physical (albeit microscopic) dimensions, and does not exactly fit our idealistic viewpoint. Euclid defined a point as "that which has no part" and a line as "breadthless length" (i.e., "length without width"). However, these statements mainly served as a reminder to a reader of the *Elements* about the obvious. Euclid made no attempt to prove properties of points and lines precisely from these definitions.

The modern axiomatic approach is to decide at the beginning what the undefined terms shall be, then set down the properties those objects are to have in the axioms. One's choice of undefined terms and axioms can be quite arbitrary, however, as long as no contradictions result. Regarding this arbitrary nature of the axioms and use of undefined terms, the great philosopher Bertrand Russell once remarked, "Mathematics may be defined as the subject in which we never know what we are talking about, nor whether what we are saying is true."

MODELS FOR AXIOMATIC SYSTEMS

A **model** for an axiomatic system is a realization of the axioms in some mathematical setting in which all the undefined terms have a specific meaning and all the axioms are true. The model may "live" or exist in coordinate geometry, or in some well-known, or even contrived, system, but it is always separate from the axioms themselves. This idea actually occurred earlier, in Example 3 of Section 2.1: The fractal of Figure 2.7 served as a *model* for the three stated axioms. This particular model could be regarded as taking place in the coordinate plane by merely assigning coordinates to all the points in the diagram of Figure 2.7 in some systematic way.

Two further examples of axioms and models are now given. In these examples each of the models illustrated by the diagrams represents a finite number of points in the Euclidean plane and certain line segments containing those points. These diagrams (models) depict only the essential features of the axioms regarding the number of points and lines and the collinearity or noncollinearity relationships. Nothing can be inferred from the pictures we do not derive from the axioms.

EXAMPLE 3

ABSTRACT SYSTEM

1. There exist two points.
2. There exists a line containing those two points.◻

MODEL

Figure 2.14

EXAMPLE 4

ABSTRACT SYSTEM

MODELS

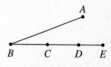

Figure 2.15

1. There exist five points.
2. Each line is a subset of those five points.
3. There exist two lines.
4. Each line contains at least two points.◻

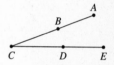

Figure 2.16

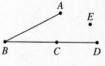

Figure 2.17

We can see that any theorem logically provable from the axioms must also be true in any model for those axioms, but not conversely. This can be seen quite clearly in Example 4, in which *three* different models were possible. A theorem true in one model but not in another (such as "every point belongs to some line" which is true in the first two models but not in the last) would amount to a theorem *not provable from the axioms*, since it is not true in all the models.

INDEPENDENCE AND CONSISTENCY IN AXIOMATIC SYSTEMS

The only things of virtue about an axiomatic system are **independence** (every axiom is essential—it is not a logical consequence of the others), and **consistency** (freedom from contradictions). Models play an important role in deciding these two issues.

If a model can be constructed in a mathematical setting which is known to be free of contradictions, then we know that no contradiction can arise from the axioms themselves. Thus, a model for a set of axioms in some concrete system always establishes (at least) **relative consistency**—the axiomatic system is as consistent as the mathematical environment of the model. On the other hand, if a model can be constructed for all but one of the axioms in a certain axiomatic system, and if the one excluded axiom is false in that model, then the excluded axiom cannot be proven from the rest; thus, it is **independent**. If this can be done for each of the axioms, then the system as a whole is independent.

One of the earliest serious debates in mathematics arose over the independence of Euclid's Fifth Postulate of Parallels. (See Section 4.1.) For centuries this postulate was thought to be unnecessary, and mathematicians, including the great Carl F. Gauss, tried to use the other Euclidean assumptions to prove it. It was not until E. Beltrami constructed a model in Euclidean geometry that the issue was completely settled. More will be said about this later—in fact, in Chapter 6 we will go through the details necessary to construct and understand one of Beltrami's models (now known as Poincaré's Model).

UNDECIDABLE ISSUES

If a property cannot be proven, and if its negation also cannot be proven, then that property is termed **undecidable** with respect to the given axiomatic system. When we say "it cannot be proven," we do not merely mean that the problem is extremely difficult and that nobody thus far has found a proof. We mean that it is *impossible* to prove; more explicitly, there exists a model in which all the axioms are true but the theorem in question is not. The Parallel Postulate in geometry is an example: it is undecidable as a property in the foundations of geometry. If within the context of a set of axioms a certain property has become a **conjecture** (statement proposed as a theorem but standing with neither a proof showing it is true, nor a counterexample showing it is false), the possibility always exists that the conjecture is undecidable.

If we can create *two models*, one in which a conjecture is valid and the other in which it is false, then that conjecture is thereby shown to be undecidable *within that system of axioms*. At this point, we have a choice. We can either include the conjecture as an axiom or not include it (and give a counterexample to show it is false). If we decide to include it as an axiom, then one of the models shows the new axiom is consistent with the others, and the other shows it is independent. If we decide to take its negation as an axiom, again the same two models show the negation to be both consistent and independent.

Long-standing conjectures, like the Four Color Conjecture, are prime suspects for being undecidable issues. The Four Color Conjecture was recently settled by two mathematicians from the University of Illinois, Wolfgang Haken and Kenneth Appel, who proved the Four Color *Theorem*, which asserts that any ordinary map may be colored in the customary manner with only four colors, however complicated the map may be. This settled an issue dating back to the early 1900s. However, the Continuum Hypothesis in set theory (involving cardinal numbers), which had also been a long-standing conjecture, was finally shown by P. J. Cohen in 1963 to be undecidable. Current examples of possible undecidable issues include Fermat's Last Theorem in number theory, and the Riemann Hypothesis in complex variables.

COMPLETENESS AND CATEGORICAL SYSTEMS

There are two other important terms associated with axiomatic systems. An axiomatic system is **complete** if any further axiom, stated within the confines of the system, would render the system either redundant or inconsistent. (That is, the additional statement either contradicts the original axioms or can be proven as a theorem.) An axiomatic system is **categorical** iff there is essentially only one model (that is, all models are the same but for names and labels). In our previous examples of simple axiom systems for points and lines, the rather trivial set in Example 3 was categorical, while the one in Example 4, possessing at least three different models, was not. We state a theorem which is easy to prove from the definitions, which we will leave as a challenge for your own possible enjoyment.

THEOREM

A categorical axiomatic system is complete.

It turns out that the ordinary axioms for Euclidean geometry (which we shall study) are categorical; all models can be shown to be equivalent to three-dimensional coordinate geometry—the model we are so familiar with from the calculus.

EXAMPLE 5   Consider the following axiomatic system:

Undefined terms: member, committee
Axioms:

1.   Every committee is a collection of at least two members.
2.   Every member is on exactly one committee.

Find two distinctly different models for this set of axioms, and discuss how it might be made categorical.

SOLUTION

Let one model be:

Members:	John, Dave, Robert, Mary, Kathy, and Jane
Committees:	Committee A: John and Robert
	Committee B: Dave and Jane
	Committee C: Kathy and Mary

Another model is:

Members:	$\{a, b, c, d, e, f, g, h\}$
Committees:	Committee A: $\{a, b, c, d\}$
	Committee B: $\{e, f\}$
	Committee C: $\{g, h.\}$

Since we have found models which are different (committees having different sizes exist), the system is noncategorical. We also can show it is not complete, since the following statement is clearly undecidable:

$$p = \text{"There exist precisely six members."}$$

The first model shows that this property is consistent with Axioms 1–2, while the second shows its negation is consistent. Hence, *p* is undecidable. Suppose we add the axiom

3.   There exist three committees and six members.

Now we have a categorical system—every model quite obviously consists of three committees, with two members on each commitee. All models would be entirely similar; only the name or identities of individual committee members could differ.❑

PROBLEMS   (§2.2)

_____ *Group A* _____

1.   The rectangle shown in the figure has an area of 5 · 13 = 65 square units. The four-piece dissection of the rectangle indicated can apparently be reassembled to form a perfect square having an area of 8 · 8 = 64 square units. Where did the missing square go? Explain the fallacy in this well-known puzzle.

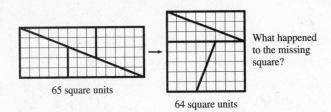

65 square units

64 square units

What happened to the missing square?

2.   A not-so-well-known puzzle is called a **Curry Triangle**, but it is even more mysterious than the one in Problem 1. However, it contains the same type of fallacy. (See if you can find it.) Start with an isosceles triangle having a base of

10 units and height 12. Hence, its area is ½ · 10 · 12 = 60 square units. When the six pieces in the dissection of the triangle are rearranged face up as shown in (a), the result is an apparent *loss* of two square units. However, in (b), some of the pieces are turned face down, and the resulting loss of area is *only one* unit square. Explain.

(a)   (b)

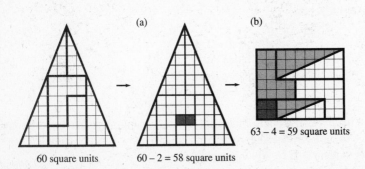

60 square units

60 – 2 = 58 square units

63 – 4 = 59 square units

3. Trace the word *noise* through a standard dictionary until a circular chain occurs.

4. Show that the following model violates the axioms in Example 5 (concerning committees):

Members:	John, Dave, Robert, Mary, Kathy, and Jane.
Committees:	Executive Committee: Mary and Robert
	Steering Committee: John and Robert
	Nomination Committee: Jane, Kathy, and Dave.

5. Find a third model for the axioms in the Example 5 (on committees) that is distinctly different from the models given in Problem 4. What is different about your model, precisely?

6. Find two different models for the following set of axioms in which "point" and "line" are undefined terms:

   Axiom 1: Every line is a set of at least two points.

   Axiom 2: Each two lines intersect in a unique point.

   Axiom 3: There are precisely three lines.

7. (a) Show that the diagram at the right constitutes a model in the Euclidean Plane for the axioms given in Example 3, Section 2.1, which has only 10 points. We repeat the axioms here for convenience:

   Axiom 1: Each line is a set of four points.

   Axiom 2: Each point is contained by precisely two lines.

   Axiom 3: Two distinct lines which intersect do so in exactly one point.

   (b) Is this axiom system categorical? Complete?

_____ *Group B* _____

8. (a) A **convex quadrilateral** is a quadrilateral having the property that its diagonals lie between the two sides adjacent to it. Show that the sum of the measures of the angles of any convex quadrilateral is 360.

   (b) Is the quadrilateral in Figure 2.12 convex? Why not? Is a parallelogram a convex quadrilateral?

9. Locate the points A, B, C, D, E in the coordinate plane, as shown in the figure, and verify that $ABDE$ is a rectangle and $BD = BC$. Find the coordinates of the midpoint L, slope of line CE, and the equations of the perpendiculars LO and MO. Then find the coordinates of the point of intersection of these two perpendiculars. Make a careful sketch revealing your computations, and draw lines OA, OE, OD, and OC. What do you observe?

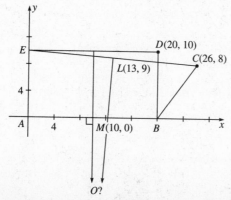

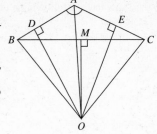

10. Find an example to explain the fallacy in the following. (Refer to the figure at the right.)

    Theorem (!) *Every triangle is isosceles.*

    Proof: Assuming that triangle $\triangle ABC$ is *NOT* isosceles, $(AB \neq AC)$, then the angle bisector at A and perpendicular bisector of side BC are not parallel and must therefore meet at some point O. Drop perpendiculars OD and OE. Then $AO = AO$ and $m\angle DAO = m\angle OAE$ so that (by Euclid) right triangles $\triangle ADO$ and $\triangle AEO$ are congruent. Therefore, $AD = AE$ and $DO = EO$. Since O is equidistant from B and C, $OB = OC$ and the right triangles $\triangle BDO$ and $\triangle CEO$ are also congruent. Hence, in addition to $AD = AE$, we also have $BD = EC$. Then, $AD + DB = AE + EC$, or, since $AB = AD + DB$ and $AC = AE + EC$, $AB = AC$.

*11. There are seven standing committees on the *Flying Aviators Club*. In drawing up a set of by-laws for committee membership, the members wanted to maximize both the distribution of its members serving on committees, and communication between committees. They decided to adopt the following rules:

    (1) There shall be seven committees of the *Flying Aviators Club*, including the Executive, Program, and Nominating Committees. The other committees shall be designated A, B, C, and D.

    (2) The Executive Committee consists of just the President, Vice-President, and Secretary/Treasurer.

    (3) The President shall serve on the three named committees mentioned in By-law 1.

    (4) Each member-at-large of the *Flying Aviators Club* must serve on at least three committees.

    (5) Two or more members are prohibited from belonging to the same two committees.

    (6) Each two committees must have a member in common.

    Show that the members have unwittingly adopted a list of rules that will work only if the club has *precisely seven members*. (*HINT:* Use "points" for members and let "lines" represent committees.)

    *See Problem 6, §2.4.

12. Prove the theorem: *A categorical axiomatic system is complete.*

———— *Group C* ————

13. Show that there are essentially only two models for the following axiom system:

    Undefined Terms: point, adjacent to, and color

    Axioms:

    (1) There are exactly five points.

    (2) If point A is adjacent to point B, then point B is adjacent to point A.

    (3) If point A is not adjacent to point B, then there exists a point C to which A and B are mutually adjacent.

    (4) Each point is assigned a color, red or green.

    (5) Any two adjacent points are assigned to different colors.

14. As an interesting excursion befitting Bertrand Russell's comment about mathematics, consider the following axiom system (adapted from Howard Eves, *Survey of Geometry, Volume 1*, p. 390).

    Undefined terms: abbas, dabbas

    Axioms:

    (1) Every abba is a collection of at least two dabbas.

    (2) There exist at least two dabbas.

(3) If d and d' are two dabbas, then there exists one and only one abba containing both d and d'.

(4) If a is an abba, then there exists a dabba d not in a.

(5) If a is an abba, and d is a dabba not in a, then there exists one and only one abba containing d and not containing any dabba that is in a.

(a) Deduce the following theorems from this postulate set:

(1)   Every dabba is contained in at least two abbas.

(2)   There exist at least four distinct dabbas.

(3)   There exist at least six distinct abbas.

(b) Find two models for this postulate set, one having four dabbas and six abbas, and the other having nine dabbas, twelve abbas, and three dabbas on each abba.

15. Find the fallacy in the following obviously incorrect reasoning (found in Graustein, *Introduction to Higher Geometry*, p. 100):

Theorem (!): Every point lying inside a given circle of radius r also lies on that circle.

Proof: Let the center of the circle be O and suppose A is an interior point, with $OA < r$. Locate point B on line OA such that

$$OB = \frac{r^2}{OA}.$$

Since $OA < r$, then

$$OB > \frac{r^2}{r} = r,$$

so B will fall outside the circle. Let M be the mid-point of segment AB and let the perpendicular at M meet the circle at C. By applications of the Pythagorean Theorem, we have, as shown in the accompanying figure,

$$OA \cdot OB = r^2$$

$$(OM - AM)(OM + MB) = r^2$$

$$(OM - AM)(OM + AM) = r^2$$

$$OM^2 - AM^2 = r^2$$

$$(OC^2 - MC^2) - (AC^2 - MC^2) = r^2$$

$$OC^2 - AC^2 = r^2$$

$$r^2 - AC^2 = r^2$$

$$-AC^2 = 0$$

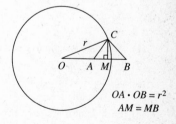

$$OA \cdot OB = r^2$$
$$AM = MB$$

Therefore, $AC = 0$ and $A = C$. That is, A lies on circle O.

*2.3   AN EXCURSION: EUCLID'S CONCEPT OF AREA AND VOLUME

Before we embark on our study of axiomatic geometry, we consider one further example of an axiomatic system and its models. We have chosen a rather elaborate example, an enduring topic for geometry and applied mathematics, yet one that is as troublesome as it is obvious. We begin with a question.

Which do you think is larger: a one-inch tall isosceles triangle with a two-inch base, or a rectangle $\frac{1}{64,000}$ of an inch wide and a mile long? Could you rely on your intuition to answer this question conclusively? Does the area even *exist* for this very long and human-hair-width rectangle?

Our intuitive concept of area and volume no doubt involves the idea of simple counting—counting the number of unit squares, or fractional parts thereof, (or unit cubes) which are contained by the region concerned. This works fine for an integer-sided or even rational-sided rectangle, which involves only a finite number of unit squares or fractional parts (like the first two examples shown in Figure 2.18), and the formula $K = bh$ is pretty obvious. But it quickly becomes problematical in the case of a rectangle with irrational sides, or a triangle, even though these figures in geometry are quite elementary (by contrast, finding the area of a circle would certainly seem insurmountable). The Greeks struggled with the problem of area for the so-called **incommensurable case** of the rectangle (when the ratio of length to width is irrational)—and prevailed, ultimately.

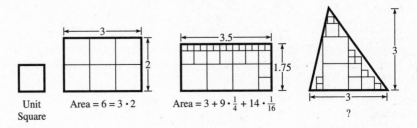

Figure 2.18

The modern approach to this idea of counting squares or cubes involves **measure theory**. This topic is devoted essentially to the study of area and volume as applied to arbitrary sets in the plane or in three-dimensional space. For a region in the plane, one essentially covers the given region by a grid of squares, "counts" the number of squares which just cover the region, and then multiplies by the area of each square in the grid (as shown in Figure 2.19). By making the grid finer and finer, one obtains a converging sequence whose limit is taken as the area of the region. A similar idea is used for the volume of a solid in three-dimensional space.

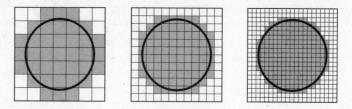

Figure 2.19

Although perhaps appealing to our basic instincts, the idea of "counting" squares remains elusive, to say the least. Since it requires using a complicated process involving limits, actual calculations are unwieldy, and useful area and volume formulas are difficult to develop. Thus we come face to face with the recurring problem in mathematics of trying to formulate a natural and workable definition, and then using the definition to derive the basic theory.

To illustrate further the dilemma facing us, we propose a short drama. It is predicated on the assumption that the only part you remember about area in geometry are the basic formulas for rectangles and triangles. As a promotion, suppose a large company has decided to sponsor a contest on mathematics. The rules involve a $500 entry fee, coming to the site of the contest to participate, and working on the problem without recourse to textbooks or calculators. The prize is a million dollars. Thus, the motivation for you to derive from scratch the formulas you have forgotten is guaranteed. Now suppose the problem is to find which of the following geometric figures (Figure 2.20)—a trapezoid, circle, regular hexagon, and a square with the four corners cut along the curve indicated in the coordinate system, as shown in the figure—has the largest area. Tie breakers consist of the degree of accuracy of the answers (exact answers are obviously desirable), and the inclusion of a mathematical development for the formulas used. No doubt the contestants will try very hard to capture the essence of the concept of area in order to derive the formulas they need.

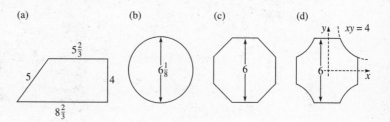

Figure 2.20

This example points up the nature of the area problem confronting ancient scholars. They, too, were trying to find formulas for the area and volume of certain geometric objects, along with proofs or plausibility arguments for those formulas. They had no books in which to look up the answer, nobody who could give them a hint. They had only their intellect and ingenuity. It is sometimes surprising to learn how extensive and ingenious those early developments of area and volume were, particularly since the tools of calculus and the theory of limits had not yet been invented. A prime example is the marvelous development given by Archimedes in his monumental derivation of the volume for a sphere.

We might note that the particular regions chosen for our imaginary contest in Figure 2.20 actually represent major plateaus of achievement in the history of mathematics. The Babylonians (2000–1600 B.C.) could have found the correct formula for the area of the trapezoid (region (a)), but they did not have the general formula for the area of a circle or regular polygon. Euclid (300 B.C.) conquered the circle and regular polygon (regions (b) and (c)), but it requires some form of calculus (A.D. 1790) to find the area of region (d), which, in terms of the usual coordinate system, requires calculating the area under the curve $y = 4/x$.

It is strangely illuminating, in modern times, to examine Euclid's ancient approach to the problem of area and volume. In the *Elements*, Euclid never bothered to define area and volume. Rather, he *identified* each object in the plane with its area and each object in space with its volume; area and volume were not regarded as real numbers, separate from the objects themselves. When he stated that two objects in the plane were "equal," he meant that they had *equal areas or volumes*. For example, his Proposition 35, Book I, states: "Parallelograms which are on the same base and in the same parallels are equal to one another."

Indeed, what Euclid did is logically equivalent to our modern approach, a development that appears in many high school geometry textbooks. One assumes the *existence* of the area and volume of all regions under consideration, and then states certain desirable laws which area and volume should obey (the *axioms*).

We present the axioms usually assumed, and show how to derive some of the basic formulas from them. In this axiom system, we take as an undefined term **region**. We may think of a region as just a set of points (like a circle or triangle and their interiors), whose boundary is not too pathological. (There do exist bounded regions in the plane whose areas do not exist, called **nonmeasurable sets** in mathematics.)

	AREA	VOLUME
1. Existence Postulate	To each region M in the plane, there corresponds a real number Area $M \geq 0$, called its area.	To each region T in space, there corresponds a real number Vol $T \geq 0$, called its volume.
2. Dominance Postulate	For regions M_1 and M_2 in a plane, if $M_1 \subseteq M_2$, then Area $M_1 \leq$ Area M_2	For regions T_1 and T_2 in space, if $T_1 \subseteq T_2$, then Vol $T_1 \leq$ Vol T_2

3. Postulate of Additivity	For any two planar regions M_1 and M_2 such that Area $(M_1 \cap M_2) = 0$, then Area $(M_1 \cup M_2) =$ Area M_1 + Area M_2	For any two regions T_1 and T_2 in space such that Vol $(T_1 \cap T_2) = 0$, then Vol $(T_1 \cup T_2) =$ Vol T_1 + Vol T_2
4. Congruence Postulate	Congruent regions in a plane have equal areas.	Congruent regions in space have equal volumes.
5. Unit of Measure	The area of the unit square is one.	The volume of the unit cube is one.
6. Cavalieri's Principle[1]	If all the lines parallel to some fixed line that meet the plane regions M_1 and M_2 do so in line segments having equal lengths, whose endpoints are the boundary points of the two regions, then Area M_1 = Area M_2 (See Figure 2.21)	If all the planes parallel to some fixed plane that meet the solid regions T_1 and T_2 do so in plane sections having equal areas, whose boundaries lie in the boundaries of the two regions, then Vol T_1 = Vol T_2

CAVALIERI'S PRINCIPLE FOR AREA

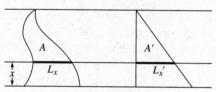

If $L_x = L_x'$ then Area A = Area A'

CAVALIERI'S PRINCIPLE FOR VOLUME

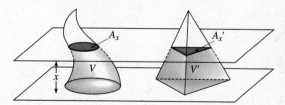

If Area A_x = Area A_x' then Volume V = Volume V'

Figure 2.21

[1]This ingenious axiom is due to one of the early pioneers of calculus, B. Cavalieri (1598–1647). Euclid and Archimedes obviously had to do without this labor-saving concept.

NOTE: Cavalieri's principle may actually be proven from the theory of integration from the axioms which precede it.

OUR GEOMETRIC WORLD

Designer scratch pads with a twist provide a perfect illustration of Cavalieri's Principle. A cubical pile of square pages is twisted to form the more interesting three-dimensional figure below. If we started with this solid and asked for its volume, the problem might seem unmanageable. But, since page-for-page the solids have equal cross-sectional areas, the volume of the twisted solid equals that of the cube, which is quite elementary.

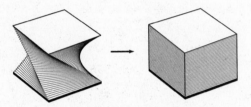

Figure 2.22

It is common in elementary treatments to assume as axioms the basic formulas for the area of a rectangle and the volume of a "box." But these may be derived by logic, using the previous axioms. To illustrate, we show this for a rectangle. Let two rectangles R and R' be given, with bases of length b and b', and altitudes h and h'.

We show first that if $h = h'$,

$$\frac{\text{Area } R'}{\text{Area } R} = \frac{b'}{b}.$$

Let m/n be a rational approximation of b'/b in positive integers m and n, where

(1)
$$\frac{m}{n} < \frac{b'}{b} \leq \frac{m+1}{n}$$

If we multiply throughout by the positive quantity nb, then

(2)
$$mb < nb' \leq (m+1)b.$$

Now consider three rectangles R_1, R_2, R_3 (Figure 2.23) each having altitude h and bases of length mb, nb', and $(m+1)b$, respectively. By the Dominance Postulate (since the rectangle plus interior having the smaller length would fit inside the one having the greater length by congruence, etc.), the inequality **(2)** implies that

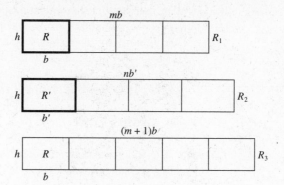

Figure 2.23

(3) $$\text{Area } R_1 < \text{Area } R_2 \leq \text{Area } R_3$$

By the Postulate of Additivity, since exactly m rectangles congruent to R make up R_1, n congruent to R' make up R_2, and $m + 1$ congruent to R make up R_3, we get

$$m \text{ Area } R < n \text{ Area } R' \leq (m + 1) \text{ Area } R$$

or

(4) $$\frac{m}{n} < \frac{\text{Area } R'}{\text{Area } R} \leq \frac{m + 1}{n}$$

Therefore, by comparing the inequalities **(1)** and **(4)** (as $n \to \infty$)

(5) $$\frac{\text{Area } R'}{\text{Area } R} = \frac{b'}{b}$$

Similarly, if the original rectangles R and R' have $b = b'$, then

(6) $$\frac{\text{Area } R'}{\text{Area } R} = \frac{h'}{h}$$

From **(5)** and **(6)** it now follows that if R' is a rectangle with height h and base 1, and U is the unit square, then

$$\frac{\text{Area } R}{\text{Area } R'} = \frac{b}{1} = b \quad \text{and} \quad \frac{\text{Area } R'}{\text{Area } U} = \frac{h}{1} = h$$

Simply multiply these two equations, and use Postulate 5 to obtain

---

THEOREM 1

If R is a rectangle with base of length b units and height of length h units, then

(7) $$\text{Area } R = bh$$

The formula Vol $P = Bh$ for a rectangular box P having base area B and height h may be derived similarly. (See Problem 18.) The formulas for the general parallelo-

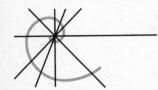

gram and parallelepiped then easily follow from Cavalieri's Principle. Triangles, trapezoids, triangular prisms, and pyramids come next. The following Discovery Unit explores an interesting way to find the area of a circle based on the axioms.

MOMENT FOR DISCOVERY

The Area of a Circle

The following diagrams depict a unique way to deduce the area of a circle of radius r, apparently making use of the postulates for area.

1. Find what number L the sequence $L_1, L_2, L_3, \ldots$ seems to be approaching. (Give a formula for L in terms of r.)

2. To what value, then, do the areas of these figures seem to be converging? (Give a formula.)

3. Was strict use of the axioms on area used, or was some other property assumed here? Discuss.

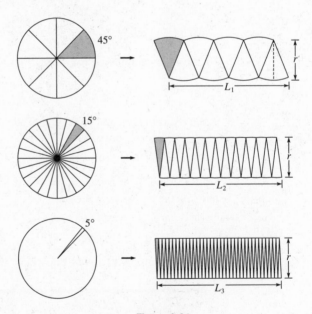

Figure 2.24

We thus have an axiom system that gives us desirable formulas, but what about its consistency? Although we may have great confidence in its validity, this does not prove consistency. One reason we introduced these axioms was so we could further illustrate the concept of models.

EXAMPLE 1   **A Model for the Axioms of Area**   Let the undefined term **region** be interpreted as any *finite set of elements*. Define **Area S** to be simply the *number of elements of S*. Two sets are said to be **congruent** iff there exists a one-to-one correspondence between their elements. Any family of finite sets then serves as a model for the first five axioms of area (or volume).

Proof of Model:

Taking the axioms in turn, we certainly have Area $S \geq 0$ (if S is the empty set, then Area $S = 0$, since S has no elements). For Axiom 2 (the Dominance Postulate), certainly if $S_1 \subseteq S_2$, then the number of elements in S_1 is not greater than the number in S_2, hence Area $S_1 \leq$ Area S_2. For Axiom 3, if Area $(S_1 \cap S_2) = 0$, then S_1 and S_2 have no elements in common. Thus, the number of elements in $S_1 \cup S_2$ is the total number in S_1 and S_2 separately, hence Area $(S_1 \cup S_2) =$ Area $S_1 +$ Area S_2. For Axiom 4, if two sets S_1 and S_2 are congruent, they have the same number of elements, and Area $S_1 =$ Area S_2. Finally, just to make it work, we agree to take a "unit square" to be any set having one element. Then Axiom 5 is valid.◻

Another model is pursued in Problem 20, which is more geometric in nature (and more elaborate). It is a model for all six axioms, including Cavalieri's Principle.

PROBLEMS   (§2.3)

_____ *Group A* _____

1. Find the areas of the first three regions proposed for the contest in Figure 2.20.
2. In the accompanying figure, one quadrant of region (d) of Figure 2.20 is shown. Part of this region is a rectangle, and the rest consists of the area under the curve $y = 4/x$ (the shaded region). Find the coordinates of point B and use calculus to find the area of the shaded region, then determine the area of region (d). Finally, find which of the regions of Figure 2.20 has the greatest area.

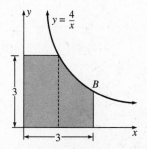

3. Use Cavalieri's Principle to find the area of the shaded region in the figure.

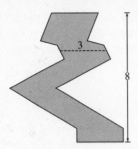

4. Use Cavalieri's Principle to find the total area covered by the letters in the word *Cavalieri* if the width of the downstroke of each letter is 2 mm and the height of the letters is 8 mm. (Assume that the gap in "C" has height equal to that of the hook in the letter "L".)

CAVALIERI

5. Which has the largest area, an isosceles triangle one inch tall with a 2-inch base, or a rectangle $\frac{1}{64,000}$ of an inch wide and one mile long?

6. **(a)** Use Cavalieri's Principle to show from Theorem 1 that if P is a parallelogram having base of length b and height h,

$$\text{Area } P = bh$$

(b) Use the axioms for area to prove the formula

$$\text{Area } T = \frac{1}{2}bh$$

for a triangle T having base of length b and height h.

(c) Prove the formula

$$K = \frac{1}{2}(b_1 + b_2)h$$

for the area of a trapezoid having bases of length b_1 and b_2 and height h.

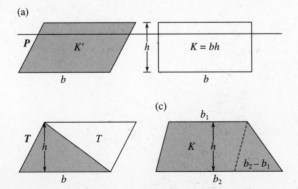

7. In the figure below, the point $P(x, y)$ varies on the ellipse $\dfrac{x^2}{25} + \dfrac{y^2}{9} = 1$. It can be

shown that the two foci of the ellipse are $F_1(4, 0)$ and $F_2(-4, 0)$ and that the characteristic property of the ellipse, $PF_1 + PF_2 = C$, a constant, holds. (Can you show this property using coordinate geometry?) Therefore, all triangles with base F_1F_2 and vertex $P(x, y)$ on the ellipse have the same perimeter.

(a) Find that common perimeter.

(b) When does $\triangle PF_1F_2$ have maximum area, and what is that maximum area?

(c) If P varies on the line $y = 3$, what properties does $\triangle PF_1F_2$ have regarding its area and perimeter?

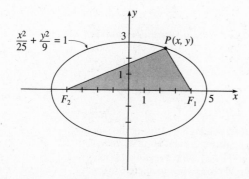

8. A window design consists of a rectangle and semicircle at the top (see figure below). In order to maximize the amount of light which passes through the window, a carpenter decided that the maximal area of a window which can fit inside a total perimeter of 18 ft. 3⅜ in. (exact value: $12 + 2\pi$) is obtained when the rectangular section is a square approximately 3.12 ft. high. Based on calculations of the two areas shown, do you think the carpenter is correct? Find the areas of the carpenter's window and the regions having the dimensions shown in the figure. (***HINT:*** The condition which gives the correct height of the rectangle in (b) is $2x + 6 + 3\pi = 12 + 2\pi$, or $x \approx 1.429$ ft.)

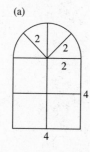

(a)

(b)

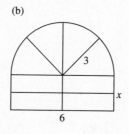

9. Use the classical area formulas to determine the ratio of the side of a square to the radius of a circle if the square and circle have the same area.

_____ *Group B* _____

10. Show by calculus that the window of Problem 8 in fact has maximal area when the rectangle is twice as long as it is high. Find the dimensions of the optimal rectangle and its exact area.

11. A farmer has 30 yds. of fencing and wants to enclose an area beside his barn. What are the dimensions of the region of maximal area, and what is that maximal area, if the region is

(a) a rectangle?

(b) an isosceles triangle with its base along the side of the barn?

(c) a semicircle with diameter along the side of the barn?

(Refer to the figure below.)

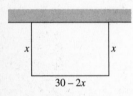
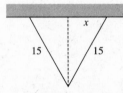
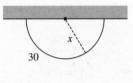

12. Calculus is not needed to solve the window problem (Problems 8 and 10) or the fence problem (Problem 11) because both involve merely a quadratic expression in some variable whose maximum can be determined by algebra. (Thus, Euclid's methods would have sufficed for such problems.) Determine whether the following analogous fence problem requires calculus: A farmer wants to enclose a plot of ground 112.5 sq. ft. in area beside his barn which uses the least amount of fencing. What are the dimensions of the region of least perimeter, and what is the length of fencing required, if the region is

(a) a rectangle?

(b) an isosceles triangle with base along the side of the barn?

(c) a semicircle with diameter along the side of the barn?

13. **Area of Regular Polygon** If a is the **apothem** of a regular *n*-sided polygon (distance from center to any side), prove that its area is given by $K = \frac{1}{2}ap$, where *p* is the perimeter. (See the figure below.)

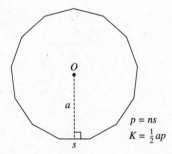

14. Find an explicit formula for the area of a regular hexagon and regular octagon in terms of the length *s* of a side.

15. Use Cavalieri's Principle to find the areas of each of the regions indicated in the figure.

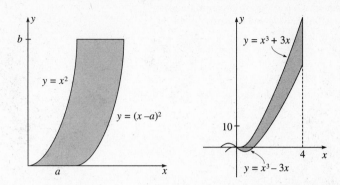

16. Find the area of the shaded region in the figure without calculus. (**HINT:** Cavalieri's Principle applies—show that at height y the generating segment of the region has length y. You will need to solve the quadratic equation $x^2 - 2xy + 2y^2 - 2y = 0$ for x.)

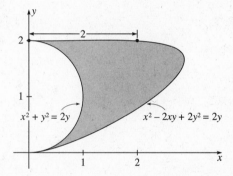

17. Area of Circle Euclid reasoned that the ratio of the circumferences of two circles are to each other as their radii. (The first two diagrams in the figure illustrate the argument.) It follows that the ratio of the circumference of any circle to its diameter is constant. This constant is defined as π, yielding the formula $C = 2\pi r$ for the circumference of a circle with radius r. Now show by the use of regular inscribed and circumscribed polygons of a circle of radius r (and the result of Problem 13) that $\frac{1}{2}a_n p_n < K < \frac{1}{2}a_n' p_n'$ for an n-sided polygon, hence, by taking the limit, $K = \frac{1}{2}rC$. Prove this, then derive the usual formula for K.

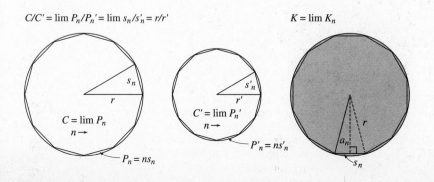

_____ *Group C* _____

18. Using the same procedure as above for rectangles, prove the volume formula $V = \ell w h$ or $V = Bh$ for a parallelepiped having dimensions ℓ, w, h, and $B = $ area

   of base $= \ell w$. (***HINT:*** First prove that $\dfrac{\text{Vol } P'}{\text{Vol } P} = \dfrac{\ell'}{\ell}$ for boxes P' and P having the

   same width and height, and lengths ℓ' and ℓ, respectively.)

19. **Alternate Euclidean Proof of Pyramid Formula** See the figure below and fill in the details of the following analysis.

   (1) Dissect the original pyramid P into two triangular prisms T_1 and T_1' and two congruent pyramids P_1 and P_1' similar to P and half its size. Why does Vol $T_1 = \frac{1}{8}Bh$ and Vol $2T_1' = \frac{1}{4}Bh$ (where $2T_1' = $ parallelepiped shown, having base $\frac{1}{2}B$)? Thus,

$$\text{Vol } P = \frac{1}{4}Bh + 2 \text{ Vol } P_1$$

   (2) Dissect the pyramid P_1 in like manner and obtain

$$\text{Vol } P = \frac{1}{4}Bh + 2\left(\frac{1}{4}B_1 h_1 + 2 \text{ Vol } P_2\right)$$

$$= Bh\left(\frac{1}{4} + \frac{1}{16}\right) + 4 \text{ Vol } P_2,$$

   where B_1 and h_1 are the base area and altitude, respectively, of P_1.

   (3) Continue the process and obtain the result

$$\text{Vol } P = Bh\left(\frac{1}{4} + \frac{1}{16} + \frac{1}{64} + \dots\right) \equiv Bh \sum_{n=1}^{\infty} \frac{1}{4^n}$$

   (4) Evaluate the geometric series in **(3)**.

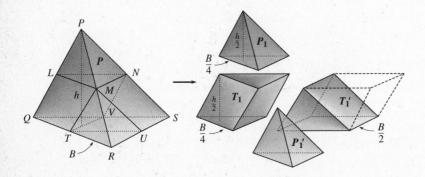

   NOTE: Euclid used this same decomposition of the pyramid to prove that two triangular pyramids having congruent bases and equal altitudes were "equal," which then enabled him to finish his proof of the relation $V = \frac{1}{3}Bh$. Cavalieri's Principle was not available to Euclid, so he had to find a more direct method of proof, like the one given in this problem.

20. **Model for Area Axioms Using Pick's Theorem** A rather unusual way to construct a model for the above axioms of area (Axioms 1–6) is to make use of Pick's Formula for area. (See Problem 7, Section 1.1.) The advantage is its great simplicity, compared to that of the usual area formula involving limits or integration theory. In the usual coordinate plane, consider the points having integer coordinates (called **lattice points**). A **lattice polygon** is an ordinary polygon (simple closed polygonal curve) whose vertices are

lattice points. Define the interior of a polygon the usual way, and use the term **degenerate lattice polygon** for a polygonal path from one point to another whose vertices are lattice points.

The Model: Let a **region** be any lattice polygon and interior, and for each such region P, suppose that B counts the number of lattice points on its boundary and I the number in its interior. Then let

$$\text{Area } P = \frac{1}{2}B + I - 1, \quad \text{if } P \text{ is not degenerate,}$$

$$= 0, \quad \text{if } P \text{ is degenerate.}$$

Show that the axioms for area are satisfied.

21. **UNDERGRADUATE RESEARCH PROJECT** An interesting approach to area is provided by the line integral $\int_C (F(x,y)\, dx + G(x,y)\, dy)$. For each simply-connected region R with boundary C, define the area of R by the formula Area $R = \int_C x\, dy$. If you are familiar with this concept, verify the area postulates above for this definition. (*HINT:* For the Congruence Postulate, use the transformation formula $x^* = ax + by$, $y^* = bx \pm ay$, $a^2 + b^2 = 1$, for an isometry that maps a given triangle to one that is congruent to it (use $\delta = \pm 1$). Using the positive direction about the triangle, show, by substitution, that $\int_C x^*\, dy^* - \int_C x\, dy = \int_C (x^*\, dy^* - x\, dy) = 0$ by showing that $\partial P/\partial y = \partial Q/\partial x$, the condition for exact differentials whose line integrals are zero.) *NOTE:* More details on isometries may be found in Section 5.5.

2.4 INCIDENCE AXIOMS FOR GEOMETRY

The axioms we adopt for geometry in this book are based on the familiar adaptation of Hilbert's axioms found in Moise, *Elementary Geometry from an Advanced Standpoint*, and used in many high school geometry texts. In order to facilitate the statement of the axioms and the later development of geometry, free use of the fundamental notions of set theory will be made. Basic set theory notation (such as $x \in S$ for set membership, $S \subseteq T$ for subsets, and $S \cap T$ for intersection will be used).

We begin with those axioms which govern how points, lines, and planes interact—the so-called **incidence axioms**. The undefined terms will be

point, line, plane, space

(the latter two indicate axioms for three-dimensional space). We let S denote the **universal set**, that is, the set of *all points in space*, and we envision lines and planes as being certain subsets of S, as illustrated in Figure 2.25. Sometimes a line

THREE-DIMENSIONAL SPACE

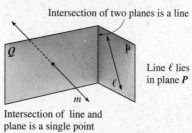

Intersection of two planes is a line

Line ℓ lies
in plane P

Intersection of line and
plane is a single point

Figure 2.25

ℓ can be a **subset** of a plane P, when all the points of ℓ belong to P ($\ell \subseteq P$). Or ℓ can meet P at just one point A ($\ell \cap P = \{A\}$), and finally, ℓ can be **parallel** to P, when there are no points at all in common ($\ell \cap P = \varnothing$).

In the interest of more colorful language and smoother discourse, we shall modify the stilted language of set theory. This will actually make the axioms shorter and easier to read. For example, instead of saying that the intersection of two lines is a "nonempty set," we shall simply say that the two lines **meet**, or **pass through** a certain point. A point which belongs to a line or plane will be said to **lie on** that line or plane. Two or more lines or planes which meet at the same point are called **concurrent** lines or planes, and two or more points are said to be **collinear** if they lie on the same line. Finally, two points are said to **determine** a line provided there exists a unique line passing through them.

The first axiom is basic to almost any geometry worth considering.

AXIOMS FOR POINTS, LINES, AND PLANES

AXIOM I-1

Each two distinct points determine a line.

Note that this axiom allows us to designate a line by any pair of its points, say A and B, and we use the common notation

$$\overleftrightarrow{AB}$$

The two-headed arrow used in this notation is suggestive of the familiar imagery of a line, consisting of an infinite number of points covering the line and extending indefinitely in both directions. Naturally, lines we draw in figures can only represent a part of the ideal line we have in mind in stating the axiom. Moreover, it is premature at this point to presume that lines consist of a continuum (i.e., "continuous infinite stream") of points, as suggested by the diagram in Figure 2.26.

Figure 2.26

Since $\overleftrightarrow{AB}$ is a set of points, we have (by definition)

$$A \in \overleftrightarrow{AB}, \quad B \in \overleftrightarrow{AB}, \quad \therefore \{A, B\} \subseteq \overleftrightarrow{AB}$$

An obvious restatement of Axiom I-1 using the notation just introduced is therefore

---

THEOREM 1

If $C \in \overleftrightarrow{AB}$, $D \in \overleftrightarrow{AB}$ and $C \neq D$, then $\overleftrightarrow{CD} = \overleftrightarrow{AB}$.

Since we are interested in axiomatic geometry of three dimensions, we need to postulate how planes are to behave. The first such axiom guarantees the uniqueness of a plane passing through three noncollinear points.

AXIOM I-2

Three noncollinear points determine a plane.

OUR GEOMETRIC WORLD

When the four legs of a chair do not rest squarely on the floor, the reason is clear from Axiom I-2: the fourth leg need not lie in the plane of the other three, and if too short or too long, creates instability. But a surveyor's transom, a camera tripod, and a child's tricycle are all highly stable objects, illustrations of Axiom I-2 at work.

Figure 2.27

A simple model at this point could consist of a set of three points, A, B, and C, three subsets for lines $\{A, B\}$, $\{B, C\}$, and $\{A, C\}$, and a single plane $\{A, B, C\}$. Note that we cannot conclude that a given plane contains three noncollinear points, nor that a line always contains two distinct points. A later axiom will take care of that.

The nature of a line to be "straight" or not "curved," or a plane to be "flat" (as in Figure 2.28) prevents a line from intersecting a plane in two different points unless the line lies completely in the plane. Another desirable property of planes is that they must intersect in a line. Thus, our next two incidence axioms are quite predictable.

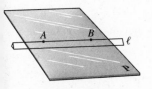

Figure 2.28

AXIOM I-3

If two points lie in a plane, then any line determined by those two points lies in that plane.

> ### AXIOM I-4
> If two planes meet, their intersection is a line.

The last axiom on incidence is of a technical nature, guaranteeing certain "obvious facts"—desirable properties we could not prove otherwise. (See Problem 11.)

> ### AXIOM I-5
> Space consists of at least four noncoplanar points, and contains three noncollinear points. Each plane contains at least three noncollinear points, and each line contains at least two distinct points.

MODELS FOR THE AXIOMS OF INCIDENCE

The model we consider first has the smallest possible number of points (four). This will establish the consistency of the axioms of incidence we have introduced so far. We use three-dimensional coordinate space to aid in its construction, and we encourage you to participate in its construction in the Discovery Unit which follows.

MOMENT FOR DISCOVERY

A Test of the Axioms of Incidence

According to Axiom I-5, there must be at least four points. Let these be labelled A, B, C, and D. We explore here whether the axioms hold with only these four points.

1. Arrange A, B, C, D in ordinary xyz-space (shown in Figure 2.29), where A is the origin, and B, C, and D are at a unit distance from A on each of the three coordinate axes. This forms a **tetrahedron** $ABCD$.

MODEL FOR AXIOMS 1.1–1.5

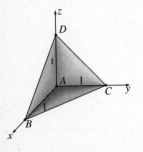

Figure 2.29

2. Using actual lines and planes in xyz-space to help you, and using set theory notation, make a list of

    (a) the lines

    (b) the planes

    (You should get, for example, that two of the *lines* are $\ell_1 = \{A, B\}$, $\ell_2 = \{A, C\}$, and one of the *planes* is $P_1 = \{A, B, C\}$.)

3. After you determine the lines, examine whether you think Axiom I-1 holds. Can you verify this without referring to the diagram? Look at your list of lines and the points you identified for them.

4. Does Axiom I-2 hold?

5.   Is Axiom I-3 obvious? Verify Axiom I-4, again using set theory, and not coordinate geometry. Does Axiom I-5 work?

6.   Can you think of a way to construct a model having five points? Six points? *n* points, where *n* is an integer > 4?

An interesting problem for the reader to work on next is to construct a model for Axioms I-2, I-3, I-4, and I-5 in which Axiom I-1 *fails to hold*. This would show the independence of Axiom I-1. (A model can be constructed with just four points.) The independence of the remaining axioms can also be considered.

The following theorem can now be proven.

THEOREM 2

If two distinct lines ℓ and m meet, their intersection is a single point. If a line meets a plane and is not contained by that plane, their intersection is a point.

Proof

I.   *Given:* Distinct lines ℓ and m, and point A on $\ell \cap m$.

   Prove: No other point, besides A, lies in $\ell \cap m$.

CONCLUSIONS	JUSTIFICATIONS
(1) Assume false. Assume point $B \neq A$ lies on both ℓ and m.	assumption for indirect proof
(2) A and B belong to both ℓ and m.	given and Step (1)
(3) $\ell = m$ $\rightarrow\leftarrow$	Axiom I-1
(4) $\therefore$ A is the only point on both ℓ and m.	Rule of Elimination

II.   *Given:* Line ℓ not lying in plane P, $A \in \ell \cap P$.

   Prove: No other point, besides A, lies in $\ell \cap P$.

(The proof will be left as Problem 8.)

PROBLEMS   (§2.4)

_____ **Group A** _____

1.   Verify that the following system is a model for the incidence axioms (Axioms I-1–I-5) (show that each axiom is true):

   Points: $S = \{1, 2, 3, 4, 5\}$

   Lines: $\{1, 2, 3\}$, $\{1, 4\}$, $\{1, 5\}$, $\{2, 4\}$, $\{2, 5\}$, $\{3, 4\}$, $\{3, 5\}$, $\{4, 5\}$

Planes: $\{1, 2, 3, 4\}$, $\{1, 2, 3, 5\}$, $\{1, 4, 5\}$, $\{2, 4, 5\}$, $\{3, 4, 5\}$

Draw a three-dimensional illustration of this model. (If done correctly, the picture serves as an immediate visual verification of Axioms I-1–I-5 in the model.)

2. Show that the following system is NOT a model for Axioms I-1–I-5:

Points: $S = \{1, 2, 3, 4, 5\}$

Lines: $\{1, 2, 3\}$, $\{1, 4\}$, $\{1, 5\}$, $\{2, 4\}$, $\{2, 5\}$, $\{3, 4\}$, $\{3, 5\}$, $\{4, 5\}$

Planes: $\{1, 2, 3, 4\}$, $\{1, 2, 3, 5\}$, $\{2, 3, 4, 5\}$, $\{1, 4, 5\}$

Which axiom is not satisfied?

3. Consider the following system of points, lines, and planes:

Points: $S = \{1, 2, 3, 4, 5\}$

Lines: $\{1, 2, 3\}$, $\{1, 4\}$, $\{1, 5\}$, $\{2, 4\}$, $\{2, 5\}$, $\{3, 4\}$, $\{3, 5\}$, $\{4, 5\}$

Planes: $\{1, 2, 3, 4\}$, $\{1, 2, 3, 5\}$, $\{1, 2, 3\}$, $\{1, 4, 5\}$, $\{2, 4, 5\}$, $\{3, 4, 5\}$

Which incidence axioms are satisfied here?

4. Consider the following system of points, lines, and planes:

Points: $S = \{1, 2, 3, 4\}$

Lines: $\{1, 2, 4\}$, $\{1, 3\}$, $\{2, 3\}$, $\{3, 4\}$

Planes: $\{1, 2, 3\}$, $\{1, 3, 4\}$, $\{2, 3, 4\}$

Which incidence axioms are satisfied in this model?

_____ Group B _____

5. Consider the infinite set of positive integers $\{1, 2, 3, ...\}$. Find an easy way to name definite subsets as lines and other definite subsets as planes in such a manner that the incidence axioms will automatically hold (without a proof being necessary) as in the manner of the previous tetrahedron model. (***HINT:*** Let one line consist of the points 3, 4, 5, ...)

6. **The 7-Point Projective Plane** Consider the following model, in which there is only one plane, the universal set **S** itself. (See the figure to the right.)

Points: $S = \{1, 2, 3, 4, 5, 6, 7\}$

Lines: $\ell_1 = \{1, 2, 3\}$, $\ell_2 = \{1, 4, 5\}$, $\ell_3 = \{3, 7, 5\}$, $\ell_4 = \{1, 6, 7\}$,
$\ell_5 = \{2, 5, 6\}$, $\ell_6 = \{3, 4, 6\}$, $\ell_7 = \{2, 4, 7\}$

Plane: $P = \{1, 2, 3, 4, 5, 6, 7\}$

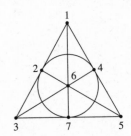

Which incidence axioms are satisfied? (Compare this model with the committee structure in the Flying Aviators Club of Problem 11, Section 2.2.)

7. **The Nine-Point Affine Plane** Show that the following system satisfies Axiom I-1, part of Axiom I-5, and the **Parallel Postulate**: *Given a point and a line not passing through it, there exists a unique line passing through the given point that is parallel to the given line.* (Parallel means "do not share a common point.")

Points: $S = \{A, B, C, D, E, F, G, H, J\}$

Lines: $\ell_1 = \{A, B, C\}$, $\ell_2 = \{D, E, F\}$, $\ell_3 = \{G, H, J\}$, $\ell_4 = \{A, D, G\}$,
$\ell_5 = \{B, E, H\}$, $\ell_6 = \{C, F, J\}$, $\ell_7 = \{A, E, J\}$, $\ell_8 = \{B, F, G\}$,
$\ell_9 = \{C, D, H\}$, $\ell_{10} = \{C, E, G\}$, $\ell_{11} = \{A, F, H\}$,
$\ell_{12} = \{B, D, J\}$

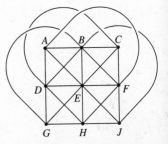

(See the figure to the right for a suggestive diagram. Note that these figures cannot represent the models faithfully in every aspect; they are intended only to show the incidence relationships, nothing else.)

8. Prove the second part of Theorem 2.

9. Construct a model having five points in which Axioms I-2, I-3, I-4, and I-5 are satisfied, but not I-1.

10. (*NOTE:* A basic knowledge of linear algebra is needed for a feasible solution of this problem.) Verify that the following system is a model for the Incidence Axioms I-1–I-5:

Points: All ordered triples of rational numbers (x, y, z)

Lines: Sets of triples (x, y, z) satisfying simultaneously two independent linear equations,

$$ax + by + cz + d = 0, \quad a'x + b'y + c'z + d' = 0$$

where a, b, c are not all zero, a', b', and c' are not all zero, and the triple (a, b, c) is not proportional to (a', b', c'). (Example: $2x + y + z + 1 = 0$ and $x - y + 3z = 5$ together represent a line, but $2x + y + z + 1 = 0$ and $6x + 3y + 3z = 0$ do not.)

Planes: Sets of all triples (x, y, z) which satisfy a single, nontrivial linear equation

$$ax + by + cz + d = 0 \quad a, b, c \text{ not all zero}$$

──────── *Group C* ────────

11. Prove explicitly from the incidence axioms that

 (a) a plane cannot be a line

 (b) each line is contained by at least two planes, whose intersection is that line.

 (Do you see the necessity for all the parts of Axiom I-5?)

12. Desargues' Theorem (Problem 12, Section 1.1) is merely an exercise in incidence geometry in the case when the two triangles $\triangle ABC$ and $\triangle A'B'C'$ *lie in different planes.* This case may then be extended to triangles lying in the same plane by projection. (See the figure below.) Prove this yourself from Axioms I-1–I-5 or look up a proof in a text from the library.

DESARGUES' THEOREM INCIDENCE GEOMETRY

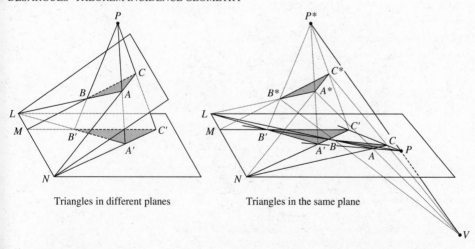

Triangles in different planes   Triangles in the same plane

13. **Axioms For Affine Space**   An **affine space** is any system of points, lines, and planes satisfying the axioms of incidence and, in addition, the following **Parallel Postulate** for space, with the same meaning of parallel as before (objects which do not meet):

AXIOM I-6: Given any two noncoplanar lines in space, there exists a unique plane through the first line that is parallel to the second.

An affine space (or affine plane) having a maximal of $n \geq 2$ points on each line is said to be of **finite order** n. Find an argument which establishes each of the following:

(a) Every plane in an affine space is an affine plane. (See Problem 7.)

(b) Given a plane P and a point A not on it, there exists one and only one plane through A that is parallel to P.

(c) An affine plane of order n has n points on each line, $n^2 + n$ lines, and n^2 points altogether.

(d) An affine space of order n has n points on each line, $n^3 + n^2 + n$ planes, and n^3 points altogether.

14. Construct a model in which Axioms I-1, I-2, I-4, and I-5 are valid, but not I-3.

15. **A 27-Point Affine Space** The model we will consider has 27 points, 117 lines, and 39 planes. In order to facilitate its construction, we resort to a series of figures which are three-dimensional drawings of cubes, with the points of the space labelled accordingly. Each horizontal and vertical face, midface, and each diagonal plane indicated in the figure below contains nine points of the geometry; each of the three cubes in the figure thereby contains 13 of the 39 planes. (See Problem 7 for the geometry of each plane.) The lines are obtained by locating any of the 27 points which lie on an *ordinary Euclidean Line* in each cube. (They may also be determined as lines of affine planes, 12 lines for each plane, as in Problem 7.) Lines will be duplicated as one passes from one cube to the next, but the planes will not. Examples are, for planes, $\{A, B, C, D, E, F, G, H, I\}$ and $\{F, J, Z, A, Q, X, H, O, S\}$; and for lines, $\{A, B, C\}$, $\{A, N, \Sigma\}$, and $\{B, K, T\}$.

Verify the properties of the model by performing the following valid geometric operations for any affine space.

(a) Find the unique line (the three points on that line) passing through each of the pairs of points (A, K), (V, H), and (S, I).

(b) Find the unique plane (the nine points on that plane) passing through each of the point triples (G, H, R) and (K, F, Σ).

(c) Find whether the planes through (R, S, T) and (B, F, L) intersect, and if they do, find their line of intersection.

(d) Find the unique plane through P parallel to the plane through (B, F, L).

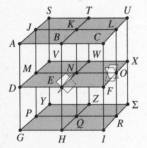 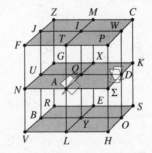 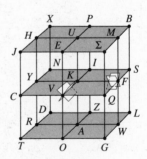

16. **Project** How many *lines* exist in an affine space of order 2? Find the general formula for an affine space of order n, and verify your answer for the case $n = 2$, and that of the case $n = 3$ in the previous problem.

2.5 METRIC BETWEENNESS, SEGMENTS, RAYS, AND ANGLES

The geometry we want to study possesses a **metric**, or *distance concept*, and, accordingly, *S* is called a **metric space**. The axioms are as follows.

> ### METRIC AXIOMS
> #### AXIOM D-1: EXISTENCE
> Each pair of points (*A*, *B*) is associated with a unique real number $AB \geq 0$, called the **distance** from *A* to *B*.
>
> #### AXIOM D-2: POSITIVE DEFINITENESS
> For all points *A* and *B*, $AB > 0$ unless $A = B$.
>
> #### AXIOM D-3: SYMMETRY
> For all points *A* and B, $AB = BA$.

In so many words, the metric axioms guarantee that we can measure the distance from one point to another, that the distance measured is positive if the points are distinct, and that the distance between two points is not affected by the order in which they occur. The usual definition of a metric includes an additional axiom, the **triangle inequality**: Given *A*, *B*, and *C*,

$$AB + BC \geq AC$$

as in Figure 2.30. But this will ultimately follow as a theorem later on. If it so happens that

$$AB + BC = AC$$

(as illustrated in Figure 2.30) then it is only natural to assume that B is somehow

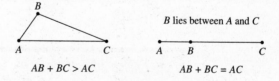

B lies between *A* and *C*

$AB + BC > AC$ $AB + BC = AC$

Figure 2.30

"between" *A* and *C*—that the ordering of the points on the line is either *A-B-C* or *C-B-A*. In the foundations of geometry, betweenness is often taken as an undefined operation, where the desired properties are postulated. This was the approach taken by David Hilbert (1862–1943) in his *Foundations of Geometry*. His axiom system was entirely geometric, altogether in the spirit and context of Euclid, but with the necessary axioms to do the job right. However, mainly for convenience, we have chosen to adopt a notion for distance first, from which we can then define the property of betweenness, which we shall now do.

*H*ISTORICAL *N*OTE

Hilbert solved outstanding problems both in mathematics and physics, and significantly extended virtually every field of mathematics. Born in 1862 in Königsberg, East Prussia (the famous city whose arrangement of seven bridges was the source of the well-known graph theory problem solved by Euler, which marked the beginning of topology), he was the son of a district judge who wanted him to become a lawyer. But Hilbert soon saw that his greatest interests, and gifts, were in mathematics. He received his doctorate from Königsberg University in May of 1885, and arrived at Göttingen University to teach mathematics in 1895, almost exactly 100 years after Gauss had been a student. Hilbert remained at Göttingen as professor of mathematics the rest of his life, and during this time directed 96 doctoral students. Hilbert's work was so far-reaching that it has been said that every significant discovery in the 20th century, if not a corollary of his

results, would have to go *through* one or more of them. Hilbert's byword, which was engraved on his tombstone, was *We Must Know; We Shall Know.* After having solved a major problem in number theory that beautifully extended Gauss's reciprocity theorem, he turned to geometry and wrote his *Foundations of Geometry* in 1898. He adopted the familiar style and terminology of Euclid. That, and its logical perfection, made it an immediate success (like the *Elements*), quickly going through seven editions. Hilbert's favorite comment was: "Instead of points, lines and planes, one must be able to say at all times tables, chairs, and beer mugs," to emphasize the abstract nature of the axiomatic development. Other notable attempts to correct the logical shortcomings of Euclid during that period included the work of Moritz Pasch, which was the basis for Hilbert's work, and the development by G. Peano.

DEFINITION OF BETWEENNESS

For any three points A, B, and C in space, we say that B is **between** A and C, and we write A-B-C, iff A, B, and C are distinct, collinear points, and $AC = AB + BC$.

Thus, if B is to lie *between* A and C, the three points must be distinct points, must lie on some line, and must exist in the explicit *order* first A, then B, finally C

(or else, first C, then B, and finally A) as we encounter them on the line. A fictitious point P, moving on the line from left to right (Figure 2.31), would encounter point A first, then point B, and finally point C (or the reversal of that if P were moving in the opposite direction). For this reason, notions of betweenness are sometimes referred to as properties of *order*. Hilbert's axioms on betweenness are called **order postulates**.

A recurring situation in geometry involves the ordering of four collinear points.

ORDER OF POINTS ON A LINE

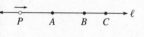

Figure 2.31

DEFINITION

If A, B, C, and D are four distinct collinear points, let the betweenness relation A-B-C-D represent the composite of all four betweenness relations A-B-C, A-B-D, A-C-D, and B-C-D.

Now we state a few theorems on betweenness which are very easy to prove and, at the same time, very useful.

THEOREM 1

If A-B-C then C-B-A, and neither A-C-B nor B-A-C.

Proof

Given: A-B-C.

Prove: C-B-A; gain a contradiction if we assume that A-C-B or B-A-C.

CONCLUSIONS	JUSTIFICATIONS
(1) $AB + BC = AC$ and A, B, C are distinct and collinear points.	definition (betweenness)
(2) $BC + AB = AC$.	algebra
(3) $CB + BA = CA$.	Symmetry Property of metric
(4) $\therefore$ C-B-A.	definition of betweenness
(5) Assume A-C-B. Then $AB = AC + CB$.	assumption for indirect argument, definition
(6) $(AC + CB) + BC = AC$.	substitute Step (5) into Step (1)
(7) $2BC = 0$ $\rightarrow$ $BC = 0$	algebra and $B \neq C$

$\rightarrow\leftarrow$

(The case B-A-C is similar, and is left as Problem 9.)

THEOREM 2

If A-B-C, A-C-D, and the inequalities $AB + BD \geq AD$ and $BC + CD \geq BD$ hold[2], then A-B-C-D.

(The proof will be left as Problem 14; the conclusion of this theorem is an easy consequence of the Ruler Postulate to be introduced in Section 2.7.)

Having introduced betweenness, we may now define the familiar geometric objects **segment**, **ray**, and **angle**. Rather than give wordy definitions for what is rather obvious, we direct the reader to a "pictorial definition" in Figure 2.32. Set theoretic definitions follow.

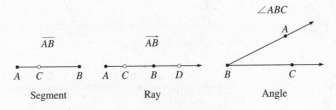

Figure 2.32

The set theoretic definitions of these concepts are as follows.

SEGMENT AB:	$\overline{AB} = \{A, B\} \cup \{C : A\text{-}C\text{-}B\}$
RAY AB:	$\overrightarrow{AB} = \{A, B\} \cup \{C : A\text{-}C\text{-}B\} \cup \{D : A\text{-}B\text{-}D\}$
LINE AB:	$\overleftrightarrow{AB} = \overrightarrow{BA} \cup \overrightarrow{BC}$, where $A\text{-}B\text{-}C$[3]
ANGLE ABC:	$\angle ABC = \overrightarrow{BA} \cup \overrightarrow{BC}$ $\qquad$ (A, B, C noncollinear)

The points A and B in the above insert which define segment $\overline{AB}$ are called **end-points**; point A of ray $\overrightarrow{AB}$ is called an **endpoint** (or sometimes, **origin**), and point B of angle $\angle ABC$ is called its **vertex**. Note that we are requiring the sides of an angle to be distinct, noncollinear rays. Thus, our development disallows degenerate angles (angles whose sides coincide) and straight angles (angles whose sides form a straight line). While it is certainly possible for two rays to form a straight line, we simply do not refer to such sets as "angles". This is a convenience, since then we do not always have to preface the statements of theorems on angles with the disclaimer "if an angle is nondegenerate and nonstraight, then ..."

---

[2]These concepts hold if the metric satisfies the Triangle Inequality mentioned previously.

[3]The result of the next axiom to be introduced.

OUR GEOMETRIC WORLD

Euclid's approach to geometry could be described by the term **caliper geometry**, a geometry in which the only means of comparing lengths of segments are by a pair of mechanical calipers. In such a geometry, one can only tell whether one segment is larger than another, or if they are equal (congruent). Euclid's *Elements* provided no concept for the "length" of a segment (in terms of a real number apart from the segment itself, as we are assuming here). One reason was that the real number system had not yet been invented in Eucild's day.

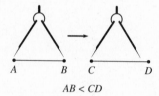

$$AB < CD$$

Figure 2.33

We introduce a few basic properties of segments and rays for you to consider, and in some cases, to prove. Remember, in order to prove that two *sets* are equal, you have to show

(1) that each element of the first set belongs to the second

(2) that each element of the second belongs to the first.

THEOREM 3

(a) $\overline{AB} = \overline{BA}$

(b) $\overline{AB} \subseteq \overrightarrow{AB}$

(c) $\overrightarrow{AB} \subseteq \overleftrightarrow{AB}$

THEOREM 4

$\overrightarrow{AB} \cap \overrightarrow{BA} = \overline{AB}.$

Proof

Given: rays $\overrightarrow{AB}$ and $\overrightarrow{BA}$.

Prove: **(a)** $\overrightarrow{AB} \cap \overrightarrow{BA} \subseteq \overline{AB}$
            (b) $\overline{AB} \subseteq \overrightarrow{AB} \cap \overrightarrow{BA}$

THEOREM 4 (CONTINUED)

CONCLUSIONS

JUSTIFICATIONS

(1) Assume $X \in \overrightarrow{AB} \cap \overrightarrow{BA}$, or that $X \in \overrightarrow{AB}$ and $X \in \overrightarrow{BA}$.

hypothesis for proving $\overrightarrow{AB} \cap \overrightarrow{BA} \subseteq \overline{AB}$

(2) Assume $X \notin \overline{AB}$. (Then $X \neq A$ and $X \neq B$.)

hypothesis for indirect argument

(3) Since $X \in \overrightarrow{AB}$ then A-B-X.

definition of ray $\overrightarrow{AB}$

(4) Since $X \in \overrightarrow{BA}$ then B-A-X.

definition of ray $\overrightarrow{BA}$

(5) Hence both A-B-X and B-A-X →←

Steps (4) and (5), and Theorem 1

(6) $X \in \overline{AB}$ ∴ (a) $\overrightarrow{AB} \cap \overrightarrow{BA} \subseteq \overline{AB}$.

Rule of Elimination

(7) $\overline{AB} \subseteq \overrightarrow{AB}$ and $\overline{AB} = \overline{BA} \subseteq \overrightarrow{BA}$.

Theorem 3 (b)

(8) ∴ (b) $\overline{AB} \subseteq \overrightarrow{AB} \cap \overrightarrow{BA}$.

set theory

Now consider this question: Suppose that A, B, and C are points of some line ℓ such that B lies between A and C (A-B-C); is it true *from the previous axioms* that the points of ℓ must belong to one of the two "opposite" rays $\overrightarrow{BA}$ or $\overrightarrow{BC}$? That is, can we actually prove from the axioms that if A-B-C then

$$\overrightarrow{BA} \cup \overrightarrow{BC} = \overleftrightarrow{AC}?$$

Obvious as this seems, we cannot; unfortunately, we will have to introduce a further axiom to take care of this. It is instructive to build a model to show *why* it is necessary to do so. We consider a more elaborate model than that given earlier, but based on the same idea. We use a diagram in three-dimensional space as a guide for lines, planes, and the pertinent relationships between them.

MODEL FOR AXIOMS I-1–I-5, D-1–D-3

Consider a tetrahedron in three-dimensional space, and take as "points" the vertices of the tetrahedron and two additional points on one edge, as shown. Thus define:

Points: $S = \{A, B, C, D, E, F\}$

Lines: $\{A, B, C, D\}$, $\{A, E\}$, $\{B, E\}$, $\{C, E\}$, $\{D, E\}$, $\{A, F\}$, $\{B, F\}$, $\{C, F\}$, $\{D, F\}$, and $\{E, F\}$

Planes: $\{A, B, C, D, E\}$, $\{A, B, C, D, F\}$, $\{A, E, F\}$, $\{B, E, F\}$, $\{C, E, F\}$, and $\{D, E, F\}$

Distance: $AE = BE = CE = DE = EF = 2$, $AF = BF = CF = DF = AC = 2$, $AB = BC = CD = AD = BD = 1$, and define $PQ = 0$ if $P = Q$ and $PQ = QP$ for all points P and Q

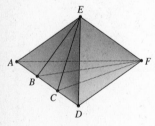

Figure 2.34

It can be observed (basically from the figure itself) that the incidence axioms are all valid here, and the metric axioms D-1, D-2, and D-3 are easy to check (symmetry was built in by requiring that $PQ = QP$ for all P and Q). In fact, the triangle inequality holds, so this model is actually a metric space (we can reason as follows: If PQ is any distance, and R is any other point, then from the definition of distance given, $PR + RQ \geq 2 \geq PQ$ in all cases). The rest of the analysis will be left to the reader for self discovery.

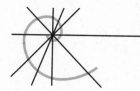

MOMENT FOR DISCOVERY

Working With the Model

Study the model just given and make sure you understand how it works. (Examples: Planes $\{A, E, F\}$ and $\{B, E, F\}$ have the line $\{E, F\}$ as intersection, and $\{A, B, C, D, E\}$ and $\{A, B, C, D, F\}$ intersect in the line $\{A, B, C, D\}$; the only line having more than two points is $\overleftrightarrow{AC} = \{A, B, C, D\}$.) See if you can identify the following sets of points, *by definition*: $\overrightarrow{BA}$ and $\overrightarrow{BC}$. (Use the following steps as a guide if you need some help.)

1. Show that B and D are the only points of segment $\overline{AC}$ besides A and C. (Note that A-B-C holds.)

2. Since there are obviously no points of this model which lie between A and B, see if there are any points P such that B-A-P (lying on ray $\overrightarrow{BA}$).

3. What points, therefore, constitute the ray $\overrightarrow{BA}$? (Did you get $\overrightarrow{BA} = \{A, B\}$?)

4. Find all points P such that B-C-P.

5. What points lie on ray $\overrightarrow{BC}$?

6. What points make up the set $\overrightarrow{BA} \cup \overrightarrow{BC}$?

What did you discover about this model?

We remedy this undesirable state of affairs by introducing a further axiom concerning distance, which is really an axiom about betweenness. However, we prefer to give a geometric statement.

AXIOM D-4

Given any three points A, B, and C on line ℓ such that A-B-C, $\overrightarrow{BA} \cup \overrightarrow{BC} = \ell$.

You should have no difficulty in verifying to your own satisfaction that Axiom D-4 is equivalent to the following restatement given strictly in terms of betweenness.

> ### AXIOM D-4'
> Given any four distinct collinear points A, B, C, and D such that A-B-C, then either D-A-B, A-D-B, B-D-C, or B-C-D.

PROBLEMS (2.5)

———— *Group A* ————

1. The points A, B, C, D, E, F, and G lying in the xy-plane are, in order, given by the respective coordinates $(0, 0)$, $(\pm 1, 2)$, $(\pm 2, 3)$, and $(\pm 3, 3)$ [i.e., $B = (+1, 2)$ and $C = (-1, 2)$, etc.]. Name the betweenness relations that exist among these points, using the ordinary concept of Euclidean distance normally assumed in coordinate geometry.

2. Suppose that in a certain metric geometry satisfying axioms D-1–D-3, points A, B, C, and D are collinear, and

$$AB = 4 = AC, \quad BC = 6, \quad BD = 3 = CD, \quad AD = 5$$

    What betweenness relations are demanded by *definition* among the four points A, B, C, and D?

3. In a certain metric geometry, the distances determined by the collinear points A, B, C, D, and E are $CE = 1$, $AB = BD = 2$, $BC = \sqrt{7}$, $AC = 3$, $BE = 2\sqrt{3}$, $CD = \sqrt{13}$, and $AD = AE = DE = 4$. Find what betweenness relations exist.

4. **Matching** The expressions on the left represent certain geometric objects in the figure on the right. For each such expression, enter the number of the correct object. The same response may be used more than once.

    (a) $\overline{RS}$ ———

    (b) $\overrightarrow{RT}$ ———

    (c) $\overrightarrow{ST}$ ———

    (d) $\overleftrightarrow{RS}$ ———

    (e) $\overline{RT}$ ———

    (f) $\overleftrightarrow{ST}$ ———

    (1) $S \quad R \quad T$

    (2) $S \qquad\qquad R$

    (3) $R \quad S \quad T$

    (4) $R \qquad\qquad T$

    (5) $T \quad S \quad R$

5. Name all segments and rays exhibited by illustration in the figure at right which do NOT contain point E.

6. Name all segments and rays exhibited by illustration in the figure at right which do NOT contain point G or point H.

7. Name all segments and rays exhibited by illustration in the figure at right which DO contain the point P but not having P as an endpoint (you need only name each set once and omit alternative symbols possible in each case).

8. If A-B-C, explain why $BC + CA > AB$.

_____ *Group B* _____

9. Complete the proof of Theorem 1. (Assume *B-A-C*, then obtain a contradiction.)

10. Prove Theorem 3: $\overline{AB} = \overline{BA}$, $\overline{AB} \subseteq \overrightarrow{AB}$, and $\overrightarrow{AB} \subseteq \overleftrightarrow{AB}$.

11. **A Line Cannot Be a Ray** If line ℓ has at least three points and has the property that given any two points A and B on ℓ there exists a third point C on ℓ such that A-B-C, then prove that $\ell \neq \overrightarrow{BA}$ for any two points $A \neq B$ on ℓ.

_____ *Group C* _____

For all problems in this section, assume that the metric obeys the triangle inequality, defined earlier, but not taken as an axiom.

12. Prove that in a metric space, if C lies between A and B, and O is any other point, then $OC \leq OA + OB$. (**HINT:** Make three applications of the triangle inequality.)

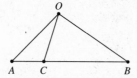

13. Prove that if $M \in \overline{AB}$, $N \in \overline{AC}$, $d = MN$, $a = BC$, $b = AC$, and $c = AB$, then $d \leq s$, where $s = \frac{1}{2}(a + b + c)$

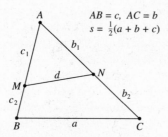

$AB = c,\ AC = b$
$s = \frac{1}{2}(a + b + c)$

14. Prove Theorem 2. (**HINT:** $BC + CD = (AC - AB) + (AD - AC) = AD - AB$ and $AD \leq AB + BD \rightarrow AD - AB \leq BD$.)

15. **Topology of Metric Spaces** In a metric space, an **open ball of radius $r > 0$** is the set of all points at a distance $< r$ from some fixed point P (called its **center**), denoted $B_r(P)$. Show that for any $r, s > 0$ and any two points P and Q, if $R \in B_r(P) \cap B_s(Q)$, and t = smaller of $r - PR$ and $s - QR$, then $B_t(R) \subseteq B_r(P) \cap B_s(Q)$. (You must prove that if $W \in B_t(R)$, then $W \in B_r(P)$ and $W \in B_s(Q)$.) This proves the topological property in metric spaces that the intersection of two open sets is an open set, where an **open set** is defined as the union of any family of open balls.

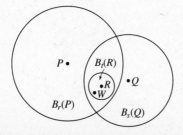

16. Suppose segment $\overline{AB}$ is as shown in the figure below, where $OA \leq r$ and $OB \leq r$. Prove that $OC \leq 2r$, where $C \in \overline{AB}$ (See Problem 12.) That is, if A and B are any two points of the open ball $B_r(O)$, then segment $\overline{AB} \subseteq B_{2r}(O)$, the best we can do in a metric space. (In Euclidean geometry, it can be shown that $\overline{AB} \subseteq B_r(O)$, or that open balls are *convex*, defined in the next section.)

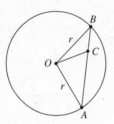

2.6   PLANE SEPARATION, POSTULATE OF PASCH, INTERIORS OF ANGLES

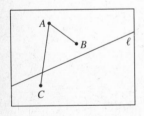

Figure 2.35

When you draw a line across a sheet of paper, you have effectively divided that sheet into two regions—the two *sides* of the line. This simple idea is very useful in geometry, but in order to develop a precise statement of what is needed, it's helpful to analyze the experiment just described.

Consider the points A, B, and C (Figure 2.35). It appears that when two points lie on the *same side* of line ℓ (as do A and B), then the segment joining them does not intersect ℓ, and all the points of that segment lie entirely on that side. Hence, that side of ℓ is what we call a *convex set*. However, if two points lie on *opposite* sides (as do A and C), then the segment joining them intersects ℓ at some "interior" point of the segment. This forms the basis for the next axiom, which establishes the idea of a *half-plane*.

CONVEX SETS

First, let's introduce formally the concept of convexity so we can use it in dealing with the sides of a line.

DEFINITION

A set K in S is called **convex** provided it has the property that for all points $A \in K$ and $B \in K$, the segment $\overline{AB} \subseteq K$. (See Figure 2.35 for illustration.)

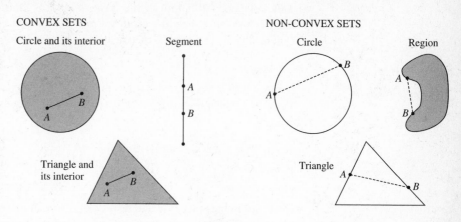

Figure 2.36

About the simplest result one can obtain directly from the definition is the most basic general property of convexity:

> If K_1 and K_2 are any two convex sets, $K_1 \cap K_2$ is also convex.

Proof: If A and B are any two points of $K_1 \cap K_2$, then A and B both belong to K_1 and K_2, and by convexity of K_1 and K_2 individually, $\overline{AB} \subseteq K_1$ and $\overline{AB} \subseteq K_2$. Hence, $\overline{AB} \subseteq K_1 \cap K_2$.

PLANE SEPARATION POSTULATE

Now we are ready to introduce the next axiom.

AXIOM H-1: PLANE SEPARATION POSTULATE

Let ℓ be any line lying in any plane P. The set of all points in P not on ℓ consists of the union of two subsets H_1 and H_2 of P such that:

(1) H_1 and H_2 are convex sets.

(2) H_1 and H_2 have no points in common.

(3)   If A lies in H_1 and B lies in H_2, line ℓ intersects the segment $\overline{AB}$.

The sets H_1 and H_2 are called the two **sides** of ℓ, or also, **half-planes** determined by ℓ. In order to identify which side of a line we are talking about, if A and B lie, respectively, in H_1 and H_2, we shall speak of the **A-side** of ℓ for H_1 and the **B-side** of ℓ for H_2.

THE HALF-PLANES
DETERMINED BY LINE ℓ

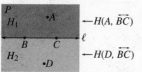

Figure 2.37

A neat notational device that will be useful (as illustrated in Figure 2.37) is

$$H(A, \overleftrightarrow{BC}) = H_1 \qquad \text{(the } A\text{-side of } \overleftrightarrow{BC})$$
$$H(D, \overleftrightarrow{BC}) = H_2 \qquad \text{(the } D\text{-side of } \overleftrightarrow{BC})$$

This notation will help us write down a kind of "formula" for the interior of an angle a little later. The first result we prove is an aid to proving other more significant results. Such a result is designated by the term **lemma** in mathematics.

PROPERTIES OF HALF-PLANES

LEMMA

Suppose that B lies on line ℓ, A lies in one of the half-planes H_1 determined by ℓ, and that A-B-C holds (Figure 2.38). Then point C will lie in the opposite half-plane, H_2.

Proof

Given: $A \in H_1$, $B \in \ell$, and A-B-C.

Prove: $C \in H_2$.

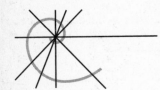

Figure 2.38

CONCLUSIONS	JUSTIFICATIONS
(1) C lies in the plane of ℓ and A.	Axiom I-3
(2) Either $C \in \ell$, $C \in H_1$, or $C \in H_2$.	all the logical cases (Plane Separation Postulate)
(3) Assume $C \in \ell$. (See Figure 2.38.)	Logical Case 1
(4) $\overleftrightarrow{BC} = \ell$.	two points determine a line
(5) $A \in \overleftrightarrow{BC} = \ell$ →←	A-B-C, $A \notin \ell$ since $A \in H_1$
(6) Assume $C \in H_1$.	Logical Case 2
(7) $B \in \overline{AC} \subseteq H_1$ and $B \in H_1$. →←	definition of segment, convexity of H_1, $B \in \ell$
(8) $\therefore C \in H_2$.	Rule of Elimination

A basic property of half-planes is taken up in the following Discovery Unit, where you will be guided through a proof. It is one of those "obvious" properties whose proof may seem unnecessary, but it is by no means trivial.

MOMENT FOR DISCOVERY

A Property of Half-Planes

Let ℓ be any line lying in plane P. Study the following argument in order to answer a few questions concerning it later. You are to supply the missing reasons. (We will borrow a property of betweenness which we will prove later from the Ruler Postulate; refer to Figure 2.39 for the argument.)

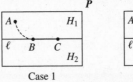
Case 1

Case 2

Figure 2.39

Conclusions	Justifications
(1) Line $\ell \subseteq$ plane P, and half-planes H_1 and H_2 determined by ℓ.	given, Axiom H-1
(2) There exists a point A in $\mathbf{P}$ not on ℓ.	Axiom I-5
(3) There exists a point B on ℓ.	_____ ?
(4) There exists a line m passing through A and B.	_____ ?
(5) There exists point C on m such that A-B-C.	Property of betweenness proven in §2.7
(6) Either $A \in H_1$ or $A \in H_2$.	_____ ?
(7) Assume $A \in H_1$.	Logical Case 1
(8) ∴ $C \in H_2$.	_____ ?
(9) Assume $A \in H_2$.	Logical Case 2
(10) ∴ $C \in H_1$.	_____ ?

What does this prove about half-planes? (See questions which follow for further guidance, if you need it.)

1. Is there any logical guarantee in the statement of the Plane Separation Postulate that half-planes are nonempty sets?

2. Does the argument in the lemma have any bearing on this? Does the argument outlined in the Discovery Unit help?

3. State the explicit theorem about half planes it proves.

Often it is necessary to refer to the part of a segment which excludes its endpoints, or a ray without its endpoint. We refer to these sets as **open segments** or **open rays**, and their individual points are called **interior points**. On the other hand, a **closed half-plane** is the set consisting of a line and one of its half-planes. To summarize,

$(\overline{AB}) = \{X : A\text{-}X\text{-}B\}$ (open segment)

$(AB] = \{X : X = B, A\text{-}X\text{-}B, \text{ or } A\text{-}B\text{-}X\}$ (open ray)

$[H_1] = H_1 \cup \ell$, where $H_1 = H(P, \ell)$ (closed half-plane)

THEOREM 1

If one point of a segment or ray lies in a half-plane H_1 determined by some line ℓ, and the endpoint of the segment or ray itself lies on ℓ, then the entire open segment or open ray lies in H_1.

(This is a key result for work with betweenness, and we will leave the proof as an interesting challenge; see Problem 10.)

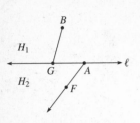

Figure 2.40

COROLLARY

Let *B* and *F* lie on opposite sides of a line ℓ and let *A* and *G* be any two distinct points on ℓ. Then segment $\overline{GB}$ and ray $\overrightarrow{AF}$ have no points in common (Figure 2.40).

Proof: By Theorem 1, $(\overline{GB}) \subseteq H_1$ and $(\overrightarrow{AF}] \subseteq H_2$, so since H_1 and H_2 are disjoint, $(\overline{GB}) \cap (\overrightarrow{AF}] = \varnothing$; since $A \neq G$, then $\overline{GB} \cap \overrightarrow{AF} = \varnothing$.

In the early development of the foundations of geometry, the following proposition was taken as an axiom and the Plane Separation Postulate proved as a theorem. It is a little easier and technically less tedious to do it the other way around, which is the approach we have taken here. This relationship is named after Moritz Pasch (1843–1930) who was the first to recognize its significance to geometry, and the first also to treat Euclid's unstated assumptions about betweenness explicitly. Either proposition, the Plane Separation Postulate or the Postulate of Pasch, may be taken as an axiom and the other proven as a theorem.

POSTULATE OF PASCH

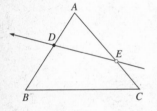

Figure 2.41

THEOREM 2: POSTULATE OF PASCH

Suppose *A*, *B*, and *C* are any three distinct noncollinear points in a plane, and ℓ is any line which also lies in that plane and passes through an interior point *D* of segment $\overline{AB}$ but not through *A*, *B*, nor *C*. Then ℓ meets either $\overline{AC}$ at some interior point *E*, or $\overline{BC}$ at some interior point *F*, the cases being mutually exclusive (Figure 2.41).

(Proof left as Problem 14.)

INTERIOR OF AN ANGLE

We all know how to identify the interior of an angle, at least by picture—even a child could do it. However, it is both interesting and useful that we now have a means of precisely defining this concept mathematically.

It is shown in Figure 2.42 how half-planes display the interior of $\angle ABC$. Informally, the interior of $\angle ABC$ consists of all points which lie between the two sides (rays $\overrightarrow{BA}$ and $\overrightarrow{BC}$). Formally, it is the intersection of the two half-planes determined by the sides of the angle and containing the other side. A point lying in this set is called an **interior point**; the **exterior**, or set of **exterior points**, is the set of all points in the plane not in the interior or sides of an angle. In symbolic form:

Formal:	Interior $\angle ABC = H(A, \overleftrightarrow{BC}) \cap H(C, \overleftrightarrow{BA})$
Informal:	The interior of $\angle ABC$ is the set of all points *X* which simultaneously lie on the *A*-side of $\overleftrightarrow{BC}$ and on the *C*-side of $\overleftrightarrow{BA}$.

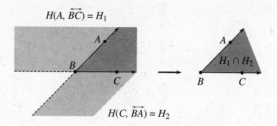

Figure 2.42

By its very definition as the intersection of two convex sets, the interior of an angle is convex—not altogether a useless fact. The next theorem is a more specialized result than this, but is the basic embodiment of the convexity property.

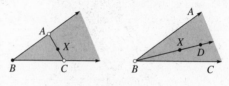

Figure 2.43

THEOREM 3

If A and C lie on the sides of $\angle B$, then, except for endpoints, segment $\overline{AC} \subseteq$ Interior $\angle B$. If $D \in$ Interior $\angle B$, then, except for B, ray $BD \in$ Interior $\angle B$ (Figure 2.43).

Proof: Let $X \in (\overline{AC})$. Then A-X-C. By Theorem 1, $X \in H(A, \overrightarrow{BC})$. Similarly, $X \in H(C, \overrightarrow{BA})$. Hence $X \in$ Interior $\angle ABC$.

(The proof of the second part is left as Problem 11.)

PROBLEMS   (§2.6)

___ *Group A* ___

1. Use the notation for half planes introduced in this section to name Regions 1 and 2 illustrated in the following figure.

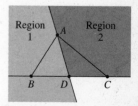

2. Use the notation for half planes introduced in this section to name Regions 1 and 2 illustrated in the following figure.

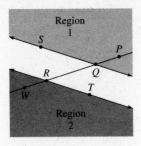

3. In the following figure, name the half-plane which
   (a) contains ray $\overrightarrow{DE}$ but does not contain points G and F
   (b) does NOT contain points D, E, and G.

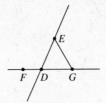

4. If $D \in H(A, \overleftrightarrow{BC})$, why is $H(D, \overleftrightarrow{BC}) = H(A, \overleftrightarrow{BC})$?

5. If two half-planes meet, where neither is a subset of the other, is their intersection always the interior of some angle?

6. Prove that if $B \notin \overleftrightarrow{AD}$ and A-B-C, then $C \in H(B, \overleftrightarrow{AD})$.

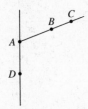

7. Explain why a line ℓ cannot meet all three sides $\overline{AB}$, $\overline{BC}$, and $\overline{AC}$ of a triangle at interior points D, E, F, respectively.

8. Why is the interior of an angle plus its sides a convex set?

9. What sorts of figures are possible from the intersection of three half-planes? Four? Five? ...?

_____ *Group B* _____

10. Prove Theorem 1. (Outline of proof to be completed: It suffices to prove the conclusion for rays only, since segment $\overline{AB}$ is a subset of ray $\overrightarrow{AB}$. So let $A \in \ell$ and suppose that $B \in H_1$, as in the following figure. Suppose also that X is any point of the open ray $(\overrightarrow{AB}]$. If $X = C$, where A-C-B, suppose $C \in \ell$ or $C \in H_2$. Then $B \in \ell$ ($\rightarrow\leftarrow$), or B-A-C ($\rightarrow\leftarrow$). $\therefore C \in H_1$. Similarly, if $X = D$, where A-B-D, then $D \in H_1$.)

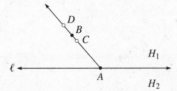

11. Prove the second part of Theorem 3: If $D \in$ Interior $\angle ABC$, then $(\overrightarrow{BD}] \subseteq$ Interior $\angle ABC$.

12. Suppose that A, B, and C are distinct, noncollinear points, and that A-D-B and C-E-D. Prove there exists a unique point F such that A-E-F and B-F-C.

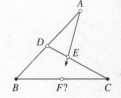

_____ *Group C* _____

13. Two convex sets K_1 and K_2 make up an entire plane P (that is, $K_1 \cup K_2 = P$). What must the two sets look like?

14. Prove the Postulate of Pasch as a theorem. (***HINT:*** Consider the two half-planes determined by line ℓ; what must be true of points A, B, and C, since they do not lie on ℓ?

2.7 ANGLE MEASURE AND THE RULER, PROTRACTOR POSTULATES

The basic axioms for angle measure assert

(a) that we can measure any angle

(b) when two angles are placed adjacent to one another and they are not too large, we can sum the measures of the smaller angles to obtain the measure of the larger angle.

EXISTENCE OF ANGLE MEASURE, ADDITION POSTULATE

AXIOM A-1: EXISTENCE OF ANGLE MEASURE

To every angle there corresponds a unique, real number θ, $0 < \theta < 180$.

The real number associated with an angle $\angle ABC$ will be called its **measure**, denoted $m\angle ABC$, or $m\angle B$ if no ambiguity results. The degree mark, as in 120°, is not used since angle measure is just a real number, needing no particular embellishment. However, we will often use degree marks in figures to distinguish angle measure from distance or lengths of segments occurring in the same figure.

NOTE: As is customary in the foundations of geometry, we assume that all angles have measure less than 180. If this seems unnatural to you, try defining what you think would be meant by an *angle of measure* 270. (Be precise.)

The next axiom for angle measure provides the basis for a betweenness concept for rays, analogous to that for collinear points.

AXIOM A-2: ANGLE ADDITION POSTULATE

If D lies in the interior of $\angle ABC$, then $m\angle ABC = m\angle ABD + m\angle DBC$, and conversely.

In the statement of the postulate, we would assert that ray $\overrightarrow{BD}$ lies **between** rays $\overrightarrow{BA}$ and $\overrightarrow{BC}$. In general,

DEFINITION

Suppose that $\overrightarrow{OA}$, $\overrightarrow{OB}$, and $\overrightarrow{OC}$ are concurrent rays, all having the same endpoint O. Then if these rays are distinct, and if

$$m\angle AOB + m\angle BOC = m\angle AOC,$$

then ray $\overrightarrow{OB}$ is said to lie **between** rays $\overrightarrow{OA}$ and $\overrightarrow{OC}$, and we write $\overrightarrow{OA}\text{-}\overrightarrow{OB}\text{-}\overrightarrow{OC}$.

The same formal properties for ray betweenness as point betweenness follow, and in the same manner as before. It is not necessary to pursue that in detail—it's enough to know that we can do it if we want. In connection with the Angle Addition Postulate, however, it is easy to become confused about what we are assuming about rays. The definition of ray betweenness, involving additivity of angle measure, is independent of the definition for the interior of an angle, *which involves the concept of half-planes*. Thus a restatement of the Addition Postulate sounds like a play on words:

If a ray whose endpoint is the vertex of an angle passes through an interior point of that angle, then the ray lies between the sides of that angle.

It is indeed fortunate that we do not have to be careful with our language here! The Addition Axiom, Axiom A-2, formally ties together the two concepts of ray betweenness and interior points of angles.

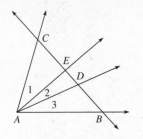

Figure 2.44

RULER AND PROTRACTOR POSTULATES

EXAMPLE 1 Suppose that in Figure 2.44 the betweenness relation C-E-D-B holds on line $\overleftrightarrow{BC}$. Show that $\overrightarrow{AC}$-$\overrightarrow{AE}$-$\overrightarrow{AD}$-$\overrightarrow{AB}$[4] holds, and that $m\angle CAB = m\angle 1 + m\angle 2 + m\angle 3$.

SOLUTION

Since E is interior to $\angle CAD$ by Theorem 3 of Section 2.6, $\overrightarrow{AC}$-$\overrightarrow{AE}$-$\overrightarrow{AD}$ follows from Axiom A-2. Similarly, $\overrightarrow{AC}$-$\overrightarrow{AE}$-$\overrightarrow{AB}$, $\overrightarrow{AC}$-$\overrightarrow{AD}$-$\overrightarrow{AB}$, and $\overrightarrow{AE}$-$\overrightarrow{AD}$-$\overrightarrow{AB}$. It follows by definition that $\overrightarrow{AC}$-$\overrightarrow{AE}$-$\overrightarrow{AD}$-$\overrightarrow{AB}$. Therefore, $m\angle CAB = m\angle CAD + m\angle 3$ and $m\angle CAD = m\angle 1 + m\angle 2$, or by substitution, $m\angle CAB = m\angle 1 + m\angle 2 + m\angle 3$.∎

As its name suggests, the Ruler Postulate affirms that the points on a line have the appearance as indicated in Figure 2.45, where the points of $\overleftrightarrow{AB}$ are marked off by the corresponding numbers on the ruler indicating their exact position on that line. Obviously, these ruler marks also serve to measure the distance between any two points. Thus, for example, $CD = 9\frac{1}{8} = 1.125$ (inches).

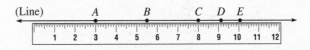

Figure 2.45

AXIOM D-5: RULER POSTULATE

The points of each line may be assigned to the real numbers x, $-\infty < x < \infty$, called **coordinates**, in such a manner that

(1) each point on ℓ is assigned to a unique coordinate

(2) each coordinate is assigned to a unique point on ℓ

(3) any two points on ℓ may be assigned to zero and a positive coordinate, respectively

(4) if points A and B on ℓ have coordinates a and b, then $AB = |a - b|$.

Likewise, any angle can be measured by a protractor, as shown in Figure 2.46. We set the edge of the protractor along one side of the angle (ray $\overrightarrow{AB}$), with the center mark at the vertex of the angle, and take the reading from the other side of the angle (ray $\overrightarrow{AC}$). Thus, we find that in this example, $m\angle CAB \approx 25.2$. Along with ray $\overrightarrow{AC}$, each of the rays with vertex A can be assigned a unique real number. Essentially, just as with distance, the difference between coordinates of two rays yields the measure of the angle defined by them. For example, in Figure 2.46, $m\angle DAC = 65 - 25.2 = 39.8$.

---

[4]Exactly analogous to A-B-C-D for points. (See definition immediately preceding Theorem 1 of §2.4.)

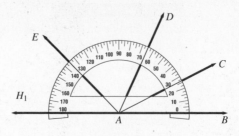

Figure 2.46

AXIOM A-3: PROTRACTOR POSTULATE

The set of rays having a common origin O and lying on one side of line $\ell = \overrightarrow{OA}$, including ray $\overrightarrow{OA}$, may be assigned to the real numbers θ for which $0 \leq \theta < 180$, called **coordinates**, in such a manner that

(1) each ray is assigned a unique coordinate θ

(2) each coordinate θ is assigned to a unique ray

(3) the coordinate of $\overrightarrow{OA}$ is 0

(4) if rays $\overrightarrow{OP}$ and $\overrightarrow{OQ}$ have coordinates θ and ϕ, then $m\angle POQ = |\theta - \phi|$.

BASIC BETWEENNESS PROPERTIES

There is a duality between the set of all collinear points on line ℓ lying on one side of some point P on ℓ, and the set of rays concurrent at O lying on one side of line $\overrightarrow{OA}$. Every property of angle measure and betweenness in the family of rays has its direct counterpart concerning distance and betweenness in this set of collinear points. For example, if we prove that points having positive coordinates a, b, and c satisfy a betweenness relation from the ruler postulate because $a < b < c$, it is clear that the protractor postulate implies the same corresponding conclusion for rays. It is therefore necessary to prove such properties for only one case. This observation will save us a great deal of work in everything that follows.

The first result is a basic tool. Its proof indicates how coordinates control betweenness relations, and makes betweenness completely predictable. For convenience, we shall use the notation $A[a]$, $B[b]$, ... to indicate that the coordinates of A and B on a line under the Ruler Postulate are a and b, respectively, and similarly for rays.

THEOREM 1

If $A[a]$, $B[b]$, and $C[c]$ are three collinear points (and $\overrightarrow{OA}[a]$, $\overrightarrow{OB}[b]$, $\overrightarrow{OC}[c]$ three concurrent rays) with their coordinates, then $A\text{-}B\text{-}C$ ($\overrightarrow{OA}\text{-}\overrightarrow{OB}\text{-}\overrightarrow{OC}$) iff $a < b < c$ or $c < b < a$.

Proof

I.   *Given:* Either $a < b < c$ or $c < b < a$.

   Prove: $A\text{-}B\text{-}C$ ($\overrightarrow{OA}\text{-}\overrightarrow{OB}\text{-}\overrightarrow{OC}$).

THEOREM 1 (CONTINUED)

CONCLUSIONS	JUSTIFICATIONS
(1) $a < b < c$	First Logical Case
(2) $AB = b - a, BC = c - b, AC = c - a$	Ruler Postulate
(3) $AB + BC = b - a + c - b = c - a = AC$	algebra
(4) $\therefore$ A-B-C (similarly, $\overrightarrow{OA}$-$\overrightarrow{OB}$-$\overrightarrow{OC}$)	definition
(5) $c < b < a$	Second Logical Case
(6) $AB = a - b, BC = b - c, AC = a - c$	Ruler Postulate
(7) The sum again produces A-B-C (or $\overrightarrow{OA}$-$\overrightarrow{OB}$-$\overrightarrow{OC}$).	algebra, definition

II.   *Given:* A-B-C ($\overrightarrow{OA}$-$\overrightarrow{OB}$-$\overrightarrow{OC}$).

   Prove: Either $a < b < c$ or $c < b < a$.

(1) By Part I, if a is numerically between b and c, then B-A-C, and if c is numerically between a and b, then A-C-B. But each of these contradicts A-B-C, so b is numerically between a and c.

COROLLARY

Suppose that four distinct collinear points are given with their coordinates: $A[a]$, $B[b]$, $C[c]$, $D[d]$. If A-B-C and A-C-D, then A-B-C-D, and similarly for rays $\overrightarrow{OA}$, $\overrightarrow{OB}$, $\overrightarrow{OC}$, and $\overrightarrow{OD}$.

It is evident that the Ruler Postulate now imposes on each line in our geometry a continuum of points, like the real number line, with betweenness of points coinciding with order relations among their corresponding coordinates. Duality extends these same results to the set of concurrent rays on one side of a line.

CONVEXITY OF SEGMENTS

The convexity of a geometric two-dimensional object is related to its shape. In the case of a one-dimensional object, such as a segment, ray, or line, convexity is related to its "straightness." The basic result, now provable with ease, is

A segment, ray, or line is a convex set.

The proof of this depends on this basic result: If $C \in \overline{AB}$ and $D \in \overline{AB}$, then $\overline{CD} \subseteq \overline{AB}$. This is easy using Theorem 1 and the definition of segment, so we leave the rest for you.

"TWO POINTS DETERMINE A RAY"

In geometry, we often take for granted that in reading an angle, such as that shown in Figure 2.47, it is immaterial whether we use $\angle BAD$ or $\angle CAD$, where C is

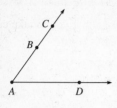

Figure 2.47

any point on ray $\overrightarrow{AB}$. This is actually another important betweenness result, addressed by Theorem 2.

LEMMA

If A and B are two distinct points, and $C \in \overrightarrow{AB}$, with $A \neq C$, then $\overrightarrow{AB} \subseteq \overrightarrow{AC}$.

Proof: From the Ruler Postulate and Theorem 1, let the coordinates of A, B, and C be, respectively, 0, $b > 0$, and c; since either A-C-B, or A-B-C, $c > 0$ (Figure 2.48). If X is any point of ray $\overrightarrow{AB}$ and x is its coordinate, then $x \geq 0$; by order properties of the reals, we must have either

$$0 \leq x \leq c \text{ (that is, } X \in \text{ segment } \overline{AC} \subseteq \text{ ray } \overrightarrow{AC})$$

or

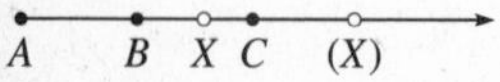

$$0 < c < x \text{ (that is, } A\text{-}C\text{-}X \text{ and } X \in \text{ ray } \overrightarrow{AC})$$

Hence, in all cases, $X \in$ ray $\overrightarrow{AC}$. Since X was any point of ray $\overrightarrow{AB}$, we have shown

$$\overrightarrow{AB} \subseteq \overrightarrow{AC},$$

as desired.

Figure 2.48

THEOREM 2

If $C \in \overrightarrow{AB}$ and $A \neq C$, then $\overrightarrow{AB} = \overrightarrow{AC}$.

(Proof left as Problem 11.)

SEGMENT/ANGLE CONSTRUCTION THEOREMS

It does not take a great deal of imagination to conclude that the next theorem will be an indispensable tool for our study of geometry, as it was throughout Euclid's *Elements*.

THEOREM 3: SEGMENT CONSTRUCTION THEOREM

If $\overline{AB}$ and $\overline{XY}$ are any two segments and $AB \neq XY$, then there is a unique point C on ray $\overrightarrow{AB}$ such that $AC = XY$, with A-C-B if $XY < AB$, or A-B-C if $XY > AB$. (See Figure 2.49.)

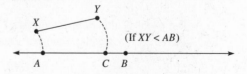

Figure 2.49

We will prove just one case of this important theorem and allow you to take care of the other, which is very similar.

Proof:

Given: $XY < AB$.

Prove: There exists $C \in \overline{AB}$ such that $AC = XY$ and A-C-B.

Conclusions	Justifications
(1) Suppose that $A[0]$ and $B[b]$, $b > 0$, are the given points, with their coordinates, on line $\overleftrightarrow{AB}$.	Ruler Postulate
(2) $AB = \lvert b - 0 \rvert = b$	Ruler Postulate
(3) Let $c = XY > 0$.	positive definiteness of metric $(X \neq Y)$
(4) Locate $C[c]$ on AB.	Ruler Postulate
(5) $0 < c < b$	$XY < AB = b$
(6) $\therefore$ A-C-B and $AC = c = XY$	Theorem 1, Ruler Postulate, Step **(3)**

Another construction which Euclid used was the location of the midpoint of a line segment, accomplished with a compass and straight-edge. In our axiomatic development, we first must define midpoint, then prove it exists.

DEFINITION

A point M on a segment $\overline{AB}$ is called a **midpoint** iff it has the property that $AM = MB$. Such a midpoint is also said to **bisect** the segment, and any line, segment, or ray passing through that midpoint is also said to **bisect** the segment.

THEOREM 4: MIDPOINT CONSTRUCTION THEOREM

The midpoint of any segment exists and is unique.

Proof

Given: $\overline{AB}$, with $A \neq B$.

Prove: **I.** A midpoint M exists for $\overline{AB}$.

 II. If M' is any other midpoint, then $M' = M$.

Conclusions	Justifications
I.	
(1) Choose a coordinate system such that $A[0]$ and $B[b]$, with $b > 0$.	Ruler Postulate
(2) Locate the point $M[\frac{1}{2}b]$ on $\overline{AB}$.	Ruler Postulate

THEOREM 4: MIDPOINT CONSTRUCTION THEOREM (CONTINUED)

(3) $AM = \left| \frac{1}{2}b - 0 \right| = \frac{1}{2}b$ and $MB =$ Ruler Postulate

   $\left| b - \frac{1}{2}b \right| = \frac{1}{2}b$, so $AM = MB$.

(4) $\therefore$ M is a midpoint of $\overline{AB}$.   definition

II.

(5) Let x be the coordinate of M'.   Ruler Postulate

(6) $0 \le x \le b$.   Theorem 1, $M' \in \overline{AB}$

(7) $AM' = x$ and $M'B = b - x$.   Ruler Postulate

(8) $AM' = M'B$.   definition

(9) $x = b - x$, hence $x = \frac{1}{2}b$ and   algebra, Step **(2)**
   $\therefore$ $M' = M$

Another of Euclid's assumptions involved "extending" a line segment by "its own length," that is, essentially to "double" a given line segment. In view of preceding arguments, its proof is clear. (See Figure 2.50.)

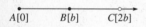

$A[0]$ $B[b]$ $C[2b]$

Figure 2.50

THEOREM 5: SEGMENT DOUBLING THEOREM

There exists a unique point C on ray $\overrightarrow{AB}$ such that B is the midpoint of $\overline{AC}$.

Each of the previous results concerning the construction of segments a certain length, midpoints, and extensions of line segments all have their counterpart with angles, using betweenness concepts for rays. The pay-off with our approach is that now we can just state the results; their proofs are the duals of the proofs we have already given for segments. The definition of **angle bisector** and a line which **bisects** an angle (a line containing the bisector) are clear from duality.

THEOREM 3': ANGLE CONSTRUCTION THEOREM

If $\angle ABC$ and $\angle XYZ$ are any two nondegenerate angles and $m\angle ABC \ne m\angle XYZ$, then there exists a unique ray $\overrightarrow{BD}$ on the C-side of $\overleftrightarrow{AB}$ such that $m\angle XYZ = m\angle ABD$, and either $\overrightarrow{BA}$-$\overrightarrow{BD}$-$\overrightarrow{BC}$ if $m\angle XYZ < m\angle ABC$, or $\overrightarrow{BA}$-$\overrightarrow{BC}$-$\overrightarrow{BD}$ if $m\angle XYZ > m\angle ABC$.

> ### THEOREM 4′: ANGLE BISECTION THEOREM
> Every angle has a unique bisector.

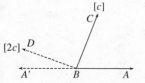

> ### THEOREM 5′: ANGLE DOUBLING THEOREM
> Given any angle $\angle ABC$ having measure < 90, there exists a ray $\overrightarrow{BD}$ such that $\overrightarrow{BC}$ is the bisector of $\angle ABD$. (See Figure 2.51.)

Figure 2.51

PROBLEMS   (§2.7)

_____ *Group A* _____

1. Suppose that four collinear points with their respective coordinates are $A[3]$, $B[4]$, $C[2]$, and $D[-4]$. Show that the origin $O[0]$ lies on segment $\overline{AD}$, but not on $\overline{AB}$, $\overline{AC}$, or $\overline{BC}$.

2. Let $A[3]$ and $B[-6]$ be two points on a line (with their coordinates). Find what values of x are allowable if point $P[x]$ is any point on
    (a) $\overleftrightarrow{AB}$
    (b) $\overrightarrow{AB}$
    (c) $\overrightarrow{BA}$.
    (You do not have to prove your answers.)

3. Let $A[5]$, $B[10]$, $C[3]$, and $D[-1]$ denote four collinear points. If $P[x]$ denotes any other point on $\overleftrightarrow{AB}$, find the range of x allowable if
    (a) $P \in \overline{CB}$
    (b) $P \in \overrightarrow{AD}$
    (c) $P \in \overrightarrow{AD} \cap \overrightarrow{CA}$.

4. Let $A[-3]$, $B[-1]$, $C[6]$, and $D[5]$ denote points on a line, with their coordinates. If $P[x]$ denotes any other point on $\overleftrightarrow{AB}$, find the range of x allowable if
    (a) $P \in \overline{AB}$
    (b) $P \in \overline{BD}$
    (c) $P \in \overrightarrow{DC}$
    (d) $P \in \overrightarrow{AB} \cup \overrightarrow{CD}$.

5. Prove that any line segment $\overline{AB}$ can be **trisected**; that is, there exist points T_1 and T_2 on $\overline{AB}$ distinct from A and B such that $AT_1 = T_1T_2 = T_2B$. (By duality, it follows that *any angle may be trisected,* axiomatically.) See Problem 7, §2.8. (**HINT:** Assume coordinates such that $A[0]$ and $B[b]$, where $b > 0$; use the Ruler Postulate.)

6. On the line indicated in the accompanying figure, if $WY = 17$, $WZ = 23$, and $XZ = 21$, find XY. (Use the betweenness relations evident from the figure at the right.) [UCSMP[5], p. 50]

[5]University of Chicago School Mathematics Project: Geometry.

7. In Problem 6, how many answers are possible for XY if no betweenness relations are assumed from a figure? (Assume X, Y, Z, and W are collinear.) Find them.

8. Let $A[a]$ and $B[b]$ designate two points on a line which are 8 units apart. What are the possible values of b?

9. If on some line ℓ we plot all points $P[x]$ such that $x < 3$, what geometric object best describes the given set? What exactly describes it?

———— *Group B* ————

10. Prove that if T_1 and T_2 are the trisectors of segment $\overline{AB}$, as defined in Problem 5, then T_1 is the midpoint of $\overline{AT_2}$, T_2 is the midpoint of $\overline{T_1B}$, and $AT_1 = \frac{1}{3}AB$.

11. Prove Theorem 2: If $C \in \overrightarrow{AB}$ and $C \neq A$, then $\overrightarrow{AB} = \overrightarrow{AC}$. (**HINT:** Use the lemma to show that both $\overrightarrow{AB} \subseteq \overrightarrow{AC}$ and $\overrightarrow{AC} \subseteq \overrightarrow{AB}$. Can the roles of B and C be reversed?)

12. Write out an explicit proof for the corollary to Theorem 1: If A-B-C and A-C-D, then A-B-C-D.

13. Prove: If A, B, and C are any three distinct, collinear points, then either A-B-C, A-C-B, or B-A-C.

14. Prove the remaining case of Theorem 3: A-B-C if $XY > AB$.

15. For practice, write out a proof of Theorem 5′, the dual of the Segment Doubling Theorem.

———— *Group C* ————

16. UNDERGRADUATE RESEARCH PROJECT How many distances among n points on a line satisfying the Triangle Inequality will uniquely determine the ordering of those points? (i.e., P_1-P_2-P_3-...-P_n, where the obvious extension of the betweenness concept A-B-C-D is indicated.) Is there a formula $d(n)$ for this number? (Start with the simplest case $n = 3$, with $AB = d_1$, $BC = d_2$, and $AC = d_3$, where $d_1 = d_2 + d_3$; will this force A-C-B? What if only two distances are given for these three points?)

2.8 CROSSBAR THEOREM, LINEAR PAIR AXIOM, PERPENDICULARITY

One of the deeper results of axiomatic geometry is the so-called **Crossbar Principle**: If a segment reaches from one side of an angle to the other and a ray lies between its sides, then the ray must intersect the segment. (See Figure 2.52.) This is certainly an obvious property, but one that requires proof if we are to derive all our results axiomatically. We encourage you to participate by supplying the missing reasons in the following two-column proof.

CROSSBAR THEOREM

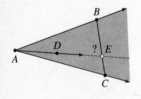

Figure 2.52

THEOREM 1: CROSSBAR THEOREM

If D is in the interior of $\angle BAC$, then ray $\overrightarrow{AD}$ meets segment $\overline{BC}$ at some interior point E. (See Figure 2.52.)

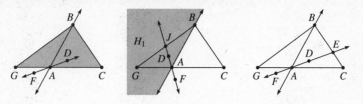

Figure 2.53

THEOREM 1: CROSSBAR THEOREM (CONTINUED)

Proof: (See Figure 2.53.)

Given: $\angle BAC$, $D \in$ Interior $\angle BAC$, ray $\overrightarrow{AD}$, and segment $\overline{BC}$.

Prove: Ray $\overrightarrow{AD}$ meets $\overline{BC}$ at some point E such that B-E-C.

CONCLUSIONS	JUSTIFICATIONS
(1) There exists F with D-A-F and G with C-A-G.	Ruler Postulate, Theorem 1, §2.7
(2) Either line $\overrightarrow{AD} \cup \overrightarrow{AF}$ meets $\overline{BG}$ at some interior point J or it meets $\overline{BC}$ at some interior point E.	_____?
(3) Assume $\overrightarrow{AD} \cup \overrightarrow{AF}$ meets $\overline{BG}$ at J.	Logical Case 1
(4) F and D, and therefore F and B, lie on opposite sides of line $\overleftrightarrow{AC}$.	F-A-D, and D lies on the B-side of $\overleftrightarrow{AC}$
(5) $J \notin \overrightarrow{AF}$	$\overline{GB} \cap \overrightarrow{AF} = \varnothing$ (corollary to Theorem 1, §2.6)
(6) Then $J \in \overrightarrow{AD}$ $\therefore D \in \overrightarrow{AJ}$	_____?
(7) D and J both lie on the G-side of $\overleftrightarrow{AB}$.	$J \in \overline{BG}$ and $\overline{BG}$ lies on the G-side of AB, hence $\overrightarrow{AJ}$ lies on that side (Theorem 1, §2.6)
(8) But D lies on the C-side of $\overleftrightarrow{AB}$. $\rightarrow\leftarrow$	_____?
(9) $\therefore \overrightarrow{AD} \cup \overrightarrow{AF}$ meets $\overline{BC}$ at point E.	Rule of Elimination
(10) $E \notin \overrightarrow{AF}$, hence $E \in \overrightarrow{AD}$, as desired.	_____?

OPPOSITE RAYS, LINEAR AND VERTICAL ANGLE PAIRS

If A-B-C, then the union of the two rays $\overrightarrow{BA}$ and $\overrightarrow{BC}$ is a line (namely, line $\overleftrightarrow{AC}$). We call any two such rays **opposite** or **opposing** rays. A basic fact, easy to prove, is that if we are given a ray $\overrightarrow{PQ}$, then there always exists a unique opposite ray $\overrightarrow{PR}$.

DEFINITION

If the sides of one angle are opposite rays to the respective sides of another angle (as shown in Figure 2.54), the angles are said to form a **vertical pair**.

VERTICAL PAIR

LINEAR PAIR

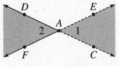

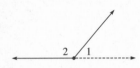

Figure 2.54

Figure 2.55

A standard fact of geometry is that vertical angles have equal measures—either taken as an axiom, or proven as a theorem from another axiom. We choose a development that follows the spirit of Euclid's *Elements*, which necessitates adding one more axiom. First, we consider another kind of angle pair.

DEFINITION

Two angles are said to form a **linear pair** iff they have one side in common and the other two sides are opposite rays. (See Figure 2.55.) We call any two angles whose angle measures sum to 180 a **supplementary pair**, or more simply, **supplementary**, and two angles whose angle measures sum to 90 a **complementary pair**, or **complementary**.

THEOREM 2

Two angles which are supplementary (or complementary) to the same angle have equal angle measures.

(To prove, use simple algebra and the definitions.)

As our final axiom on angles, we assume

AXIOM A-4

A linear pair of angles is a supplementary pair.

Now we are in a position to prove the so-called **Vertical Pair Theorem**.

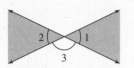

Figure 2.56

THEOREM 3: VERTICAL PAIR THEOREM

Vertical angles have equal measures.

Proof: By the previous definitions, the angles marked 1 and 3 in Figure 2.55 form a linear pair, as are those marked 3 and 2; thus by Axiom A-2, angles 1 and 2 are each supplementary to a common angle 3, and must have equal angle measures by Theorem 2.

The intuitive idea of perpendicularity comes from a concept of "erect" or "prone position," or one line lying "vertical" relative to another. This means that the line must form a linear pair of angles having equal measures with the other line. The definition suggested by this idea is asymmetric; that is, it cannot be immediately concluded (at least, not from the definition) that if line ℓ is perpendicular to line m, then m is perpendicular to ℓ.

DEFINITION

If line ℓ intersects another line m at some point A and forms a supplementary pair of angles at A having equal measures, then ℓ is said to be **perpendicular** to m, and we write $\ell \perp m$. (In Figure 2.57, we have $\ell \perp m$ because $\angle CAB$ and $\angle CAB'$ have equal angle measures.)

Simple algebra shows that if $\ell \perp m$ at point A, as in Figure 2.57, the supplementary pair of angles (e.g., $\angle CAB$ and $\angle CAB'$) must each have (degree) measure 90. (Do you see why?)

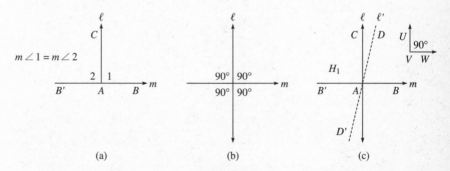

Figure 2.57

DEFINITION

An angle having measure 90 is called a **right angle**. Angles having measure less than 90 are **acute angles**, and those with measure greater than 90, **obtuse angles**.

> ### THEOREM 4
> One line is perpendicular to another line iff the two lines form four right angles at their point of intersection. (See Figure 2.57.)

We shall leave the proofs of the remaining theorems to you for problems.

COROLLARY A

Line ℓ is perpendicular to line m iff ℓ and m contain the sides of a right angle.

COROLLARY B

The relation of perpendicularity is symmetric, that is, if $\ell \perp m$, then $m \perp \ell$.

> ### THEOREM 5: EXISTENCE AND UNIQUENESS OF PERPENDICULARS
> Suppose that in some plane line m is given and an arbitrary point A on m is located. Then there exists a unique line ℓ that is perpendicular to m at A.

PROBLEMS (§2.8)

_____ *Group A* _____

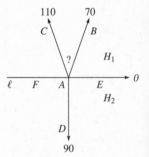

1. In the figure to the right, the rays in both half-planes H_1 and H_2 have been assigned coordinates according to the Protractor Postulate. Rays $\overrightarrow{AB}$ and $\overrightarrow{AC}$ have coordinates 70 and 110 in H_1 and ray $\overrightarrow{AD}$ has coordinate 90 in H_2 (as illustrated). Using betweenness relations evident from the figure, determine

    (a) $m\angle BAC$

    (b) $m\angle CAD$

    (c) $m\angle BAD$

    (d) What betweenness relations among rays $\overrightarrow{AB}$, $\overrightarrow{AC}$, and $\overrightarrow{AD}$ hold, if any?

2. As in Problem 1, rays $\overrightarrow{AE}$ and $\overrightarrow{AF}$ have coordinates 31 and 150, as shown in the figure below, with line $\ell = \overrightarrow{AM} \cup \overrightarrow{AN}$.

    (a) Find $m\angle EAF$.

    (b) Which of the rays $\overrightarrow{AM}$ or $\overrightarrow{AN}$ lies between $\overrightarrow{AE}$ and $\overrightarrow{AF}$?

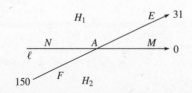

3. As in Problem 1, the rays shown in the figure to the right have the coordinates as indicated. Find

  (a) $m\angle RAT$

  (b) $m\angle RAW$

  (c) $m\angle SAW$

  (d) $m\angle TAW$

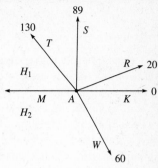

4. Using your answers in Problem 3, verify the following formula for $m\angle XAY$, when $\overrightarrow{AX}$ and $\overrightarrow{AY}$ lie on opposite sides of line ℓ—you do not have to provide proofs: If x and y are the coordinates of $\overrightarrow{AX}$ and $\overrightarrow{AY}$, respectively, then

$$m\angle XAY = x + y, \quad \text{if } x + y < 180$$

$$= 360 - (x + y), \quad \text{if } x + y > 180$$

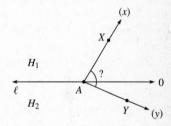

_____ **Group B** _____

5. In Problem 3 (see the top right figure) prove that $m\angle SAW = 149$.

6. Prove that if B-D-E-C and $A \notin \overrightarrow{BD}$, then $\overrightarrow{AB}$-$\overrightarrow{AD}$-$\overrightarrow{AE}$-$\overrightarrow{AC}$.

*7. Define angle trisectors and prove that any angle can be trisected in axiomatic geometry.

  *See Problem 11, §4.6.

  NOTE: A famous problem going back to ancient times is finding a compass/straightedge construction for the trisectors of a given angle. It is known that no construction exists. The proof of this fact involves methods of abstract algebra.

8. Prove the corollary of Theorem 4 (Symmetry of Perpendicularity).

9. Prove Theorem 4. (Use Figure 2.57c.)

10. Prove Theorem 5. (Use Figure 2.57c.)

_____ **Group C** _____

*11. Converse of Vertical Pair Theorem Prove the following *converse* of the Vertical Pair Theorem: If B-A-C, rays $\overrightarrow{AD}$ and $\overrightarrow{AE}$ lie on opposite sides of line $\overleftrightarrow{AB}$, and if $m\angle BAD = m\angle CAE$, then D-A-E and ($\angle BAD$, $\angle CAE$) form a vertical pair of angles.

  *See Problem 21, §4.1, and Problem 12, §6.2.

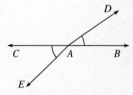

*12. Prove the well-known geometric principle: *The sum of the measures of any set of adjacent angles about a point, whose sides and interiors cover then entire plane, is 360.*

  *See Problem 11, §3.8.

This chapter began with a discussion of axiomatic mathematics, logic and proof, and how models play a role in determining whether a set of axioms is consistent and independent. This paved the way for beginning our study of axiomatic geometry, with *point*, *line*, *plane*, and *space* as undefined terms.

The axioms in the first group, I-1 through I-5, describe the interaction of points, lines and planes in three-dimensional space, called *incidence* properties. There are no surprises here—points, lines and planes behave just like we imagine them: A line will either "pierce" a plane at precisely one point, or be wholly contained by it (unless the line and plane are parallel), and given a line and point in space such that the point does not lie on the line, there is precisely one plane that contains both the line and point. Finally, two nonparallel planes will always intersect in a straight line.

The next group of axioms, D-1 through D-3, establishes the metric concept, or distance between points. From this, betweenness ideas were explored, and *segments*, *rays* and *angles* were defined. Axiom D-4 provides us with a desirable property of lines as the union of two opposite rays at any point of the line.

The Plane Separation Postulate, Axiom H-1, makes explicit the assumptions we are making regarding the behavior of the two regions into which a line divides a plane. Everything we take for granted about such concepts as the "side of a line" is given a formal basis here in terms of *half-planes*. The concept of half-planes makes it easy to define, and deal with, the interior of an angle (without using angle measure). Angle measure emerges from the basic existence axiom, Axiom A-1, and the Angle Addition Postulate, Axiom A-2, provides a connection between the interior of an angle and betweenness of rays, which would not otherwise exist.

The Ruler/Protractor Postulates (Axioms D-5 and A-3) provide us with the basis for thinking of lines as a *continuum* of points, as we normally do; analogously, a continuum of rays with a common endpoint exists. In particular, there is a system of coordinates for the points on each line ℓ and, dually, one for the set of all "concurrent" rays in a half-plane H such that

$$m\overline{AB} \equiv AB = |a - b|,$$

where a and b are the respective coordinates of points A and B on ℓ, and

$$m\angle AOB = |a - b|,$$

where a and b are the respective coordinates of rays $\overrightarrow{OA}$ and $\overrightarrow{OB}$ in H. In this part of our development, all the familiar ideas concerning betweenness for collinear points and concurrent rays were established. The construction theorems were proven, and the existence of midpoints and angle bisectors established.

Finally, in the last section, the Crossbar Principle was established and the final axiom on angle measure introduced which enabled us to deal with perpendicular lines, right angles, and vertical angles in the customary way. (You are to be reminded that we have not yet proven the existence of a line perpendicular to a given line passing through a given *external* point not on the line.)

This completes the presentation and development of 15 of the 17 axioms needed for the development of Euclidean geometry.

Answer each of the following questions True (T) or False (F):

1. The logical purpose of a model for an axiomatic system is to enlighten our understanding of the axioms.

2. An axiomatic system is independent provided that, for each axiom, there exists a separate model in which that axiom is false and all the rest are true.

3. Axiom I-4 ("if two planes meet, they do so in a line") can be logically proven from Axiom I-3 ("if two points of a line lie in a plane, the line lies entirely in the plane").

4. The Triangle Inequality was not assumed because it would be redundant to do so.

5. The following was assumed as an axiom: Given any four distinct collinear points A, B, C, and D, either D-A-B, A-D-B, B-D-C, or B-C-D.

6. Angles of arbitrary positive measure are possible in our system.

7. The interior of an angle with measure greater than 180 is defined as a nonconvex region.

8. Half-planes, segments, rays, and angles are all examples of convex sets.

9. Perpendicular lines were defined as lines which meet at an angle of measure 90 (a right angle).

10. Except for the case of perpendicularity, one of the angles in a linear pair of angles is always acute.

3

*F*OUNDATIONS OF *G*EOMETRY *2*: *T*RIANGLES, *Q*UADRILATERALS, *C*IRCLES

OVERVIEW

In his organization of geometry, Euclid developed the basic properties of triangles before those of quadrilaterals and rectangles. Consequently, the postulate of parallels was postponed until late in the development. We consider here that same body of material, which provides the foundation for two distinctly different geometries. This development is known as **absolute geometry**[1], and it consists of that part of Euclidean geometry which excludes any reference to parallel lines or related notions. As we shall see, the character of geometry is profoundly affected by the kind of parallelism properties it has.

---

[1]This term was first used by J. Bolyai to describe a general development of three-dimensional space, or what any basic geometric study must include without a commitment regarding parallelism. Any logical consequence thereof would be an *absolute truth* (about physical space), thus the term.

It is remarkable that so much can be done without assumptions about parallelism, and it may come as a shock to know that a totally different yet consistent
geometry based on a non-Euclidean concept for parallel lines is possible (which we
study in Chapter 6). In Euclid's day, the study of geometry was synonymous with
the study of physical space, and philosophers maintained that any other geometry
was impossible or contradictory. Today, we know that this narrow view of consistency in axiomatic geometry is false. Because of the discovery of non-Euclidean
geometry in the 1800s, geometry, as well as mathematics in general, was liberated
from its old bonds to the natural world. The resulting proliferation of new areas of
mathematics and the new applications to science *and* nature spawned by this revolution have greatly surpassed the wildest dreams of the early Euclidean geometers.

3.1   TRIANGLES, CONGRUENCE RELATIONS, SAS HYPOTHESIS

TRIANGLES   The terminology commonly associated with triangles will be introduced formally.

DEFINITION

A **triangle** is the union of three segments (called its **sides**), whose endpoints
(called its **vertices**) are taken, in pairs, from a set of three noncollinear
points. Thus, if the vertices of a triangle are A, B, and C, then its sides are $\overline{AB}$
$\overline{BC}$, and $\overline{AC}$, and the triangle is then the set defined by $\overline{AB} \cup \overline{BC} \cup \overline{AC}$,
denoted by $\triangle ABC$. The **angles** of $\triangle ABC$ are $\angle A \equiv \angle BAC$, $\angle B \equiv \angle ABC$, and
$\angle C \equiv \angle ACB$.

(See Figure 3.1 for a pictorial definition of these and other terms often associated with triangles; the formal statement of the remaining definitions will
be left as an option for the reader.)

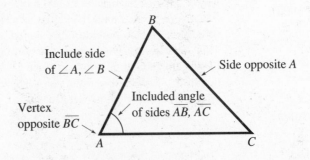

Figure 3.1

OUR GEOMETRIC WORLD

The triangle is a convenient and sturdy form used in engineering and construction. The photograph shows a common method for constructing a very strong support. Due to the construction used (i.e., supports of equal lengths), the congruence theorems we study in Section 3.3 show that many pairs of triangles in this photo are congruent.

EUCLID'S CONCEPT FOR CONGRUENCE

Euclid had no word reserved exclusively for what we today call **congruence**. He used the term "equal" to describe not only congruence of triangles, but equality of areas. Two polygons were said to be "equal" if they had the same area. Euclid attributed to congruent (equal) triangles the property that one triangle could be placed *precisely on top of another.* The act of placing one figure on top of another has been called "superposition" (a term not used by Euclid).

If we were to use superposition as a criterion for congruence, then somehow the properties of motions or isometries (distance-preserving maps) would have to be dealt with in the axioms. This can, in fact, be done, making isometries the basis for congruence, and this is not a bad way to study geometry. We shall, however, take the more traditional approach here. (In Chapter 5, we will discuss Euclid's superposition proof in the context of transformation theory.)

CONGRUENCE AS AN EQUIVALENCE RELATION

Modern usage of the term "equal" in mathematics is restricted to mean "identically the same as," as in two sets, two numbers, or two algebraic expressions. For example, if we say that two points are equal, as in $A = B$, then we mean they coincide, or are the *same* point. We write $\overline{AB} = \overline{CD}$ only if the *set* of points $\overline{AB}$ is the exact same set of points denoted by $\overline{CD}$(which can presumably happen only if either $A = C$ and $B = D$, or $A = D$ and $B = C$). In arithmetic, $\dfrac{6}{(-3)^2} = \dfrac{2}{3}$ because these two numbers are identically the same, or have the *same value,* although their representations differ. We could not say that "Bill equals Sam" in the mathematical sense unless the person named Bill were the same identical person as the one named Sam. Now, "Bill = Sam" could mean, however, that Bill and Sam have the same height, or that they have the same weight, or that they belong to the same personality type. But this usage of *equals* is entirely different, and gives rise to a general type of equality in mathematics called an **equivalence relation,** a concept of equality that applies to many different situations. Its defining properties are just the three familiar properties of equality: *Reflexive Law, Law of Symmetry,* and the *Transitive Property* (explained later).

In geometry, if we want "=" to be used as it is in all other branches of mathematics (and we shall), then when a segment can be made to fit exactly on top of another (as in Euclid) but the segments are *different sets of points,* we need a new symbol and a different concept for equality. We shall use ≅ for this and call it **congruence.** Under this kind of equality, it is necessary for the two segments to have the *same length,* but they do not have to be identically the same set of points. Similarly, two *angles* are said to be congruent if and only if their measures are equal.

Thus, we require

CONGRUENCE FOR SEGMENTS AND ANGLES

$\overline{AB} \cong \overline{XY}$	iff	$AB = XY$
$\angle ABC \cong \angle XYZ$	iff	$m\angle ABC = m\angle XYZ$

CONGRUENCE FOR TRIANGLES

The concept for congruence in triangles is more complicated, simply because there is no single measure that adequately describes both the size and shape of a triangle. It is clear that in order for two triangles to be considered "congruent," one triangle must "fit exactly on top of" the other (as in two identical cardboard models representing those triangles). Thus, as in Figure 3.2, all the sides and all the angles of the two triangles are, respectively, congruent (in the sense of congruence already defined for segments and angles). We might, of course, require less in our definition of congruence, but it is usually better mathematics to require all we shall ever need in a definition, and then use theorems to show what lesser powers suffice.

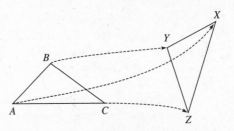

Figure 3.2

Note that in order to state precisely what we mean by congruent triangles, we must decide how to make the vertices correspond. In the example illustrated in Figure 3.2, we obviously could not let vertex A correspond to vertex Y because $m\angle A \neq m\angle Y$. We shall define a correspondence between the vertices of two triangles by writing

$$ABC \leftrightarrow XYZ \quad \text{iff} \quad A \leftrightarrow X, B \leftrightarrow Y, \text{ and } C \leftrightarrow Z$$

In words, this means that A, B, and C *correspond* to X, Y, and Z, respectively. Under this correspondence, therefore, corresponding angles and sides are thereby generated (Figure 3.3):

$$\overline{AB} \leftrightarrow \overline{XY}, \overline{BC} \leftrightarrow \overline{YZ}, \overline{AC} \leftrightarrow \overline{XZ}$$

and

$$\angle A \leftrightarrow \angle X, \angle B \leftrightarrow \angle Y, \angle C \leftrightarrow \angle Z.$$

CORRESPONDENCE BETWEEN
TWO NONCONGRUENT TRIANGLES

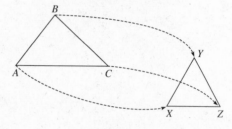

Figure 3.3

(Obviously, triangles do not have to be congruent to have such a correspondence.)

Note that there are potentially six ways for one triangle to be congruent to another, simply because there are six ways to make the vertices correspond:

$$ABC \leftrightarrow XYZ \quad ABC \leftrightarrow YXZ \quad ABC \leftrightarrow ZXY$$

$$ABC \leftrightarrow XZY \quad ABC \leftrightarrow YZX \quad ABC \leftrightarrow ZYX$$

> **DEFINITION**
>
> If, under a certain correspondence between the vertices of two triangles, corresponding sides and corresponding angles are congruent, the triangles are said to be **congruent**.

It is useful, if only to save on verbosity and writing, to use a sort of universal notation and brief restatement of this definition that succinctly states the property which congruent figures must have:

CPCF

Corresponding parts of congruent figures are congruent.

We shall make frequent use of this abbreviation in proofs.

It will be convenient to use a somewhat stronger notation for congruence than normally used. Using it properly requires some care, but it will save time in the long run. *We shall denote the congruence of* $\triangle ABC$ *and* $\triangle XYZ$ *by* $\triangle ABC \cong \triangle XYZ$ *only when the correspondence which defines the congruence is* $ABC \leftrightarrow XYZ$. Thus, in symbols,

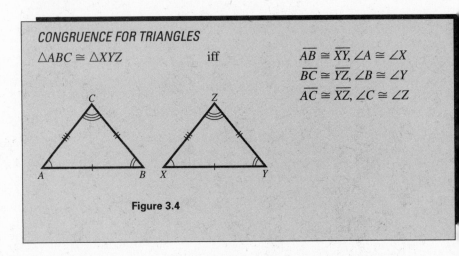

CONGRUENCE FOR TRIANGLES

$\triangle ABC \cong \triangle XYZ$ iff $\overline{AB} \cong \overline{XY}, \angle A \cong \angle X$
$\overline{BC} \cong \overline{YZ}, \angle B \cong \angle Y$
$\overline{AC} \cong \overline{XZ}, \angle C \cong \angle Z$

Figure 3.4

The correct *order* of vertices in naming the triangles which are congruent is essential; changing the order changes the congruence.

Finally, the three basic properties of congruence can be stated and easily established (left as an exercise). We state these properties in terms of congruence for triangles, but they also hold for congruence of segments and angles as well.

PROPERTIES OF CONGRUENCE

1. Reflexive Law: $\triangle ABC \cong \triangle ABC$
2. Law of Symmetry: If $\triangle ABC \cong \triangle XYZ$, then $\triangle XYZ \cong \triangle ABC$.
3. Transitive Law: If $\triangle ABC \cong \triangle XYZ$ and $\triangle XYZ \cong \triangle UVW$, then $\triangle ABC \cong \triangle UVW$.

QUESTION: In view of the Reflexive Law, can a triangle be congruent to itself if the correspondence of vertices is different from the identity correspondence, that is, $A \leftrightarrow A$, $B \leftrightarrow B$, $C \leftrightarrow C$? In order to explore this properly, we recommend that you work through the next Discovery Unit.

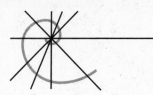

MOMENT FOR DISCOVERY

The Reflexive Law of Congruence

Suppose $\triangle ABC$ is a given triangle. Since the triangles $\triangle ABC$ and $\triangle BAC$ are identical sets, $\triangle ABC = \triangle BAC$. Therefore, the Reflexive Law of Congruence demands that $\triangle ABC$ be congruent to $\triangle BAC$. Can one write this congruence as $\triangle ABC \cong \triangle BAC$? Let's explore the implications of this congruence in detail.

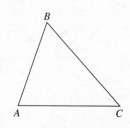

Figure 3.5

1. What correspondence between A, B, C and A, B, C is indicated *by definition* if $\triangle ABC \cong \triangle BAC$?

2. Name the corresponding angles (i.e., $\angle A \leftrightarrow \angle B$, etc.).

3. Name the corresponding sides.

4. Using *CPCF*, write down the pairs of *congruent* sides and angles under this correspondence.

5. Do you still think that $\triangle ABC \cong \triangle BAC$ in general? If not, what kind of triangle would lead to this situation?

SAS HYPOTHESIS

Earlier we raised the issue of how much information would be needed to guarantee the congruence of triangles. Can we get by with fewer than six sets of congruent pairs, as required in the definition? Obviously, we could not merely require that the three sets of *angles* in two triangles be congruent, because similar, noncongruent triangles have this feature. Euclid's *Elements* provides some clues. Euclid presents an argument that two triangles are congruent if the following hypothesis is satisfied:

THE SAS HYPOTHESIS

Under the correspondence $ABC \leftrightarrow XYZ$, let two sides and the included angle of $\triangle ABC$ be congruent, respectively, to the corresponding two sides and the included angle of $\triangle XYZ$. That is, for example, $\overline{AB} \cong \overline{XY}$, $\overline{BC} \cong \overline{YZ}$, and $\angle ABC \cong \angle XYZ$.

Euclid's familiar result actually cannot be established with the current set of axioms (those presented in the last chapter). In the next section, we develop an interesting model which shows this. Thus, if we want this property for congruence, we must assume it as an axiom.

PROBLEMS   (§3.1)

_____ *Group A* _____

1. In the figure to the right, $RS = VW$ and $RT = UW$.
   (a) Name all congruent pairs of segments as guaranteed *by definition*.
   (b) Prove that $\overline{ST} \cong \overline{UV}$.

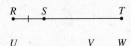

2. (a) Name the congruent pairs of angles as indicated in the figure below, and as guaranteed *by definition*.
   (b) Prove that $\angle SVW \cong \angle UVR$.

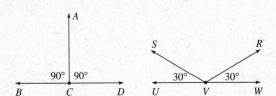

3. In the figure below, name the congruent pairs of segments and angles justified by our axioms thus far. (Segments have their measures as indicated in the figure; perpendicularity is indicated by small squares in the customary manner.)

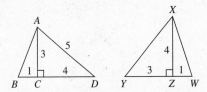

4. In the following figure, name the congruent pairs of segments if $\triangle RST \cong \triangle LMN$.

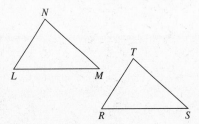

5. In the figure at the right, name the congruent pairs of (distinct) angles if $\triangle XYZ \cong \triangle XZY$.

6. Under the correspondence $LMN \leftrightarrow VUW$, the triangles $\triangle LMN$ and $\triangle UVW$ in the figure below are congruent, and $\triangle LMN$ has $LN = LM$. Name the congruent segments.

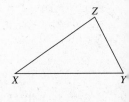

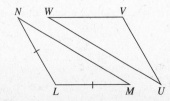

7. If $\triangle ABC \cong \triangle XYZ$, under what circumstances is $\triangle BCA \cong \triangle XZY$?

8. Which pairs of triangles in the figure below (with measures as indicated) would satisfy the SAS Hypothesis? (Do not assume knowledge of congruent triangles.)

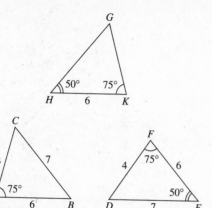

9. Which pairs of triangles in the following figure would satisfy the SAS Hypothesis? (Do not assume knowledge of congruent triangles.)

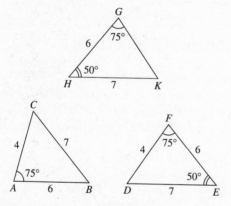

10. What reason(s) would you give for the (valid) general statements

    (a) $AB = BA$

    (b) $\overline{AB} \cong \overline{BA}$

11. What reason(s) would you give for the (valid) general statements

    (a) $\triangle ABC = \triangle BCA$

    (b) $\triangle ABC \cong \triangle ABC$

    (c) Under what circumstances is $\triangle ABC \cong \triangle BCA$?

12. In the figure to the right, what reason(s) would you give for the statement $\angle DAB \cong \angle DAC$?

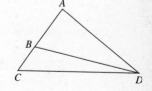

_____ *Group B* _____

13. Prove that if $\overline{AB} = \overline{CD}$ and $A \neq B$, then either $A = C$ and $B = D$ or $A = D$ and $B = C$.

14. Using the definition of angle, prove that if $\angle ABC = \angle XYZ$, then $B = Y$ and either $\overrightarrow{BA} = \overrightarrow{YX}$ and $\overrightarrow{BC} = \overrightarrow{YZ}$, or $\overrightarrow{BA} = \overrightarrow{YZ}$ and $\overrightarrow{BC} = \overrightarrow{YX}$.

15. Under what circumstances would you regard the two configurations in below as "congruent" if $\overrightarrow{PQ} \perp \overrightarrow{RS}$?

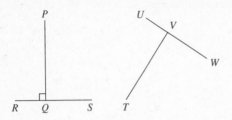

16. Prove the Transitive Law for triangles using the Transitive Law for equality in the real number system. (If two angles are congruent, then they have *equal measures* in the real number system.)

———— *Group C* ————

17. Discuss and analyze critically the following conjecture: Two triangles are congruent if any 5 out of the 6 parts of the two triangles are congruent, respectively, in some order (e.g., three pairs of angles and two pairs of sides, or two pairs of angles and three pairs of sides, etc.).

---

*3.2   TAXICAB GEOMETRY: GEOMETRY WITHOUT SAS CONGRUENCE

We are now going to look at a rather unusual geometry. While it is something you can have some fun with, the purpose is entirely serious. This geometry serves as a model for all the axioms thus far assumed, but violates the SAS Postulate which we shall introduce in Section 3.3. What does this say logically about the relationship of SAS to our other axioms?

In order to construct a true model for the axioms thus far assumed, which were three-dimensional in character, the points of our model will be the ordinary points of three-dimensional coordinate geometry: $P(x, y, z)$, where x, y and z are real numbers. The lines and planes will be defined exactly as in coordinate/vector geometry. That is, a plane is the set of points whose coordinates satisfy a linear equation $ax + by + cz = d$ for a, b, and c not all zero, and a line is the intersection of two planes. In this setting, we know that the axioms of incidence (Axioms I-1–I-5) will all be satisfied.

Moving on to the concept of distance and angle measure, instead of taking the distance between two points $P(x_1, y_1, z_1)$ and $Q(x_2, y_2, z_2)$ to be the usual metric given by the distance formula

$$PQ = \sqrt{(x_1 - x_2)^2 + (y_1 - y_2)^2 + (z_1 - z_2)^2}$$

take the following metric

(1) $$PQ^* = |x_1 - x_2| + |y_1 - y_2| + |z_1 - z_2|$$

$\mathcal{H}$ISTORICAL $\mathcal{N}$OTE

The Taxicab Metric was originally discovered by Hermann Minkowski (1864–1909) as a special case of a metric defined in terms of an arbitrary convex set centered at the origin, whose boundary is a "circle" of the geometry. Minkowski, like Hilbert, grew up in Königsberg. In 1879, when he won a prize for a competition in mathematics, he met Hilbert for the first time, and they became close friends. At Hilbert's urging, Minkowski accepted a teaching position at Göttingen in 1902, and they were colleagues there until Minkowski's untimely death of appendicitis at the age of 45. Besides

Minkowski's monumental work in the theory of numbers (using his metric), he is also known for his mathematical foundation for the Theory of Relativity, which Einstein—a student of Minkowski—used extensively in his work. The idea of the four-dimensional space/time framework commonly used for relativity is due to Minkowski. The physicist Max Born, also a student of Minkowski's, once said: "Minkowski laid out the whole arsenal of relativity mathematics...as it has been used ever since by every theoretical physicist."

TAXICAB METRIC

When restricted to the plane, this metric is commonly called the **Taxicab**, or **Manhattan**, **Metric**. We shall soon learn why.

The definition for betweenness will be the same as that for Euclidean betweenness: If A, B, and C are distinct collinear points, and $AB^* + BC^* = AC^*$, then A-B-C. It will be seen later that even though the two metrics for the plane are very different, taxicab betweenness and Euclidean betweenness coincide. If we adopt Euclidean angle measure for this geometry, all the axioms on angle measure (Axioms A-1, A-2, A-3, and A-4) hold, and this geometry becomes a model for the previous axioms, assuming we can verify the Ruler Postulate.

At this point, let's confine our attention to a single plane. For that purpose, we can assume that z = 0 and that in **(1)**, $z_1 = z_2 = 0$. Hence, the distance formula reduces to

$$\textbf{(2)} \qquad PQ^* = |x_1 - x_2| + |y_1 - y_2|$$

In Figure 3.6, where we have illustrated the components of this formula, $|x_1 - x_2|$ is the length of one leg of the right triangle $\triangle PRQ$ shown, and $|y_1 - y_2|$ is that of the other. That is, the "distance" from P to Q is the Euclidean distance *around the corner* along the two legs of the triangle, as if to get from P to Q, you had to go through point R!

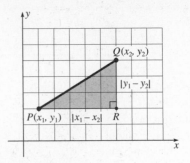

Figure 3.6

To make this more graphic, Figure 3.7 shows a portion of a fictitious city map. The route we must take to get from point A to B is $AB^* = AC + CB$. But a variety of other routes would also take us from A to B with the same total distance travelled, such as

$$AD + DE + EF + FB$$

Point E is the only point of the route $ADEFB$ that lies *between* A and B, but many other points, such as C, D, and F, are **metrically between** A and B, in the sense that equality in the triangle inequality holds for them. For example, since $AD^* = AD$ and $DB^* = DE + EF + FB$, we have

$$AB^* = AD^* + DB^*$$

THE TAXICAB METRIC:
$AB^* = AC^* + CB^*$

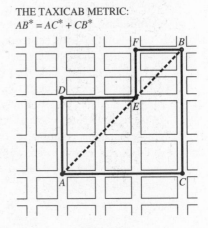

Figure 3.7

If we are given a line ℓ in this geometry, such as $y = mx + b$, then for any two points along this line, say $P(x_1, y_1)$ and $Q(x_2, y_2)$, we have, by substitution into the taxicab distance formula **(2)**,

$$(3) \qquad PQ^* = |x_1 - x_2| + |(mx_1 + b) - (mx_2 + b)|$$

$$= |x_1 - x_2| + |m| \, |x_1 - x_2| = k |x_1 - x_2|,$$

where $k = 1 + |m|$. Since k is a constant, the Ruler Postulate will be cleared up if we take as the correspondence between real numbers and points on ℓ:

(4) $$P(x, y) \leftrightarrow kx \qquad (k = 1 + |m|)$$

That is, the taxicab **line coordinate** of $P(x, y)$ on ℓ is kx, or $P[kx]$ as in the earlier notation. The value k depends on the line chosen, but it is a *fixed constant for each line*. (Compare this with the corresponding result using the Euclidean metric; instead of $k = 1 + |m|$, we would have $k = \sqrt{1 + m^2}$. The same basic results follow for both metrics, but with a different value for k.)

Thus, if $A[a]$ and $B[b]$ are any two points on ℓ with line coordinates $a = kx_1$ and $b = kx_2$, where $A = (x_1, y_1)$ and $B = (x_2, y_2)$ as points in the plane, $x_1 \neq x_2$, then by **(3)**

(5) $$AB^* = k|x_1 - x_2| = |kx_1 - kx_2| = |a - b|$$

and the Ruler Postulate works. If $x_1 = x_2$, then ℓ is a vertical line, with equation $x = c$ constant, **(3)** reduces to $PQ^* = |c - c| + |y_1 - y_2| = |y_1 - y_2|$, and we can use the simpler correspondence $P(x, y) \leftrightarrow y$. Again, we have $AB^* = |y_1 - y_2| = |a - b|$.

Thus, we see that the Ruler Postulate is valid. Were we to switch over to the Euclidean metric, then **(3)**, **(4)**, and **(5)** become, respectively,

$$PQ = k'|x_1 - x_2|, \qquad P(x, y) \leftrightarrow k'x,$$
$$AB = |k'x_1 - k'x_2| \equiv |a' - b'|,$$

where $A[a']$ and $B[b']$ are any two points on ℓ, with their Euclidean line coordinates. Since $a = kx$ and $a' = k'x$, or $a/a' = k/k' = $ a constant, it is clear that by Theorem 1 of Section 2.6

$$A\text{-}B\text{-}C^* \text{ iff } a < b < c \text{ or } c < b < a$$
$$\text{iff } a' < b' < c' \text{ or } c' < b' < a'$$
$$\text{iff } A\text{-}B\text{-}C$$

This proves

THEOREM 1
Taxicab betweenness coincides with Euclidean betweenness.

Although *geometric* betweenness for the Euclidean and Taxicab Metrics are the same, metric betweenness is an entirely different matter. As mentioned before, we could have $AB^* = AD^* + DB^*$ under the Taxicab Metric without having $AB = AD + DB$ under the Euclidean Metric as shown in Figure 3.7. Consequently, there are some unusual features of the Taxicab Metric, as we might imagine. For starters, circles are actually diamond-shaped, ellipses are crystal-like hexagons or octagons, and "perpendicular bisectors" of line segments can be irregular curves (see Figure 3.8 for some illustrations).

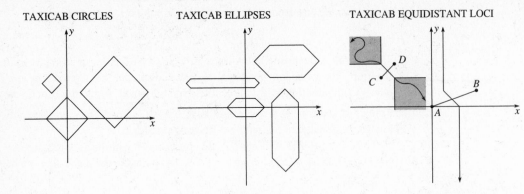

Figure 3.8

All of these facts can be verified without difficulty; we leave the interesting analysis for the ellipse and other "conic sections" as problems (Problems 12 and 15–17). The theory of equidistant loci, which in Euclidean geometry results in the idea of "perpendicular bisectors," is particularly bizarre, or so it seems. Again pursued in the problems, such a locus can be, in certain cases and in certain regions of the plane, a completely *arbitrary curve* and not uniquely defined. One such instance of this is depicted in Figure 3.8 (points "equidistant" from C and D). The Discovery Unit will allow you to find out for yourself what taxicab circles look like.

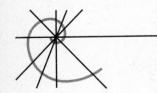

MOMENT FOR DISCOVERY

Square Circles

In Figure 3.9, consider the square centered at the origin with its diagonals on the coordinate axes. Calculate these taxicab distances:

1. OQ^* (which equals $OM + MQ$), where $Q = (2, 2)$
2. OR^* (which equals $ON + NR$), where $R = (3, 1)$
3. What did you discover?
4. For any point $P(x, y)$ on the line $x + y = 4$, for $0 \le x \le 4$, find OP^*. Any conclusions?

In order to prove that the square is a taxicab circle, we need to start with the geometric definition for a circle centered at O ($OP^* = 4$) and use the coordinate formula for the Taxicab Metric (2).

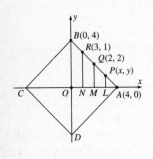

Figure 3.9

5. Write down the coordinate formula for OP^* if $O = (0, 0)$ and $P = (x, y)$, and simplify. Now set $OP^* = 4$. This will give you a coordinate equation for the taxicab circle considered above. (This equation will involve absolute values.)

6. Analyze this equation in each of the four quadrants to see what you get. (e.g., in Quadrant II, it follows that $x < 0$ and $y > 0$, so $|x| = -x$ and $|y| = y$; substitute these into the coordinate equation from Step 5.)

We are now going to consider the problem of congruent triangles in the taxicab geometry. Specifically, does the ordinary SAS Theorem work or not? Could you decide whether two triangles must be congruent if the sides of one have the same taxicab lengths as the corresponding sides of the other? What do "equilateral" triangles look like?

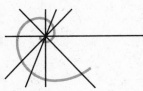

MOMENT FOR DISCOVERY

Taxicab Triangles

Consider the two triangles in Figure 3.10, $\triangle ABC$ and $\triangle XYZ$. They are both "right" triangles, with right angles at B and Y, respectively.

1. Find the taxicab lengths of sides $\overline{BA}$, $\overline{BC}$, and $\overline{AC}$.

2. Find the taxicab lengths of sides $\overline{XY}$, $\overline{YZ}$, and $\overline{XZ}$.

3. Under the correspondence $ABC \leftrightarrow XYZ$, are corresponding angles congruent?

4. Under the correspondence $ABC \leftrightarrow XYZ$, are corresponding sides congruent?

5. Is the *SAS* Hypothesis satisfied?

6. Are the triangles congruent?

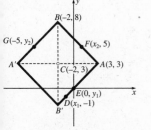

Figure 3.10

EXAMPLE 1 A taxicab circle centered at $C(-2, 3)$ is indicated in Figure 3.11.

(a) Find x_k and y_k ($k = 1$ and 2) which are components of coordinate pairs for certain points in the figure, as indicated, using the coordinate method and equations of the "circle."

(b) Find the taxicab distances CD^*, CE^*, CF^*, and CG^*.

Figure 3.11

SOLUTION

(a) AB' (slope 1): $y - 3 = 1 \cdot (x - 3) = x - 3 \rightarrow y = x$
$\therefore D = (-1, -1)$ and $E = (0, 0)$
AB (slope -1): $y - 3 = -(x - 3) \rightarrow y = -x + 6$
$\therefore F = (1, 5)$
BA' (slope 1): $y - 8 = x + 2 \rightarrow y = x + 10$
$\therefore G = (-5, 5)$

(b) $CD^* = |-2 + 1| + |3 + 1| = 5$,
$CE^* = |-2 - 0| + |3 - 0| = 5$,

$$CF^* = |-2 - 1| + |3 - 5| = 5,$$
$$CG^* = |-2 + 5| + |3 - 5| = 5 \square$$

E X A M P L E   2   The triangle whose vertices are $A(2, 2)$, $B(-1, 1)$, an
$C(1, -1)$ in the xy-plane is equilateral under the Taxicab Metric. Verify this. (Se
Figure 3.12.)

SOLUTION

(a) $AB^* = |2 - (-1)| + |2 - 1| = 3 + 1 = 4,$
$BC^* = |-1 - 1| + |1 - (-1)| = 2 + 2 = 4,$
$AC^* = |2 - 1| + |2 - (-1)| = 1 + 3 + 4$

Thus $AB^* = BC^* = AC^*$ under the Taxicab Metric, and the triangle is a taxica
equilateral triangle having sides of measure 4 units.$\square$

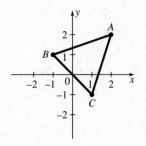

Figure 3.12

E X A M P L E   3   The analogue of the Euclidean parabola $y = \frac{1}{4}x^2$ in th
taxicab plane (having "focus" $F(0, 1)$ and "directrix" $y = -1$) is illustrated in Figu
3.13. Find the general location of a variable point $P(x, y)$ on this parabola using th
defining property $PF^* = PQ^* = |y + 1|$.

SOLUTION

By formula (2), $PF^* = |x| + |y - 1|$. Thus, we must have $|x| + |y - 1|$
$|y + 1|$. There are three possibilities for y: $y < -1$, $-1 \leq y < 1$, and $y \geq 1$. We tak
each case in turn.

(1) $y < -1 \rightarrow y + 1 < 0$, hence $y - 1 < 0$: $|x| - (y - 1) = -(y + 1)$ or $|x| = -2$
$\rightarrow\leftarrow$ (because $|x|$ cannot be negative). Hence, no points exist in this case.

(2) $-1 \leq y < 1 \rightarrow y + 1 \geq 0$, $y - 1 < 0$: $|x| - (y - 1) = y + 1 \rightarrow |x| = 2y$ $\therefore$ $y \geq$
Hence, P lies on either of the two lines $y = \pm \frac{1}{2}x$, $0 \leq y \leq 1$

(3) $y \geq 1 \rightarrow y - 1 \geq 0$ and $y + 1 > 0$: $|x| + y - 1 = y + 1 \rightarrow |x| = 2$
Hence, P lies on either of the two lines $x = \pm 2$, $y \geq 1$.$\square$

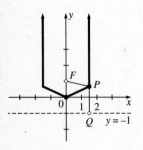

Figure 3.13

PROBLEMS   (§3.2)

_____ *Group A* _____

1. Is the triangle △ABC of Example 2 equilateral under the Euclidean metric? Isosceles?

2. The triangle whose vertices are $A(0, 0)$, $B(2, 0)$, and $C(1, \sqrt{3})$ in the xy-plane is easily found to be equilateral under the Euclidean distance formula. (See the figure to the right.) In the taxicab geometry, find

   (a) the lengths of the sides of this triangle

   (b) the angle measures

   (c) Is △ABC equilateral in the taxicab geometry? Equiangular? Isosceles?

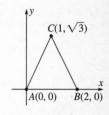

3. Is the Pythagorean Theorem valid in the taxicab geometry? If not, give a specific counterexample.

4. In the figure below, a taxicab circle centered at $C(3, 5)$ is indicated.

   (a) Using only the coordinates of the corner points, deduce the coordinate equations of the sides and complete the components of the missing coordinates of the points D, E, F, and G.

   (b) Find the taxicab distances CD^*, CE^*, CF^*, and CG^*.

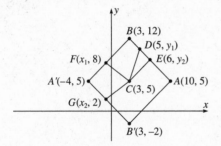

5. (a) A man has a newspaper stand at $W = (1, 0)$, eats regularly at a cafeteria located at $E = (8, 3)$, and does his laundry a laundromat at $L = (7, 2)$. If he wants to find a room R using the Taxicab Metric so as to be at the same walking distance away from each of these points, where should the room R be located? (Give the coordinates.)

   (b) How many blocks does he have to walk from his room to each of the three points, assuming he finds such a room?

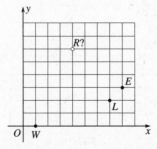

6. (a) Referring to Problem 5, if all conceivable shortcuts are possible (i.e., using the Euclidean Metric), where should the man's room be?

(b) Answer part (b) of Problem 5 if his room is located using the Euclidean metric. (Note that the distance here is $8\frac{1}{2}$ blocks compared to 5 blocks in Problem 5(b)!)

7. Do three noncollinear points always determine a unique taxicab circle? Consider all cases.

8. In the figure below, verify to your own satisfaction that *all points* in the shaded region are equidistant from A and B in taxicab geometry.

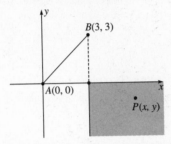

Group B

9. (a) Show that any point $P(x, y)$ on the line $x = 5$ for $y < 0$ is equidistant from $A(0, 0)$ and $B(6, 4)$ using the Taxicab Metric

$$PA^* = |x - 0| + |y - 0| = |x| + |y|$$
$$PB^* = |x - 6| + |y - 4|$$

(Remember that $|a| = -a$ if $a < 0$.)

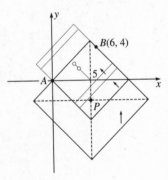

(b) In particular, note that $P(5, -2)$ is such a point, and is the center of a taxicab circle passing through A and B (hence, $PA^* = PB^* = $ radius of circle).

(c) By sliding the taxicab circle at P upwards and shrinking its size, always requiring that it pass through A and B, find the path of its center P. (This gives the set of all points equidistant from A and B under the Taxicab Metric.) Sketch this path, and show that it consists of the broken line $x = 5$ when $y < 0$, $x + y = 5$ when $0 \leq y \leq 4$, and $x = 1$ when $y > 4$. (These equations come from (a), with $0 < x < 6$, substituting $|x - 6| = -(x - 6)$.)

10. In a town having perfect square blocks and equally-spaced streets running north and south, east and west, two police stations are to be located at $A(0, 0)$ and $B(30, 20)$ (in the figure to the right). The town officials want to divide the town into two precincts—Precinct 1 served by Station A and Precinct 2 served by Station B. How should the boundary be drawn? (See Problem 9 for ideas.)

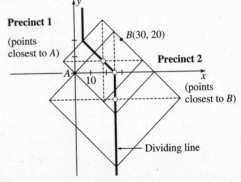

11. Given an angle $\angle ABC$, what sort of locus is obtained for the set of all points "equidistant" from its sides? Is it a straight line? Answer in all cases, if you can.

12. In the crystal-like figures shown in the figure to the right, the pointed ends are isosceles right triangles and the "axis" is either a vertical or horizontal line. Show that if F_1 and F_2 are the centers of the taxicab semicircles indicated, every point P on the boundary in each case satisfies the relation $PF_1{}^* + PF_2{}^* = c$, a constant.

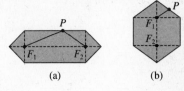

(a)   (b)

13. Find a pair of noncongruent triangles in the taxicab geometry (besides right triangles) which satisfy the SAS Hypothesis.

_____ *Group C* _____

14. Rework Problem 10 if there are to be *three* precincts: At $A(0, 0)$, $B(30, 20)$, and $C(10, 40)$.

15. Investigate the parabola and the focus/directrix property (as in Example 3) for an oblique axis. What do such curves look like? Explore how the computer might be of use in this problem.

16. Show that a regular octagon with two horizontal sides is an ellipse having foci inside the octagon and on a line making a 45° angle with the *x*-axis. (Use the two-focus property of the ellipse $PF_1{}^* + PF_2{}^* = c$, a constant, to make your analysis.)

17. Using the two-focus property of hyperbolas in Euclidean geometry, investigate taxicab hyperbolas. What would the "asymptotes" of such curves look like?

NOTE: Readable references to Problems 15–17 include E. Krause, "Taxicab Geometry", *Mathematics Teacher,* Vol. 66 (1973), p. 695–706, and R. Laatsch, "Pyramidal Sections in Taxicab Geometry", *Mathematics Magazine,* Vol. 55, No. 4 (1982), p. 205–212.

3.3 *SAS, ASA, SSS Congruence, and Perpendicular Bisectors*

The postulate anticipated in the preceding section represents the last axiom we need for absolute geometry. This postulate leads to all the familiar properties of Euclidean geometry that do not involve the concept of parallel lines. Thus, all the results we obtain are also true for non-Euclidean geometry as well. We can be sure that this new axiom is independent from the rest due to the validity of taxicab geometry as a model for the axioms we have already assumed.

AXIOM C-1: SAS POSTULATE

If the SAS Hypothesis holds for two triangles under some correspondence between their vertices, then the triangles are congruent.

A major application of the SAS Postulate is to establish the other two fundamental congruence criteria for triangles, namely the ASA and SSS theorems. (Secondary school texts frequently take these results as axioms also.) This, and the basic properties of isosceles triangles, will be our goal for this section.

THE ASA CONGRUENCE CRITERION

Euclid's proof of the ASA Theorem relied on the undefined operation of superposition—placing or "applying" one triangle onto another. However, a logical proof based on the development thus far is not difficult to construct.

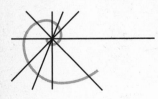

MOMENT FOR DISCOVERY

ASA Congruence

Suppose that $\angle A \cong \angle X$, $\overline{AB} \cong \overline{XY}$, and $\angle B \cong \angle Y$. If it is assumed that $AC > XZ$, we can locate D on $\overrightarrow{AC}$ so that A-D-C and $\overline{AD} \cong \overline{XZ}$, by the Segment Construction Theorem.

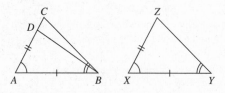

Figure 3.14

1. What do you observe about $\triangle ABD$ and $\triangle XYZ$? (Use the SAS Postulate.)
2. Does this make $\angle ABD \cong \angle Y$?
3. We are given that $\angle Y \cong \angle ABC$. Then what about $m\angle ABD$ and $m\angle ABC$? Do you reach a contradiction? What properties are used?
4. Does this argument prove anything significant?

If you were successful with the preceding Discovery Unit, then you have completed the proof of the following theorem.

THEOREM 1: ASA THEOREM

If, under some correspondence, two angles and the included side of one triangle are congruent to the corresponding angles and included side of another, the triangles are congruent under that correspondence.

EXAMPLE 1   (See Figure 3.15.)
Given: $\angle EBA \cong \angle CBD$, $\overline{AB} \cong \overline{BC}$, $\angle A \cong \angle C$.
Prove: $\overline{EB} \cong \overline{DB}$.

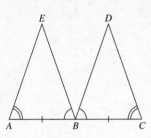

Figure 3.15

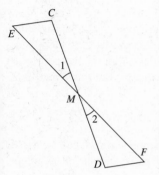

Figure 3.16

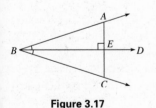

Figure 3.17

CONCLUSIONS	JUSTIFICATIONS
(1) $\triangle ABE \cong \triangle CBD$.	ASA Theorem
(2) $\therefore \overline{EB} \cong \overline{DB}$.	CPCF

[UCSMG, p. 317]❏

EXAMPLE 2   (See Figure 3.16.)

Given: M is the midpoint of $\overline{CD}$ and $\overline{EF}$.

Prove: $\angle C \cong \angle D$.

CONCLUSIONS	JUSTIFICATIONS
(1) $MC = MD$, $ME = MF$.	definition of midpoint
(2) Angles 1 and 2 form a vertical pair, hence $\angle 1 \cong \angle 2$.	definition and Vertical Pair Theorem
(3) $\triangle CME \cong \triangle DMF$.	SAS Postulate
(4) $\therefore \angle C \cong \angle D$	CPCF

[UCSMG, p. 318]❏

EXAMPLE 3   Prove that a line which bisects an angle also bisects any segment perpendicular to it which joins two points on the sides of that angle.

Proof

Given: $\angle ABC$, line $\overleftrightarrow{BD}$ bisects $\angle ABC$, $\overleftrightarrow{AC} \perp \overleftrightarrow{BD}$.

Prove: Line $\overleftrightarrow{BD}$ bisects segment $\overline{AC}$.

CONCLUSIONS	JUSTIFICATIONS
(1) Ray $\overrightarrow{BD}$ intersects segment $\overline{AC}$ at some interior point E.	Crossbar Theorem
(2) $\angle ABE \cong \angle EBC$.	definition (angle bisector)
(3) $\overline{BE} \cong \overline{BE}$.	Reflexive Property
(4) $\angle BEA \cong \angle BEC$.	definition ($\perp$ lines)
(5) $\triangle ABE \cong \triangle CBE$.	ASA
(6) $\overline{AE} \cong \overline{EC} \rightarrow AE = EC$.	CPCF
(7) $\therefore E$ is the midpoint of segment $\overline{AC}$ and line $\overleftrightarrow{BD}$ bisects $\overline{AC}$.❏	definition of midpoint

CAN A TRIANGLE HAVE TWO RIGHT ANGLES?

Before we proceed to the next result, a preliminary matter must be tended to which will be needed in the development. The question of whether a triangle can have two right angles is clearly equivalent to whether two lines can be perpendicular to a third line from the same external point. Or, put another way, is the perpendicular to a line from an external point unique? (Recall that the uniqueness of the perpendicular to a line through a point *on that line* has already been established: Theorem 5 and Problem 10 in Section 2.8.)

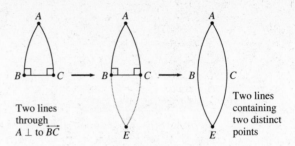

Two lines through A ⊥ to $\overleftrightarrow{BC}$

Two lines containing two distinct points

Figure 3.18

Suppose we assume, for sake of argument, that a triangle △ABC exists which has right angles at B and C (Figure 3.18). In Problem 22, you will be guided in a proof that uses the SAS Postulate to establish the existence of a second point E on the two distinct lines $\overleftrightarrow{AB}$ and $\overleftrightarrow{AC}$, as indicated in the figure. But this would be in direct violation of Axiom I-1. Hence, this proves

LEMMA A

Given a point A and a line ℓ, at most one line m perpendicular to ℓ can be passed through A, and point B = ℓ ∩ m, called the **foot** of perpendicular m, is unique.

COROLLARY

A triangle can have at most one right angle.

Note that the *existence* of the perpendicular m in Lemma A is another technical matter, and it has not yet been established in the affirmative.

ISOSCELES TRIANGLE
THEOREM

A simple result which makes use of SAS and ASA involves an **isosceles triangle**—a triangle having two sides congruent. The congruent sides are called the **legs** of the triangle, and the third side is called the **base**, with the vertex opposite the base called simply the **vertex**, and the angle at that vertex, called the **vertex angle**. The result may be stated as follows: If two sides of a triangle are congruent, the angles opposite are congruent.

Euclid went to a great deal of trouble to prove this using a special construction method which later became a test of geometric ability (the *pons asinorum*— "bridge of asses"). Any student who could not "cross this bridge" was considered unable to go further in the study of geometry. (See Problem 24.)

However, there is a very efficient, streamlined proof, first given by Pappus, used by Hilbert, and reportedly rediscovered by a computer in the 1950s when programmers were conducting early experiments in artificial intelligence. The reader has only to look back to the Discovery Unit of Section 3.1 and examine the correspondence ABC ↔ BAC in connection with that. Hence, there is a one-line proof!

LEMMA B

In △ABC, if $\overline{AC} \cong \overline{BC}$, then ∠A ≅ ∠B.

Proof: By the SAS Postulate, $\triangle ABC \cong \triangle BAC$.

LEMMA C

In $\triangle ABC$, if $\angle A \cong \angle B$, then $\overline{AC} \cong \overline{BC}$.

Proof: By the ASA Theorem, $\triangle ABC \cong \triangle BAC$.

EXAMPLE 4 (See Figure 3.19.)

Given: $GJ = KM = 2$, $JK = HJ = HK = 10$, with betweenness relations as evident from the figure.

Prove: $HG = HM$.

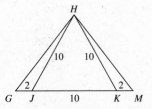

Figure 3.19

CONCLUSIONS	JUSTIFICATIONS
(1) $GK = GJ + JK = 12$ and $MJ =$ $MK + KJ = 12$.	betweenness properties
(2) $\overline{GK} \cong \overline{MJ}$, $\overline{KH} \cong \overline{JH}$, and $\overline{HJ} \cong \overline{HK}$.	definition of $\cong$
(3) $\angle HJM \cong \angle HKG$.	Lemma B
(4) $\triangle HJM \cong \triangle HKG$.	SAS Postulate
(5) $\therefore \overline{HG} \cong \overline{HM}$ (and $HG = HM$).	CPCF☐

OUR GEOMETRIC WORLD

The center beam in the photograph is used to support the beams at the top. Depending on whether certain other supporting beams are of equal length, it may or may not follow that the center support is erect ($\perp$ to the base). A theoretical basis for this question in provided by Theorem 2.

The line passing through the vertex of an isosceles triangle and the midpoint of its base is evidently a line of symmetry for the triangle (Figure 3.20). In addition to bisecting the base, it is perpendicular to the base, and it bisects the vertex angle.

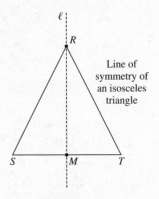

Figure 3.20

Several propositions involving this line of symmetry imply that if a line has any two of the four possible symmetry properties, it has the other two as well. The symmetry properties we are referring to are: perpendicular to the base (**PB**), bisecting the base (**BB**), bisecting the vertex angle (**BV**), and passing through the vertex (**PV**). Since obviously **BV** → **PV**, the six possible results, corresponding to six pairs among four objects, reduce to only four.

THEOREM 2: ISOSCELES TRIANGLE THEOREM

A triangle is **isosceles** iff the base angles are congruent. Furthermore, in any isosceles triangle, if line ℓ satisfies any two of the four symmetry properties mentioned above, it satisfies all four, and ℓ is a line of symmetry for the triangle.

Proof: The first part of the theorem incorporates Lemmas A and B, already proven. The second part has four cases, as mentioned above. They are

I. Given PV and BB, to prove **PB** and **BV**. This result may be stated as follows: The line joining the vertex of an isosceles triangle and the midpoint of the base is also perpendicular to the base and bisects the vertex angle.

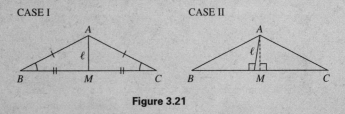

Figure 3.21

THEOREM 2 (CONTINUED)

Proof: (See Figure 3.21; you are asked to supply the missing reasons.)

Given: $\triangle ABC$ with $\overline{AB} \cong \overline{AC}$, M the midpoint of base $\overline{BC}$.

Prove: $\overrightarrow{AM} \perp \overline{BC}^2$ and $\overrightarrow{AM}$ bisects $\angle BAC$.

CONCLUSIONS		JUSTIFICATIONS
(1)	$\overline{BM} \cong \overline{MC}$, B-M-C	definition of midpoint
(2)	$\angle ABM \cong \angle ACM$	Lemma B
(3)	$\triangle ABM \cong \triangle ACM$.	_____ ?
(4)	$\angle AMB \cong \angle AMC$	_____ ?
(5)	$\therefore \overline{AM} \perp \overline{BC}$.	_____ ?
(6)	$\angle BAM \cong \angle CAM$.	_____ ?
(7)	$\overrightarrow{AB}$-$\overrightarrow{AM}$-$\overrightarrow{AC}$	_____ ?
(8)	$\therefore \overrightarrow{AM}$ bisects $\angle BAC$.	definition

II. Given **PV** and **PB**, to prove **BB** and **BV**. Draw the line joining the vertex A and midpoint M of the base $\overline{BC}$ (Figure 3.21). Then, as in Case I, $\overline{AM} \perp \overline{BC}$, hence, by the uniqueness of the perpendicular through A (Lemma A), the lines $\overleftrightarrow{AM}$ and ℓ coincide. Hence **BB** and **BV** clearly follow.

III. Given **PV** and **BV** (that is, just **BV**), to prove **PB** and **BB**. The argument is to be completed as a problem (Problem 12).

IV. Given **PB** and **BB**, to prove **PV** and **BV**. Join the vertex and midpoint of the base (a previous case) and prove that this line must coincide with the given line that is perpendicular to the base at its midpoint, as in Case II. (Argument to be completed in Problem 15.)

PERPENDICULAR BISECTORS, LOCUS

The first instance (or among the first) in which one encounters the notion of *locus* in geometry is in the following result, an obvious corollary of the Isosceles Triangle Theorem (specifically, Case IV). We define the **perpendicular bisector** of a segment $\overline{AB}$ to be the line which both bisects $\overline{AB}$ and is perpendicular to it.

PERPENDICULAR BISECTOR THEOREM

The set of all points equidistant from two fixed points A and B is the perpendicular bisector of segment $\overline{AB}$.

The usual language is, "The locus of points equidistant from two fixed points ..." The term **locus** is an old-fashioned expression for a *set of points*. The proof of any

[2]It is customary in geometry to speak of two *segments* (or two *rays*) being perpendicular if the lines containing them are perpendicular.

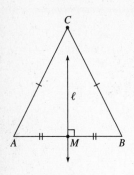

Figure 3.22

such "locus" problem in geometry requires two parts:

(1) Prove that each point of the given set, or locus, satisfies the properties.

(2) Prove, conversely, that any point satisfying the properties lies in the set.

These two steps constitute the standard set-theoretic argument proving that two sets S_1 and S_2 are identical; first prove that $S_1 \subseteq S_2$, then $S_2 \subseteq S_1$.

Applying this to the proposition just mentioned, first suppose that C is equidistant from A and B (Figure 3.22). Then $CA = CB$ and $\triangle ABC$ is isosceles; by Case IV of Theorem 2, the perpendicular bisector of $\overline{AB}$ passes through C. Conversely, if C lies on the perpendicular bisector ℓ of C, clearly the two triangles on either side of ℓ are congruent by *SAS*, hence $\overline{CA} \cong \overline{CB}$. The proof is thereby complete.

SSS CONGRUENCE CRITERION

The results on isosceles triangles and perpendicular bisectors make the SSS Theorem an easy exercise, which we have formulated as a Discovery Unit.

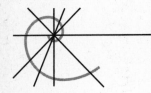

MOMENT FOR DISCOVERY

SSS Theorem Via Kites and Darts

Two geometric figures, the **kite** and the **dart**, though simple, are quite useful. These are four-sided figures which have a line of symmetry passing through two opposite vertices. That is, in Figure 3.23, $\overline{AB} \cong \overline{AD}$, and $\overline{BC} \cong \overline{CD}$. The dart is distinguished from the kite by virtue of both angles at A (and C) being acute for the kite and obtuse for the dart. Since $\overline{AC} \cong \overline{AC}$, the two triangles on either side of line $\overleftrightarrow{AC}$ would be congruent by SSS. However, this can be proved without using SSS, thereby providing a means of *proving* SSS from kites and darts.

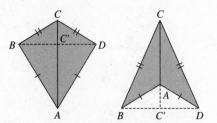

Figure 3.23

1. Given $\triangle ABC$ and $\triangle XYZ$ having congruent sides under the correspondence $ABC \leftrightarrow XYZ$, construct $\triangle ADC$ congruent to $\triangle XYZ$ on the other side of line $\overleftrightarrow{AC}$ from point B, forming a kite or dart. (Show that a kite or dart is formed; that is, show that $\overline{AB} \cong \overline{AD}$ and $\overline{BC} \cong \overline{CD}$.)

2. We now have A equidistant from B and D, and C equidistant from B and D. What do you conclude about line $\overleftrightarrow{AC}$?

3. What theorem tells us that ray $\overrightarrow{AC}$ bisects $\angle BAD$ (if C and C' lie on the same side of A), or ray $\overrightarrow{CA}$ bisects $\angle BCD$ (if C and C' lie on opposite sides of A)? Do not use SSS here.

4. Finish the argument that $\triangle ABC \cong \triangle XYZ$.

If you were successful with discoveries concerning the kite and dart, then you have proven the following result.

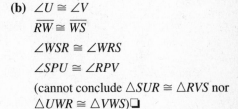

THEOREM 3: SSS THEOREM

If, under some correspondence between their vertices, two triangles have the three sides of one congruent to the corresponding three sides of the other, then the triangles are congruent under that correspondence.

E X A M P L E   5 Name all congruent pairs of distinct segments and angles in Figure 3.24, which follow logically from previous concepts, or from the congruence theorems thus far developed.

SOLUTION

(a) $\overline{PT} \cong \overline{TQ}$

$\angle P \cong \angle Q$

$\overline{PW} \cong \overline{WQ}$

$\angle PWT \cong \angle QWT$

$\angle PTW \cong \angle WTQ$

(b) $\angle U \cong \angle V$

$\overline{RW} \cong \overline{WS}$

$\angle WSR \cong \angle WRS$

$\angle SPU \cong \angle RPV$

(cannot conclude $\triangle SUR \cong \triangle RVS$ nor $\triangle UWR \cong \triangle VWS$)❏

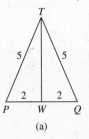

(a)

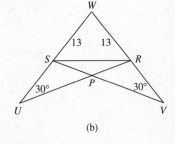

(b)

Figure 3.24

E X A M P L E   6 In Figure 3.25, we have a four-sided figure with the congruent sides and angles as marked.

(a) Prove that $\overline{AD} \cong \overline{CD}$.

(b) If $\overline{BC} \cong \overline{CD}$, prove $\angle D$ is a right angle. (Presumably, $ABCD$ is then a square). We shall see later that it is impossible with what we know to prove that $\angle D$ is a right angle without the assumption $\overline{BC} \cong \overline{CD}$.

SOLUTION

(a)

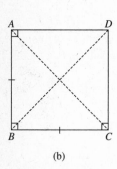

(a)

CONCLUSIONS	JUSTIFICATIONS
(1) Construct $\overline{AC}$.	Axiom I-1
(2) $\triangle ABC$ is isosceles, so $\angle BAC \cong \angle BCA$.	definition, Isosceles Triangle Theorem
(3) $m\angle CAD = 90 - m\angle BAC$ $= 90 - m\angle BCA$ $= m\angle ACD$.	betweenness considerations
(4) $\therefore \overline{AD} \cong \overline{DC}$.	Isosceles Triangle Theorem

(b)

CONCLUSIONS	JUSTIFICATIONS
(1) Construct $\overline{AC}$ and $\overline{BD}$.	Axiom I-1
(2) $\triangle ABC \cong \triangle DCB$.	SAS Postulate
(3) $\angle ACB \cong \angle DBC$, $\overline{AC} \cong \overline{BD}$.	CPCF
(4) $m\angle ABD = 90 - m\angle DBC$ $= 90 - m\angle ACB$ $= m\angle DCA$.	betweenness considerations
(5) $\triangle ABD \cong \triangle DCA$.	SAS Postulate
(6) $m\angle D = m\angle A = 90$ $\therefore \angle D$ is a right angle.❏	CPCF, definition

Figure 3.25

We now take care of a problem in axiomatic geometry that we have mentioned before. Although we use it constantly, its proof can be regarded as an optional exercise since the result itself is intuitively obvious, sometimes taken as an axiom in elementary treatments.

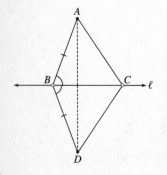

Figure 3.26

THEOREM 4: EXISTENCE OF PERPENDICULAR FROM AN EXTERNAL POINT

Let line ℓ and point A not on ℓ be given. Then there exists a unique line m perpendicular to ℓ and passing through A.

Proof: Given A not on line ℓ locate arbitrary points B and C on ℓ, as shown in Figure 3.26, and let $\angle DBC \cong \angle ABC$, $\overline{BD} \cong \overline{BA}$. It follows that B and C are both equidistant from A and D. (Why?) Hence, $AD \perp BC$ by the Perpendicular Bisector Theorem. (The uniqueness has already been established in Lemma A.)

PROBLEMS (§3.3)

———— Group A ————

1. Name all congruent pairs of distinct segments in each figure which follow logically either from previous concepts, or from the congruence theorems thus far developed.

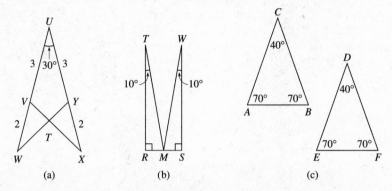

2. Name all congruent pairs of distinct angles in the figure of Problem 1 which follow logically either from previous concepts, or from the congruence theorems thus far developed.

3. Using the figure, find the missing lengths *x* and *y*, and state which congruence criterion you used.

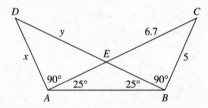

4. In the figure to the right, find the measure of angle ∠*QVW* and state what congruence criterion you used. (Assume that *U-V-W*, $\overrightarrow{VW}$-$\overrightarrow{VQ}$ -$\overrightarrow{VR}$-$\overrightarrow{VS}$, and $\overrightarrow{VR}$-$\overrightarrow{VS}$-$\overrightarrow{VT}$-$\overrightarrow{VU}$.)

5. Find the missing angle measures *x*, *y*, *z* in the figure below using the congruence criteria of this section. Explain your solution.

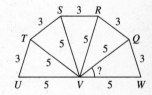

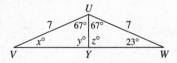

6. (a) List the eight triangles shown in the figure to the right.

   (b) From the marked information, there are three pairs of triangles which can be proved congruent. Name them, *with vertices in corresponding order*.

   [UCSMP, p. 321]

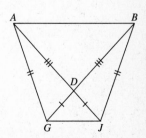

7. The following proof applies to the figure to the right, with the information as indicated. Supply the missing reasons.

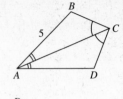

CONCLUSIONS	JUSTIFICATIONS
(1) $\overline{AC} \cong \overline{AC}$	(a) _____?
(2) $\triangle ABC \cong \triangle ADC$	(b) _____?
(3) $\overline{AD} \cong \overline{AB}$	(c) _____?
(4) $AD = 5$	(d) _____?

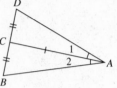

8. Explain why the SAS Postulate cannot be used to prove $\triangle ADC \cong \triangle ABC$ in the figure to the right. (Assume $\angle 1 \cong \angle 2$.)

9. A support brace has the shape illustrated in the figure below, with the approximate dimensions as indicated in the figure. What principle in geometry requires that line $\overleftrightarrow{ST}$ be perpendicular to line $\overleftrightarrow{QR}$?

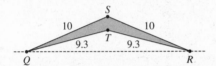

10. In the figure to the right, we are given $\overline{AB} \cong \overline{CD}$, $\overline{AB} \perp \overline{BC}$, and $\overline{CD} \perp \overline{BC}$.

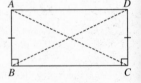

   (a) Why must $\overline{AC} \cong \overline{BD}$? (What congruence criterion guarantees this?)

   (b) Must $\overline{AD}$ be congruent to $\overline{BC}$? (Be careful not to use properties of parallelism or parallelograms you may be familiar with; try to picture this on the surface of a sphere to get a feel for what is provable from the axioms.)

   (c) If you are further given that $m\angle A = m\angle D = 90$, can it now be proven that $\overline{AD} \cong \overline{BC}$?

11. Triangle $\triangle ABG$ is an isosceles triangle, as indicated in the figure to the right. Segments of equal length have been measured off on the base. Fill in the missing parts of the following proof, assuming all betweenness relations that are obvious from the figure.

   Given: $\overline{AB} \cong \overline{AG}$, $\overline{BC} \cong \overline{CD} \cong \overline{DE} \cong \overline{EF} \cong \overline{FG}$.

   Prove: $\angle CAD \cong \angle EAF$.

CONCLUSIONS	JUSTIFICATIONS
(1) $\angle ABD \cong \angle AGE$	(a) _____?
(2) $BD = BC + CD$ and $EG = EF + FG$.	(b) _____?
(3) $BC + CD = EF + FG$, so $\overline{BD} \cong \overline{GE}$.	(c) algebra and _____?
(4) (d) _____?	(e) _____?
(5) $\overline{AD} \cong \overline{AE}$, $\angle ADC \cong \angle AEF$	(f) _____?
(6) $\triangle CAD \cong \triangle FAE$	(g) _____?
(7) $\therefore \angle CAD \cong \angle FAE$	(h) _____?

12. Prove Case III of the Isosceles Triangle Theorem: The line passing through the vertex and bisecting the vertex angle of an isosceles triangle is the perpendicular bisector of the base.

_____ *Group B* _____

13. For this proof, assume all betweenness relations evident in the figure to the right.

   Given: $\overline{AB} \perp \overline{BC}$, $\overline{CD} \perp \overline{BC}$, and $BE = EC$.

   Prove: $AB = CD$.

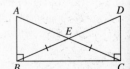

14. Use the figure for Problem 13 and give a proof for the following.

Given: $\overline{AB} \perp \overline{BC}$, $\overline{CD} \perp \overline{BC}$, and $AB = CD$.

Prove: $BE = EC$.

15. Prove Case IV of the Isosceles Triangle Theorem: The perpendicular bisector of the base of an isosceles triangle passes through the vertex and bisects the vertex angle. (**HINT:** This uses the uniqueness of the perpendicular to a line at a point on that line, established in Chapter 2. Do not use the Perpendicular Bisector Theorem, because that result uses this one.)

16. In this problem, you are to establish all betweenness relations needed. (See figure to the right.)

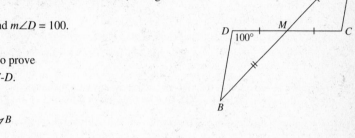

Given: $\overline{AB}$ and $\overline{CD}$ bisect each other at M and $m\angle D = 100$.

Prove: $m\angle C = 100$.

17. Establish all betweenness relations needed to prove

Given: $DF = EF$, $BF = FC$, A-D-B, and C-F-D.

Prove: $AD = AE$.

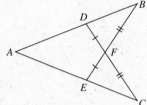

18. Prove the following.

Theorem: The two segments joining the vertex of an isosceles triangle with the trisection points of the base are congruent.

19. A common misconception related to Problem 18 is assuming that joining the vertex of an isosceles triangle with the trisection points on the base will trisect the vertex angle. Create a convincing argument to dispel this notion based on results in this section. (You may use the fact that the interior segments are shorter than the legs of the given triangle—a result of later theorems, but intuitively clear.)

20. Write out a complete proof of

Theorem: The perpendicular bisectors of the sides of a triangle are concurrent if any two of them meet.

(**HINT:** If two perpendicular bisectors meet at O, what relation does O have with respect to the three vertices?)

21. Prove the following.

Given: A-B-C, D and E are on the same side of line $\overleftrightarrow{AC}$, $\angle ABD \cong \angle CBE$, $AB = BD$, and $BC = BE$.

Prove: $AE = CD$.

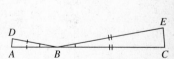

(Establish all betweenness relations.)

22. Fill in the missing reasons for the proof of Lemma A. (See figure on next page.)

Given: $\overleftrightarrow{AB} \perp \ell$

Prove: $\overleftrightarrow{AB}$ is the only perpendicular to ℓ through A.

CONCLUSIONS	JUSTIFICATIONS
(1) Suppose there is a second line $\overleftrightarrow{AC} \perp \ell$, $B \neq C$.	assumption for indirect proof

(2) With B the midpoint of $\overline{AE}$, consider $\overleftrightarrow{CE}$. $\overline{BC} \cong \overline{BC}$, so $\triangle ABC \cong \triangle EBC$.

(a) _____ ?

(3) $m\angle BCA = m\angle BCE = 90$

(b) _____ ?

(4) $\overleftrightarrow{CE} \perp \ell$ at C.

(c) _____ ?

(5) $\overleftrightarrow{AC}$ coincides with $\overleftrightarrow{CE}$.

(d) _____ ?

(6) $\overleftrightarrow{AB}$ and $\overleftrightarrow{AC}$ have A and E in common. $\rightarrow\!\!\leftarrow$

(e) _____ ?

(7) $\therefore$ $\overleftrightarrow{AB}$ is the only perpendicular to ℓ.

(f) _____ ?

23. (a) Prove that an isosceles triangle exists on a given segment as base.

(b) Prove that equilateral triangles exist. (Do not start with a given base.)

NOTE: Later we develop the tools necessary for proving that an equilateral triangle exists with a given base—Euclid's *first* proposition in Volume I of the *Elements*.

24. In the figure below, $\triangle ABC$ is given, with $\overline{AB} \cong \overline{AC}$. Can you construct a proof based on the diagram shown that $\angle ABC \cong \angle ACB$? (This was Euclid's *pons asinorum*.)

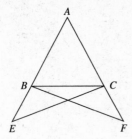

_____ *Group C* _____

25. See if you can prove the *Triangle Inequality* using the isosceles triangle theorems of this section: *If A, B, C are noncollinear, then AB + AC > BC*. What other facts do you need in order to prove this? (The figure below provides a hint for one case.)

(Assume $AB + AC = BC$)

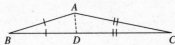

3.4 EXTERIOR ANGLE INEQUALITY

A result which plays a major role in the development of absolute geometry is the Exterior Angle Theorem. Our proof is virtually the same as that found in the *Elements*. Surprisingly, it is valid for absolute geometry.

EXTERIOR ANGLE OF A TRIANGLE

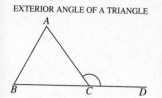

Figure 3.27

DEFINITION

Let $\triangle ABC$ be given, and suppose D is a point on $\overrightarrow{BC}$ such that the betweenness relation B-C-D holds (Figure 3.27). Then $\angle ACD$ is called an **exterior angle** of the given triangle. The interior angles at A and B are called **opposite interior angles** of $\angle ACD$.

OUR GEOMETRIC WORLD

When sailors use *Polaris* (the North Star) for sighting, it is routinely assumed that their line of sight is parallel to the **polar axis** (the line through the north and south poles, which points directly to *Polaris*). This assumption implies that the angle of elevation gives their precise lattitude θ (since $\theta = 90 - \phi_1 = 90 - \phi_2$). Although the error is small enough to justify the procedure, these lines are, in fact, not parallel. The inequality we study in this section shows that $\phi_2 > \phi_1$. (See Problem 14.)

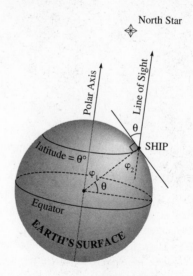

Figure 3.28

You may remember that in Euclidean geometry the measure of an exterior angle of any triangle in Euclidean geometry equals the sum of the measures of the other two opposite interior angles. In absolute geometry, this relationship is not valid, in general.

*SPHERICAL GEOMETRY
AND THE EXTERIOR
ANGLE OF A TRIANGLE*

Spherical geometry is the geometry which takes place on the surface of any sphere (often, for convenience, the **unit sphere**—a sphere with radius = 1). We take as "lines" the great circles on the sphere, as "distance" the length of the minor arc along a great circle joining any two points, and as "angle measure" the measure of the angle between the tangents to two circular arcs which are the sides of a spherical angle. (See Figure 3.29.)

It is pretty clear (by a standard result from spherical trigonometry) that the SAS Postulate holds for all spherical triangles, and everything we have thus far assumed for a single plane, except for Axiom I-1 and the Ruler Postulate, is true for the sphere. A "half-plane" on the sphere would be the set of all points in one hemisphere; thus each line (great circle) has two well-defined sides obeying the properties of the Plane Separation Axiom.

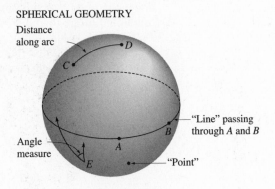

SPHERICAL GEOMETRY

Figure 3.29

We thus find that substantial parts of the theory we have developed up to this point are valid for spherical geometry. Even the principle "two points determine a line" is *almost true*. Except for the case when the points are the endpoints of a diameter (i.e., "poles"), there is only one great circle ("line") which passes through them. If we were to revise Axiom I-1 slightly, and revamp the Ruler Postulate so that bounded distances were possible, our axioms would indeed become valid on the sphere in their entirety. This is, in fact, the very problem addressed in Section 6.8, where axiomatic spherical geometry is introduced.

Since the sphere is thus a model for the previous axioms, as modified, we can use it to test the validity of theorems in absolute geometry. Anything provable in absolute geometry without making explicit use of unbounded distance is valid on the sphere, so if we know something is *not* true on the sphere, then it is also likely not true in absolute geometry.

A case in point is the familiar result in Euclidean geometry we mentioned previously: $m\angle ACD = m\angle A + m\angle B$. On the sphere, we could have the situation depicted in Figure 3.30, where A is the north pole and $\overleftrightarrow{CD}$ is the equator. Here, both $\angle ACD$ and $\angle B$ are right angles, and $m\angle ACD$ cannot be the "sum" of $\angle B$ and $\angle A$. This is an indication that we cannot prove this theorem in absolute geometry without introducing further ideas. The best we can do is an inequality.

$$m \angle ACD \neq m \angle A + m \angle B$$

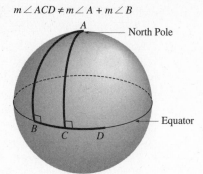

Figure 3.30

EXTERIOR ANGLE OF A TRIANGLE IN ABSOLUTE GEOMETRY

We now present the theorem and its proof.

THEOREM 1: EXTERIOR ANGLE INEQUALITY

An exterior angle of a triangle has angle measure greater than that of either opposite interior angle.

Proof:

Given: Exterior $\angle ACD$ of $\triangle ABC$, as in Figure 3.31.

Prove: (a) $m\angle ACD > m\angle A$.

(b) $m\angle ACD > m\angle B$.

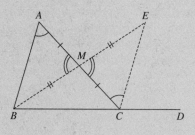
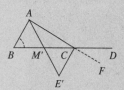

Figure 3.31

(a)

CONCLUSIONS	JUSTIFICATIONS
(1) Locate M, the midpoint of $\overline{AC}$.	existence of midpoint
(2) Extend $\overline{BM}$ its own length to point E such that M is the midpoint of $\overline{BE}$.	Segment Doubling Theorem
(3) $A\text{-}M\text{-}C$ and $AM = MC$; $B\text{-}M\text{-}E$ and $BM = ME$.	definition of midpoint

---

THEOREM 1: EXTERIOR ANGLE INEQUALITY (CONTINUED)

(4) $\angle AMB$ and $\angle CME$ form a vertical pair.	definition
(5) $\angle AMB \cong \angle CME$	Vertical Pair Theorem
(6) $\triangle AMB \cong \triangle CME$	SAS
(7) $m\angle ACE \equiv m\angle MCE = m\angle MAB \equiv m\angle A$	CPCF
(8) A, M and E are on the same side of $\overleftrightarrow{BC} = \overleftrightarrow{CD}$, hence $E \in H(A, \overleftrightarrow{CD})$.	Theorem 1, § 2.6
(9) Similarly, $E \in H(D, \overleftrightarrow{CA})$ and hence $E \in$ Interior $\angle ACD$.	Definition of the interior of an $\angle$
(10) $m\angle ACD = m\angle ACE + m\angle ECD$ $> m\angle ACE$	Angle Addition Postulate, algebra
(11) $\therefore\ m\angle ACD > m\angle A$	Step 7

(b)

For the second part, merely locate F so that A-C-F; then $m\angle BCF = m\angle ACD$ by the Vertical Pair Theorem, and from the first part, $m\angle BCF > m\angle B$. By substitution, $m\angle ACD > m\angle B$.

It is instructive to observe how this proof breaks down on the sphere. Consider in Figure 3.32 the same triangle $\triangle ABC$ as in the proof of Theorem 1, only with A the "north pole" and $\overleftrightarrow{BC}$ the "equator." The first step in the proof was to locate the midpoint M, then to *double* segment $\overline{BM}$ by its own length to form segment $\overline{BE}$. On the sphere, point E ends up *on line BC itself*, so is not in the interior of $\angle ACD$! Other situations could cause E not to remain in the northern hemisphere at all, but to lie on the *opposite side of* $\overleftrightarrow{BC}$ *from A*. (Do you see a situation on the sphere in which the above proof does apply, using hemispheres for half-planes and the other basic geometric properties?)

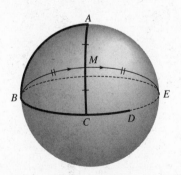

Figure 3.32

APPLICATIONS

The Exterior Angle Inequality has far-reaching effects in geometry. For example, it puts a limit on the number of obtuse or right angles which a triangle can have. A short list of the simpler consequences follows. (Figure 3.33 indicates how the proof of the first can be derived; the others are corollaries.)

- The sum of the measures of any two angles of a triangle is less than 180.
- A triangle can have at most one right or obtuse angle.
- The base angles of an isosceles triangle are acute.

METHOD FOR PROVING $m \angle A + m \angle B < 180$

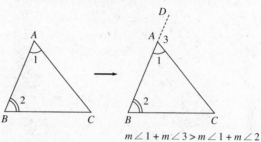

$$m \angle 1 + m \angle 3 > m \angle 1 + m \angle 2$$

Figure 3.33

A more important application is the famous Saccheri-Legendre Theorem: *Every triangle in absolute geometry has angle sum ≤ 180.* This can be proven with what we now know—no additional assumptions have to be made. The idea of the proof can be dramatized in a discovery project specifically designed for this purpose.

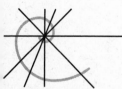

MOMENT FOR DISCOVERY

Central Idea in Proof of Saccheri-Legendre Theorem

Suppose the angles of $\angle ABC$ are as marked in the figure: $m\angle A = 56$, $m\angle B = 50$, and $m\angle C = 70$. Points M and E are constructed as in Euclid's proof of the Exterior Angle Inequality, where M is the midpoint of both $\overline{AC}$ and $\overline{BE}$. Thus, $\triangle AMB \cong \triangle CME$ and $m\angle MCE = m\angle A$, $m\angle E = m\angle ABM$.

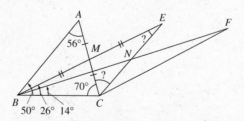

Figure 3.34

1. Calculate the angle sum for $\triangle ABC$: $m\angle A + m\angle B + m\angle C = ?$ (Note that this answer is not 180, which is quite possible in non-Euclidean geometry.)

2. Suppose that $m\angle EBC = 26$. Find the measures of the angles at C and E. (Did you get $m\angle BCE = 126$?)

3. Record the following angle measures thus far determined:

$$m\angle EBC, \; m\angle BCE, \; m\angle BEC$$

4. Find the angle sum in $\triangle BCE$ (the sum of the measures of the angles in Step 3). Did anything happen?

5. $\triangle BCF$ is constructed from $\triangle BCE$ in the same way that $\triangle BCE$ was constructed from $\triangle ABC$ (N is the midpoint of $\overline{CE}$ and $\overline{BF}$). Suppose $m\angle FBC = 14$, for example. Then calculate

$$m\angle FBC, \; m\angle BCF, \; m\angle BFC$$

6. Find the angle sum of $\triangle BCF$. Did you notice anything?

Saccheri and Legendre both observed in their own developments of non Euclidean geometry, by using an argument based on the ideas which you should have found evident in the Discovery Unit, that the angle sum of any triangle is numerically the same as that of a sequence of other triangles constructed in the manner of the proof of the Exterior Angle Inequality. Since the argument is so clever and the resulting theorem so remarkable (all derived from the proof of the Exterior Angle Inequality), let's take some time to go through it.

Suppose $\triangle ABC$ has angle sum exceeding 180. That is,

$$m\angle A + m\angle B + m\angle C = 180 + t,$$

where t is some positive real number (in Figure 3.35, $\angle C \equiv \angle ACB$). Proceed to replace the original triangle by a sequence of other triangles, $\triangle BCE$, $\triangle BCF$, $\cdots$ each having $\overline{BC}$ as base, and having angle sum equal to that of the previous triangle (and to that of $\triangle ABC$). That is, in Figure 3.35,

$$m\angle E + m\angle EBC + m\angle BCE = m\angle F + m\angle FBC + m\angle BCF = \cdots = 180 + t$$

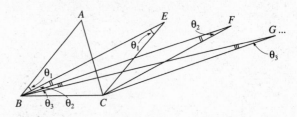

Figure 3.35

This is easily proved, since in $\triangle BCE$ we have

$$m\angle ABE = m\angle E \quad \text{and} \quad m\angle A = m\angle ACE$$

or

$$m\angle E + m\angle EBC + m\angle BCE = m\angle ABE + m\angle EBC + m\angle BCA + m\angle ACE = m\angle ABC + m\angle BCA + m\angle A = 180 + t$$

This argument can be repeated to show that each triangle in the sequence has angle sum equal to that of $\triangle ABC$, or $180 + t$. We would now like to show that the angles at E, F, G, ... are approaching zero as limit. Let

$$\theta_1 = m\angle E = m\angle ABE, \quad \theta_2 = m\angle F = m\angle EBF, \quad \theta_3 = m\angle G = m\angle FBG, \cdots$$

Then we have

$$\theta_1 + \theta_2 + \cdots + \theta_n = m\angle ABE + m\angle EBF + m\angle FBG + \cdots < m\angle ABC \equiv m\angle B$$

or

$$\sum_{n=1}^{\infty} \theta_n = \theta < m\angle B$$

and thus $\lim\limits_{n \to \infty} \theta_n = 0$. Therefore, the angle opposite side $\overline{BC}$ of the nth triangle for sufficiently large n is smaller than any preassigned real number, such as t, and thus for large enough n the nth triangle, call it $\triangle BCW$, has $m\angle W < t$, with angle sum $180 + t$. That is,

$$180 + t = m\angle WBC + m\angle BCW + m\angle W < m\angle WBC + m\angle BCW + t$$

or, canceling t from both sides,

$$180 < m\angle WBC + m\angle BCW,$$

a contradiction, since the sum of the measures of two angles of a triangle is *always less than 180*. Hence, we have proven

---

THEOREM 2: SACCHERI-LEGENDRE THEOREM
The angle sum of any triangle cannot exceed 180.

Note that we have not yet proven that the angle sum of a triangle *equals* 180; this result depends on properties of parallelism which we have not yet introduced.

PROBLEMS ($\S3.4$)

_____ **Group A** _____

1. In the figure at the right, $\angle ECD$ is an exterior angle of $\triangle ACE$ and B is a variable point on $\overrightarrow{AC}$. What can be said about the range of values possible for x and y when

   (a) A-B-C?

   (b) A-C-B-D?

2. Using the figure for Problem 1, suppose $\overrightarrow{EB}$ bisects $\angle AEC$. What can be said about the range of values for x and y for all possible positions of the points if $m\angle A = 30$?

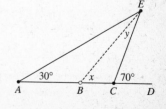

3. In the accompanying figure, $\triangle ABC$ has a right angle at C. What can be said about the type of angles the triangle has at A and B, and why?

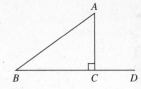

4. Must the base angles of an isosceles triangle be acute in absolute geometry? Try to write a formal proof of this, or give specific reasons for your answer. (See figure below.)

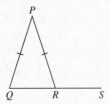

*5. In the following figure, $\triangle WKT$ has acute angles at K and T. If $\overleftrightarrow{WR} \perp \overleftrightarrow{KT}$, show that R lies between K and T. (If R does *not* lie between K and T, then either K-T-R or R-K-T; show either case leads to a contradiction of the Exterior Angle Inequality.)

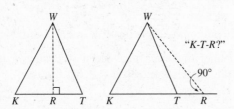

*An important result for absolute geometry which is used repeatedly from now on.

6. Suppose $\overleftrightarrow{AM} \perp \overleftrightarrow{BC}$ in the figure below and that $m\angle B = m\angle C = \frac{1}{2}$. A point P moves on $\overleftrightarrow{BC}$ and assumes 179 locations on $\overleftrightarrow{BC}$ (P_1, P_2, ... , P_{179}) such that

$$m\angle AP_1C = 1,\ m\angle AP_2C = 2,\ m\angle AP_3C = 3, ...$$

(In general, $m\angle AP_nC = n$.) Find the approximate locations of these points in terms of their order on $\overleftrightarrow{BC}$. Does P_5 follow P_4, etc.? Does the Exterior Angle Inequality play a role?

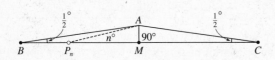

7. Use the Exterior Angle Inequality to prove that we cannot have two perpendiculars to the same line from the same external point.

8. Explain how to construct a spherical triangle (prove that it exists on the unit sphere) having two angles each of measure 120. Is this possible in absolute geometry? Why, or why not?

9. Prove that the sum of the measures of any two angles of a triangle is less than 180 without using the Saccheri-Legendre Theorem. (See Figure 3.33.)

_____ *Group B* _____

10. If a quadrilateral $ABCD$ has right angles at A, B, and C, show that $m\angle D \leq 90$.

11. Prove that the hypotenuse of a right triangle has length greater than that of either leg.

12. Prove in △*ABC* that if *AB* < *AC*, then *m∠C* < 90. (See figure to the right.)

13. An indirect consequence of the Exterior Angle Inequality is the "obvious" fact that for any three distinct points *A*, *B*, *C*, if *AB* + *BC* = *AC*, then *A-B-C*. Prove this by assuming that *B* does not lie on line $\overleftrightarrow{AC}$ and gaining a contradiction, then showing that if $B \in \overleftrightarrow{AC}$, we cannot have *B-A-C* or *A-C-B*. (See figure below for a hint.)

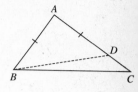

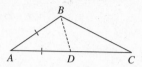

*14. Prove as a corollary to the Exterior Angle Inequality: If two lines make equal angles with another line, and if those angles are oriented in the same direction, the lines are parallel. (**HINT:** Suppose that the lines are *not* parallel, but meet at some point *P*, as shown in the figure. Use the Exterior Angle Inequality to gain a contradiction.)

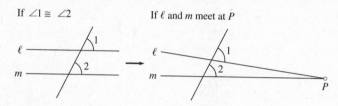

*See lemma, §4.1.

3.5 THE INEQUALITY THEOREMS

The Exterior Angle Inequality leads to numerous other inequality theorems for triangles. These, in turn, are very useful in proving certain "equality" theorems, identities, and congruences. We shall continue our list of properties which logically follow from this inequality. The first one lends itself to an easy "proof by contradiction," and the other two are obvious corollaries. (See Problem 12, Section 3.4 and Problem 10, Section 3.5.)

- If one side of a triangle has greater length than another side, then the angle opposite the longer side has the greater angle measure, and conversely. (**Scalene Inequality**)

- If a triangle has an obtuse or right angle, then the side opposite that angle has the greatest length.

- The hypotenuse of a right triangle has length greater than that of either leg.

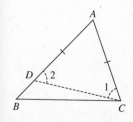

Figure 3.36

To prove the Scalene Inequality, observe Figure 3.36: If *AB* > *AC* and *AD* = *AC*, then by the Exterior Angle Inequality, *m∠ACB* > *m∠1* = *m∠2* > *m∠B*. The converse then follows easily: Following Euclid's old argument (quoted in Section 2.1), suppose that *m∠C* > *m∠B*; then by the result just obtained, we cannot have *AB* ≤ *AC* or else *m∠C* ≤ *m∠B*, a contradiction. Therefore, *AB* > *AC*.

Now we show how to obtain the important triangle inequality with the theory we now have.

THEOREM 1: TRIANGLE INEQUALITY

In any triangle, the sum of the measures of two sides is greater than that of the third side. Consequently, for any three distinct points A, B, and C, $AB + BC \geq AC$, with equality if and only if A-B-C.

Proof

CONCLUSIONS	JUSTIFICATIONS
(1) Extend $\overline{BC}$ to point D so that C-B-D and $BD = AB$ (Figure 3.37)	Segment Construction Theorem
(2) $DC = DB + BC = AB + BC$	C-B-D, substitution
(3) $m\angle 1 = m\angle 2$	Isosceles Triangle Theorem
(4) $B \in$ Interior $\angle DAC$	Theorem 3, § 2.6
(5) $m\angle DAC = m\angle 2 + m\angle 3 > m\angle 2$ $= m\angle 1 \equiv m\angle ADC$	substitution, Steps 3 and 4
(6) $DC > AC$	Scalene Inequality
(7) $AB + BC > AC$	substitution (Step 2)
(8) $\therefore$ In all cases for three distinct points A, B, and C, $AB + BC \geq AC$. If $AB + BC$ $= AC$, then A-B-C.	If B is not on line $\overrightarrow{AC}$, use the above argument; if A-B-C, then $AB + BC = AC$, and if C-A-B or A-C-B, then $AB + BC > AC$

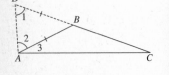

Figure 3.37

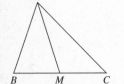

Figure 3.38

COROLLARY: MEDIAN INEQUALITY

Suppose that $\overline{AM}$ is the median to side $\overline{BC}$ of $\triangle ABC$ (Figure 3.38). Then $AM < \frac{1}{2}(AB + AC)$.

(The proof will be left as Problem 14.)

The next result is sometimes referred to as the "Hinge" Theorem, or "Alligator" Theorem (the wider the angle, the larger the opening). We shall give it a more formal name, however.

THEOREM 2: SAS INEQUALITY THEOREM

If in $\triangle ABC$ and $\triangle XYZ$ we have $AB = XY$, $AC = XZ$, but $m\angle A > m\angle X$, then $BC > YZ$, and conversely, if $BC > YZ$, then $m\angle A > m\angle X$. (See Figure 3.39.)

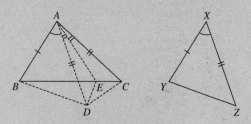

Figure 3.39

Proof

Given: $AB = XY$, $AC = XZ$, and $m\angle A > m\angle X$.

Prove: $BC > YZ$.

CONCLUSIONS	JUSTIFICATIONS
(1) Construct ray $\overrightarrow{AD}$ such that $\overrightarrow{AB}$-$\overrightarrow{AD}$-$\overrightarrow{AC}$ and $\angle BAD \cong \angle X$, with $AD = XZ = AC$.	Angle and Segment Construction Theorems
(2) $\triangle ABD \cong \triangle XYZ$	SAS
(3) $\overline{BD} \cong \overline{YZ}$	CPCF
(4) Construct the bisector of $\angle DAC$; this bisector will cut segment $\overline{BC}$ at an interior point E.	Angle Construction Theorem, Crossbar Theorem
(5) $\angle DAE \cong \angle EAC$, $\overline{AE} \cong \overline{AE}$, $\overline{AD} \cong \overline{AC}$	definition of bisector, Reflexive Property
(6) $\triangle DAE \cong \triangle CAE$	SAS
(7) $DE = EC$	CPCF
(8) $BC = BE + EC = BE + DE > BD$, $\therefore BC > YZ$	B-E-C, substitution from Step 7, Triangle Inequality and Step 3

EXAMPLE 1 Name the longest and shortest segments shown in Figure 3.40, and give the reasons for your answers.

SOLUTION

$PT > PW$, $PT > WT > QW$ (Scalene Inequality)

$PT > PQ$, $PT > QT$ (betweenness property)

$\quad \therefore \overline{PT}$ is the longest segment.

$PQ < QW < PW$, $QT = QW < WT < PT$ (Scalene Inequality)

$\quad \therefore \overline{PQ}$ is the shortest segment. ◻

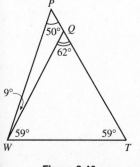

Figure 3.40

E X A M P L E   2   In Figure 3.41, a circle (to be introduced formally in Section 3.8) with diameter $\overline{QR}$ and center O is shown. If P varies on the circle on either side of $\overrightarrow{QR}$, and if $\theta = m\angle POQ$, define the function

$$f(\theta) = PQ, \quad 0 < \theta < 180$$

Explain why $f(\theta)$ is an increasing function. (That is, if $\theta_1 < \theta_2$, then $f(\theta_1) < f(\theta_2)$.) Observe points P and P' in the figure.

SOLUTION

In $\triangle OPQ$ and $\triangle OP'Q$, $\theta_1 = m\angle POQ < m\angle P'OQ = \theta_2$ and $f(\theta_1) = PQ$, $f(\theta_2) = P'Q$. Since $OP = OP'$ and $OQ = OQ$, by the SAS Inequality Theorem, $PQ < PQ'$, or $f(\theta_1) < f(\theta_2)$.□

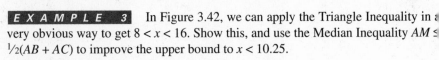

P
P'
R
θ₂ θ₁
O
Q

Figure 3.41

E X A M P L E   3   In Figure 3.42, we can apply the Triangle Inequality in a very obvious way to get $8 < x < 16$. Show this, and use the Median Inequality $AM \le \frac{1}{2}(AB + AC)$ to improve the upper bound to $x < 10.25$.

SOLUTION

Using the Triangle Inequality for $\triangle ABD$,

$$x < 4 + 12 = 16 \text{ and } x + 4 > 12 \rightarrow 8 < x < 16$$

To improve this inequality, use the fact that the midpoint M of $\overline{BC}$ will define a segment $\overline{BM}$ of length $\frac{1}{2}(4 + 12) = 8$, hence D is the midpoint of $\overline{BM}$. By the Median Inequality applied to the two resulting triangles,

$$x < \frac{1}{2}(12 + AM) \text{ and } AM < \frac{1}{2}(12 + 5) = 8.5$$

$$\therefore x < \frac{1}{2}(12 + 8.5) = 10.25,$$

a considerable improvement over $x < 16$. Thus, $8 < x < 10.25$.□

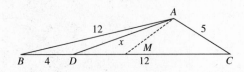

A
12   5
x
M
B   4   D   12   C

Figure 3.42

PROBLEMS   (§3.5)

_____ *Group A* _____

1. Name the longest and shortest segments shown in the figure to the right and state your reasons.

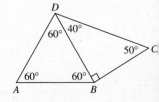

D
60° 40°
50° C
60°   60°
A   B

2. If $QR = QT$, which segment is the shortest, and why? (See the figure to the right.)

3. Which angle in the following figure is smallest, and why? (Do not use Euclidean geometry and properties we have not yet introduced axiomatically.)

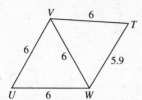

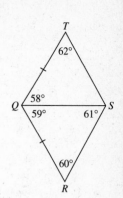

4. In right triangles $\triangle ABC$ and $\triangle DEF$, the legs $\overline{AC}$ and $\overline{DF}$ are congruent, but $BC > EF$. Show that $AB > DE$. (***HINT:*** Construct $\overline{AG}$ so that $GC = EF$.)

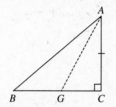

5. The base of the triangle in the figure at right has been divided into four congruent segments, each of length 21 units, and R, the foot of the perpendicular from O on that base forms noncongruent segments with C and D, as indicated. List the five segments from O to points A, B, C, D, and E in order, according to their lengths, from largest to smallest. (Use the result of Problem 4.)

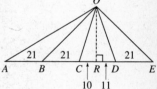

6. Which of the angles at the base of the triangle in the figure for Problem 5 is the

(a) largest?

(b) smallest?

(***HINT:*** Use Problem 4 to produce a similar result involving angles; show $m\angle B < m\angle E$.)

7. Suppose that $\triangle ABC$ has sides $AB = 31$, $AC = 35$, and M is the midpoint of $\overline{BC}$.

(a) Find $y = CE$ if M is the midpoint of $\overline{AE}$.

(b) Show that $2 < x < 33$, where $x = AM$.

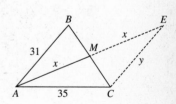

8. As in the preceding problem, M is the midpoint of $\overline{BC}$, $x = AM = ME$, and $y = CE$, but $AB = 60$ and $AC = 40$ as shown in the figure below. Find

(a) y and z

(b) the appropriate bounds on x.

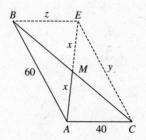

9. In $\triangle RST$, what is the largest value possible for $m\angle S$?

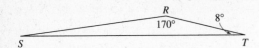

_____ *Group B* _____

10. Prove that in a triangle with a right or obtuse angle at A, the side opposite angle A has the greatest length.

11. (a) Extend the result of Problem 4 to obtuse triangles $\triangle ABC$ and $\triangle DEF$, where instead of $m\angle C = m\angle F = 90$, $m\angle C = m\angle F > 90$.

 (b) Is the result true if $m\angle C = m\angle F < 90$?

12. Prove that if A, B, and C are distinct, collinear points and A-B-C does not hold, then $AB + BC > AC$.

13. Give a proof of the following inequality: For any three points A, B, and C,

$$AB - BC \leq AC \leq AB + BC,$$

with equality in either case if and only if A, B, and C are collinear.

14. Prove the Median Inequality : $AM < \frac{1}{2}(AB + AC)$. (**HINT:** As shown in the figure, double segment $\overline{AM}$ to D, and use the Triangle Inequality for $\triangle ACD$.)

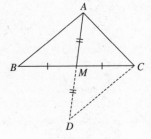

15. Prove that the total angle sum of two adjacent triangles with their bases lying on a common line exceeds the angle sum of the large triangle by 180. That is, if A-D-C and k_1 and k_2 are the angle sums of $\triangle BAD$ and $\triangle BDC$, respectively, then

$$k_1 + k_2 = m\angle A + m\angle ABC + m\angle C + 180$$

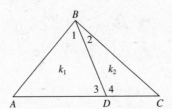

_____ *Group C* _____

16. Using the result of Problem 15, prove that if the angle sum of triangles in absolute geometry is constant, then the angle sum of all triangles equals 180. (**HINT:** Let $k_3 = m\angle A + m\angle B + m\angle C$; by hypothesis, $k_1 = k_2 = k_3$.)

17. Prove the SSS Congruence Theorem using inequality theorems.

18. Prove that if A, B, and C are any three points, and M is the midpoint of $\overline{BC}$, then

$$\frac{1}{2}(AB - AC) \leq AM \leq \frac{1}{2}(AB + AC)$$

*19. Prove that in absolute geometry, open balls are convex (and circular disks are convex, as a corollary).

*See Problems 15, 16 in §2.5.

20. UNDERGRADUATE RESEARCH PROJECT Make several applications of the
Median Inequality to prove in $\triangle ABC$ that

$$d < \frac{3}{8}a + \frac{5}{8}b$$

What would it take to generalize this to

$$d < pa + qb,$$

where $p = \dfrac{AP}{AB}$ and $q = \dfrac{PB}{AB}$? (See accompanying figure.)

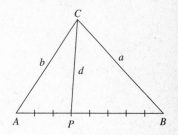

21. UNDERGRADUATE RESEARCH PROJECT The lengths 2 in., 2 in., 2 in.,
and 2 in. can be the sides of a quadrilateral, but 2 in., 4 in., 8 in., and 16 in. cannot. Is there a "quadrilateral inequality"? That is, is there some general way you
can tell when four positive numbers a, b, c, d can be the lengths of the sides of a
quadrilateral? [UCSMP, p. 51] Generalize and prove all results in either absolute
or Euclidean geometry. Is there a difference if we allow the vertices to be noncoplanar, that is, lying in three-dimensional space?

3.6   ADDITIONAL CONGRUENCE CRITERIA

Two further congruence criteria for arbitrary triangles can now be established,
symbolized by AAS, and a special one to be explained later, SsA. Special cases of
these for right triangles, the HA and HL criteria, are quite useful. At this point in
our development, it is time to cut back on the strict outline mode for arguments and
be on our way to learning the skill of writing mathematical proofs.

**AAA AND SSA
CONGRUENCE CRITERIA**

We begin with the situation where a pair of triangles have two pairs of congruent
angles and *any* pair of congruent sides (not necessarily the "included sides"). It
seems apparent that when we have two triangles $\triangle ABC$ and $\triangle XYZ$ with two pairs
of angles congruent, $m\angle A = m\angle X$ and $m\angle B = m\angle Y$, then the third pair will be
congruent also, $m\angle C = m\angle Z$. Indeed, the conclusion is correct under certain circumstances, but it is clear only in Euclidean geometry where the angle sum of all
triangles is 180. The proof here is valid for absolute geometry, and for spherical
geometry for triangles which are sufficiently small. The following Discovery Unit
explores a special situation on the sphere in this regard.

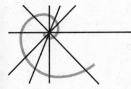

MOMENT FOR DISCOVERY

Spherical Triangles

On a unit sphere, consider the north pole N and the equator ℓ.

1.   Choose points A and B on ℓ at a distance $\pi/4$ apart. This will make $\angle AOB$
have measure 45 since the radii are of unit length.

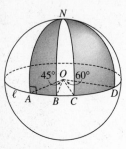

Figure 3.43

2. Draw the meridians through A and B (and passing through N). This forms a spherical triangle $\triangle ABN$.

3. How large is $m\angle ANB$? (The tangents to arcs $\overset{\frown}{NA}$ and $\overset{\frown}{NB}$ at N are parallel to $\overline{OA}$ and $\overline{OB}$.)

4. Repeat for two other points C and D on ℓ, but at a distance $\pi/3$ apart. (Thus $m\angle COD = 60$.)

5. How large is $m\angle CND$?

6. Do $\triangle ANB$ and $\triangle CND$ have a pair of congruent angles? Two pairs? What about the third pair?

The correct argument depends on the inequality theorems.

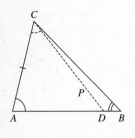

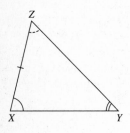

Figure 3,44

THEOREM 1: AAS THEOREM

If, under some correspondence, two angles and a side opposite one of the angles of one triangle are congruent, respectively, to the corresponding two angles and side of a second triangle, then the triangles are congruent.

Proof

Given: $\angle A \cong \angle X$, $\angle B \cong \angle Y$, and $AC \cong XZ$ (Figure 3.44).

Prove: $\triangle ABC \cong \triangle XYZ$.

(1) We want to prove that $m\angle C = m\angle Z$. Hence, assume for the sake of argument that $m\angle C > m\angle Z$. (The argument is the same if $m\angle C < m\angle Z$.)

(2) Construct ray $\overrightarrow{CP}$ such that $m\angle ACP = m\angle Z$ and $\overrightarrow{CA}\text{-}\overrightarrow{CP}\text{-}\overrightarrow{CB}$. (Angle Construction Theorem)

(3) $P \in$ Interior $\angle ACB$ and $\overrightarrow{CP}$ meets $\overline{AB}$ at some point D. (Angle Addition Postulate and Crossbar Principle)

(4) $\triangle ADC \cong \triangle XYZ$ (ASA Theorem)

(5) $m\angle ADC = m\angle Y$ (CPCF)

(6) But $m\angle ADC > m\angle B = m\angle Y$ →← (Exterior Angle Inequality)

(7) Hence our assumption that $\angle C \neq \angle Z$ was false, and $\therefore m\angle C = m\angle Z$ (the triangles are congruent by ASA).

You might recall from trigonometry the so-called "ambiguous" case for solving a triangle given various information. This particular case involves solving for the remaining parts of a triangle when the measures of two sides and an angle opposite one of them are presented; it is ambiguous because there can be either two solutions, one solution, or no solution, depending on circumstances. It is because of this that we do not get the likely congruence criterion indicated by SSA (see Figure 3.45 for a specific counterexample). However, if we require the pair of congruent

NONCONGRUENT TRIANGLES SATISFYING
SSA HYPOTHESIS

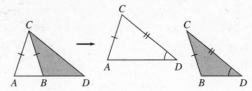

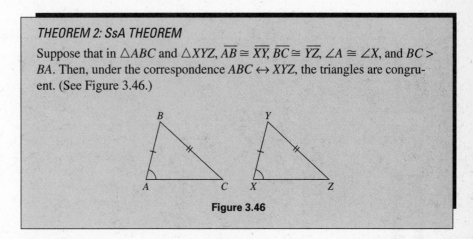

Figure 3.45

sides opposite the congruent angles to have greater length than the other pair, then congruence *will* follow. How would you like to try your hand at proving this important tool? (See Problem 9 for some guidance if you need it.)

THEOREM 2: SsA THEOREM

Suppose that in $\triangle ABC$ and $\triangle XYZ$, $\overline{AB} \cong \overline{XY}$, $\overline{BC} \cong \overline{YZ}$, $\angle A \cong \angle X$, and $BC > BA$. Then, under the correspondence $ABC \leftrightarrow XYZ$, the triangles are congruent. (See Figure 3.46.)

Figure 3.46

NOTE: The restriction to the case where one of the congruent pairs of sides is the longer side is essential. Using the theorem without this property can very well lead to a false conclusion.

TWO CONGRUENCE THEOREMS FOR RIGHT TRIANGLES

Essentially, we need only two pairs of congruent parts, in addition to the right angles, in order to conclude that two right triangles are congruent. Since these results are trivial consequences of SAS, AAS, etc., we merely list the two most useful criteria.

For convenience, we let H, L, and A designate the hypotenuse, leg, and acute angle, respectively, of a right triangle in stating these results. The first is an obvious corollary of the AAS Theorem; the second is a corollary of the SsA theorem. Proving them are nice exercises for you.

COROLLARY A: HA THEOREM

If, under some correspondence between the vertices of two right triangles, the hypotenuse and an acute angle of one triangle are congruent, respectively, to the corresponding hypotenuse and acute angle of the other, then the triangles are congruent. (See Figure 3.47.)

COROLLARY B: HL THEOREM

If, under some correspondence between the vertices of two right triangles, the hypotenuse and leg of one triangle are congruent, respectively, to the corresponding hypotenuse and leg of the other, then the triangles are congruent. (See Figure 3.47.)

THE HA HYPOTHESIS

THE HL HYPOTHESIS

Figure 3.47

PROBLEMS   (§3.6)

_____ *Group A* _____

1. Draw two separate noncongruent triangles of your own which satisfy the SSA Hypothesis. Can you draw these triangles in such a way that one of them has a right angle? Can they both be right triangles?

2. Provide the missing reasons in the following proof of the HL Congruence Theorem. (See figure below.)

   Given: Right triangles $\triangle ABC$ and $\triangle XYZ$, with right angles at C and Z, and with $AC = XZ$ and $AB = XY$.

   Prove: $\triangle ABC \cong \triangle XYZ$.

   (1) $m\angle C = 90$.   (a)_____ ?
   (2) $m\angle B < 90$, or $m\angle B < m\angle C$.   (b)_____ ?
   (3) $AB > AC$.   (c)_____ ?
   (4) $\therefore \triangle ABC \cong \triangle XYZ$ (d)_____ ?

3. In elementary geometry texts where the SsA theorem is *not* included, a different proof of the HL Congruency Theorem is required. Find such a proof. (See figure for hint.)

 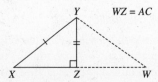

4. From the information given in the figure at the right, determine what triangle congruence theorem tells you which pair of triangles is congruent, and then write down the congruence with vertices correctly corresponding, and the congruence criterion used.

5. From the information given in the figure below, determine what triangle congruence theorem tells you which pair of triangles is congruent, and then write down the congruence with vertices correctly corresponding, with the congruence criterion used.

(a)

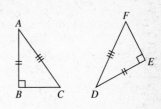

(a)   (b)   (c)

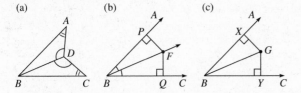

(b)

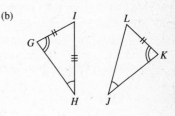

6. **(a)** List all eight triangles occurring in the figure below.

   (b) From the marked information, there are three pairs of triangles which can be proved congruent. Name them, with vertices in corresponding order.

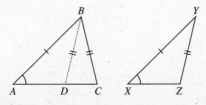

7. Follow these steps to draw (construct) a certain triangle.

   a. Draw a ray $\overrightarrow{XY}$.

   b. Draw $\angle WXY$ with measure (approximately) 50.

   c. Draw $\overline{XZ}$ on ray $\overrightarrow{XW}$ such that $ZX = 11$ cm.

   d. Draw circle Z with radius 9 cm. Let P be a point of intersection, Circle $Z \cap \overleftrightarrow{XY}$.

   e. Consider $\triangle XPW$. Will everyone else who does this correctly have a $\triangle XPW$ congruent to yours? Discuss.

   [UCSMP, p. 331]

8. Explain why the SsA Theorem cannot be used to prove that $\triangle ADC \cong \triangle ABC$ in the accompanying figure.

   [UCSMP, p. 321]

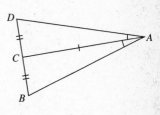

_____ *Group B* _____

9. Prove the SsA congruence criterion. (***HINT:*** In $\triangle ABC$ and $\triangle XYZ$, with $AB = XY < YZ = BC$ and $m\angle A = m\angle X$, suppose side $AC > XZ$; then $\overline{AD}$ may be constructed on $\overline{AC}$ so that A-D-C and $AD = XZ$. Draw segment $\overline{BD}$. Now work with this figure to get a contradiction of the hypothesis.)

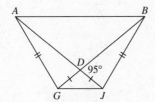

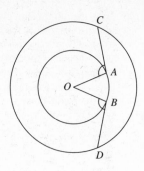

10. Assume the usual definition for circles (to be discussed later in this chapter), and concentric circles. In the figure, prove that if $\overline{OA}$ and $\overline{OB}$ are radii of the small circle and $\angle OAC \cong \angle OBD$, then $\overline{AC} \cong \overline{BD}$.

11. Prove that if I is any point on the bisector $\overrightarrow{BD}$ of $\angle ABC$, then I is equidistant from the sides of the angle ("perpendicular distance" understood).

12. State and prove the *converse* of the proposition of Problem 11.

*13. Prove that the angle bisectors of any triangle are concurrent at a point that is equidistant from the three sides of the triangle. (See Problems 11 and 12.)

   *This result is needed for Problem 8, §3.8 and in the proof of Theorem 1, §6.5.

14. Prove that if two triangles $\triangle ABC$ and $\triangle XYZ$ are known to have all acute angles, and under $ABC \leftrightarrow XYZ$ two pairs of corresponding sides and a pair of corresponding angles are congruent, then $\triangle ABC \cong \triangle XYZ$.

_____ *Group C* _____

15. **Steiner-Lehmus Theorem** If two angle bisectors of a triangle are congruent, the triangle is isosceles. Prove, using the inequality theorems. (**HINT:** In $\triangle ABC$, suppose bisectors $\overline{BD}$ and $\overline{CE}$ are congruent, but that $AC > AB$ and hence, $m\angle ABC > m\angle ACB$. Construct F on $\overline{AD}$ so that $\angle FBD \cong \angle ACE$, and gain a contradiction.)

$$FC > BF \rightarrow CG > BD$$

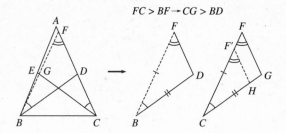

3.7 *QUADRILATERALS*

Some of the more common types of parallelograms, such as the square and rectangle, do not generally exist in absolute geometry. This section will explore this question, as well as develop the subject of quadrilaterals formally. The symbol $\diamond$ or $\square$ is often used to denote quadrilaterals; we shall adopt the former for the general case and reserve the latter for squares and rectangles.

DEFINITION

If A, B, C, and D are any four points lying in a plane such that no three of them are collinear, and if the points are so situated that no pair of open segments $(\overline{AB})$, $(\overline{BC})$, $(\overline{CD})$, and $(\overline{DA})$ have any points in common (Figure 3.48), then the set

$$\diamond ABCD \equiv \overline{AB} \cup \overline{BC} \cup \overline{CD} \cup \overline{DA}$$

is a **quadrilateral**, with **vertices** A, B, C, D; **sides** $\overline{AB}, \overline{BC}, \overline{CD}, \overline{DA}$; **diagonals** $\overline{AC}, \overline{BD}$; and **angles** $\angle DAB, \angle ABC, \angle BCD$, and $\angle CDA$.

QUADRILATERALS

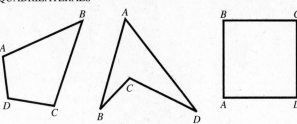

Figure 3.48

A few more terms associated with quadrilaterals are used frequently. Sides like $\overline{BC}$ and $\overline{CD}$ with a common endpoint are called **adjacent** (or sometimes **consecutive**), and angles containing a common side, like $\angle BCD$ and $\angle CDA$ (where $\overrightarrow{CD} \subseteq \overrightarrow{CD} \cap \overrightarrow{DC}$), are also called **adjacent** (or **consecutive**). Two sides (or angles) which are not adjacent are called **opposite** (such as $\overline{AB}$ and $\overline{CD}$).

CONVEX QUADRILATERALS

In Figure 3.48, one of the quadrilaterals has the peculiar property that its diagonals do not intersect, and two vertices (A and C) lie on one side of the diagonal joining the other two vertices (B and D). In most work with quadrilaterals in elementary geometry, we want the diagonals to intersect and to lie between opposite vertices. Such quadrilaterals are called **convex**. We leave fundamental details concerning this, as well as the formal definition, as problems. A few useful properties of convex quadrilaterals may be listed.

- The diagonals of a convex quadrilateral intersect at an interior point on each diagonal.

- If $\diamondsuit ABCD$ is a convex quadrilateral, then D lies in the interior of $\angle ABC$ (and similarly for the other vertices).

- If A, B, C, and D are consecutive vertices of a convex quadrilateral, then $m\angle BAD = m\angle BAC + m\angle CAD$.

CONGRUENCE CRITERIA FOR CONVEX QUADRILATERALS

A few basic congruence properties will be used repeatedly throughout the rest of this text.

DEFINITION

Two quadrilaterals $\diamondsuit ABCD$ and $\diamondsuit XYZW$ are **congruent** under the correspondence $ABCD \leftrightarrow XYZW$ iff all pairs of corresponding sides and angles under the correspondence are congruent (i.e., CPCF). Such congruence will be denoted by $\diamondsuit ABCD \cong \diamondsuit XYZW$. For a display of these details, see the following table.

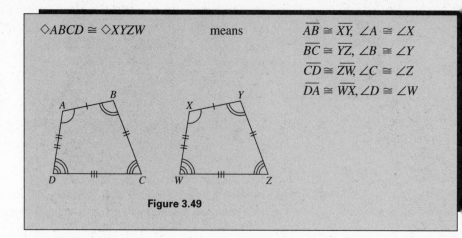

$\Diamond ABCD \cong \Diamond XYZW$ means

$\overline{AB} \cong \overline{XY}, \angle A \cong \angle X$
$\overline{BC} \cong \overline{YZ}, \angle B \cong \angle Y$
$\overline{CD} \cong \overline{ZW}, \angle C \cong \angle Z$
$\overline{DA} \cong \overline{WX}, \angle D \cong \angle W$

Figure 3.49

THEOREM 1: SASAS CONGRUENCE

Suppose that two convex quadrilaterals $\Diamond ABCD$ and $\Diamond XYZW$ satisfy the **SASAS Hypothesis** under the correspondence $ABCD \leftrightarrow XYZW$. That is, three consecutive sides and the two angles included by those sides of $\Diamond ABCD$ are congruent, respectively, to the corresponding three consecutive sides and two included angles of $\Diamond XYZW$. Then $\Diamond ABCD \cong \Diamond XYZW$.

Proof: Draw a pair of corresponding diagonals in the two quadrilaterals and use congruence criteria for triangles; complete the details as Problem 5.

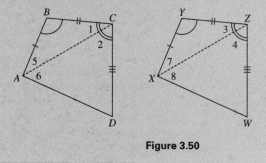

Figure 3.50

Other congruence theorems for convex quadrilaterals, symbolized in the usual manner, may also be derived (they also make interesting exercises for you):

• ASASA Theorem

• SASAA Theorem

• SASSS Theorem

QUESTION: Do you think ASAA is a valid congruence criterion for convex quadrilaterals?

OUR GEOMETRIC WORLD

Unlike triangles, quadrilaterals are unstable geometric designs. This is reflected in the fact that a quadrilateral can be made rigid only by fixing one of its angles, in addition to its four sides (the SASSS congruence criterion is valid, but SSSS is not). If a single brace is added (dotted line in Figure 3.51), then the whole assembly becomes (theoretically) stable.

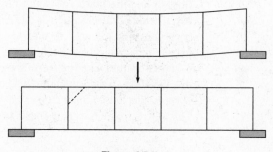

Figure 3.51

SACCHERI, LAMBERT QUADRILATERALS

Our main goal now is to develop the basic properties of the so-called **Saccheri Quadrilateral.** This figure is named after the first geometer to pursue seriously the consequences of a non-Euclidean hypothesis for parallels, Girolamo Saccheri (pronounced "Sack-er'-ee") (1667–1733), a Jesuit priest. The Saccheri Quadrilateral is about the nearest thing we can get to a rectangle in absolute geometry. In defining a rectangle, we have to be careful not to presume the Euclidean Postulate for Parallels; our definition must be meaningful in absolute geometry.

> **DEFINITION**
>
> A **rectangle** is a convex quadrilateral having four right angles.

If you think you can prove rectangles exist at this point, be careful—you are probably using properties of parallelograms we have not yet introduced. Once again, a good way to get a feel for what can or cannot be proven in absolute geometry is to consider spherical geometry. It's pretty clear you cannot draw a quadrilateral having four right angles on a sphere. Do you agree?

SACCHERI
QUADRILATERAL

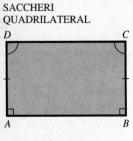

Figure 3.52

DEFINITION

Let $\overline{AB}$ be any line segment, and erect two perpendiculars at the endpoints A and B. (See Figure 3.52.) Mark off points C and D on these perpendiculars so that C and D lie on the same side of line $\overleftrightarrow{AB}$, and $BC = AD$. Join C and D. The resulting quadrilateral is a **Saccheri Quadrilateral**. Side $\overline{AB}$ is called the **base**, $\overline{BC}$ and $\overline{AD}$ the **legs**, and side $\overline{CD}$ the **summit**. The angles at C and D are called the **summit angles**.

To see what a Saccheri Quadrilateral looks like on a sphere, consider the equator (line ℓ) and the north and south poles (points N and S) in Figure 3.53. By taking two meridian lines through N and S perpendicular to the equator, and constructing congruent arcs $\overset{\frown}{AD}$ and $\overset{\frown}{BC}$ on those meridians meeting the equator at A and B, we may exhibit a Saccheri Quadrilateral $\Diamond ABCD$ in spherical geometry. The angles at C and D can be observed to be congruent. (Do you see that $\triangle NCD$ is isosceles, so that the supplementary angles at C and D are congruent?) In fact, it is not hard to observe that the summit angles of this Saccheri Quadrilateral are *obtuse angles*. This is true for all Saccheri Quadrilaterals on a sphere.

SACCHERI QUADRILATERAL
IN SPHERICAL GEOMETRY

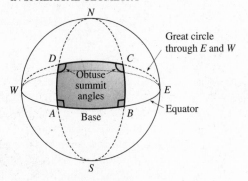

Figure 3.53

By reviewing the previously stated properties of convex quadrilaterals, and the fact that because of the uniqueness of perpendiculars from an external point in absolute geometry, lines $\overleftrightarrow{BC}$ and $\overleftrightarrow{AD}$ cannot meet, we conclude that

LEMMA

A Saccheri Quadrilateral is convex.

THEOREM 2

The summit angles of a Saccheri Quadrilateral are congruent.

Proof:

(1) In Figure 3.54, $\overline{DA} \cong \overline{CB}$, $\angle A \cong \angle B$, $\overline{AB} \cong \overline{BA}$, $\angle B \cong \angle A$ and $\overline{CB} \cong \overline{DA}$ (definition)

(2) $\Diamond DABC \cong \Diamond CBAD$, under the correspondence $DABC \leftrightarrow CBAD$. (SASAS Theorem)

(3) $\therefore \angle C \cong \angle D$. (CPCF)

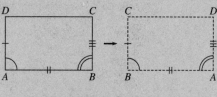

Figure 3.54

Numerous properties of Saccheri Quadrilaterals may be easily proven as corollaries to Theorem 2, which we simply list for possible additional exercises. They are all intuitively obvious if we appeal to geometric symmetry.

- The diagonals of a Saccheri Quadrilateral are congruent.

- The line joining the midpoints of the base and summit of a Saccheri Quadrilateral is the perpendicular bisector of both the base and summit.

- If each of the summit angles of a Saccheri Quadrilateral is a right angle, the quadrilateral is a rectangle, and the summit is congruent to the base.

NOTE: The existence of a common perpendicular bisector of the base and summit of a Saccheri Quadrilateral proves the existence of a quadrilateral in absolute geometry having *three* right angles, called a **Lambert Quadrilateral,** after J.H. Lambert (1728–1777), another early pioneer in the development of non-Euclidean geometry.

One further corollary is not a result of symmetry, and is more challenging to prove. (See Problem 13.)

COROLLARY

If the summit angles of a Saccheri Quadrilateral are acute, the summit has greater length than the base.

ATTEMPTS TO REPAIR
EUCLID'S ELEMENTS

Saccheri dreamed of providing a proof of Euclid's Parallel Postulate. His method was to investigate systematically the three possible cases for a given Saccheri Quadrilateral.

Hypothesis of the Obtuse Angle:	Summit angles of a Saccheri Quadrilateral are obtuse angles.
Hypothesis of the Right Angle:	Summit angles of a Saccheri Quadrilateral are right angles.
Hypothesis of the Acute Angle:	Summit angles of a Saccheri Quadrilateral are acute angles.

Saccheri titled his work *Euclides ab omne naevo vindicatus* ("Euclid freed of every flaw"), which was published in 1733, the year of his death. However, had he succeeded in his effort, namely, to eliminate the Hypothesis of the Acute Angle, he

$\mathcal{H}$ISTORICAL $\mathcal{N}$OTE

Saccheri and Lambert (pictured) were the forerunners of non-Euclidean geometry. They were the first to start with a hypothesis contrary to Euclidean geometry and derive results from that hypothesis. Having shown that the Hypothesis of the Obtuse Angle is impossible, Saccheri next assumed the Hypothesis of the Acute Angle, hoping to reach another contradiction. (This would have shown that the Hypothesis of the Right Angle was the only valid one.)

His work in 1733 established important facts about absolute geometry, some of which appear in this section,

but the sought-for contradiction never showed up. In desperation, thinking there had to be a contradiction, Saccheri appealed to the "intuitive nature of the straight line." Thirty years later, in 1766, Lambert undertook a similar investigation, revealing unusual flashes of insight. Among them was a plausibility argument showing that if the angle sums of triangles are not 180, then there exists an *absolute unit of measure*. This was to be proven 60 years later by both Bolyai and Lobachevski, the founders of true non-Euclidean geometry.

would have destroyed rather than repaired Euclidean geometry on logical grounds. For this would have meant that Euclidean geometry is self-contradictory due to the existence of models for non-Euclidean geometry which can be constructed with the Euclidean axioms. (In Chapter 6, we shall study one of those models in detail.)

ELIMINATING THE HYPOTHESIS OF THE OBTUSE ANGLE

We can prove with our present assumptions that the hypothesis of the obtuse angle cannot hold. To accomplish this, we introduce a special construction which makes every triangle correspond to a Saccheri Quadrilateral in such a way that the *angle sum of the triangle equals the sum of the measures of the two summit angles of the quadrilateral*. Since we know that the angle sum of a triangle is not greater than 180, this will disprove the hypothesis of the obtuse angle.

Let $\triangle ABC$ be given. Locate the midpoints M and N of sides $\overline{AB}$ and $\overline{AC}$, and drop perpendiculars from B and C to line $\overleftrightarrow{MN}$ at B' and C', respectively, as illustrated in Figure 3.55. Since B and C lie on the opposite side of line MN as A, they lie on the same side of $\overleftrightarrow{MN}$, and as in the case of the Saccheri quadrilateral, we can show that $\Diamond BCC'B'$ is convex. If we can show that $BB' = CC'$, then $\Diamond BCC'B'$ will be a Saccheri Quadrilateral—the **Saccheri Quadrilateral associated with $\triangle ABC$**.

SACCHERI QUADRILATERAL
ASSOCIATED WITH A TRIANGLE

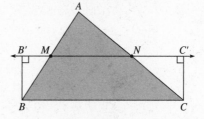

Figure 3.55

THEOREM 3

The quadrilateral $\Diamond BCC'B'$ (Figure 3.55) is a Saccheri Quadrilateral with base $\overline{B'C'}$ and summit $\overline{BC}$. Moreover, the angle sum of $\triangle ABC$ equals twice the measure of either summit angle of the quadrilateral, and $MN = \frac{1}{2}B'C'$.

Proof: Drop the perpendicular $\overline{AQ}$ to $\overleftrightarrow{MN}$. There are three cases.

   I. $\angle AMN$ and $\angle ANM$ are both acute (the case shown in Figure 3.56).

   II. $\angle AMN$ (or $\angle ANM$) is a right angle.

   III. $\angle AMN$ (or $\angle ANM$) is obtuse.

In these three cases, the ordering of the points on $\overleftrightarrow{MN}$ is, respectively

   I. B'-M-Q-N-C' (by the Exterior Angle Inequality)

   II. $B' = M = Q$ and M-N-C'

   III. Q-M-B'-C' and M-N-C'

THEOREM 3 (CONTINUED)

We shall give the proof of Case I only, and leave the other cases as a problem. (See Problem 8.)

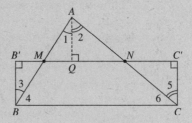

Figure 3.56

(1) $\overline{AM} \cong \overline{MB}$, $\angle AMQ \cong \angle BMB'$, $\overline{AN} \cong \overline{NC}$, and $\angle ANQ \cong \angle CNC'$. (definition of midpoint, Vertical Pair Theorem)

(2) $\triangle AQM \cong \triangle BB'M$ and $\triangle AQN \cong \triangle CC'N$. (HA Theorem)

(3) $\overline{BB'} \cong \overline{AQ} \cong \overline{CC'}$. $\therefore \lozenge BCC'B'$ is a Saccheri Quadrilateral. (CPCF, Transitive Law)

(4) $B'M = MQ$, $QN = NC'$. Then B'-M-Q-N-C' implies that $B'C' = B'M + MQ + QN + NC' = 2MQ + 2QN = 2MN$, and $\therefore MN = \frac{1}{2}B'C'$.

(5) $m\angle MAQ = m\angle B'BM$ and $m\angle QAN = m\angle NCC'$. (CPCF)

(6) $Q \in$ Interior $\angle MAN$, $M \in$ Interior $\angle B'BC$, and $N \in$ Interior $\angle BCC'$. (M and B' lie on the same side of $\overleftrightarrow{BC}$, and M and C lie on the same side of $\overleftrightarrow{BB'}$.)

(7) $\therefore$ Angle sum of $\triangle ABC = m\angle 1 + m\angle 2 + m\angle 4 + m\angle 6 = m\angle 3 + m\angle 5 + m\angle 4 + m\angle 6 = m\angle B'BC + m\angle C'CB = 2 \cdot m\angle B'BC$. (Angle Addition Postulate, algebra)

As a direct result of this construction and the preceding theorem, we find that since $m\angle B'BC = \frac{1}{2} \cdot$ angle sum of $\triangle ABC$,

- The summit angles of a Saccheri Quadrilateral in absolute geometry are either acute angles or right angles (Saccheri-Legendre Theorem).

- The summit of a Saccheri Quadrilateral in absolute geometry has length greater than or equal to that of the base. (See Problem 13.)

We state a third interesting corollary which results from the fact that $MN = \frac{1}{2}B'C'$ in Figure 3.56. Since $B'C' \leq BC$ by the second corollary just mentioned, $MN \leq \frac{1}{2}BC$, or

- The line joining the midpoints of two sides of a triangle has length less than or equal to one-half that of the third side.

This result is the best we can do in absolute geometry. While we are quite familiar with the Euclidean result $MN = \frac{1}{2}BC$, this requires properties of parallelograms, not generally valid in absolute geometry.

PROBLEMS   (§3.7)

_____ *Group A* _____

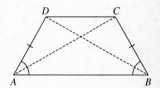

1. In quadrilateral $\Diamond ABCD$, $\overline{AD} \cong \overline{BC}$ and $\angle A \cong \angle B$.
   (a) Prove $\overline{AC} \cong \overline{BD}$.
   (b) Use this result to prove that the diagonals of a Saccheri Quadrilateral are congruent.

2. Prove that if the summit and base of a Saccheri Quadrilateral are congruent, the quadrilateral is a rectangle.

3. State the ASASA Congruence Theorem in terms of $\Diamond ABCD \leftrightarrow \Diamond XYZW$.

4. State the SASAA Congruence Theorem in terms of $\Diamond ABCD \leftrightarrow \Diamond XYZW$

5. Prove Theorem 1, the SASAS Congruence Theorem for convex quadrilaterals. (See Figure 3.50.)

6. In the figure, $\square ABCD$ is a rectangle, and segments $\overline{AE}$ and $\overline{DF}$ are the doubling of $\overline{AB}$ and $\overline{DC}$, respectively. Prove that $\Diamond AEFD$ is also a rectangle.

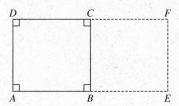

7. Write out an argument for the following indicated proposition.
   Given: $\Diamond ABCD$ is a Lambert Quadrilateral and $AB = BC$.
   Prove: $AD = CD$ and ray $\overrightarrow{BD}$ bisects $\angle ABC$.

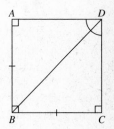

_____ *Group B* _____

8. Prove Theorem 3 for
   (a) Case II
   (b) Case III.

9. Prove that the opposite sides of a rectangle are congruent. (See definition for *rectangle*, given earlier.) (***HINT:*** Assume that one side, say $\overline{AB}$, has length less than that of the opposite side, $\overline{CD}$. What construction can you make? Use the Exterior Angle Inequality.)

10. Prove the ASASA Congruence Theorem for convex quadrilaterals.

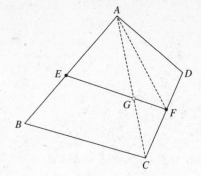

11. Prove that the segment joining any two points of opposite sides of a convex quadrilateral forms two adjacent convex quadrilaterals. (Prove that the new quadrilaterals have the property that each diagonal lies between opposite vertices. See figure to the right.)

12. Prove the SASSS Theorem for convex quadrilaterals.

13. Give a complete proof of the related corollaries of Theorem 2 and Theorem 3, that the line segment joining the midpoints of two sides of a triangle has length less than or equal to half that of the third side ($MN = \frac{1}{2}BB' \leq \frac{1}{2}BC$ in Figure 3.56). (*HINT:* In the following figure, show that $m\angle 1 + m\angle 2 = 90 \geq m\angle 2 + m\angle 3 \rightarrow m\angle 1 \geq m\angle 3$; apply the Hinge Theorem to $\triangle CAD$ and $\triangle ACB$. Prove that $AB \leq CD$.)

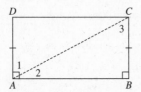

14. Show that, unlike triangles, specifying the lengths of four sides does not uniquely determine a quadrilateral. Give a specific example. (Thus the saying, *quadrilaterals are not rigid* becomes meaningful.)

15. Show that, unlike triangles, specifying the angle measures and the length of one side does not uniquely determine a quadrilateral. (Hence there is no ASAAA Congruence Criterion.)

16. Prove that the line joining the midpoints of the summit and base of a Saccheri Quadrilateral is the perpendicular bisector of the summit and base.

17. Determine whether ASAASA denotes a valid congruence criterion for convex quadrilaterals.

Group C

18. Determine whether SSASA is a valid congruence criterion for convex quadrilaterals.

19. Determine whether SSSAA is a valid congruence criterion for convex quadrilaterals.

20. **UNDERGRADUATE RESEARCH PROJECT** Write a paper on all the possible congruence criteria for convex quadrilaterals

   (a) in absolute geometry

   (b) in Euclidean geometry (involving, for example, Problem 23, § 4.2).

3.8 CIRCLES

Many familiar properties of circles carry over to absolute geometry. We need a definition to introduce the subject formally.

> **DEFINITION**
>
> A **circle** is the set of points in a plane which lie at a positive, fixed distance r from some fixed point O. The number r is called the **radius** (as well as any line segment joining point O to any point on the circle), and the fixed point O is called the **center** of the circle. A point P is said to be **interior** to the circle, or an **interior point**, whenever $OP < r$; if $OP > r$, then P is said to be an **exterior point**.

The numerous other terms commonly associated with circles, such as **diameter**, **chord**, and **tangent**, *will be defined by a pictorial glossary.* From this you should be able to write formal definitions of these terms whenever necessary. (See Figure 3.57.)

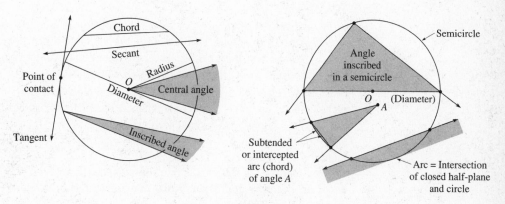

Figure 3.57

THE ELEMENTARY PROPERTIES

All the basic theorems for circles may be proven with ease by the use of the congruence theorems given so far, so most of these results will be left as problems. A few examples are stated as follows.

- The center of a circle is the midpoint of any diameter.

- The perpendicular bisector of any chord of a circle passes through the center.

- A line passing through the center of a circle and perpendicular to a chord bisects the chord.

- Two congruent central angles subtend congruent chords, and conversely.

- Two chords equidistant from the center of a circle have equal lengths, and conversely.

E X A M P L E   1   Prove the property of circles that two congruent central angles subtend congruent chords, and conversely.

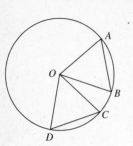

Figure 3.58

SOLUTION

Let the circle be as shown in Figure 3.58. By hypothesis, we have $\angle AOB \cong \angle COD$ and must prove that $\overline{AB} \cong \overline{CD}$. But, by the definition of a circle, $\overline{OA} \cong \overline{OB} \cong \overline{OC} \cong \overline{OD}$, so $\triangle AOB \cong \triangle COD$ by SAS. Hence $\overline{AB} \cong \overline{CD}$ by CPCF. For the converse, we are given that $\overline{AB} \cong \overline{CD}$ and must prove that $\angle AOB \cong \angle COD$. This time SSS applies to prove $\triangle AOB \cong COD$, and therefore $\angle AOB \cong \angle COD$.❏

There are, however, a few fundamental facts about Euclidean circles which are not provable in absolute geometry.

EXAMPLE 2   On the sphere, find an example of a circle and two intersecting chords cutting off segments of lengths a and b on one chord, and c and d on the other, such that $ab \neq cd$.

SOLUTION

For convenience, we are first going to look for an example of a right triangle violating the Pythagorean Theorem. Observe the three-dimensional coordinate system as indicated in Figure 3.59. The unit sphere with center O has equation $x^2 + y^2 + z^2 = 1$, and we locate points $P(1, 0, 0)$ and $Q(\sqrt{1/2}, \sqrt{1/2}, 0)$ on the equator (thus, $m\angle POQ = 45$ and arc $\overset{\frown}{PQ} = \pi/4$), and point $R(\sqrt{1/2}, 0, \sqrt{1/2})$ in the xz-plane halfway up to the "north pole" (hence, $m\angle POR = 45$ and arc $\overset{\frown}{PR} = \pi/4$). From the distance formula for xyz-space, $QR = \sqrt{0 + 1/2 + 1/2} = 1$, which means $m\angle QOR = 60$ or arc $\overset{\frown}{QR} = \pi/3$. Thus, if $r =$ arc $\overset{\frown}{QR}$, $a =$ arc $\overset{\frown}{PR}$, and $s =$ arc $\overset{\frown}{PQ}$, then $r^2 \neq a^2 + s^2$.

Now observe the spherical circle shown in the figure with radius r, the spherical chord $\overline{AB}$ of length $2a$, perpendicular to diameter $\overline{CD}$ of length $c + d$, where $c = r - s$ and $d = r + s$. Then, with $a = b$,

$$ab = a^2 \quad \text{and} \quad cd = (r-s)(r+s) = r^2 - s^2$$

Hence

$$ab = a^2 \neq r^2 - s^2 = cd❏$$

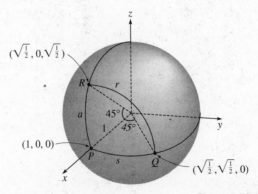

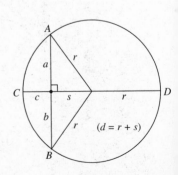

Figure 3.59

CIRCULAR ARC MEASURE

A natural idea in geometry, and a useful one, is that of circular arc measure and its property of additivity (the measure of the union of two nonoverlapping arcs equals the sum of their measures). This can be easily introduced into absolute geometry in the conventional way. This development will be used later when we want to establish the properties of inscribed angles of circles in Euclidean geometry (Chapter 4).

DEFINITION

As shown in Figure 3.60, consider the three types of arcs of a circle with center O, along with their measures. A **minor arc** is the intersection of the circle with a central angle and its interior, a **semicircle** is the intersection of the circle with a closed half-plane whose edge passes through O, and a **major arc** of a circle is the intersection of the circle and a central angle and its exterior (that is, the complement of a minor arc, plus endpoints). If the endpoints of an arc are A and B, and C is any other point of the arc (which must be used in order to uniquely identify the arc), then we define the **measure** $m\widehat{ACB}$ of the arc as follows.

MINOR ARC	SEMICIRCLE	MAJOR ARC
$m\widehat{ACB} = m\angle AOB$	$m\widehat{ACB} = 180$	$m\widehat{ACB} = 360 - m\angle AOB$

THE THREE TYPES OF ARCS AND THEIR MEASURES

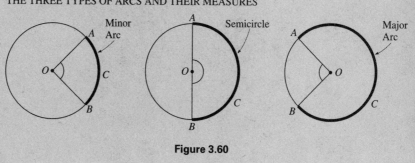

Figure 3.60

Having separate definitions for the measure of a major and minor arc poses a problem for proving the additivity property. The best way to handle it is to introduce coordinates. In Figure 3.61, circle O is given, and we choose an arbitrary, fixed half-plane H_1 with edge passing through O, determined by opposite rays $\overrightarrow{OP}$ and $\overrightarrow{OP'}$. The Protractor Postulate assigns the real numbers θ, $0 < \theta < 180$, to the rays $\overrightarrow{OA}$ and $\overrightarrow{OX}$ lying in H_1, and 0 to ray $\overrightarrow{OP}$. Thus, in particular, if $\overrightarrow{OA}$ lies in H_1, $\overrightarrow{OA}$ is assigned the number a such that $m\angle AOP = a$. Now consider any ray lying in the opposite half-plane H_2, such as $\overrightarrow{OB}$ in Figure 3.61. We assign the *negative* of the real number which would have been assigned to $\overrightarrow{OB}$ by the Protractor Postulate applied to $\overrightarrow{OP}$ and all rays in H_2. That is, $\overrightarrow{OB}$ is assigned the number $b < 0$ such that

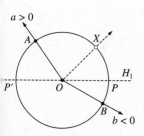

Figure 3.61

$b = -m\angle BOP$. Finally, we assign the number 180 to the ray $\overrightarrow{OP'}$, the ray opposite $\overrightarrow{OP}$. Thus we see that all rays with vertex O are assigned unique real numbers θ, where $-180 < \theta \le 180$

LEMMA

For any arc $\overset{\frown}{ACB}$ on circle O, if P' lies in the complementary arc of $\overset{\frown}{ACB}$ and $a > b$ are the coordinates of rays $\overrightarrow{OA}$ and $\overrightarrow{OB}$, respectively, relative to the half-planes determined by line $\overleftrightarrow{PP'}$, then

$$m\overset{\frown}{ACB} = a - b$$

The proof of this lemma is straightforward, but involves many tedious details. The three cases when arc $\overset{\frown}{ACB}$ is a minor arc, a semicircle, or a major arc must be treated separately. We choose to pass over these details at the present time.

THEOREM 1: ADDITIVITY OF ARC MEASURE

Suppose arcs $A_1 = \overset{\frown}{ADC}$ and $A_2 = \overset{\frown}{CEB}$ are any two arcs of circle O having just one point C in common, and such that their union, $A_1 \cup A_2 = \overset{\frown}{ACB}$, is also an arc. Then $m(A_1 \cup A_2) = mA_1 + mA_2$.

Proof (Figure 3.62): Choose point P' any point of the circle not on arc $\overset{\frown}{ACB}$, and take $\overrightarrow{OP}$ the opposite ray of $\overrightarrow{OP'}$ as the origin of a coordinate system for the rays from O. If the coordinates for $\overrightarrow{OA}, \overrightarrow{OB}$, and $\overrightarrow{OC}$ are a, b, and c, respectively, it is clear that c lies numerically between a and b (actual proof involves Theorem 1, §2.6, and a few other details involving betweenness for rays). We may assume without loss of generality that $b < c < a$. Then, by the lemma,

$$m\overset{\frown}{ADC} + m\overset{\frown}{CEB} = (a - c) + (c - b) = a - b = m\overset{\frown}{ACB}$$

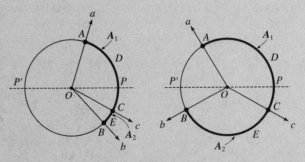

Figure 3.62

E X A M P L E   3 The angles $\angle ROT$, $\angle ROS$, and $\angle SOP$ have measures as indicated by Figure 3.63.

(a) Deduce the arc measures $m\overset{\frown}{RVS}$, $m\overset{\frown}{SPT}$, and $m\overset{\frown}{RST}$.

(b) Verify the additivity property for this case.

(c) Find the coordinates of rays $\overrightarrow{OR}$, $\overrightarrow{OS}$, and $\overrightarrow{OT}$.

(d) Verify the formula of the above lemma for the three arcs involved here.

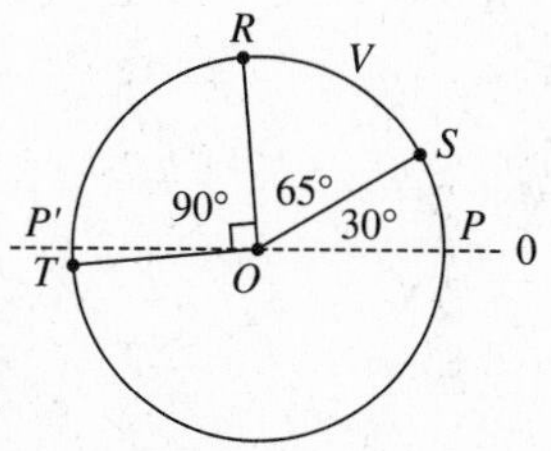

SOLUTION

(a) Observe that $m\angle TOS = m\angle TOR + m\angle ROS = 90 + 65 = 155$. By definition,

$$m\overset{\frown}{RVS} = 65,$$

$$m\overset{\frown}{SPT} = 360 - 155 = 205,$$

$$m\overset{\frown}{RST} = 360 - 90 = 270$$

(b) $m\overset{\frown}{RVS} + m\overset{\frown}{SPT} = 65 + 205 = 270$, in agreement with (a).

(c) The coordinates of rays $\overrightarrow{OR}$, $\overrightarrow{OS}$, and $\overrightarrow{OT}$ are found to be 95, 30, and −175, respectively.

(d) $m\overset{\frown}{RVS} = 95 - 30 = 65$, $m\overset{\frown}{SPT} = 30 - (-175) = 205$, and $m\overset{\frown}{RST} = 95 - (-175) = 270$, in agreement with (a). ❑

Figure 3.63

THE TANGENT THEOREM

We define formally both a secant and tangent of a circle at this time.

DEFINITION

A line which meets the circle in two distinct points is a **secant** of the circle. A line that meets a circle at only one point is called a **tangent** to that circle, and the point in common between them is the **point of contact or point of tangency**.

In elementary geometry, the obvious way to construct a tangent to a circle at a given point on that circle is to construct the perpendicular to the radius drawn to that point (Figure 3.64). The actual proof of this is simple enough to leave to your own inclinations. If you are unfamiliar with the proof, you may want to follow through the next Discovery Unit for that purpose before proceeding further.

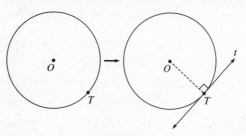

Figure 3.64

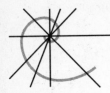

MOMENT FOR DISCOVERY

Constructing Tangents to Circles

We have a circle, a point A of that circle, and a line t passing through A. Let's explore what perpendicularity to the radius at A has to do with the tangency of line t.

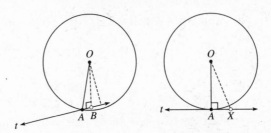

Figure 3.65

1. First, if t is *not* perpendicular to $\overrightarrow{OA}$, let $\overrightarrow{OB} \perp t$. Can you show that there must be a second point in common between the line and the circle? (Think symmetry.) A proof requires that you actually show how to locate such a point.

2. Then is t tangent to the circle in this case? (Use the definition, not your intuition.) What does this prove so far? (If t is not perpendicular to $\overrightarrow{OA}$, then ... ?)

3. Put the statement you found in Step 2 in the positive form (contrapositive). (If t *is* tangent to circle O at A, then ... ?)

4. Now suppose $t \perp \overrightarrow{OA}$. Let X be any other point on t besides A. Is there any definite, proven, relation between the distances OX and OA? What?

5. In this case, can t have any other point besides A in common with the circle? What have you discovered about t this time?

6. Based on what you have discovered, how may one characterize the tangent to a circle at a given point?

If you met with success in the preceding Discovery Unit, then the proof of the following theorem is clear.

THEOREM 2: TANGENT THEOREM

A line is tangent to a circle iff it is perpendicular to the radius at the point of contact.

One of the useful corollaries of the tangent theorem is the following.

COROLLARY

If two tangents $\overline{PA}$ and $\overline{PB}$ to a circle O from a common external point P have A and B as the points of contact with the circle, then $\overline{PA} \cong \overline{PB}$ (Figure 3.66).

(For the proof, just consider the radii $\overline{OA}$, $\overline{OB}$, and the line of center $\overleftrightarrow{PO}$.

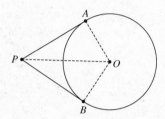

Figure 3.66

THE SECANT THEOREM

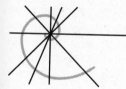

We shall concentrate our remaining efforts on a somewhat deeper result for absolute geometry. If a line passes through an interior point of a circle, it should be obvious that it must intersect that circle and be one of its secants. We may either take this as an additional axiom, or prove it as a theorem. We naturally prefer the latter approach.

It is interesting to explore this idea about a line and circle in more detail.

MOMENT FOR DISCOVERY

The Secant Theorem

If a line passes through an interior point of a circle, the question is whether that line can, perhaps, "roam around" inside the circle and never "get out". (See Figure 3.67.) Even if we could prove that there exist points on the line which are arbitrarily far from the center, we still have not proven that the line actually *intersects the circle*. Obviously, the line cannot just "slip" through the circle, can it? (Actually, it *can* if we are working in the rational plane. See Problem 13 in this connection.) To be specific, consider a circle of radius 5, with A on line ℓ inside the circle.

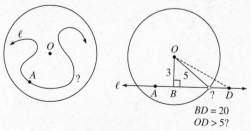

$$BD = 20$$
$$OD > 5?$$

Figure 3.67

1. Locate the foot of the perpendicular, B, on the given line, from A. In the example, $OB = 3$ units. How far from B should the points of intersection of the line and circle be? If this were the Euclidean plane, how might the Pythagorean Theorem be applied?

2. Since we cannot use the Pythagorean Theorem in absolute geometry, locate point D on ℓ so that $BD = 20$. Can we be certain that $OD > 5$? Why?

3. Generalize Step 2 to any circle of radius r and any line ℓ passing through any interior point A.

4. This gives us two points on ℓ, one in the interior of the circle (point A), and one exterior to it (point D). It remains to show that $\overline{AD}$ meets circle O.

The proof mentioned in the Discovery Unit (that $\overline{AD}$ meets circle O) involves, for the first time, an important, fundamental property of the real numbers, the property of **completeness**. The fact that a line segment has no "gaps" or "missing points" forces it to intersect the circle, and this is the intuitive meaning of *completeness*. It can be shown that this property of the line implies that of the existence of least upper bounds (the same concept you were introduced to in calculus). We shall remind you of the essentials of this concept in a moment.

We are then exploring a way to prove the so-called **Secant Theorem** in absolute geometry. The issue we have raised actually involves a very fundamental topological problem: Given a simple, closed curve C (that is, a curved or broken line, like a polygon with a million sides, which comes back to itself but never intersects itself to form a figure 8) and the arc L of another curve joining a point A inside C to a point B outside, *then must L intersect C?* The example in Figure 3.68 shows how difficult it is even to determine which points of the plane lie interior to C and which lie outside of C in certain cases. (This diagram consists of a map of the hedge that forms the famous maze in the gardens at Hampton Court Palace, England; *two separate closed curves* can be found!)

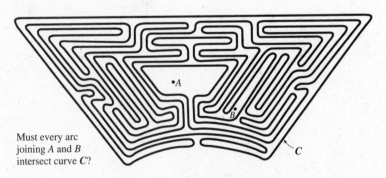

Must every arc joining A and B intersect curve C?

Figure 3.68

The tools of geometry do not seem, at first, to be powerful enough to handle such a problem. It took the clever thinking of C. Jordan (1838–1922) to solve it

His theorem is a landmark for topologists, called the **Jordan Closed Curve Theorem**, and it answers the question we asked in the affirmative.

In the case of a circle, we do have a well-defined interior and exterior, but we still cannot prove the desired property unless we use the least upper bound property of the real line, or some equivalent property. The **least upper bound property** asserts that every nonempty, bounded set *M* of real numbers (hence, points on a line) has a **least upper bound** (also called the **supremum**) which may be written

$$c = \sup M$$

and which satisfies the following properties.

(1) If $x \in M$, then $x \le c$ (that is, c is an upper bound of *M*).

(2) If $x \le u$ for all $x \in M$, then $u \le c$ (that is, c is least among all upper bounds u of *M*).

This property is often used in calculus when the sum of a bounded infinite series having positive terms is first considered and the property is invoked to ensure that such a series always converges.

THEOREM 3: SECANT THEOREM

If a line ℓ passes through an interior point *A* of a circle, it is a secant of the circle, intersecting the circle in precisely two points.

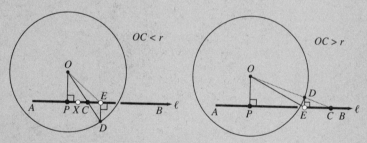

Figure 3.69

Proof:

(1) We are given that $OA < r$, where r is the radius of the circle. Let P be the foot of the perpendicular from O to line $\ell = \overrightarrow{AB}$. (See Figure 3.69.)

(2) For all points $X \in$ ray $\overrightarrow{PB}$, define the set of reals

$$M = \{PX : OX < r\}$$

By the Triangle Inequality, $PX \le PO + OX \le OA + OX < 2r$, so *M* is bounded and has a least upper bound, say $c = \sup M$.

(3) Choose C on ray $\overrightarrow{PB}$ such that $PC = c$.

(4) If $OC < r$, let $\overline{OD}$ be the radius through C. It is clearly possible, as indicated in Figure 3.69, to locate point E on ray $\overrightarrow{PB}$ such that P-C-E and $OE < OD = r$. But then the real number $p = PE$ belongs to *M*, yet $p > c$.
$\rightarrow\leftarrow$

(This would deny that c is an *upper bound* of *M*.)

THEOREM 3: SECANT THEOREM (CONTINUED)

(5) If $OC > r$, again locate points D and E as shown. Here we have $q = PE$ as an upper bound for M (proof?), yet $c > q$.→←

(This denies that c is the *least* upper bound.)

(6) The only remaining possibility is $OC = r$. Hence, C lies on both ℓ and the circle. The other point of intersection may be located by a simple construction, completing the proof.

PROBLEMS   (§3.8)

———— *Group A* ————

1. Prove that the perpendicular bisector of any chord of a circle passes through the center.

2. Prove that the perpendicular from the center of a circle to any chord bisects that chord.

3. Prove the following elementary facts about circles.

   (a) Two chords of a circle are congruent iff they are equidistant from the center of the circle.

   (b) Two chords of a circle are congruent iff they subtend arcs of equal measure.

4. (For this problem, you may use basic properties of similar triangles and the Pythagorean Theorem.) In the figure, the circles are 13 cm apart, having respective radii of 2 cm and 7 cm, and A and B are the points of contact of the common external tangent, while C and D are those of the common internal tangent. Find

   (a) AB

   (b) CD

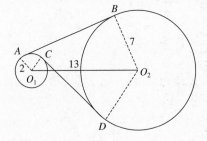

5. An equilateral octagon $PQRSTUVW$ is inscribed in a circle, as shown in the figure to the right. Prove that $\diamond PRTV$ is equilateral and equiangular. (Note that in Euclidean geometry, $\diamond PRTV$ would be a square.)

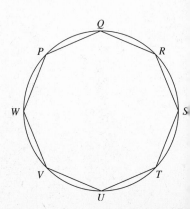

6. A circle passes through the vertices of $\Diamond ABCD$, and $AB = BC = CD$. (See figure to the right.) Prove that $m\angle B = m\angle C$. (**HINT:** Use (**b**) of Problem 3)

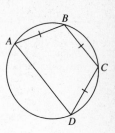

7. In the following figure, $\overrightarrow{PA}$ and $\overrightarrow{PB}$ are tangent lines at A and B, and O is the center of the circle. Prove that ray $\overrightarrow{PO}$ bisects $\angle APB$.

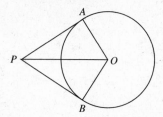

8. The Incircle of a Triangle A circle is called the **incircle** of a triangle iff it is tangent to all three sides (and, accordingly, lies inside the triangle). The radius and center of the incircle is called the **inradius** and **incenter** of the original triangle.

(**a**) Prove that the incircle of any triangle exists. (See Problem 13, §3.6.)

(**b**) Using the notation in the figure, where a, b, c is standard notation for the lengths of the sides of the triangle and $s = \frac{1}{2}(a + b + c)$ is the semiperimeter, deduce the formulas

$$x = s - a, \quad y = s - b, \quad z = s - c$$

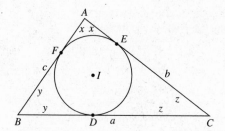

9. Two circles are tangent externally at C, with the common external and internal tangents $\overleftrightarrow{AB}$ and $\overleftrightarrow{CD}$ drawn. Prove that $\overleftrightarrow{CD}$ bisects $\overline{AB}$ at M, the point of intersection.

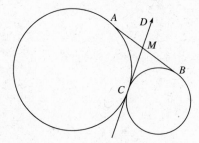

_____ *Group B* _____

10. Observe the Euclidean construction of the tangent to a circle from a given outside point. What well-known theorem in Euclidean geometry do you need in order to justify it? (If you do not know the answer, see §4.5.)

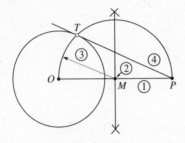

11. Use circles and circular arc measure to prove:

    Theorem: The sum of the measures of all the angles about a point is 360 provided those angles and their interiors contain the entire plane and are mutually disjoint.

12. Prove that the set of interior points of any circle is convex.

13. If it were possible for the **rational plane** to be a model for our geometry, then we could have a circle of rational radius and a line passing through an interior point of that circle which do not meet. To be specific, consider the coordinate plane with the restriction that the coordinates x, y of all points $P(x, y)$ must be rational numbers ($\sqrt{2}$, $\sqrt{3}$, π, etc. are disallowed). Consider the circle whose equation is $x^2 + y^2 = 25$ and the line whose equation is $y = 1$ (all variables x and y are rational). Show that this line and this circle can have no points in common.

_____ *Group C* _____

14. Prove the **Two Circle Theorem:** If circles O and O' having radii $r \geq r'$ respectively have their centers at a distance d apart, where $r - r' < d < r + r'$, then the circles will meet at two distinct points. (***HINT:*** First prove that circle O' has a diameter $\overline{AB}$ with A interior to circle O and B exterior to it. Then define the set of reals

$$M = \{m\angle XO'A : X \in \text{circle } O', OX < r\}$$

and work with $c = \sup M$.

15. **Converse of Triangle Inequality Theorem** Use the Two Circle Theorem (Problem 14) to prove in absolute geometry that if three numbers a, b, and c satisfy the (strict) Triangle Inequality (that is, $a + b > c$, $a + c > b$, and $b + c > a$), then there exists a triangle $\triangle ABC$ having a, b, and c as the lengths of its sides. Thus,

    Corollary: Given a segment $\overline{AB}$, there exists an equilateral triangle $\triangle ABC$ having $\overline{AB}$ as base.

The heart of absolute geometry is the development in this chapter resulting from the notion of congruent triangles. We saw at first, after looking at taxicab geometry, that in order for congruence to play a role it would be necessary to add one additional axiom—the SAS Postulate, Axiom C-1 (thus bringing the total number of axioms to 16 so far). The early consequences of the SAS Postulate were then established: the ASA Congruence Theorem and the standard results concerning isosceles triangles and related theorems on equidistant loci. The remaining congruence theorem SSS was proved as a consequence, using darts and kites.

The primary inequality theorem in absolute geometry is the Exterior Angle Inequality in triangles. Euclid's proof was revised to acceptable standards of rigor, which then led to many other important inequality theorems, notably the Triangle Inequality ($AB + BC > AC$ if A, B, and C are noncollinear), and the so-called Hinge or Alligator Theorem (in $\triangle ABC$, $m\angle A > m\angle B$ iff $BC > AC$). Further congruence criteria then result, including the very useful HL Theorem for right triangles.

Finally, quadrilaterals and circles were introduced and basic results proven. The congruence criterion SASAS for convex quadrilaterals was observed to be analogous to SAS for triangles, leading to an important concept, the Saccheri Quadrilateral having two consecutive right angles at the base and a line of symmetry at the midpoint of the base. This line of symmetry divides a Saccheri Quadrilateral into two congruent quadrilaterals, each having three right angles, which were studied by Lambert (and called Lambert Quadrilaterals). Saccheri's Hypothesis of the Acute Angle was seen to be equivalent to a "deficient" angle sum for all triangles (sum of angle measures < 180). The basic properties of circles were established, culminating in the Tangent/Secant Theorems: A line is tangent to a circle iff it is perpendicular to the radius at the point of contact, and a line is a secant (intersects the circle in two distinct points) iff it passes through an interior point of the circle.

Answer each of the following questions True (T) or False (F).

1. Absolute geometry is a study of the consequences of only the 16 axioms, or their equivalent, which have been introduced thus far.

2. In order for two triangles to be congruent, it is sufficient for two sides and an angle of one triangle to be congruent, respectively, to the corresponding two sides and angle of the other.

3. In order for two triangles to be congruent, it is sufficient for two angles and a side of one triangle to be congruent, respectively, to the corresponding two angles and side of the other.

4. In order for two right triangles to be congruent, it is sufficient for two sides of the first to be congruent to two sides of the second.

5. The Triangle Inequality states that for any three distinct points A, B, and C, $AB + BC \leq AC$, with equality only when A-B-C.

6. The SsA congruence criterion requires the congruent sides opposite the congruent angles to have greater length than the other pair of congruent sides.

7. Saccheri and Lambert were both early pioneers in non-Euclidean geometry, but are not credited for its discovery.

8. The existence of a rectangle involves extra geometric assumptions we have not yet made.

9. A line cannot be a secant of a circle if it is perpendicular to a radius of that circle.

10. In absolute geometry, the measure of an exterior angle of any triangle equals the sum of the measures of the opposite interior angles.

4

EUCLIDEAN GEOMETRY: TRIGONOMETRY, COORDINATES AND VECTORS

OVERVIEW

It is the purpose of this chapter to adopt the Parallel Postulate for Euclidean geometry and to develop the basic concepts of classical geometry—rectangles, regular polygons, and the circle theorems, including coordinates and vectors. All this is accomplished by adding to our list of 16 axioms one further axiom, the parallel postulate for Euclidean geometry. The effect which properties of parallelism have on absolute geometry is dramatic, as we shall see.

4.1   EUCLIDEAN PARALLELISM, EXISTENCE OF RECTANGLES

Before we get started, we need to establish what we mean by parallel lines. Euclid's Definition 23 in Book I of the *Elements* is quite descriptive:

Parallel lines are lines which, being in the same plane and being produced indefinitely in both directions, do not meet one another in either direction.

Our statement is similar, but will omit the unnecessary references to "extending a line" and "direction." Such features have already been taken care of by the axioms.

DEFINITION

Two distinct lines ℓ and m are said to be **parallel** (and we write $\ell \parallel m$) iff they lie in the same plane and do not meet (Figure 4.1).

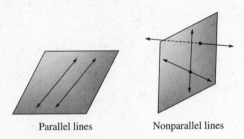

Parallel lines   Nonparallel lines

Figure 4.1

Another preliminary detail is introducing the usual terminology associated with the angles created when one line, a **transversal**, intersects two other lines (parallel or not). We let Figure 4.2 speak for itself—definitions using betweenness considerations can be formulated from the figure if need be.

Alternate interior angles   Corresponding angles   Interior angles on same side of transversal

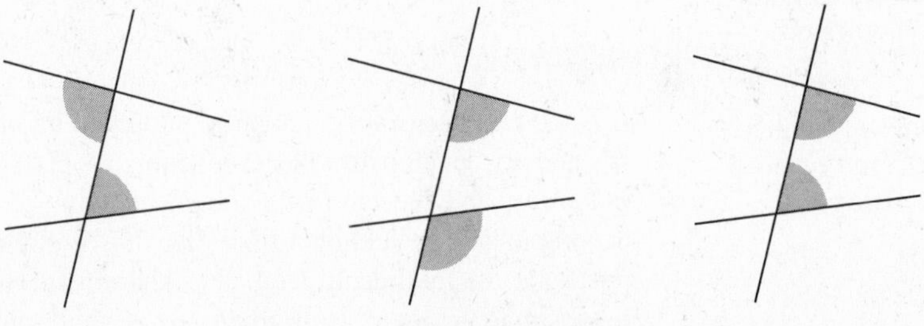

Figure 4.2

**A USEFUL FACT
ABOUT PARALLELS**

The following result was substantially the last theorem proved by Euclid before he began using his parallel postulate. Although it involves parallel lines, it in no way uses the parallel postulate and is valid in absolute geometry.

LEMMA: PARALLELISM IN ABSOLUTE GEOMETRY

If two lines in the same plane are cut by a transversal so that a pair of alternate interior angles are congruent, the lines are parallel.

Proof

Given: Lines ℓ and m, transversal t meeting line ℓ at A and m at B, and $\angle 1 = \angle 2$ (Figure 4.3).

Prove: $\ell \parallel m$.

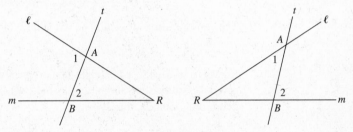

Figure 4.3

(1) Suppose that ℓ is not parallel to m and that, therefore, ℓ and m meet at some point R.

(2) Then $\angle 1$ (or $\angle 2$)is an exterior angle of $\triangle ABR$. (definition of exterior angle)

(3) $m\angle 1 > m\angle 2$ (or $m\angle 2 > m\angle 1$). →← (Exterior Angle Inequality)

(4) ∴ $\ell \parallel m$.

It is the *converse* of this proposition which is the Euclidean Parallel Postulate, a rather curious phenomenon. We now know that this converse cannot be proven, which is in itself of great significance. This fact is the legacy of a long, unprecedented struggle in mathematics. It represents the defeat of lifetime efforts of many first-rate mathematicians, including Gauss and Legendre, who attempted to prove the postulate, but failed. Euclid is obviously not among those who failed because he stated the property as a postulate and made no attempt to prove it. Whether this was a matter of luck, mere convenience, or deep insight on his part can only be speculated.

**EUCLID'S FIFTH
POSTULATE OF
PARALLELS**

We state Euclid's parallel postulate for later reference (as paraphrased).

EUCLID'S FIFTH POSTULATE OF PARALLELS

If two lines in the same plane are cut by a transversal so that a pair of interior angles on the same side of the transversal have a total measure less than 180, the lines will meet on that side of the transversal..

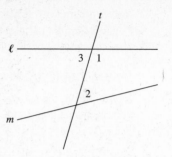

Figure 4.4

In effect, this says that in Figure 4.4:

If $m\angle 1 + m\angle 2 \neq 180$, then ℓ is not parallel to m.

The contrapositive, logically equivalent to this statement, is

If $\ell \parallel m$, then $m\angle 1 + m\angle 2 = 180$ (or $m\angle 2 = m\angle 3$).

In this form, Euclid's postulate is more easily spotted as an equivalent form of the parallel postulate we adopt, and as the *converse* of the above lemma (which stated that if $m\angle 2 = m\angle 3$, then $\ell \parallel m$).

THREE POSSIBLE NOTIONS OF PARALLELISM

Consider in a single fixed plane a line ℓ and a point P not on it (Figure 4.5). There are obviously three logical cases:

(1) There exists no line through P parallel to ℓ.
(2) There exists exactly one line through P parallel to ℓ.
(3) There exists more than one line through P parallel to ℓ.

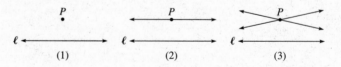

Figure 4.5

Which of these do you think is valid? The obvious answer, no doubt, is **(2)**. That is what most people who are only familiar with Euclidean geometry would say. But, in fact, both projective geometry and spherical geometry satisfy case **(1)**, and hyperbolic geometry satisfies case **(3)**, which we study later, in Chapter 6. Since we want to study Euclidean geometry in this and the following chapter, we shall adopt hypothesis **(2)** at this time, formally stated as our last axiom for Euclidean geometry.

AXIOM P-1: EUCLIDEAN PARALLEL POSTULATE

If ℓ is any line and P any point not on ℓ, there exists in the plane of ℓ and P one and only one line m that passes through P and is parallel to ℓ.

This form of the parallel postulate is due to John Playfair (1748–1819), an English mathematician who made important contributions to the foundations of geometry in several editions of a book on geometry first published in 1795.

NOTE: The remaining major results of this text will be restricted to a *single plane*, so we do not always make references to three-dimensional space in statements of definitions or theorems.

THEOREM 1

If two parallel lines are cut by a transversal, then either pair of interior angles on the same side of the transversal are supplementary, and any pair of alternate interior angles are congruent.

We are going to let you "discover" the validity of this result for yourself in the next discovery project.

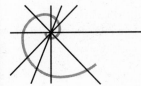

MOMENT FOR DISCOVERY

Using the Parallel Postulate

Consider two parallel lines, ℓ and m, and a transversal t, with alternate interior angles $\angle 2$ and $\angle 3$ in the figure. Apparently, $m\angle 2 = m\angle 3$. Suppose this is *not* true.

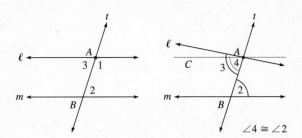

Figure 4.6

1. Assume $m\angle 3 > m\angle 2$ as shown in the second diagram. Then we may construct ray $\overrightarrow{AC}$ interior to $\angle 3$ such that $m\angle 4 \equiv m\angle BAC = m\angle 2$.

2. Since $\angle 2$ and $\angle 4$ are congruent alternate interior angles along transversal t, what does this tell us about lines $\overleftrightarrow{AC}$ and m? (Lemma?)

3. Does your conclusion in Step **(2)** give you a contradiction? (How many parallels to *m* passing through *A* are there?)

4. If $m\angle 3 < m\angle 2$, can you construct ray $\overrightarrow{AC'}$ so that $m\angle 4 = m\angle 2$ and get a contradiction for this case?

5. Does all this prove anything?

THE Z, F, AND U THEOREMS: TRANSVERSALS OF PARALLELISM

Theorem 1 has three important corollaries concerning parallel lines which can be more easily remembered (and applied) if we use suggestive terms to designate them. (Refer to Figure 4.7.) Their proofs will be left as problems.

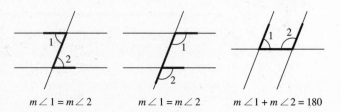

$m\angle 1 = m\angle 2$ $m\angle 1 = m\angle 2$ $m\angle 1 + m\angle 2 = 180$

Figure 4.7

COROLLARY A: THE Z PROPERTY

If two lines in the same plane are cut by a transversal, the two lines are parallel iff a pair of alternate interior angles are congruent.

COROLLARY B: THE F PROPERTY

If two lines in the same plane are cut by a transversal, the two lines are parallel iff a pair of corresponding angles are congruent.

COROLLARY C: THE U PROPERTY

If two lines in the same plane are cut by a transversal, the two lines are parallel iff a pair of interior angles on the same side of the transversal are supplementary.

In working geometry problems involving parallel lines, it is often helpful to look for configurations involving the letters Z, F, or U in the figures involved.

EXAMPLE 1   Prove in Euclidean geometry:

Euclidean Exterior Angle Theorem
If $\angle DCA$ is an exterior angle of $\triangle ABC$ (Figure 4.8), then $m\angle ACD = m\angle A + m\angle B$.

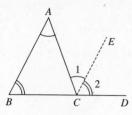

Figure 4.8

Proof:

(1) Construct $\overleftrightarrow{CE} \parallel \overleftrightarrow{BA}$ and choose E on the A-side of $\overleftrightarrow{BC}$. (Euclidean Parallel Postulate, Axiom P-1)

(2) By the Z Property, $m\angle 1 = m\angle A$.

(3) By the F Property, $m\angle 2 = m\angle B$.

(4) Since A and B lie on the same side of $\overleftrightarrow{CE}$, and B-C-D, A and D lie on opposite sides of $\overleftrightarrow{CE}$. Hence E lies on the D-side of $\overleftrightarrow{AC}$ and $E \in$ Interior $\angle ACD$.

(5) $\therefore m\angle ACD = m\angle 1 + m\angle 2 = m\angle A + m\angle B$, as desired.❏

COROLLARY D

If a line is perpendicular to one of two parallel lines, it is perpendicular to the other also.

Proof: Problem 17.

EXISTENCE OF RECTANGLES AND CONSEQUENT EFFECTS

Corollary C (the U Property of Parallelism) implies that rectangles (convex quadrilaterals having four right angles) exist in Euclidean geometry. In fact, every Saccheri Quadrilateral (and every Lambert Quadrilateral) is a rectangle, as the following reasoning shows: If $\lozenge ABCD$ is a Saccheri Quadrilateral with right angles at A and B (Figure 4.9), then by Corollary C, $\overleftrightarrow{AD} \parallel \overleftrightarrow{BC}$. Hence, again by Corollary C, the angles at C and D are supplementary. But those angles are also congruent by Theorem 2, Section 3.7, hence they are right angles and $\lozenge ABCD$ is a rectangle.

Practically the same argument shows that any Lambert Quadrilateral (a quadrilateral having three right angles) is also a rectangle. This is an interesting little exercise you might enjoy working out on your own. (See Problem 13.)

The next theorem may be proven in any number of ways. One way is to use the fact (proven in Section 3.7) that the angle sum of a triangle equals the sum of the measures of the summit angles of the associated Saccheri quadrilateral, which was just shown to be a rectangle. Or, we could use the Euclidean Exterior Angle Theorem (Example 1). We will leave this as Problem 20.

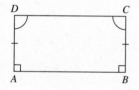

Figure 4.9

THEOREM 2

The sum of the measures of the angles of any triangle equals 180.

COROLLARY

The acute angles of a right triangle are complementary.

OUR GEOMETRIC WORLD

The great mathematician Carl F. Gauss, who in the 1800s was an official surveyor for the German government, once led an incredible expedition. He set out to test the Euclidean Hypothesis by measuring the angles of a triangle formed by the lines of sight between three distant mountain peaks. Beyond experimental error, no discrepancy from 180° was found. Of course, as Gauss himself realized, this experiment does not *prove* that our world is Euclidean, and no experiment of this type could ever prove it because of experimental error.

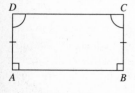

Figure 4.10

EXAMPLE 2   Prove that the summit and base of a Saccheri Quadrilateral are congruent (that is, prove $AB = DC$ in Figure 4.10).

SOLUTION

In Figure 4.10, there is given a Saccheri Quadrilateral $\diamond ABCD$, with $BC = AD$. Since $\diamond ABCD$ is a rectangle, the angles at C and D are also right angles, so $\diamond ABCD$ is a Saccheri Quadrilateral with $\overline{CD}$ as base. By the result of Problem 13, Section 3.7, the base of a Saccheri Quadrilateral is less than or equal to the summit. Hence we have $AB \le CD$ (with $\overline{AB}$ as base) and $CD \le AB$ (with $\overline{CD}$ as base). Therefore, $AB = DC.\ \square$

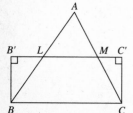

Figure 4.11

THEOREM 3: THE MIDPOINT CONNECTOR THEOREM

The segment joining the midpoints of two sides of a triangle has length one-half that of the third side and is parallel to it.

Proof: Recall that if L and M are the midpoints of $\overline{AB}$ and $\overline{AC}$ (Figure 4.12), then, by an earlier construction, line $\overleftrightarrow{LM}$ contains the summit of a Saccheri Quadrilateral $\diamond BCC'B'$ with base $\overline{B'C'}$, and it was proven that $LM = \frac{1}{2}B'C'$. (See Section 3.7.) Since Example 2 shows that $B'C' = BC$, we have $LM = \frac{1}{2}BC$. Since the base and summit of a Saccheri quadrilateral have a common perpendicular, $\overleftrightarrow{LM} \parallel \overleftrightarrow{BC}$ by Corollary C of Theorem 1.

COROLLARY

If a line bisects one side of a triangle and is parallel to the second, it also bisects the third side.

(To be proven as Problem 18.)

PROBLEMS   (§4.1)

_____ **Group A** _____

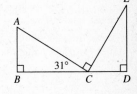

1. In the figure, $\overline{AB} \perp \overline{BD}$, $\overline{ED} \perp \overline{BD}$, and $\overline{AC} \perp \overline{CE}$. If $m\angle ACB = 31$, find the measures of the remaining nonright angles in the figure.

2. In the figure, $\Diamond ABCD$ and $\Diamond RSTU$ are rectangles. If $m\angle CBT = 10$, find the remaining nonright angles.

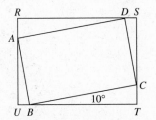

3. Given right triangle $\triangle ABC$ with angle measures as indicated in the figure, find x, y, and z.

4. Use Playfair's Postulate (Axiom P-1) and Corollary D to give another proof that the perpendicular to a line ℓ from a given external point A is unique.

5. In the slanted block letter A, $m\angle 1 = m\angle 2$. Prove: $m\angle 3 = m\angle 4$.

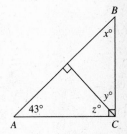

6. Prove from the theory in this section that parallel lines are everywhere equidistant.

7. Prove your choice of at least one of the corollaries A, B, or C of Theorem 2.

*8. Transitivity of Parallelism in Euclidean Plane Geometry Prove that in all cases, if $\ell \parallel m$ and $m \parallel n$, then $\ell \parallel n$. (You will need to assume that a line is *parallel to itself*.)

*See Chapter 6, §6.6.

9. In the figure at the right, $\triangle ABC$ is equilateral and $\overrightarrow{AE}$ bisects exterior angle $\angle CAD$. Prove that $\overrightarrow{AE} \parallel \overrightarrow{BC}$. Generalize this result.

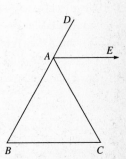

10. We are given that $\overrightarrow{UY}$ bisects $\angle WUV$ and $UW = WY$. Prove that $\overrightarrow{UV} \parallel \overrightarrow{YW}$.

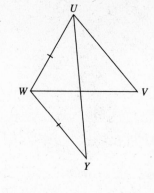

_____ **Group B** _____

11. If two angles have their respective sides parallel, the angles are either congruent or supplementary. Prove.

12. If two angles have their respective sides perpendicular, the angles are either congruent or supplementary. Prove.

13. Prove that if a convex quadrilateral has three right angles, then the fourth angle is a right angle and the quadrilateral is a rectangle.

14. Carefully draw (or construct) the internal bisector $\overrightarrow{AD}$ of $\angle BAC$ of $\triangle ABC$, and then construct the external bisector $\overrightarrow{AE}$ of exterior angle $\angle FAB$ at A. Does it appear that $\overrightarrow{AD} \perp \overrightarrow{AE}$? Replace your experimental observation by one grounded in logic. Is this result valid in absolute geometry?

15. Prove from the results of this section that if $PQ = PR = PS$ in the accompanying figure, then $\triangle RQS$ is a right triangle.

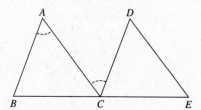

16. If $\triangle ABC \cong \triangle DCE$ and B-C-E, prove that $\overrightarrow{AB} \parallel \overrightarrow{CD}$.

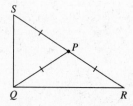

17. Prove Corollary D of Theorem 1 using the U property.

18. Prove the corollary to Theorem 3. (***HINT:*** If D is the midpoint of $\overline{AB}$ and $\overrightarrow{DE} \parallel \overrightarrow{BC}$, we must prove that E is the midpoint of $\overline{AC}$. If not, let F be that midpoint. Now apply Theorem 3 and obtain a contradiction to Axiom P-1.)

19. Prove that Euclid's Fifth Postulate is logically equivalent to Axiom P-1. You must prove

   (a) that Euclid's Postulate implies Axiom P-1,

   (b) that Axiom P-1 implies Euclid's Postulate.

   Of course, any of the results of this section are legitimate tools for this problem.

20. Prove Theorem 2 using your choice of methods.

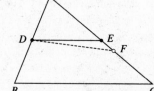

_____ **Group C** _____

21. Work the following problem in absolute geometry: If lines ℓ and m are cut by a transversal so that a pair of alternate interior angles are congruent, then ℓ and m have a common perpendicular. Do not use any of the results of this section that depend on the

Euclidean Parallel Postulate. (*HINT:* In the accompanying figure, the angles at *A* and *B* are congruent, as indicated. Drop perpendiculars from the midpoint *M* to ℓ and *m* at *C* and *D*. See Problem 11, §2.8.)

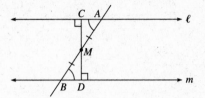

4.2 PARALLELOGRAMS AND TRAPEZOIDS: PARALLEL PROJECTION

We consider here the basic ratio-preserving property of parallel projections in Euclidean geometry, which follows directly from the basic properties of parallelograms and trapezoids.

PARALLELOGRAMS AND RHOMBI

The sequence of theorems in this section are elementary applications of the basic results of parallelism which appeared in the last section. After we prove the first of these, you should have no difficulty following suit. Our intention here is to cover only the highlights and major ideas in an area containing literally hundreds of intricate relationships and theorems.

> **DEFINITION**
>
> A convex quadrilateral ◇*ABCD* is called a **parallelogram** if the opposite sides $\overline{AB}$, $\overline{CD}$ and $\overline{BC}$, $\overline{AD}$ are parallel (Figure 4.12). A **rhombus** is a parallelogram having two adjacent sides congruent. A **square** is a rhombus having two adjacent sides perpendicular.

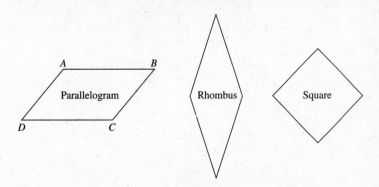

Figure 4.12

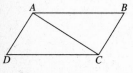

Figure 4.13

THEOREM 1

A diagonal of a parallelogram divides it into two congruent triangles.

Proof: (Figure 4.13) We want to show that $\triangle ABC \cong \triangle CDA$.

(1) $\angle BAC \cong \angle DCA$ (Z-Property of Parallelism)

(2) $\angle ACB \cong \angle CAD$ (Z-Property of Parallelism)

(3) $AC = AC$ (Reflexive Property of =)

(4) $\therefore$ $\triangle ABC \cong \triangle CDA$ (ASA)

The following familiar properties are all corollaries of the preceding theorem, or to each other. Any of them make good exercises (taken in the order given).

• The opposite sides of a parallelogram are congruent.

• The diagonals of a parallelogram bisect each other.

• If a convex quadrilateral has opposite sides congruent, then it is a parallelogram.

• If a convex quadrilateral has a single pair of opposite sides which are both parallel and congruent, it is a parallelogram.

• A parallelogram is a rhombus iff its diagonals are perpendicular.

• A parallelogram is a rectangle iff its diagonals are congruent.

• A parallelogram is a square iff its diagonals are both congruent and perpendicular.

OUR GEOMETRIC WORLD

Carpenters and construction crews routinely use one of the geometric principles in this section to serve as a check on their measurements. When a rectangular figure is called for, such as a doorway or cabinet top, the lengths of diagonals $\overline{AC}$ and $\overline{BD}$ are measured. If they are off by a half inch or more, workers know the angles are not true.

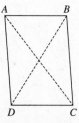

Figure 4.14

The following discovery project will lead you to a rather unexpected result concerning some of the above ideas.

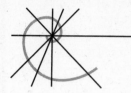

MOMENT FOR DISCOVERY

A Quadrilateral Within a Quadrilateral

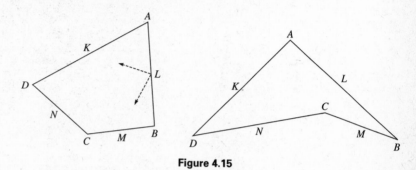

Figure 4.15

1. Draw an arbitrary quadrilateral ◇ABCD, one convex, and one nonconvex.
2. Locate, as carefully as you can, or by actual construction, the midpoints K, L, M, and N of the respective sides.
3. Join the points K, L, M, and N in order, forming a quadrilateral. Do you observe anything in particular?
4. Draw a kite ◇ABCD (with AB = AD and BC = CD), and locate the midpoints K, L, M, and N as before. Join these points, as before, forming ◇KLMN. Do you observe anything different? Can you prove what you observed?
5. Draw isosceles trapezoid ◇ABCD (see definition which follows) and draw ◇KLMN determined as before. Go one step further, and draw ◇PQRS, where P, Q, R, and S are the midpoints of the sides of ◇KLMN. Do you observe anything? Can you prove it?
6. What theorems have you discovered? Write a concise statement of each.

    (*NOTE:* You can say that you have *discovered* a theorem only if it is your idea and you can prove it.)

TRAPEZOIDS, MEDIANS, AND PARALLEL PROJECTIONS

DEFINITION

A **trapezoid** is a (convex) quadrilateral with at least two opposite sides parallel, called the **bases**, with the other two sides called the **legs**. (See Figure 4.16.) The segment joining the midpoints of the legs of a trapezoid is called the **median** (also the term used for the *line* passing through those midpoints). A trapezoid is said to be **isosceles** iff its legs are congruent *and it is not a parallelogram.*

PARTS OF A TRAPEZOID

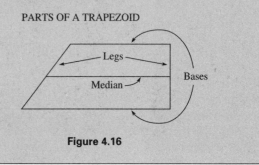

Figure 4.16

NOTE: Our definition of a trapezoid allows parallelograms to be trapezoids. This means that any result proven for a trapezoid is automatically true for a parallelogram, and we do not have to give separate proofs.

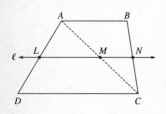

Figure 4.17

THEOREM 2: MIDPOINT-CONNECTOR THEOREM FOR TRAPEZOIDS

If a line segment bisects one leg of a trapezoid and is parallel to the base, then it is the median and its length is one-half the sum of the lengths of the two bases. Conversely, the median of a trapezoid is parallel to the bases.

Proof: (Figure 4.17) Let line ℓ bisect leg $\overline{AD}$ at L and be parallel to base $\overline{CD}$; hence $\ell \parallel \overleftrightarrow{AB}$ by definition of trapezoid and Transitivity of Parallelism (Problem 8, Section 4.1). Construct diagonal $\overline{AC}$.

(1) ℓ bisects $\overline{AC}$ at some point M. (corollary to Midpoint Connecter Theorem)

(2) $\therefore$ ℓ bisects $\overline{BC}$ at some point N (same reason).

(3) $LM = \frac{1}{2}DC$ and $MN = \frac{1}{2}AB$ (Midpoint Connector Theorem)

(4) L-M-N (ray $\overrightarrow{AC} \subseteq$ Interior $\angle BAD$)

(5) $\therefore$ $LN = LM + MN = \frac{1}{2}(AB + DC)$ (Steps (3) and (4))

(The converse will be left as Problem 16.)

OUR GEOMETRIC WORLD

The base of the Statue of Liberty has the shape of a **truncated square pyramid** (a square pyramid with the top cut off). The outer edge of each lateral face is an isosceles trapezoid.

The basis for ideas about ratio and proportion needed for the theory of similar polygons is an extension of Theorem 2. A certain mapping from one line to another will play a prominent role in later developments.

In Figure 4.18, we are given two lines, ℓ and m, and arbitrary points A and A' are located on the two lines. We set up a correspondence $P \leftrightarrow P'$ between the points of ℓ and m by requiring $\overleftrightarrow{PP'} \parallel \overleftrightarrow{AA'}$ for all P on ℓ. It is clear that this mapping is **one-to-one** (each distinct P and Q on ℓ leads to distinct **images** P' and Q' on m). This correspondence has a number of **invariant properties** (properties that are unchanged by the mapping). One of these is betweenness on ℓ (obvious because parallel lines cannot cross), and the other, which we want to focus on here, is *ratios of segments* on ℓ. We claim that, in general,

$$\frac{PQ}{PR} = \frac{P'Q'}{P'R'}$$

PARALLEL PROJECTION

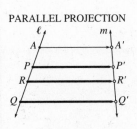

Figure 4.18

It suffices to prove that in a trapezoid $\Diamond ABCD$, if E lies on $\overleftrightarrow{AB}$ and F on $\overleftrightarrow{CD}$ such that $\overleftrightarrow{EF} \parallel \overleftrightarrow{BC}$, then $AE/AB = DF/DC$ (Figure 4.19). Our method of proof is to *assume that $\overleftrightarrow{EF}$ is not parallel to $\overleftrightarrow{BC}$*, then show that $AE/AB \neq DF/DC$. This is the contrapositive of what we are trying to prove, and hence is logically equivalent to it.

LEMMA

If $\overleftrightarrow{EF}$ is **not parallel to** $\overleftrightarrow{BC}$, then $AE/AB \neq DF/DC$.

Proof: Construct the parallels $\overleftrightarrow{EH}$ and $\overleftrightarrow{GF}$ to $\overleftrightarrow{BC}$ through E and F, respectively. This forms either a Z or backward Z depending on whether A-G-E or A-E-G (the case actually shown in Figure 4.19). Let's assume for sake of argument that A-E-G holds. Hence, since betweenness is preserved under parallel projection, D-H-F.

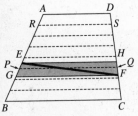

Figure 4.19

(1) Bisect segments $\overline{AB}$ and $\overline{DC}$, then bisect the segments which these midpoints determine, and continue the bisection process indefinitely. It is clear that at some stage we will have one of these midpoints, say P, falling on segment $\overline{EG}$, and, correspondingly, midpoint Q falls on $\overline{HF}$.[1] (An illustration of this is shown in Figure 4.19, where it takes *three* bisections to achieve the desired point P.)

(2) The lines joining corresponding midpoints are parallel to $\overleftrightarrow{BC}$ and $\overleftrightarrow{AD}$. (Midpoint Connector Theorem for Trapezoids)

(3) For some integer k, $AP = k \cdot AR = k \cdot AB/2^n$ (n = number of bisections required). Since betweenness is preserved, $DQ = k \cdot DS = k \cdot DC/2^n$.

(4) $\therefore$ $AP/DQ = AB/DC$, or $AP/AB = DQ/DC$ (algebra)

(5) Also, by betweenness, $AE < AP$ and $DQ < DF$, so

$$\frac{AE}{AB} < \frac{AP}{AB} = \frac{DQ}{DC} < \frac{DF}{DC}$$

(6) The case A-G-E leads to the result $AE/AB > DF/DC$ by the same argument.

(7) $\therefore$ $AE/AB \neq DF/DC$.

COROLLARY

If $AE/AB = DF/DC$, then $\overleftrightarrow{EF} \parallel \overleftrightarrow{BC}$.

THEOREM 3: PARALLEL PROJECTION THEOREM

The mapping from ℓ to m described previously preserves ratios of line segments.

Proof: We need to prove that in Figure 4.20, if $\overleftrightarrow{EF} \parallel \overleftrightarrow{BC}$, then $AE/AB = DF/DC$. Locate F' on $\overline{DC}$ such that $DF' = DC \cdot (AE/AB)$, or $AE/AB = DF'/DC$, and construct line $\overleftrightarrow{EF'}$. By the preceding corollary, $\overleftrightarrow{EF'} \parallel \overleftrightarrow{BC}$. But $\overleftrightarrow{EF} \parallel \overleftrightarrow{AD}$ by hypothesis, hence $\overleftrightarrow{EF} = \overleftrightarrow{EF'}$. (Why?) Therefore, $F' = F$ and $AE/AB = DF/DC$, as desired.

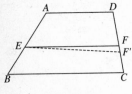

Figure 4.20

---

[1]This depends on the so-called **Archimedean Principle** of the real numbers, which states that there exists an integer n such that $n \cdot EG > AB$, or, since $2^n > n$, $2^n \cdot EG > AB$, and $AB/2^n < EG$.

NOTE: The above proofs apply also to triangles. If $A = D$ and $\diamond ABCD = \triangle ABC$, the above proof is still valid and we have the result that if $\overrightarrow{EF} \parallel \overrightarrow{BC}$, then $AE/AB = DF/DC \equiv AF/AC$. This, of course, is the key to ratio properties in similar triangles which will be dealt with in the following section. Many elementary texts take this property as axiomatic, but we have shown that is not necessary to do so.

COROLLARY: THE SIDE-SPLITTING THEOREM

If a line parallel to the base $\overline{BC}$ of $\triangle ABC$ cuts the other two sides $\overline{AB}$ and $\overline{AC}$ at E and F, respectively, then

$$\frac{AE}{AB} = \frac{AF}{AC}$$

and by algebra

$$\frac{AE}{EB} = \frac{AF}{FC}$$

(See Figure 4.21.)

Figure 4.21

PROBLEMS   *(§4.2)*

_____ *Group A* _____

1. Prove that a rhombus is an equilateral quadrilateral.

2. Is an equilateral quadrilateral a rhombus? Prove or disprove.

3. Prove that the diagonals of a rhombus are perpendicular.

4. If the diagonals of a parallelogram are perpendicular, the parallelogram is a rhombus. Prove.

5. In the accompanying figure, $\diamond WXYZ$ is an isosceles trapezoid with base angles X and Y (Problem 19 shows that $\angle X \cong \angle Y$). If $m\angle X = -2q + 71$ and $m\angle Y = -5q + 32$, find $m\angle X$.

   [UCSMP, p. 239]

6. You have a given line segment $\overline{AB}$ of length > 6 in. drawn on an ordinary $8\frac{1}{2}$ by 11 in. sheet of lined paper, and you want to use the lines of the paper to divide it into 5 congruent segments. How should you proceed? Prove your method in terms of the properties given in this section.

7. Prove that a parallelogram is a square iff its diagonals are both congruent and perpendicular.

8. In trapezoid $\diamond ABCD$, we are given $AB = AD = DC$. Find $m\angle DBC$ in terms of θ and show that $\overrightarrow{BD}$ bisects $\angle ABC$.

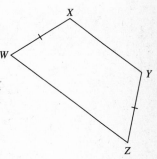

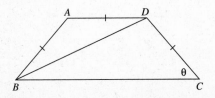

9. The diagonals of a rhombus have lengths 10 and 24. Find the lengths of the sides. (You may use the Pythagorean Theorem.)

10. In the following figure, the lines ℓ, m, n, p, and q are parallel, cutting off segments as shown. Find x, y, and z.

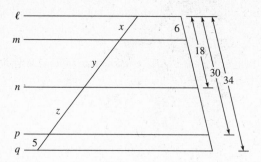

_____ *Group B* _____

11. In the accompanying figure, two congruent isosceles triangles having angles of measure 30, 75, and 75, respectively, are placed end-to-end at point B, with the two congruent sides opposite B lying on parallel lines. Lines $\overleftrightarrow{CD}$ and $\overleftrightarrow{AE}$ are drawn. Prove:

(a) $\triangle CBD$ is an equilateral triangle

(b) $\Diamond ACDE$ is a square.

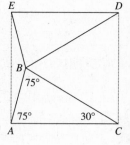

12. If two opposite sides of a convex quadrilateral are congruent and parallel, prove that the quadrilateral is a parallelogram.

13. Perpendiculars $\overline{AE}$ and $\overline{CF}$ are dropped from the vertices to the diagonal $\overleftrightarrow{BD}$ of parallelogram $\Diamond ABCD$. Prove: $AE = CF$.

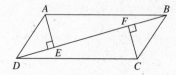

14. Euclidean Construction of Parallel Lines Given line ℓ and point A not on it, explain, with proofs, how the sequence of four steps illustrated in this figure will produce the unique parallel m to ℓ through A.

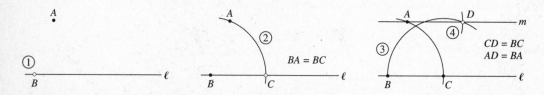

15. Give a Euclidean construction that will trisect a given line segment. (Use the Parallel Projection Theorem.)

16. Prove the converse part of the Midpoint Connector Theorem for Trapezoids: The median of a trapezoid is parallel to either base. (***HINT:*** Suppose the median is *not* parallel to the base; construct a line passing through the midpoint of one leg that is parallel to the base. What must happen?)

17. If two sets of parallel lines at equal distances apart are superimposed on one another, are rhombi formed, or just parallelograms? Prove your answer.

18. Prove the Side-Splitting Theorem (Corollary to Theorem 3). (***HINT:*** $AB/AE = 1 + EB/AE$.)

19. The base angles of an isosceles trapezoid are congruent, and conversely, if the base angles of a trapezoid are congruent it is isosceles. Prove.

20. Prove: The diagonals of an isosceles trapezoid are congruent. (See Problem 19.)

———— *Group C* ————

21. Examine, with proofs, the validity of the following conjecture in
   (a) Euclidean geometry
   (b) absolute geometry.

   Three perpendiculars to a line ℓ and equally spaced will cut off congruent segments on every line they intersect.

   (You may assume that if the Euclidean Parallel Postulate be denied then *no rectangles exist.*)

22. In a previous construction in absolute geometry (§3.4), a sequence of triangles *ABC*, *BCE*, *BCF*, ... was constructed on a common base $\overline{BC}$, such that the pairs of segments $\overline{BE}$ and $\overline{AC}$, $\overline{BF}$ and $\overline{CE}$, ... bisect each other at *L, M, N, O,* ... Prove that the two sets of points {*A, E, F, G,* ...} and {*L, M, N, O,* ...} lie on parallel lines. (Note that this is true only in Euclidean geometry.)

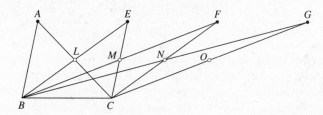

23. State and prove the SSAAA Congruence Criterion for convex quadrilaterals in Euclidean geometry.

4.3   SIMILAR TRIANGLES, PYTHAGOREAN THEOREM, TRIGONOMETRY

The development of Euclidean geometry will continue. Here we consider Euclidean similarity, the Pythagorean Theorem, and a development of the (Euclidean) trigonometry of the triangle.

> **DEFINITION**
> Two polygons P_1 and P_2 are said to be **similar**, denoted $P_1 \sim P_2$, iff under some correspondence of their vertices, corresponding angles are congruent, and the ratio of the lengths of corresponding sides is constant $(= k)$. The number k is called the **constant of proportionality**, or **scale factor**, for the similarity.

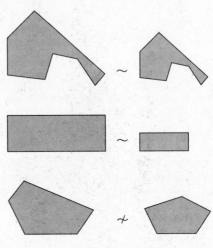

Figure 4.22

SIMILAR POLYGONS

The diagrams in Figure 4.22 illustrate a few pairs of similar polygons, and one pair that is not. The basic idea of similarity, as opposed to congruence, is that of **shape**. Polygons having both the same size and shape are, of course, congruent. But they are merely similar if they have the same shape, but not the same size.

The same care that was necessary for correspondence in a congruence will also be needed for similarity. Thus, if k is the constant of proportionality, then

$$\triangle ABC \sim \triangle XYZ \qquad \text{iff} \qquad \begin{aligned} \angle A &\cong \angle X, & AB &= k \cdot XY \\ \angle B &\cong \angle Y, & BC &= k \cdot YZ \\ \angle C &\cong \angle Z, & AC &= k \cdot XZ \end{aligned}$$

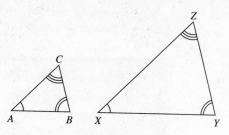

Figure 4.23

(Note that $k = 1$ corresponds to congruent triangles.)

The first similarity criterion we consider is basic, and all others depend on it: The AA Similarity Theorem. You will be given a chance to "discover" it.

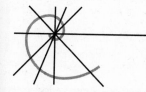

MOMENT FOR DISCOVERY

Exploring a Criterion for Similar Triangles

Suppose that under the correspondence $ABC \leftrightarrow XYZ$ we have $\angle A \cong \angle X$, $\angle B \cong \angle Y$, and, therefore, $\angle C \cong \angle Z$.

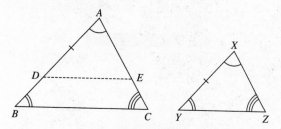

Figure 4.24

1. If $AB = XY$, what must be true regarding the lengths of the remaining two sides of the triangles?

2. Assume $AB > XY$. Locate D on $\overline{AB}$ such that $AD = XY$. What will make $AD/AB = AE/AC$ for E on $\overline{AC}$?

3. Is $\triangle XYZ \cong \triangle ADE$? Can you see how to prove $XY/AB = XZ/AC$ for this case?

4. What about the case $AB < XY$? What is your conclusion in general? (State this as a general lemma.)

5. Does your lemma imply that $XZ/AC = YZ/BC$?

6. What have you proven now, in general?

If you were successful in the preceding Discovery Unit, then the proof of the following theorem is evident.

THEOREM 1: AA SIMILARITY CRITERION

If, under some correspondence, two triangles have two pairs of corresponding angles congruent, the triangles are similar under that correspondence.

The remaining similarity theorems are now mere exercises, so we state them without proof and leave them as problems. (See Problems 14 and 15.)

THEOREM 2: SAS SIMILARITY CRITERION

If in $\triangle ABC$ and $\triangle XYZ$ we have $AB/XY = AC/XZ$ and $\angle A \cong \angle X$, then $\triangle ABC \sim \triangle XYZ$.

THEOREM 3: SSS SIMILARITY CRITERION

If in $\triangle ABC$ and $\triangle XYZ$ we have $AB/XY = BC/YZ = AC/XZ$, then $\angle A \cong \angle X$, $\angle B \cong \angle Y$, $\angle C \cong \angle Z$, and $\triangle ABC \sim \triangle XYZ$.

EXAMPLE 1   Using Figure 4.25, determine as many missing angles or sides (their measures) as possible without using trigonometry. [UCSMP, p. 607]

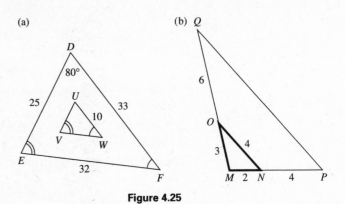

Figure 4.25

SOLUTION

(a) The AA Similarity Criterion applies to $\triangle DEF$ and $\triangle UVW$, with scale factor $k = {}^{10}\!/_{33}$.

$$\therefore UV = k \cdot DE = (\frac{10}{33}) \cdot 25 = \frac{250}{33}$$

$$VW = (\frac{10}{33}) \cdot 32 = \frac{320}{33}$$

$$m\angle U = m\angle D = 80$$

(b) For $\triangle MPQ$ and $\triangle MNO$, the SAS Similarity Criterion applies, with MP/MN $= k = \frac{6}{2} = 3$. Therefore, $PQ = k \cdot ON = 3 \cdot 4 = 12$. No actual angle measures can be determined without trigonometry.❏

E X A M P L E   2   In Figure 4.26, $\triangle ABC$ and $\triangle ADB$ are both isosceles triangles ($AB = AC$ and $AD = BD$).

(a) Prove the two triangles are similar and establish the relation $AB^2 = BC \cdot BD$.

(b) If $m\angle B = 70$, find $m\angle D$.

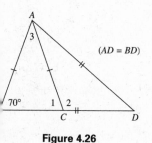

$(AD = BD)$

Figure 4.26

SOLUTION

(a) Since $\angle B$ is in common between the two triangles, they have one pair of congruent angles; since they are isosceles, $\angle 1 \cong \angle B \cong \angle BAD$ ∴ $\triangle ABC \sim$ $\triangle DBA$ by the AA Similarity Criterion. Hence

$$\frac{AB}{DB} = \frac{BC}{BA}, \quad \text{or} \quad AB^2 = DB \cdot BC$$

(b) $m\angle D = m\angle 3 = 180 - (m\angle B + m\angle 1) = 180 - 140 = 40.$❏

THE PYTHAGOREAN THEOREM DERIVED FROM SIMILAR TRIANGLES

The Pythagorean Theorem, trigonometry, and Cartesian coordinates can all be based on the concept of similarity in Euclidean geometry. We encourage you to make the former your own discovery in the following project.

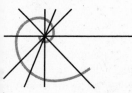

MOMENT FOR DISCOVERY

Applying the AA Similarity Criterion

Let $\triangle ABC$ be any right triangle with right angle at C, and with side lengths $a = BC$, $b = AC$, and $c = AB$. Drop perpendicular CD to AB. Since the angles at A and B are acute, we must have A-D-B.

1. Show that $\triangle BDC \sim \triangle BCA$.
2. Show that $\triangle ADC \sim \triangle ACB$.
3. Observe the corresponding sides of the three similar triangles mentioned in Steps **(1)** and **(2)**:

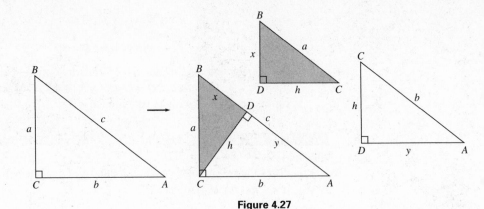

Figure 4.27

$$\frac{x}{a} = \frac{a}{c} \quad \text{and} \quad \frac{y}{b} = ?$$

(Find the correct ratio.)

4. Solve for x and y in these two equations, then use the relation $x + y = c$, and simplify. What did you prove?

EXAMPLE 3 Using similar triangles, prove the classical relationship between the altitude to the hypotenuse of a right triangle and the segments formed on the hypotenuse:

$$h^2 = c_1 c_2$$

(See Figure 4.28 and the preceding Discovery Unit.)

SOLUTION

Since $\triangle BCD \sim \triangle CAD$,

$$\frac{h}{c_1} = \frac{c_2}{h} \qquad \text{or} \qquad h^2 = c_1 c_2 \;\square$$

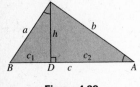

Figure 4.28

THE TRIGONOMETRY OF RIGHT TRIANGLES

If each of the three right triangles in Figure 4.29 has an acute angle congruent to $\angle A$, they are similar, and the ratio of a particular pair of side lengths, say the *opposite over the hypotenuse*, is a constant for the three right triangles pictured, and for all other right triangles having acute angle congruent to $\angle A$. The same is true for other pairs of sides. Hence, we define the following **trigonometric ratios** (which depend only on the measure of the acute angle A):

$$\sin A = \frac{a}{c} = \frac{a'}{c'} = \frac{a''}{c''} \qquad \cos A = \frac{b}{c} \qquad \tan A = \frac{a}{b}$$

called **sine**, **cosine**, and **tangent**, respectively. Thus, the **cosine** of an acute angle of measure A in any right triangle is the ratio of the lengths of the *adjacent leg to the hypotenuse*, and its **tangent** is the ratio of the lengths of the *opposite leg to the*

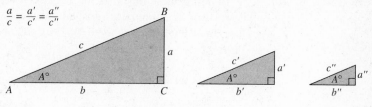

$$\frac{a}{c} = \frac{a'}{c'} = \frac{a''}{c''}$$

Figure 4.29

adjacent leg, etc. The other three ratios are the reciprocals of those already defined, and are denoted csc A, sec A, and cot A (which are abbreviations of their full names, **cosecant**, **secant**, and **cotangent**).

NOTE: We have introduced the six trigonometric functions from the viewpoint of similar triangles. The form for these definitions should be reminiscent of your first introduction to trigonometry. The only reason for repeating this here is to show the fundamental importance of similar triangles to trigonometry now that you have seen a more sophisticated treatment of geometry.

The following table summarizes the definitions and results involving right triangles introduced so far.

NOTE: To illustrate the geometric nature of these definitions, you should be able to use geometry to find the exact values of sin 30, cos 45, and tan 60. (*HINT:* For $A = 30$, consider an equilateral triangle, etc.)

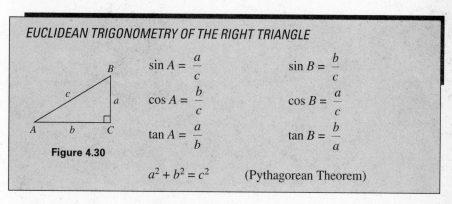

EUCLIDEAN TRIGONOMETRY OF THE RIGHT TRIANGLE

Figure 4.30

$$\sin A = \frac{a}{c} \qquad \sin B = \frac{b}{c}$$

$$\cos A = \frac{b}{c} \qquad \cos B = \frac{a}{c}$$

$$\tan A = \frac{a}{b} \qquad \tan B = \frac{b}{a}$$

$$a^2 + b^2 = c^2 \qquad \text{(Pythagorean Theorem)}$$

These definitions are obviously valid only for $0 < A < 90$. In order to make the usual applications to geometry, it is necessary to extend them to the range $0 \le A \le 180$. To that end, we make the following definitions for $0 < A < 180$.

(1) $$\sin (180 - A) = \sin A$$

(2) $$\cos (180 - A) = -\cos A$$

(3) $$\tan A = \frac{\sin A}{\cos A} \qquad (A \ne 90)$$

(similarly for the remaining three functions)

To define these functions for the special values 0 and 90, since it is desirable to have them be continuous (gradual changes in A produce gradual changes of the functions), let A be close to 90 ($A \approx 90$), and observe that

$$\cos 90 \approx \cos A = -\cos(180 - A) \approx -\cos(180 - 90) = -\cos 90$$

so that $\cos 90 \approx -\cos 90$, or $2 \cos 90 \approx 0$. This gives us a reason to define

(4) $$\cos 90 = 0$$

Since the Pythagorean Theorem yields

$$\frac{a^2}{c^2} + \frac{b^2}{c^2} = 1$$

we get

(5) $$\sin^2 A + \cos^2 A = 1$$

and continuity extends this to $A = 90$. Thus,

$$\sin^2 90 + \cos^2 90 = 1 \qquad \text{or} \qquad \sin^2 90 + 0 = 1$$

Hence, define

(6) $$\sin 90 = 1$$

Also, for acute angles, $\sin A = a/c = \cos B = \cos(90 - A)$, and continuity extends this to $A = 0$. Thus if $A \approx 0$, we have $\sin 0 \approx \cos 90 = 0$. It follows from **(5)** that if $\sin 0 = 0$, then $\cos^2 0 = 1$, so define

(7) $$\sin 0 = 0 \qquad \text{and} \qquad \cos 0 = 1$$

If we set $A = 0$ in the identities involving $180 - A$, we deduce that

(8) $$\sin 180 = \sin 0 = 0 \qquad \text{and} \qquad \cos 180 = -\cos 0 = -1$$

TRIGONOMETRY FOR ARBITRARY TRIANGLES

For completeness, we outline the proofs for the Law of Sines and Law of Cosines for any triangle $\triangle ABC$ (using standard notation). From Figure 4.31 and the three cases shown, we have

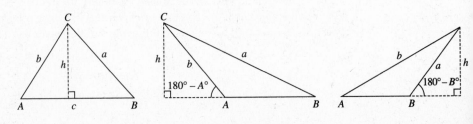

Figure 4.31

Case 1: $A, B < 90$	Case 2: $A \geq 90$	Case 3: $B \geq 90$

$$\sin A = \frac{h}{b}, \; \sin B = \frac{h}{a} \qquad \sin(180 - A) = \frac{h}{b} \qquad \sin A = \frac{h}{b}$$

$$\frac{\sin A}{\sin B} = \frac{\frac{h}{b}}{\frac{h}{a}} = \frac{a}{b} \qquad\qquad \sin B = \frac{h}{a} \qquad\qquad \sin(180 - B) = \frac{h}{a}$$

(Same results as in Case 1, since $\sin(180 - x) = \sin x$ for all x)

Hence, we have

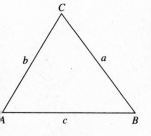

Figure 4.32

LAW OF SINES FOR ANY TRIANGLE

$$\frac{a}{\sin A} = \frac{b}{\sin B} = \frac{c}{\sin C}$$

The Law of Cosines involves a little more algebra, but the proof is standard. You may have seen a different proof of this in your college trigonometry course involving the Distance Formula. Our proof is intended to be a direct application of the definitions and development given here.

Again, there are three cases, as illustrated in Figure 4.33.

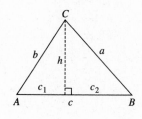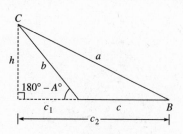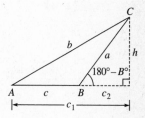

Figure 4.33

Case 1: $A, B < 90$	Case 2: $A \geq 90$	Case 3: $B \geq 90$

$$\cos A = \frac{c_1}{b} \qquad\qquad \cos(180 - A) = \frac{c_1}{b} \qquad\qquad \cos A = \frac{c_1}{b}$$

Case 1:
$$2bc \cos A = 2cc_1$$
$$a^2 = h^2 + c_2{}^2$$
$$= h^2 + (c - c_1)^2$$
$$= h^2 + c_1{}^2 + c^2 - 2cc_1$$
$$= b^2 + c^2 - 2bc \cos A$$

Case 2:
$$-2bc \cos A = 2cc_1$$
$$a^2 = h^2 + c_2{}^2$$
$$= h^2 + (c + c_1)^2$$
$$= h^2 + c_1{}^2 + c^2 + 2cc_1$$
$$= b^2 + c^2 - 2bc \cos A$$

Case 3:
$$2bc \cos A = 2cc_1$$
(You should try to complete this case yourself to test your understanding.)

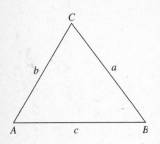

Figure 4.34

Thus,

LAW OF COSINES FOR ANY TRIANGLE

$$a^2 = b^2 + c^2 - 2bc \cos A$$
$$b^2 = a^2 + c^2 - 2ac \cos B$$
$$c^2 = a^2 + b^2 - 2ab \cos C$$

E X A M P L E   4   Use the Law of Sines to prove that if $\overrightarrow{AD}$ is the angle bisector of $\angle A$ in $\triangle ABC$ (Figure 4.35), then

$$\frac{AB}{AC} = \frac{BD}{DC}$$

SOLUTION

From the Law of Sines in $\triangle ADB$, we have

(a) $\dfrac{AB}{\sin \theta} = \dfrac{BD}{\sin A/2}$

From the Law of Sines in $\triangle ADC$,

(b) $\dfrac{AC}{\sin (180 - \theta)} = \dfrac{AC}{\sin \theta} = \dfrac{CD}{\sin A/2}$

Take reciprocals in **(b)** and multiply:

$$\frac{AB}{\sin \theta} \cdot \frac{\sin \theta}{AC} = \frac{BD}{\sin A/2} \cdot \frac{\sin A/2}{CD}$$

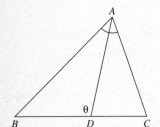

Figure 4.35

or

$$\frac{AB}{AC} = \frac{BD}{DC}$$

The result in the next example is a useful formula, and will appear frequently from now on.

E X A M P L E   5   **The Cevian Formula**

Find the following formulas for the length d of a **cevian** $\overline{CD}$ in $\triangle ABC$ (Figure 4.36), with $p = AD/AB$ and $q = DB/AB$ in each case:

(9) $d^2 = pa^2 + qb^2 - pqc^2$ if A-D-B

(10) $d^2 = -pa^2 + qb^2 + pqc^2$ if D-A-B

(11) $d^2 = pa^2 - qb^2 + pqc^2$ if A-B-D

SOLUTION

For the case *A-D-B*, use the Law of Cosines in $\triangle CDB$ and $\triangle ADC$:

$$a^2 = d^2 + DB^2 - 2d \cdot DB \cos \theta$$

$$b^2 = d^2 + AD^2 - 2d \cdot AD \cos (180 - \theta)$$

Multiply both sides of the first equation by *p*, the second by *q*, and use $DB = qAB = qc$, $AD = pAB = pc$, and $\cos(180 - \theta) = -\cos \theta$

$$pa^2 = pd^2 + pq^2c^2 - 2pqcd \cos \theta$$

$$qb^2 = qd^2 + qp^2c^2 + 2pqcd \cos \theta$$

Sum the last two equations, using $p + q = 1$:

$$pa^2 + qb^2 = (p + q)d^2 + pq(p + q)c^2 = d^2 + pqc^2$$

$$\therefore d^2 = pa^2 + qb^2 - pqc^2$$

For the other two cases, simply use the formula just obtained for cevian $\overline{CA}$ in $\triangle DBC$ or cevian $\overline{CB}$ in $\triangle ADC$ and solve for d^2, using $p + q = 1$ where appropriate. This will be left as a problem.❏

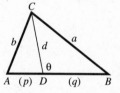
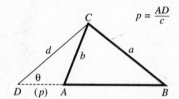
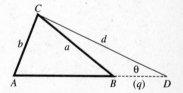

Figure 4.36

The cevian formula is quite useful in work with triangles in geometry. For example, suppose we wanted to find the length of the median to side $\overline{AB}$ of $\triangle ABC$ (the side of length *c*). Since $p = q = \frac{1}{2}$, we immediately find the general formula, as illustrated in Figure 4.37.

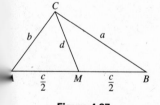

Figure 4.37

(12)
$$d = \sqrt{\frac{1}{2}a^2 + \frac{1}{2}b^2 - \frac{1}{4}c^2}$$

E X A M P L E   6   In right triangle $\triangle RST$, a cevian $\overline{RP}$ of length 12 is perpendicular to side $\overline{ST}$, cutting off segments of length 9 and 16, as shown in Figure 4.38

(a) Find *RT* using the Pythagorean Theorem

(b) Find $RS = d$ using the Pythagorean Theorem.

(c) Find *d* by use of equation (**10**).

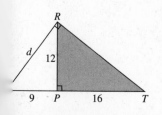

Figure 4.38

SOLUTION

(a) $RT^2 = 12^2 + 16^2 = 144 + 256 = 400$ $\therefore RT = 20$

(b) $d^2 = 12^2 + 9^2 = 144 + 81 = 225$ $\therefore d = 15$

(c) With $p = PS/PT = \frac{9}{16}$ and $q = ST/PT = \frac{25}{16}$, we have

$$d^2 = -\frac{9}{16} \cdot 20^2 + \frac{25}{16} \cdot 12^2 + \frac{9}{16} \cdot \frac{25}{16} \cdot 16^2$$

$$= \frac{-9 \cdot 400 + 25 \cdot 144}{16} + 9 \cdot 25 = 225$$

$$\therefore d = 15\square$$

PROBLEMS   (§4.3)

___ **Group A** ___

1. Consider $\triangle ABC$ and $\triangle XYZ$, with the sides and approximate angle measures as indicated in the figure.

    (a) Determine whether the triangles are similar.

    (b) Assuming the triangles are similar, then $\triangle ABC \sim$ _____? (Use the proper notation.)

    (c) If possible, determine the measures of angles X, Y, and Z.

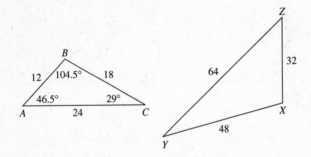

2. Given: X is the midpoint of $\overline{WY}$. V is the midpoint of $\overline{WZ}$.

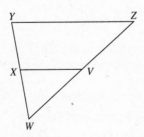

    Prove: $\triangle WXV \sim \triangle WYZ$. (See figure below.)

    [UCSMP, p. 608]

3. Show that the lengths of corresponding altitudes of two similar triangles are in the same ratio as the scale factor between the two triangles.

4. An altitude of △ABC is k times the altitude of △PQR and the base of △ABC is k times the base of △PQR. Explain why the two triangles might not be similar (give a numerical example to illustrate your answer).

In Problems 5 and 6, the two triangles in each of the figures are similar.

   (a) Determine the correspondence for the vertices.

   (b) Find the scale factor.

   (c) Determine the measures of as many unknown angles and segments as possible. [UCSMP, p. 612]

5.

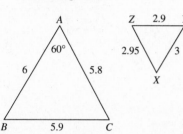

6.

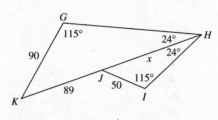

7. In Problem 6, find $HJ = x$.

8. The figure to the right contains two triangles.

   (a) Are they similar?

   (b) If so, what similarity theorem guarantees their similarity?

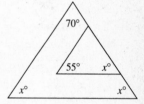

9. Algebraic Lemma for Ratio and Proportion

   (a) Prove that if $\dfrac{a}{b} = \dfrac{c}{d}$ then $\dfrac{a + c}{b + d} = \dfrac{a}{b}$ and $\dfrac{a - c}{b - d} = \dfrac{a}{b}$

   where $b \neq 0$, $d \neq 0$, $b \neq d$.

   (b) Use this in the figure to the right to prove that if

   $\dfrac{AD}{DB} = \dfrac{AE}{EC}$ $\left(\text{or } \dfrac{AD}{AE} = \dfrac{DB}{EC}\right)$

   then

   $\dfrac{AB}{AC} = \dfrac{AD}{AE}$ $\left(\text{or } \dfrac{AD}{AB} = \dfrac{AE}{AC}\right)$

   (c) Using the indicated distances, find DE.

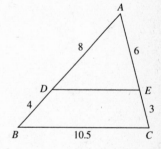

10. In △PQS, the cevian $\overline{PW}$ cuts off segments of length 6 and 15 on side $\overline{QS}$, as shown in the accompanying figure. Using the Cevian Formula, find $d \equiv PW$.

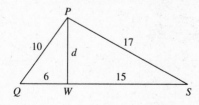

_____ *Group B* _____

11. (a) Use similar triangles to prove the result of Example 4, that in any △ABC, $\overrightarrow{CD}$

bisects $\angle C$ iff $BD/DA = a/b$. (**HINT:** Draw $\overrightarrow{BE} \parallel \overleftrightarrow{CA}$.)

(b) Use this to solve the following geometry problem: A rectangle $\square PQRS$ which is twice as long as it is wide is inscribed in right $\triangle ABC$, having its longest side $\overline{PQ}$

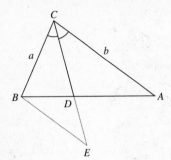

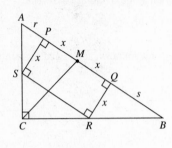

on the hypotenuse. If M is the midpoint of $\overline{PQ}$, prove that $\overrightarrow{CM}$ bisects $\angle C$.

12. In a sequence of three squares, the second and third squares are symmetrically inscribed within their predecessors, as indicated. Find the ratio of the area of the largest square to the smallest.

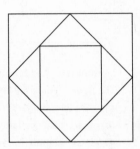

13. **Characterization of the Orthic Triangle** Use similar right triangles to prove

(a) In any acute-angled triangle $\triangle ABC$, if D, E, and F are the feet of the altitudes to the sides $\overline{BC}$, $\overline{AC}$, and $\overline{AB}$, respectively, then $\overrightarrow{DA}$ bisects $\angle FDE$ and $m\angle BDF = m\angle CDE = A$. (**HINT:** Several right triangles have acute angles in common; show that $\triangle DEC \sim \triangle ABC$.)

(b) Conversely, if altitude $\overrightarrow{DA}$ bisects $\angle FDE$ and $m\angle BDF = m\angle CDE = A$ for certain

THE ORTHIC TRIANGLE

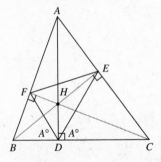

points E and F on sides $\overline{AC}$ and $\overline{BC}$, respectively, then $\overline{BE}$ and $\overline{CF}$ are altitudes of $\triangle ABC$.

14. Prove the SAS Similarity Criterion.

15. Prove the SSS Similarity Criterion.

16. In trapezoid $\diamondsuit ABCD$, the median $\overline{MN}$ cuts diagonals $\overline{AC}$ and $\overline{BD}$ at P and Q, respectively, with $BC = 6$ and $AD = 10$. Find PQ.

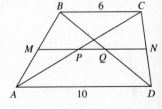

*17. **Golden Isosceles Triangles** In the accompanying figure, one isosceles triangle is constructed adjacent to another (with $AB = AC = CD$ and B-C-D). It is discovered that, in addition, $AD = BD$.

(a) Show that this can only happen (and will happen) when $\triangle ABC$ is a **Golden Isosceles Triangle**, that is, when $AB = AC = \tau BC$, where $\tau = \frac{1}{2}\left(1 + \sqrt{5}\right)$ (the **Golden Ratio**—see Problems 5, 6, and 7, §1.2).

(b) Show that $m\angle B = 72$.

(c) Using $\triangle ABC$, show that $\cos 72 = \frac{1}{2}\tau^{-1} \equiv \left(1 + \sqrt{5}\right)^{-1}$.

(d) Show that $\cos 36 = \frac{1}{2}\tau$.

*Used in the construction of a regular pentagon, Problem 13, §4.4

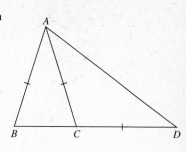

18. Let $\ell \parallel m$ and consider a **central projection** from ℓ to m with **center** P (P not on ℓ or m): For each Q on ℓ, define Q' on m such that

$$Q' = \vec{PQ} \cap m$$

Use similar triangles to prove that *central projections between parallel lines are ratio-preserving*. That is, in the accompanying figure, show that $AB/BC = A'B'/B'C'$.

19. State and prove the *converse* of the Pythagorean Theorem.

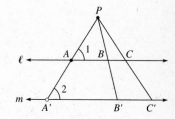

_____ *Group C* _____

20. **Directed Distance** The formulas of geometry are often greatly simplified by the use of directed distances: If ℓ is any line coordinatized as in the Ruler Postulate, for any two points $A[a]$, $B[b]$ with their coordinates, define the **directed distance**

$$AB = b - a$$

Prove the following fundamental properties of directed distance (for arbitrary A, B, C on ℓ):

(a) $AB = -BA$ (or $AB + BA = 0$)

(b) $AB + BC = AC$ (or $AB + BC + CA = 0$)

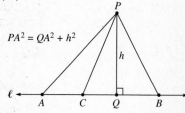

$$PA^2 = QA^2 + h^2$$

21. **Stewart's Theorem** Given 3 collinear points A, B, and C, P any other point, and Q the foot of the perpendicular from P to line $\vec{AB}$ (using directed distance as introduced in Problem 20), show that

(a) $QA^2 \cdot BC + QB^2 \cdot CA + QC^2 \cdot AB + BC \cdot CA \cdot AB = 0$

(b) $PA^2 \cdot BC + PB^2 \cdot CA + PC^2 \cdot AB + BC \cdot CA \cdot AB = 0$

(*HINT:* For (a), replace QB by $QA + AB$ and QC by $QA + AC$ and expand; for (b), let $h = PQ$ and replace PA^2 by $QA^2 + h^2$, etc.)

22. **Comprehensive Cevian Formula** Assuming directed distance on line $\vec{AB}$, define $p = AD/AB$ and $q = DB/AB$ (where p can be an arbitrary real number), and $CD = d$. Thus, $p + q = 1$ in all cases. Use Stewart's Theorem (Problem 21) to prove *without trigonometry* that for all points D on line $\vec{AB}$, $d^2 = pa^2 + qb^2 - pqc^2$.

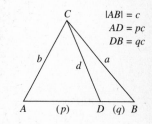

$|AB| = c$
$AD = pc$
$DB = qc$

23. **Cevian Proof of Heron's Formula** Using the result of Problem 22, let X represent a variable point on line $\vec{AB}$, with $p = x$ and $q = 1 - x$. The formula for the cevian becomes a quadratic in x.

(a) Fix the value $d = h_c$ = length of the altitude $\vec{CF}$ to $\vec{AB}$ (F can fall exterior to side $\vec{AB}$), and show that $AF/AB = p$ must satisfy the equation

$$c^2x^2 + (a^2 - b^2 - c^2)x + (b^2 - h_c^2) = 0$$

What must be true of the possible values for x?

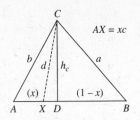

(b) That being the case, the discriminant Δ of the quadratic equation in x must equal zero. Write this out and solve the resulting equation $\Delta = 0$ for d (or h_c) in terms of a, b, and c.

(c) Put the formula you found in **(b)** in the form

$$h_c = \frac{2}{c}\sqrt{s(s-a)(s-b)(s-c)}$$

(d) What would be the corresponding forms for h_a and h_b?

(e) Prove Heron's Formula.

24. Undergraduate Research Project Find a formula for the *sides* of a triangle in terms of

(a) its *medians* (Can these lengths be given arbitrarily to determine a triangle, and is the triangle uniquely determined?).

(b) its *altitudes* (see Problem 23).

(c) Is there a characteristic relationship for the medians of a right triangle? For the altitudes of a right triangle?

(d) Establish a Heron-like formula for the area of a triangle *in terms of its three altitudes*.

---

*4.4   REGULAR POLYGONS AND TILING

An old problem in geometry, going back to Gauss, is how to construct a 17-sided regular polygon with the Euclidean tools, the compass and straight-edge. We consider the problem of both the existence of regular polygons, and their construction, and then we discuss briefly the interesting concept of tiling.

REGULAR POLYGONS IN ABSOLUTE GEOMETRY

We first tackle the problem of existence for regular polygons with an arbitrary number of sides (≥ 3). Since our construction does not use the Parallel Postulate or any of its consequences, it is valid in absolute geometry.

DEFINITION

An n-sided **polygon**, or n-gon, for any integer $n \geq 3$, is the closed union of line segments

$$\overline{P_1P_2} \cup \overline{P_2P_3} \cup \overline{P_3P_4} \cup \dots \cup \overline{P_{n-1}P_n} \cup \overline{P_nP_1}$$

with **vertices** P_1, P_2, P_3, $\dots$, P_n, and **sides** $\overline{P_iP_{i+1}}$, $\overline{P_nP_1}$ for $1 \leq i \leq n-1$, where no three **consecutive vertices** P_i, P_{i+1}, P_{i+2} (or P_{n-1}, P_n, P_1 and P_n, P_1, P_2) are collinear, and no pair of **nonadjacent sides** P_iP_{i+1}, P_jP_{j+1} $(i+1 < j)$ meet. If no vertex P_i for $1 \leq i \leq n$ lies in the interior of the triangle formed by any other three vertices (or, equivalently, all the vertices P_j lie on, or on the same side of the line $\overleftrightarrow{P_iP_{i+1}}$ containing a given side), then the polygon is said to be **convex**. An **angle** of polygon $P_1P_2P_3{\dots}P_n$ is any of the n angles formed by three consecutive vertices P_{i-1}, P_i, and P_{i+1}. A **regular polygon** is a convex polygon having congruent sides and congruent angles.

It is clear that since we are free to "construct" angles with any given measure (that is, such angles *exist*), then a regular n-gon may be constructed for each integer $n \geq 3$, and with any given length as radius. For example, to prove that a regular decagon (10-sided polygon) exists, with radius $r > 0$, let O be any point and P_1 any other point such that $OP_1 = r$ (Figure 4.39). Construct ray $\overrightarrow{OP_2}$ such that $m\angle P_1OP_2 = 360/n \equiv \theta \ (= 36)$ and $OP_2 = OP_1 = r$. If $n \geq 3$ (the smallest value for n permitted), $m\angle P_1OP_2 \leq 120 < 180$, so this construction is valid for any n. If $2\theta < 360$, construct $\overrightarrow{OP_3}$ so that $\overrightarrow{OP_2}$ bisects $\angle P_1OP_3$ and $OP_3 = r$. If $3\theta < 360$, construct $\overrightarrow{OP_4}$ so that $\overrightarrow{OP_3}$ bisects $\angle P_2OP_3$ and $OP_4 = r$. The process continues, and we can locate points $P_1, P_2, \dots, P_k$ with $\overrightarrow{OP_{k-1}}$ the bisector of $\angle P_{k-2}OP_k$ and $OP_k = r$, as long as $(k - 1)\theta < 360$, which will be true as long as $k \leq n \ (= 10)$.

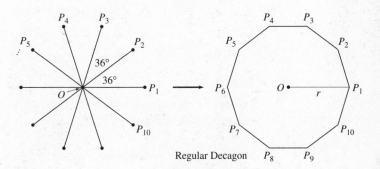

Regular Decagon

Figure 4.39

It seems clear enough that the resulting polygon $P_1P_2P_3\dots P_{10}$ is a regular 10-sided polygon. The details for actually proving the symmetry and congruence properties needed, as well as the convexity of the resulting polygon, are numerous, but routine. As an example of a detail that needs to be verified in order to justify this construction, we have the question of how to prove that the *last* angle, $\angle P_1OP_{10}$, has the same measure θ as all the rest, and that $\overrightarrow{OP_{10}}$ bisects $\angle P_1OP_9$. The reader can see that the betweenness properties normally taken for granted are anything but obvious *without recourse to figures*. Such matters are usually not even mentioned in elementary treatments, and here, we just mention them!

THEOREM 1

Any regular polygon having n sides, $n \geq 3$, exists, and may be inscribed in a circle (has its vertices lying on a circle), or may be circumscribed about a circle (has its sides tangent to a circle).

REGULAR POLYGONS IN EUCLIDEAN GEOMETRY

The above ideas are valid in absolute geometry. In Euclidean geometry, we can say a lot more, including exact formulas for the interior and exterior angles of regular polygons—formulas not valid in spherical or hyperbolic geometry.

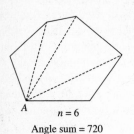

A

$n = 6$

Angle sum = 720

Figure 4.40

THEOREM 2

The angle sum of a convex n-gon is $180(n - 2)$. (See Figure 4.40.)

Proof: Since the polygon is convex, the diagonals from any vertex $P_1 = A$ to the remaining $n - 2$ vertices not adjacent to A will occur in order, and betweenness considerations allow us to sum the angles of the $n - 2$ triangles thus formed, which is obviously $180(n - 2)$, to give us the total sum of the measures of the angles of the polygon.

COROLLARY A

Each interior angle of a regular polygon (Figure 4.40) has measure

$$\phi = \frac{180(n - 2)}{n}$$

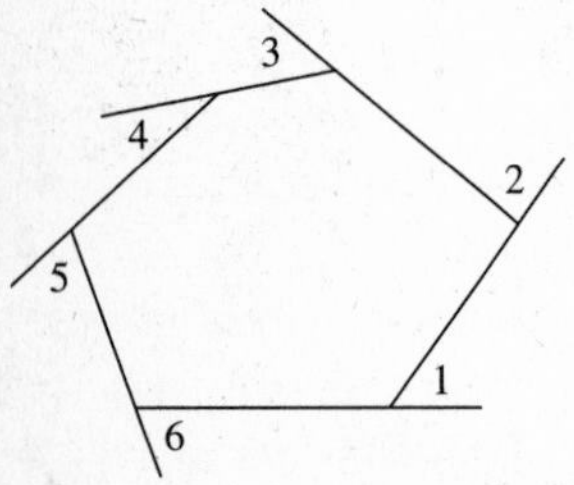

COROLLARY B

The sum of the measures of the exterior angles of any convex n-gon, taken in the same direction, is 360. (See Figure 4.41.)

Sum of exterior angles = 360

Figure 4.41

(The last corollary is intuitively clear. As you proceed around the polygon, the measure of each exterior angle gives the amount of rotation necessary in passing from one side to the next; since the total rotation necessary to return to your initial starting point is 360, then the sum of the exterior angles must be 360.)

CLASSICAL CONSTRUCTION OF REGULAR POLYGONS

Although we have discussed how to prove the *existence* of a regular polygon having $n \geq 3$ sides, the problem of actually constructing such a polygon with only a compass and straight-edge is quite another matter. Only those regular polygons which belong to a select group are so constructible. The following theorem is quite remarkable; it was discovered by Gauss when he was only 19.

THEOREM 3: THEOREM OF GAUSS ON REGULAR POLYGONS

A regular n-gon may be constructed with the Euclidean tools iff n is either a power of two, or the product of a power of two and distinct **Fermat primes**, that is, distinct primes of the form

$$F_m = 2^{2^m} + 1$$

NOTE: Fermat Primes were named after the great number theorist Pierre de Fermat (1601–1665) because he conjectured around 1640 that F_m is prime regardless of the value of m. These numbers become incredibly large very quickly with increasing m. For example, $F_5 = 4,294,967,297 = 641 \cdot 6,700,417$—a product of two primes (thus, Fermat's conjecture is false). It is known that F_m is composite for $5 \leq m \leq 16$, and it is believed that for $m > 16$, F_m is never prime. The largest values for which F_m is known to be prime are $F_3 = 257$ and $F_4 = 65{,}537$, which therefore satisfy Gauss's theorem. According to H.S.M. Coxeter, *Introduction to Geometry*, p. 27, Richelot and Schwendennwein constructed the 257-gon about 1898. J. Hermes spent 10 years on the regular 65,537-gon and deposited the manuscript in a large box at the University of Göttingen for others to inspect. (In view of this discussion, you would be advised not to start working on constructing a 4,294,967,297-sided regular polygon!)

$\mathcal{H}$ISTORICAL $\mathcal{N}$OTE

Carl Frederick Gauss, known as the *prince of mathematicians*, is regarded by some scholars to be the greatest mathematician of all time, making outstanding contributions to all fields of mathematics. His work fills many volumes, including his desk drawer collections, where he kept unpublished work. The latter was a point of contention with several prominent mathematicians of his day, notably Jacobi and Legendre. The story is told that Jacobi once was describing to Gauss the details of a recent discovery of his, whereupon Gauss reached in his desk drawer and pulled out his own work which contained the essentials of Jacobi's results. Jacobi responded angrily, in essence, "It is a pity you did not publish this paper, for it is so much better than many other papers which you have published!" Gauss was a perfectionist, however, and he would not publish material that did not meet his own high standards of rigor and thorough checking. Gauss was born in Germany in 1777 into a poor and unlettered family. His father was, from time to time, a bricklayer, gardener, and construction worker. Having noticed Gauss's genius, the Duke of Brunswick helped pay for Gauss to attend Caroline College, then later, the

University of Göttingen, from 1795–1798. In spite of his other accomplishments of greater importance, Gauss seemed proudest of the result concerning the 17-sided regular polygon (the first advance since the time of Euclid), which he had discovered while he was at Göttingen. He asked that this figure be carved on his tombstone, a request that was never carried out. In 1806, he was appointed director of the newly built observatory at the University of Göttingen, a position he held until his death nearly a half-century later. His interests had, by that time, turned to astronomy, and he devoted three years to observing and calculating the orbit of Ceres— an asteroid (mini-planet)—with amazing accuracy. His least squares method for minimizing experimental error allowed him to accurately calculate the orbits of a series of planetoids that lie between Mars and Jupiter as soon as they were discovered by astronomers, and he made significant new contributions to the theory of astronomy. Many scholars feel that this activity amounted to genius wasted, and that such efforts might have been more effectively devoted to mathematics and theoretical physics.

Since 3, 5, and 17 are Fermat primes (corresponding to $m = 0, 1,$ and 2), $n = 34 = 2 \cdot 17$ and $n = 15 = 3 \cdot 5$ are of the required form, and we can construct a 34-sided and 15-sided regular polygon with the Euclidean tools. But $n = 45 = 3^2 \cdot 5$ is not (it is not the product of *distinct* Fermat primes), so we cannot construct a 45-sided regular polygon. For $n \leq 20$, the values of n which yield constructible n-gons are

$$n = 3, 4, 5, 6, 8, 10, 12, 15, 16, 17, 20$$

We can get through this list fairly quickly with what we already know. For $n = 3$ (corresponding to the equilateral triangle), we merely swing two circular arcs centered at A and B, each having radius $\overline{AB}$ as shown in Figure 4.42. These two arcs will meet at the third vertex C of the desired triangle (Euclid's original construction, which also yields the construction for an angle of measure 60). From this we can construct a regular hexagon ($n = 6$) and, after bisecting an angle of measure 60, we can construct a regular dodecagon ($n = 12$). The regular n-gons for $n = 4, 8, 16$ are obtained by constructing a right angle, then bisecting a right angle and the half-angle obtained from that. The construction of the regular pentagon involves the Golden Ratio (introduced in Problems 5, 6, and 7 of Section 1.2), and was also given by Euclid. The essence of this construction is the fact that $\cos 36 = \frac{1}{2}\tau$ (Problem 17, Section 4.3). From that we can obtain the regular pentagon, regular decagon, and the regular 20-sided polygon. This leaves the values 15 and 17. For $n = 17$, we refer the reader to Eves, *Survey of Geometry, Volume I*, pp. 218–224, or Coxeter's *Introduction to Geometry*, p. 27. The value $n = 15$ can be based on the numerical fact

$$\frac{1}{15} = \left(\frac{2}{5} - \frac{1}{3}\right).$$

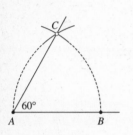

Figure 4.42

Using Figure 4.43, construct, with compass/straight-edge, an equilateral triangle with side $\overline{AC}$, and a regular pentagon with sides $\overline{AB}$ and $\overline{BD}$. Then

$$m\overset{\frown}{CD} = m\overset{\frown}{ABD} - m\overset{\frown}{ABC} = 360\left(\frac{2}{5} - \frac{1}{3}\right) = 360 \cdot \frac{1}{15} = 24$$

the required angle measure for the central angle of a 15-sided regular polygon.

Note that $n = 7$ (corresponding to the **regular heptagon**) is the first case for which there is no Euclidean construction. This was recognized in antiquity by Archimedes, who showed it could be constructed by the aid of a special transcendental curve.

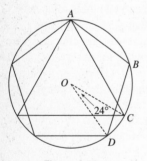

Figure 4.43

TILING AND PLANE TESSELATIONS

The subject of **tiling theory** is very old, which will be apparent from some of our illustrations. However, recently this area has had a surge of new interest, aroused in part by the publication of the definitive book by B. Grünbaum and G.C. Sheppard, *Tiling and Patterns*, which has brought a semblance of order to an unwieldy collection of seemingly unrelated ideas (viz, wallpaper patterns, and ancient classical designs), and this work is a monumental achievement in terms of both number of results and the high quality of the art work. Here, we will just introduce you to this topic.

OUR GEOMETRIC WORLD

An unusual tiling pattern decorates a museum in Grenada, Spain. Built in the 1300s, the artist is unknown. The structure is used today as a mosque.

Johann Kepler (1571–1630) was the first to apply geometry systematically to the subject of plane tilings when he observed in 1619 that the only regular polygons which can be used to tile the plane are the equilateral triangle, square, and regular hexagon, as shown in Figure 4.44. He also discovered the unusual tiling shown in Figure 4.45, which uses the regular pentagon and decagon, and includes a "fused" double decagon and five-pointed star.

(Tiles continue indefinitely in all directions)

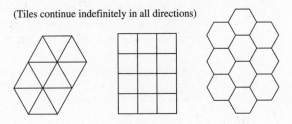

Figure 4.44

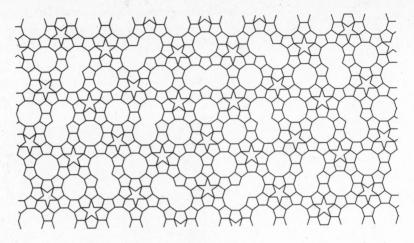

Figure 4.45

In general, a **tiling** or **tessellation** of the plane is a collection of regions T_1, T_2, ..., T_n, ... , called **tiles** (usually polygons and their interiors), such that

(1) no two of the tiles have any interior points in common

(2) the collection of tiles completely covers the plane

Here, we shall require that all tilings by polygons be **edge-to-edge**, that is, each edge of a polygon in the tiling must be an edge of one of its neighbors.

In order to make it more interesting, one normally requires a tiling to have certain regularity properties, such as restricting all tiles to be regular polygons, requiring the tiles to be congruent to a single tile, etc. The question of the *existence* of such tilings (or proving they do not exist) becomes a very interesting, and sometimes very challenging, problem.

DEFINITION

If all the tiles in a plane tiling are congruent to a single region, the tiling is said to be **elementary of order one** (or, simply, **elementary**) and the single region is called the **fundamental region** or **prototile** of the tiling. If the tiles are each congruent to one of n tiles T_1, T_2, ..., T_n (also called **fundamental regions**), the tiling is called **elementary of order n**. If the adjective **semi-regular** is added to any of the preceding terms, it means that all the fundamental regions are regular polygons[2].

A major problem in the theory of tiling is to determine whether a given two-dimensional figure, such as a polygonal region, can serve as a fundamental region

---

[2]The term **regular tiling** is restricted to tilings whose tiles are each congruent to a single regular polygon. Our "elementary tilings of order one" are termed **monohedral tilings** by Grünbaum and Sheppard.

for an elementary tiling (a tiling of order one). For example, we could ask what regular polygons can serve as the fundamental region for such a tiling. Suppose that an n-sided regular polygon, $n \geq 3$, is such a fundamental region. If there are k of these polygons at each vertex, then, with θ the measure of each interior angle of the regular polygon, we have

$$k\theta = 360, \qquad \text{where } \theta = \frac{n-2}{n}180$$

Then, by substitution,

$$k \cdot \frac{n-2}{n}180 = 360, \qquad \text{or} \qquad k = \frac{2n}{n-2}$$

Since we are seeking integer solutions for k, observe that the right side of the last equation above must be an integer. But integer values for

$$\frac{2n}{n-2}$$

occur only for $n = 3$, 4, and 6—that is, when the regular polygon has either 3, 4, or 6 sides, precisely the cases found by Kepler.

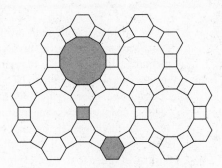

Figure 4.46

The next logical place to look for tilings involving regular polygons is in higher order semi-regular tilings (having more than one fundamental region). Many more types of regular polygons can occur in such tilings of the plane. For example, the regular dodecagon (12-sided), together with a regular hexagon and square, are the fundamental regions for the tiling illustrated in Figure 4.46. A few other examples are shown in Figures 4.47 and 4.48. A systematic way to find all possible tessellations of this type is pursued in Problem 20.

(a)

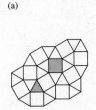

(b)

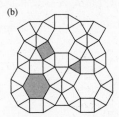

Figure 4.47

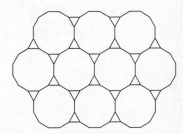

Figure 4.48

The next Discovery Unit will lead you to a construction of one of these tilings for yourself.

MOMENT FOR DISCOVERY

Exploring Semi-Regular Tilings

1. Draw a regular octagon (stop sign) on a sheet of paper.

2. If you now draw another regular octagon next to it, edge-to-edge, will another octagon fit in the space next to the common edge?

3. Draw as many edge-to-edge regular octagons as you can on your sheet of paper.

4. What spaces are not covered by the octagons? Do you see a semi-regular tiling of order two? What are the fundamental regions?

We still have not found a tiling that utilizes the pentagon, heptagon, nonagon, and decagon. In order to involve these, we must allow a different type of fundamental region other than regular polygons. The pentagon can be used in a tiling if we also allow a rhombus to be a fundamental region, thus providing us with a tiling of order two. (See Figure 4.49.) It is more difficult to obtain a tiling that involves a regular decagon; this one requires not only a rhombus and regular pentagon to serve as fundamental regions, but also the regular **pentagonal star** (the famous five-pointed star which the Pythagoreans wore as a badge of identity). The result is a tiling of order four. (See Figure 4.50.)

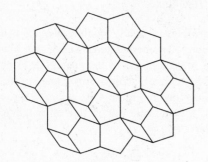

Figure 4.49

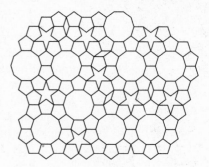

Figure 4.50

PROBLEMS   (§4.4)

_____ Group A _____

For Problems 1 and 2, determine which of the polygonal regions (a), (b), and (c) in the figures shown can be used as a fundamental region for an elementary tiling of the plane.

1.

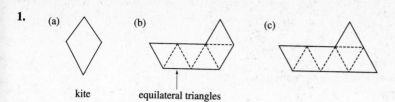

(a)   (b)   (c)

kite   equilateral triangles

2. (a)   (b)   (c)

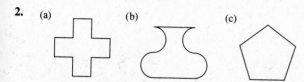

3. Verify with a sketch that an isosceles trapezoid is the fundamental region of an elementary tiling of the plane.

4. Show that any convex pentagon and interior which has a pair of opposite sides parallel and congruent, as shown here, may be used to tile the plane.

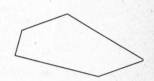

5. **(a)** The nonconvex quadrilateral ◇*ABCD* and interior shown in the figure having a line of symmetry through *A* and *C* (a dart) is the fundamental region of an elementary tiling of the plane. Make a sketch of this tiling.

   (b) Generalize to an arbitrary nonconvex quadrilateral.

6. If the two nontriangular polygons are regular, find the measures of the angles of △*PQR*.

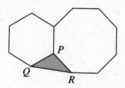

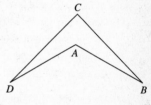

7. If the two nontriangular polygons are regular, find the measures of the angles of △*WST*.

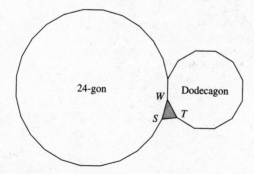

24-gon   *W* Dodecagon

   S   *T*

8. If the pentagon, hexagon, and octagon are regular polygons, find *x*. (**HINT:** Draw a diagonal, and use some trigonometry.)

9. Verify the validity of the tiling shown in Figure 4.47a.

10. Verify the validity of the tiling shown in Figure 4.48.

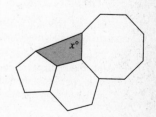

_____ *Group B* _____

11. Find a distinctly different second-order tiling of the plane from that of Figure 4.49 (different arrangement of pentagons and rhombuses).

12. Show that each of the regions illustrated in the following figure is a fundamental region for an elementary regular tiling of the plane.

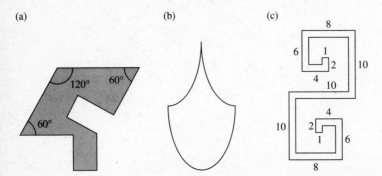

(a)   (b)   (c)

13. **Euclidean Construction of the Regular Pentagon** Show that in the figure, cos θ = ¹/2τ, so that θ = 36, m∠FOB = 72, and the construction yields a regular pentagon. The steps of the construction are as follows:

(1) Draw a unit circle, with diameter $\overline{AB}$, and erect perpendicular $\overline{OC}$.

(2) Construct the midpoint D of $\overline{AO}$.

(3) Swing arc CE centered at D, making DE = DC.

(4) Swing are EF centered at A, making AF = AE.

(*HINT:* Let M be the midpoint of base $\overline{AF}$ of isosceles △AOF, where AO = 1. Show that AM = ¹/2τ and cos θ = ¹/2τ.

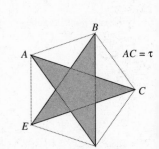

14. The midpoints of the sides of a regular pentagon are joined, in order, forming a second regular pentagon. Show that the lengths of the sides of the new pentagon are precisely ¹/2τ times those of the original polygon.

15. **The Star of Pythagoras** The Pythagoreans were known by the regular five-pointed pentagonal star which they wore as a badge for identification in their elite society. As part of their initiation rites, members were required to know how to demonstrate the Pythagorean Theorem. Show that this star, derived from the diagonals of a regular pentagon, has congruent sides, each having length τ times the side of the regular pentagon which defines it, and that the points form an angle of measure 36 each.

16. The regular hexagon can be used in a third-order semi-regular tiling of the plane. Find the other two regular polygons, and explain how this may be accomplished.

_____ *Group C* _____

17. Use Gauss's Theorem to determine which of the regular *n*-gons for 21 ≤ *n* ≤ 100 may be constructed with compass and straight-edge.

18. The 24 regions illustrated in the following figure were proposed by T.H. O'Bierne as possible fundamental regions for elementary (order one) plane tesselations. Each region proposed consists of the union of 7 equilateral triangles and interior. One of the 24 regions is not such a fundamental region. Can you find it?

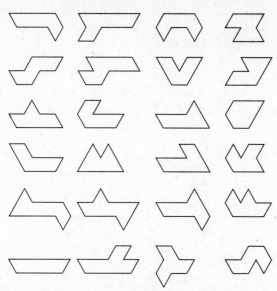

19. Niven's Theorem A 1978 theorem due to I. Niven states that if the areas of all tiles of an arbitrary plane tiling are to be greater than some positive constant α, and if the tiles consist of convex polygons having 7 or more sides, then there exist tiles whose perimeters β are *arbitrarily large* (see Klamken and Liu, "Note On a Result of Niven on Impossible Tesselations", *Amer. Math. Monthly*, Vol. 87, No. 8 (1980), pp. 651–653). Show that if the convexity requirement be dropped, the polygons in the accompanying figure are the fundamental regions of a tiling of order two, for which the perimeters of the tiles are bounded. (This would be a counterexample to the theorem if convexity requirements were deleted.)

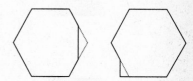

20. UNDERGRADUATE RESEARCH PROJECT Make a list of the regular polygons and interiors which can be used in a tiling of the plane of order three, where other regular polygons and semi-regular nth-order **star polygons** (star polygons with n points distributed uniformly on a circle—including the rhombus of order two used in Figures 4.49 and 4.50) are allowed. (**HINT:** If $n_1, n_2, \dots, n_k$ are the numbers of sides of regular polygons permitted at a vertex (k need not be the same for all vertices), and if θ is the common measure of the angles of a semi-regular star polygon, assuming only one of these is permitted at each point, then show that these numbers must satisfy

$$\frac{2}{n_1} + \frac{2}{n_2} + \dots + \frac{2}{n_k} = \frac{\theta}{360} + k - 1$$

4.5 THE CIRCLE THEOREMS

Much of the power of synthetic geometry derives from the study of circles. We present here the basic development of circles for Euclidean geometry. Before we begin, an interesting project for discovery is proposed.

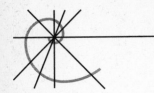

MOMENT FOR DISCOVERY

Inscribed Angles

Draw a large circle and any of its chords $\overline{AB}$, as shown. Locate three points C, C' and C'' at random on the circle and on the same side of the chord.

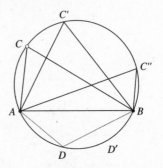

Figure 4.51

1. With a protractor (or using SKETCHPAD), carefully measure $\angle ACB$ and write down your result.

2. Repeat Step 1 for $\angle AC'B$ and $\angle AC''B$.

3. What did you discover?

4. If D and D' are points on the circle on the opposite side of chord $\overline{AB}$ from C, what do you think must be true of $m\angle ADB$ and $m\angle AD'B$? Verify by direct measurement (or by using SKETCHPAD).

5. Did you make any further discoveries?

The fact you should have discovered from the preceding experiment is one of the most amazing, and useful, theorems in Euclidean geometry. It defies intuition, and, indeed, it is totally false in non-Euclidean geometry. It is also difficult to prove by coordinates or vectors. The synthetic geometry of Euclid remains the most appropriate method of proof, even today.

THEOREM 1: INSCRIBED ANGLE THEOREM

The measure of an inscribed angle of a circle equals one-half that of its intercepted arc.

Proof: We must show that in Figure 4.52, $m\angle ABC = \frac{1}{2}m\widehat{AC}$ in all three cases.

THEOREM 1: INSCRIBED ANGLE THEOREM (CONTINUED)

Case I: When center O lies on side $\overleftrightarrow{BC}$ of $\angle ABC$

(1) Since $B\text{-}O\text{-}C$, $\angle AOC$ is an exterior angle of $\triangle AOB$. Hence, from Figure 4.52, $\theta = x + y$ (Euclidean Exterior Angle Theorem, §4.1)

(2) $\overline{OA}$ and $\overline{OB}$ are radii of circle O, so $OA = OB$ and $x = y$. (Isosceles Triangle Theorem)

(3) $\theta = 2x$ or $x = \frac{1}{2}\theta$ (algebra)

(4) $\therefore m\angle ABC = x = \frac{1}{2}\theta = \frac{1}{2}m\widehat{AC}$ (definition of circular arc measure)

Case II: When center O lies interior to $\angle ABC$

(1) Consider the diameter $\overline{BD}$ passing through B. Then

$$m\angle ABC = m\angle ABD + m\angle DBC$$

(Angle Addition Axiom)

(2) $\therefore m\angle ABC = \frac{1}{2}m\widehat{AD} + \frac{1}{2}m\widehat{DC} = \frac{1}{2}m\widehat{AC}$ (by Case 1 and Additivity of Circular Arc Measure)

Case III: When center O lies exterior to $\angle ABC$

(1) Let $\overline{BD}$ be the diameter through B and construct the tangent to circle O at B; the circle and points A, D, and C all lie on the same side of the tangent line $\overleftrightarrow{BE}$, and, using coordinates for the rays through B and Theorem 1 of §2.5, either $\overrightarrow{BA}\text{-}\overrightarrow{BD}\text{-}\overrightarrow{BC}$, $\overrightarrow{BA}\text{-}\overrightarrow{BC}\text{-}\overrightarrow{BD}$, or $\overrightarrow{BC}\text{-}\overrightarrow{BA}\text{-}\overrightarrow{BD}$. Since O is not in the interior of $\angle ABC$, we rule out the case $\overrightarrow{BA}\text{-}\overrightarrow{BD}\text{-}\overrightarrow{BC}$; the other two cases are logically the same, but for notation.

(2) Assume, therefore, that $\overrightarrow{BA}\text{-}\overrightarrow{BC}\text{-}\overrightarrow{BD}$. Then

$$m\angle ABC = m\angle ABD - m\angle CBD$$

(3) $\therefore m\angle ABC = \frac{1}{2}m\widehat{AD} - \frac{1}{2}m\widehat{CD} = \frac{1}{2}m\widehat{AC}$ (Case 1 and Additivity of Arc Measure)

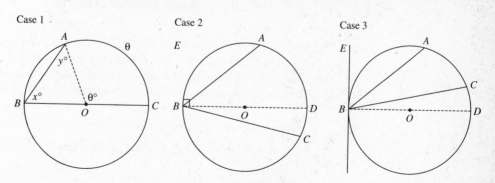

Figure 4.52

EXAMPLE 1   In Figure 4.53, find $m\angle ABC$.

SOLUTION

Observe: $m\widehat{AC} = 360 - 126 - 110 = 124$. Hence, by Theorem 1, $m\angle ABC = \frac{1}{2}(124) = 62.\square$

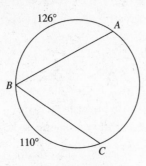

Figure 4.53

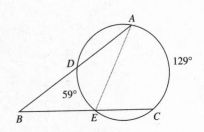

Figure 4.54

EXAMPLE 2   Find $m\angle ABC$ in Figure 4.54.

SOLUTION

Construct chord $\overline{AE}$. Then $\angle AEC$ is an exterior angle of $\triangle ABE$, so $m\angle AEC = m\angle B + m\angle BAE$. But angles $\angle AEC$ and $\angle BAE$ are both inscribed angles of the circle, so

$$m\angle B = m\angle AEC - m\angle BAE$$
$$= \frac{1}{2}(m\widehat{AC} - m\widehat{DE})$$
$$= \frac{1}{2}(129 - 59) = 35\square$$

A special case of Theorem 1 is important enough to state separately.

COROLLARY

Any angle inscribed in a semicircle is a right angle.

The converse of this result has already been mentioned several times; it is a simple property of right triangles: The midpoint of the hypotenuse of a right triangle is the center of a circle passing through its vertices. Hence, a right angle is always an inscribed angle of a semicircle with diameter joining any two points on its sides.$\square$

OUR GEOMETRIC WORLD

A carpenter's square is hung on a wall between two nails, as shown. As the square slides along the two nails, the corner will trace an arc of a perfect circle. The reason is the converse of the corollary of Theorem 1.

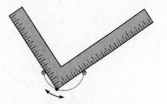

Figure 4.55

The Inscribed Angle Theorem also has several other important corollaries, which we will leave to you to prove as exercises (based on the idea advanced in examples. (See Figure 4.56.)

- An angle whose vertex lies inside a circle and whose sides are intersecting chords of the circle (intercepting arcs of measure x and y) has measure $\theta = \frac{1}{2}(x + y)$. (See Problem 7.)

- An angle whose vertex is exterior to a circle and whose sides are intersecting secants of the circle (intercepting arcs of measure x and y) has measure $\theta = \frac{1}{2}|x - y|$. (See Problem 7.)

- An angle formed by a chord and tangent of a circle, with its vertex at the point of tangency and intercepting an arc of measure x on that circle, has measure $\theta = \frac{1}{2}x$. (See Problem 8.)

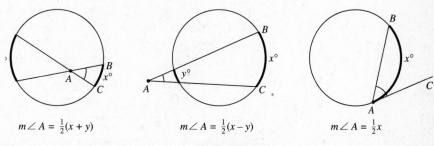

$$m\angle A = \tfrac{1}{2}(x + y) \qquad m\angle A = \tfrac{1}{2}(x - y) \qquad m\angle A = \tfrac{1}{2}x$$

Figure 4.56

Yet another important application of the inscribed angle theorem is the so-called **Two Chord Theorem**. You can discover it for yourself by working through the following Discovery Unit.

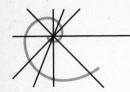

MOMENT FOR DISCOVERY

The Two Chord Theorem

Point P is an interior point of a circle having center O and radius $\sqrt{34}$, as shown. In addition, $\overline{OP}$ is the base of an isosceles right triangle. Chord $\overline{AB}$ is determined, which contains a leg of this isosceles right triangle.

(a) (b) 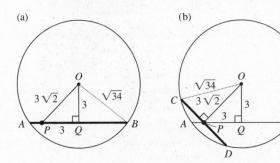 (c)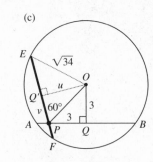

Figure 4.57

1. Determine, by the Pythagorean Theorem, the values for AP and PB. (Did you get AP = 2?)

2. A second chord $\overline{CD}$ through P is perpendicular to $\overline{OP}$ (second diagram in figure). Find the values CP and DP. (**HINT:** P is the midpoint of $\overline{CD}$ in this case.)

3. A third chord $\overline{EF}$ passes through P and makes an angle of measure 60 with $\overline{OP}$. Find the values EP and FP, using trigonometry. (**HINT:** Let Q' be the midpoint of $\overline{EF}$, and let u, v be the sides of the 30–60 right triangle in the third diagram; solve for u, v, and EQ. You should get $u = \frac{3}{2}\sqrt{6}$.)

4. Find $EP = EQ + v$ and $PF = EQ - v$.

5. Check all calculations before you proceed.

6. Evaluate the three products $AP \cdot PB$, $CP \cdot PD$, and $EP \cdot PF$. Did anything unusual happen?

NOTE: This experiment may be more efficiently handled by the computer software GEOMETRIC SUPPOSER.

THEOREM 2: TWO-CHORD THEOREM

When two chords of a circle intersect, the product of the lengths of the segments formed on one chord equals that on the other chord. That is, in Figure 4.58,

$$AP \cdot PB = CP \cdot PD$$

THEOREM 2: TWO-CHORD THEOREM (CONTINUED)

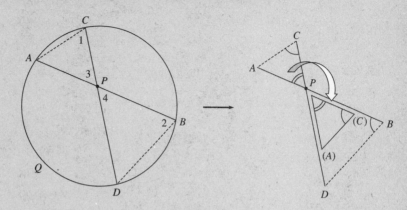

Figure 4.58

Proof: Let P be the point of intersection of chords $\overline{AB}$ and $\overline{CD}$ of the circle.

(1) $m\angle 1 = m\angle 2$ (Inscribed Angle Theorem—both angles subtend the same arc $\overset{\frown}{AQD}$)

(2) $m\angle 3 = m\angle 4$ (Vertical Pair Theorem)

(3) $\triangle APC \sim \triangle DPB$ (AA Similarity Criterion)

(4) $PA/PD = PC/PB \therefore AP \cdot PB = CP \cdot PD$ (ratios of corresponding sides of similar triangles are equal)

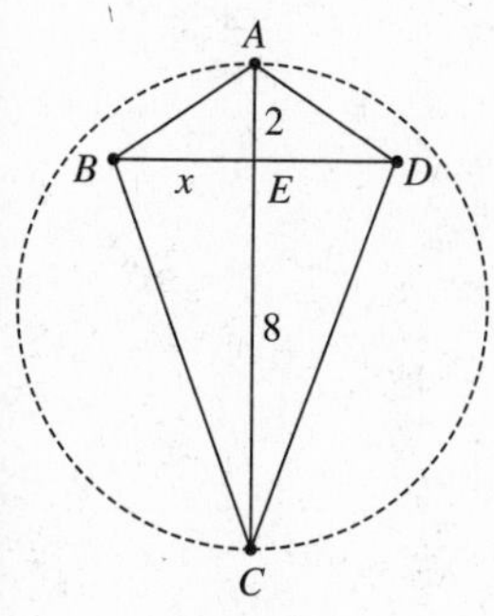

EXAMPLE   3   A kite $\diamondsuit ABCD$ has perpendicular struts meeting at E such that $AE = 2$ and $EC = 8$ (Figure 4.59). What must the length of strut $\overline{BD}$ equal in order for the corners of the kite to exactly fit on a circle?

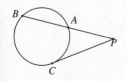

Figure 4.59

SOLUTION

According to Theorem 2, if the four corners lie on a circle, we must have $BE \cdot ED = AE \cdot EC$, or $x^2 = 16$. Therefore, $x = 4$ and $BD = 2x = 8.$ ❏

THE SECANT/TANGENT
THEOREM

Figure 4.60

THEOREM 3: SECANT/TANGENT THEOREM

If a secant $\overrightarrow{PA}$ and tangent $\overrightarrow{PC}$ meet a circle at the respective points A, B, and C (point of contact), then (Figure 4.60)

$$PC^2 = PA \cdot PB$$

(The proof is very similar to that of Theorem 2; see Problem 14.)

COROLLARY: TWO SECANT THEOREM

If two secants $\overrightarrow{PA}$ and $\overrightarrow{PC}$ of a circle meet the circle at A, B, C, and D, respectively (Figure 4.61), then

$$PA \cdot PB = PC \cdot PD$$

Proof: Draw a tangent from P and apply the Secant/Tangent Theorem to both secants. (See Figure 4.62.)

THE TWO-SECANT THEOREM

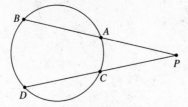

Proof of the Two-Secant Theorem

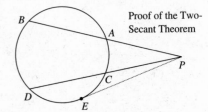

Figure 4.61

Figure 4.62

PROBLEMS   (§4.5)

_____ *Group A* _____

1. A circle has radius 6 units. A tangent is drawn from an external point 10 units from the center. If P is the external point and A the point of contact with the circle, find PA.

2. In the figure shown, Q is the center of the circle, $\overline{SQ} \perp \overline{PR}$, and $m\angle QSR = 60$. Find $m\overset{\frown}{PT}$.

3. Find the length of chord $\overline{PT}$ in Problem 2 if the radius of the circle equals 4 units.

4. *Given:* Isosceles $\triangle ABC$, with base $\overline{BC}$. (See the following figure.)

   Prove: If P is any point on minor arc $\overset{\frown}{BC}$, then ray $\overrightarrow{PA}$ bisects $\angle BPC$.

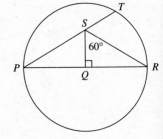

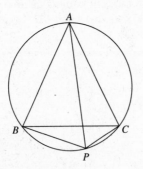

5. Using standard area formulas, find the area of each of the following figures. (In **(e)**, find the total area inside the outer boundary of the two circles.)

(a)

(b)

(c)

(d)

(Regular hexagon)

(e)

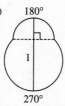

6. Two perpendicular secants from P meet a circle at A, B, C, and D, as shown, with $PA = 2$, $PB = 6$, $PC = 3$ and $PD = 4$. Find the radius r.

7. Prove the two corollaries of the Inscribed Angle Theorem which involve an angle formed by

(a) two intersecting chords

(b) two intersecting secants.

(See figure below for hints.)

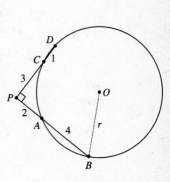

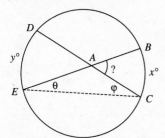

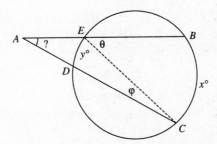

8. Prove that an angle formed by a chord and tangent of a circle has measure one-half the intercepted arc (in the figure, $m\angle BAC = \frac{1}{2}x$). (**HINT:** Let $\overleftrightarrow{BD}$ be constructed parallel to $\overleftrightarrow{AC}$, drop the perpendicular $\overline{AE}$ to $\overline{BD}$, and consider $\triangle ABD$. Why is this triangle isosceles?)

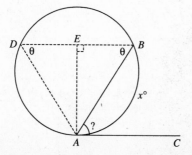

9. Polygon FANCYPROBLEM is a regular dodecagon inscribed in a circle.

   (a) Find the measure of each of the arcs $\overset{\frown}{FA}$, $\overset{\frown}{AN}$, $\overset{\frown}{NC}$, ...

   (b) What argument would prove these measures are equal?

   (c) Find $m\angle RPY$ in two different ways, using the concepts from this section, and those from the previous one involving properties of regular polygons. Compare your answers.

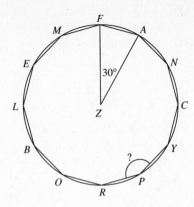

10. In the figure, $\overline{MT} \perp \overline{RW}$, $VN = VM = VW = 3$, $RN = 1$, and $m\overset{\frown}{MW} = 75$. Find

   (a) VT

   (b) NS

   (c) $m\overset{\frown}{RS}$

   (d) $m\overset{\frown}{ST}$.

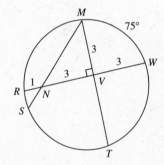

11. A circle P of radius 10 units passes through the center Q of another circle, of radius $r > 10$, cutting off an arc of measure 90 on the larger circle. Find r. (See the figure at the right.)

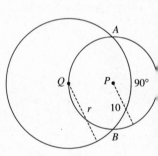

_____ **Group B** _____

12. In the figure below, the center of one circle lies on the other.

   (a) What is the relationship between the measures of the two minor arcs having endpoints at A and B?

   (b) Find $m\overset{\frown}{AB}$ if the circles have equal radii.

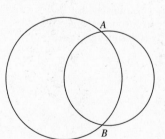

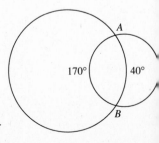

13. Two circles intersect at A and B. If $m\overset{\frown}{AB} = 40$ on the larger circle and $m\overset{\frown}{AB} = 170$ on the smaller, and if the radius of the larger circle is 12, find the radius of the smaller circle. Use trigonometry and geometric facts about circles.

14. In the accompanying figure, $\overrightarrow{PC}$ is tangent to the circle at point C and line $\overrightarrow{PA}$ is a secant.

   (a) Using similar triangles, prove that $PC^2 = PA \cdot PB$. (**HINT:** Why is $\angle B \cong \angle ACP$? What two triangles are similar?)

   (b) Prove the Two Secant Theorem stated as a corollary of Theorem 3.

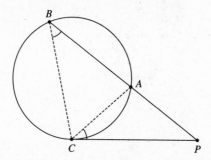

15. In right $\triangle ABC$, $\overrightarrow{MR}$ is the perpendicular bisector of the hypotenuse $\overline{AB}$, and $MR = MA$. Prove: Ray $\overrightarrow{CR}$ bisects $\angle ACB$. (**HINT:** Draw a circle through A, B, and C.)

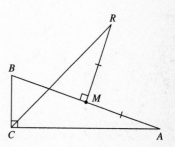

The next sequence of problems is devoted to the interesting relationship between the *power* of a point P with respect to a circle (defined in Problem 16), and the constant products $AP \cdot PB$ relative to variable chords and secants through P.

16. The Power of a Point The **power** of point P with respect to a circle with center O and radius r is the real number

$$\text{Power }(P) = PO^2 - r^2$$

Note that this number is positive if P lies outside the circle, is zero if P lies on the circle, and is negative if P lies interior to the circle.

   (a) Prove that if P lies outside circle O and $\overrightarrow{PT}$ is tangent at B, then

   $$\text{Power }(P) = PT^2$$

   (b) Identify the set of all points P for which

   $$\text{Power }(P) = k \text{ (a constant)}$$

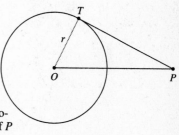

17. (a) In the following figure, use the Cevian Formula and directed distance (as introduced in Problems 20 and 22, §4.3) to prove that, regardless of the location of P inside or outside the circle,

   $$\text{Power }(P) = PA \cdot PB$$

   (b) Prove Theorems 2, 3, and the corollary to Theorem 3 as immediate corollaries of this result.

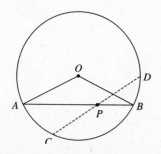
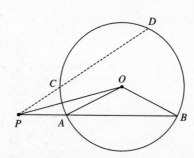

18. Using the accompanying figure and the result of Problem 17, but *independent of the other results of this section,* prove that $m\angle C = m\angle D = \frac{1}{2}m\angle AOB$, where C is any point on the circle for which $\angle ACB$ subtends the minor arc $\overset{\frown}{AB}$, and D is the point of intersection of the circle and the perpendicular from O to chord $\overline{AB}$. (What happens if arc $\overset{\frown}{ASB}$ is a semicircle or a major arc?)

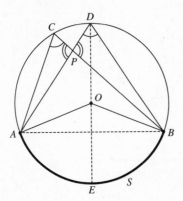

19. The Coast Guard has determined that the waters near a certain point O on shore are treacherous due to rocks hidden below the surface. The warning has been issued to all ships sailing in these waters to stay outside of a region marked by two warning lights at A and B (on shore) at a distance of 10 miles on either side of O and all other points within 10 miles of O. What simple method of sighting could be used by the captain of a ship in order to guarantee that he obeys this order? Prove your answer. (Assume a relatively straight shore line.)

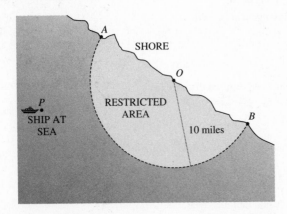

20. (a) In the accompanying figure, $\triangle ABC$ is an isosceles triangle inscribed in the circle, with $AB = AC$. If $\overline{CD}$ is a diameter, find the measures of $\angle 1$, $\angle 2$, $\angle 3$, and $\angle 4$ in terms of $\theta \equiv m\angle ABC$.

(b) What is the value of θ if $\triangle OBC$ is equilateral?

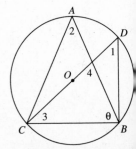

_____ *Group C* _____

21. In the figure at right, $\overline{BC}$ and $\overline{CD}$ are congruent chords of circle O, $\triangle ABC$ is isosceles with base $\overline{BC}$. Find the angles of $\triangle OAE$ in terms of $\theta \equiv m\angle OCD$ and $\phi \equiv m\angle ADC$.

22. How To Find the Radius of a Circle In the figure which follows, we are given circle O, $\triangle ABC$ is equilateral, $BC = CD$, and line $\overrightarrow{DA}$ is drawn, cutting the circle again at E. Prove that $AE = r$ (radius of circle O). (***HINT:*** With $m\angle OCD = \theta$, obtain the angles of $\triangle OAE$ in terms of θ, as in Problem 21.)

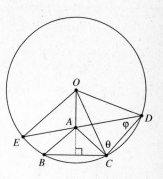

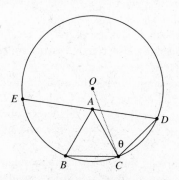

23. A point P varies on the circumcircle of $\triangle ABC$, and for each position of P a point X is located on segment $\overline{PC}$ such that $PX = PB$.

   (a) Find the locus of point X.

   (b) Do you notice anything peculiar about the configuration when $m\angle A = 60$? Generalize.

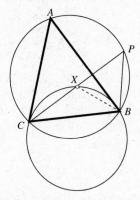

24. Problem of Appolonius Let three circles having equal radii be given. Find a Euclidean construction for a fourth circle that is tangent to the other three.

   NOTE: The general problem of constructing a circle tangent to three given circles is the famous **Problem of Appolonius**. Appolonius was a Greek geometer who lived around 250 B.C.

25. Computer Problem With the appropriate computer software, explore the following geometric problem: Let two circles C_1 and C_2 be given. Find the locus of the center C of a circle C tangent (externally or internally) to both C_1 and C_2.

4.6 COORDINATE GEOMETRY AND VECTORS

In 1637, the great mathematician/philosopher René Descartes made a discovery that would revolutionize geometry. He found that geometric configurations could be described entirely by coordinated real number pairs, and two-variable equations. From this concept ultimately emerged the entire field of real analysis, vectors, linear algebra, and matrix theory. Without it, calculus as we know it could not have developed. No doubt Descartes was one of the "giants" Newton was referring to in his famous quote.

We shall run quickly through the major steps in creating what are called **Cartesian coordinates**—sometimes also referred to as **coordinate geometry**—and **vectors**, or **vector geometry**. The previous development is distinguished from the present one by the term **synthetic**, versus the **analytic method** we are about to introduce. It is expected that you will be familiar with most of these ideas. As you read through this material, you should focus on the deeper question of validity and the existence of a "linear" coordinate system for Euclidean geometry.

THE COORDINATE PLANE:
THE EQUATION OF A LINE

Let ℓ_1 and ℓ_2 be two perpendicular lines intersecting at point O, and on each line take a coordinate system as guaranteed by the Ruler Postulate so that O has the zero coordinate on each line. (See Figure 4.63.) These lines are, respectively, the **coordinate axes** for the system. For any point P in the plane, drop perpendiculars to each line ℓ_1 and ℓ_2; if the feet of the perpendiculars meet ℓ_1 and ℓ_2 at the points $Q[x]$ and $R[y]$, respectively, then assign to P the **coordinate pair** (x, y). Conversely, every ordered pair of real numbers (x, y) is a coordinate pair of a unique point P. We designate this relation by

$$P(x, y)$$

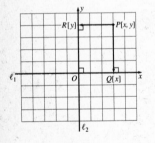

Figure 4.63

Thus, ℓ_1 is called the **x-axis** and ℓ_2 the **y-axis**.

One major result at this point is that the lines of the plane are sets of number pairs whose coordinates satisfy first-degree equations. To see this, let ℓ be any **non-vertical line** which makes an angle of measure $\theta \neq 90$ with the x-axis (Figure 4.64), and suppose ℓ intersects the y-axis at $B[b] \equiv B(0, b)$. If $P(x, y)$ is any point on ℓ, then, in right triangle $\triangle PQB$ (since $\theta \neq 90$),

$$\tan \theta = m = \frac{y - b}{x}$$

Conversely, if the coordinates of P' satisfy this relation, then

$$\tan \theta' = \frac{y - b}{x} = \tan \theta$$

hence $m\angle P'BQ = \theta = m\angle PBQ$ so that $P' = P$ and P' lies on ℓ. Solving the preceding equation for y, we deduce that ℓ has the equation $y = mx + b$, where $m = \tan \theta$ (the **slope** of ℓ), θ the **inclination** of ℓ, and b the **y-intercept** of ℓ. The equation of a **vertical line**, one that is parallel to the y-axis (with inclination $\theta = 90$) is clearly $x = a$, where a is some real constant. Thus, all lines have an equation of the form $ax + by + c = 0$, and hence the term **linear equation**.

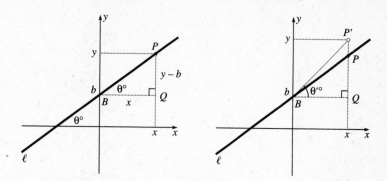

Figure 4.64

SLOPES AND GEOMETRIC RELATIONSHIPS BETWEEN LINES

It is easy to establish by geometry that *two nonvertical lines are parallel iff they have equal slopes.* This is because if $m_1 = m_2$ and hence $\tan \theta_1 = \tan \theta_2$, or $\theta_1 = \theta_2$, then by Property F, $\ell_1 \parallel \ell_2$. A good problem for you is to prove, by geometry, that *two nonvertical, nonhorizontal lines are perpendicular iff the product of their slopes is –1.* (See Figure 4.65 for a hint; if $P = (1, m_1)$, $Q = (1, 0)$, and $R = (1, m_2)$, where $m_2 < 0$, why is $PQ \cdot QR = 1$ and $m_1 m_2 = -1$?)

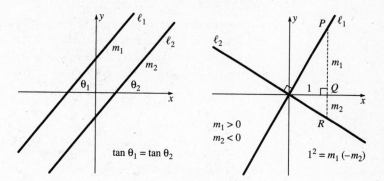

Figure 4.65

FORMULAS FOR SLOPE AND DISTANCE

Finally, we derive the analytic formulas for slope and distance. Let $P(x_1, y_1)$ and $Q(x_2, y_2)$ be two points in the plane (Figure 4.66). Construct the lines parallel to the coordinate axes through P and Q, as shown, which then determines a third point $R(x_2, y_1)$. Note that since $\vec{PR}$ and $\vec{QR}$ are, respectively, horizontal and vertical lines, $PR = \left| x_2 - x_1 \right|$ and $QR = \left| y_2 - y_1 \right|$. Since $\triangle PQR$ is a right triangle

$$PQ^2 = PR^2 + RQ^2$$

or

(1)
$$PQ = \sqrt{\left(x_1 - x_2\right)^2 + \left(y_1 - y_2\right)^2}$$

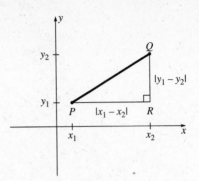

Figure 4.66

We leave the proof of the following standard slope formula as a problem: If $x_1 \neq x_2$, the slope of the line $\overleftrightarrow{PQ}$ (with the same coordinates as before) is given by

(2)
$$m = \frac{y_1 - y_2}{x_1 - x_2}$$

EXAMPLE 1   Use coordinate geometry to prove that the diagonals of a rhombus are perpendicular.

SOLUTION

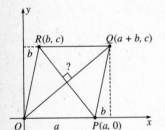

Figure 4.67

It is immaterial where we locate the coordinate system, so we can suppose that the vertices of the rhombus have the coordinates as shown in Figure 4.67. Since the sides of the pararallelogram have equal lengths, it follows from (1) that $a^2 = OP^2 = PQ^2 = (a + b - a)^2 + (c - 0)^2$ or $a^2 = b^2 + c^2$. Also, by (2),

$$m_1 = \{\text{Slope } \overleftrightarrow{OQ}\} = \frac{c - 0}{a + b - 0} = \frac{c}{a + b}$$

$$m_2 = \{\text{Slope } \overleftrightarrow{PR}\} = \frac{0 - c}{a - b} = -\frac{c}{a - b}$$

We must show that $m_1 m_2 = -1$.

$$m_1 m_2 = -\frac{c^2}{(a + b)(a - b)} = -\frac{c^2}{a^2 - b^2} = \frac{-c^2}{c^2} = -1$$

Therefore, $\overleftrightarrow{OQ} \perp \overleftrightarrow{PR}.$ ❏

VECTORS

Coordinate geometry provides the basis for introducing another tool having tremendous power in the study of geometry, with far-reaching applications in mathematics and science. We define a **vector** v to be a *column of two real numbers*, called its **components**, as in

$$v = \begin{bmatrix} x \\ y \end{bmatrix}$$

We concentrate first on just the simple algebraic properties of vectors.

VECTOR SUM, LINEAR COMBINATION OF VECTORS

We can define a **zero** for vectors, we can **add** two vectors, and **multiply** a real number times a vector, and form arbitrary **linear combinations** of vectors, according to the following definitions (let $u = [x, y]$ and $v = [z, w]$):

ZERO VECTOR:
$$0 = \begin{bmatrix} 0 \\ 0 \end{bmatrix}$$

SUM OF TWO VECTORS:
$$u + v = \begin{bmatrix} x \\ y \end{bmatrix} + \begin{bmatrix} z \\ w \end{bmatrix} = \begin{bmatrix} x + z \\ y + w \end{bmatrix}$$

SCALAR MULTIPLICATION:
$$au = a\begin{bmatrix} x \\ y \end{bmatrix} = \begin{bmatrix} ax \\ ay \end{bmatrix}$$

LINEAR COMBINATION:
$$au + bv = a\begin{bmatrix} x \\ y \end{bmatrix} + b\begin{bmatrix} z \\ w \end{bmatrix} = \begin{bmatrix} ax + bz \\ ay + bw \end{bmatrix}$$

INNER PRODUCT OF TWO VECTORS, LENGTH OR NORM OF A VECTOR

We can also multiply two vectors, although the kind of **product** we consider produces a *real number* (a **scalar**) instead of another vector: If $u = [x, y]$ and $v = [z, w]$, then

INNER PRODUCT:
$$u \cdot v = \begin{bmatrix} x \\ y \end{bmatrix} \cdot \begin{bmatrix} z \\ w \end{bmatrix} = xz + yw$$

For convenience, we shall use the notation uv for $u \cdot v$ and u^2 for $\mathbf{u} \cdot \mathbf{v}$ when $\mathbf{u} = \mathbf{v}$.

The resulting **commutative** and **distributive** laws are obvious: For arbitrary vectors u, v, w:

$$uv = vu$$

and

$$u(v + w) = uv + uw$$

Also, we have the obvious **associative** laws for scalar and dot products, $(au)v = u(av) = a(uv) \equiv auv$. It then follows that

(3) $$(au + bv)^2 = (au + bv) \cdot (au + bv) = a^2u^2 + 2abuv + b^2v^2$$

This rule has the same appearance as the ordinary algebraic rule for squaring a binomial, but do not be misled! There is a lot more to this rule than appears, and it has some geometric power built into it, as will soon be evident.

Finally, we introduce the **length** or **norm** of a vector:

LENGTH OF A VECTOR: $$\|v\| = \sqrt{v^2}$$

This too, is misleading; it looks like the familiar rule for absolute values of real numbers $|a| = \sqrt{a^2}$. But a closer examination shows that, in reality, there is also hidden information in this equation, for it can be expanded into the form

$$v^2 = vv = \begin{bmatrix} x \\ y \end{bmatrix} \cdot \begin{bmatrix} x \\ y \end{bmatrix} = x^2 + y^2, \quad \text{or} \quad \|v\| = \sqrt{x^2 + y^2}$$

Thus, from the Distance Formula, we find that $\|v\|$ is just the distance from $O(0, 0)$ to $P(x, y)$. Note the following basic properties, obvious from the definition.

$$\|v\| \geq 0, \text{ with equality only when } v = 0$$

$$\|av\| = |a| \, \|v\| \text{ for any scalar } a$$

The first inequality implies an important property of the dot product:

$$v^2 > 0 \text{ unless } v = 0.$$

VECTORS AS DIRECTED SEGMENTS; TRIANGLE ADDITION LAW

In order to actually use vectors in geometry, we must show how they are to be connected to objects in geometry.

DEFINITION

A **directed line segment** from A to B, denoted AB, is the ordinary segment $\overline{AB}$ with a **direction**, defined as the ordered pair (A, B), where point A is called the **initial point** and B the **terminal point** of the directed segment.

(Thus, each segment $\overline{AB}$ as previously defined has two directed segments associated with it, namely AB and BA.)

The **vector representation** of the directed line segment AB, where $A(a, b)$ and $B(c, d)$ are any two points, is now defined as the vector v whose components are precisely the numbers $c - a$ and $d - b$. (See Figure 4.68.) That is,

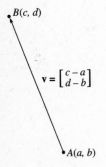

$$v = \begin{bmatrix} c - a \\ d - b \end{bmatrix}$$

Figure 4.68

VECTOR REPRESENTATION OF DIRECTED LINE SEGMENT

$$(4) \quad v = v(AB) = \begin{bmatrix} c - a \\ d - b \end{bmatrix} \qquad \text{where } A = (a, b), B = (c, d)$$

In reality, we are defining a mapping or function *from the set of all directed line segments to the set of all vectors*. We could ask if the mapping is one-to-one, that is, for two distinct segments AB and CD, is it possible to have

$$v(AB) = v(CD)?$$

It is easy to convince yourself that this is true iff the quadrilateral $\Diamond ABDC$ is a parallelogram (perhaps degenerate), as indicated in Figure 4.69. We omit the proof although simple, it is in no way a trivial matter—this is the basis for virtually all the applications of vectors to engineering mathematics and physics.

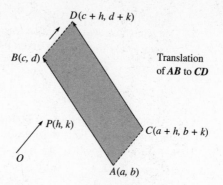

Figure 4.69

Although it is inaccurate to do so, it is customary to *identify* the vector with its geometric representation, as in

$$v = AB$$

No great harm is done as long as we understand that equality here is taken as an *equivalence up to translation*. Under this identification, note that if

$$AB = u \qquad \text{and} \qquad BC = v$$

then

(5) $$AB + BC = u + v = AC$$

(*Proof:* Suppose that $A = (a, b)$, $B = (c, d)$, $C = (e, f)$; if $AB = u$ and $BC = v$, then $u = [c - a, d - b]$ and $v = [e - c, f - d]$, and therefore $u + v = [c - a + e - c, d - b + f - d] = [e - a, f - b] = AC$.)

The relation **(5)** is called the **Triangle Law for Addition**, and may also be used to generate an equivalent **Parallelogram Law for Addition**, as indicated in Figure 4.70, along with the geometric interpretation for scalar multiplication.

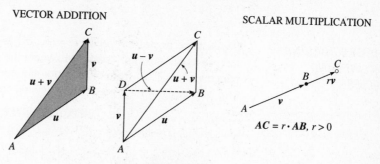

Figure 4.70

We have noted that each vector may be associated with an *infinite number* of directed line segments, parallel to one another and having equal lengths. If we consider only directed segments of the form *OP*, where *O* is the origin, then there is a *one-to-one correspondence* between vectors and such directed segments. These are called the **position vectors**, a misnomer because they are not vectors at all but directed line segments. The association between position vectors and the algebraic vectors they represent is given by

$$v = \begin{bmatrix} x \\ y \end{bmatrix} \leftrightarrow OP, \text{ where } P = (x, y)$$

LAW OF COSINES

An important question in both geometry and analysis is: What does $uv = 0$ mean geometrically? First we note from the Law of Cosines that for any triangle $\triangle ABD$ (Figure 4.70), where we let $u = AB$, $v = AD$, and $DB = DA + AB = -v + u$, with $\theta = m\angle BAD$,

$$BD^2 = AB^2 + AD^2 - 2AB \cdot AD \cdot \cos \theta$$

or

$$\|u - v\|^2 = \|u\|^2 + \|v\|^2 - 2\|u\| \|v\| \cos \theta$$

and, from **(3)**,

$$u^2 - 2uv + v^2 = u^2 + v^2 - 2\|u\| \|v\| \cos \theta$$

This last expression simplifies to

(6)   **LAW OF COSINES FOR VECTORS**: $uv = \|u\| \|v\| \cos \theta$

where θ is the measure of the angle between u and v (as represented geometrically). Thus, after setting $\theta = 90$, we deduce the following result.

THEOREM 1

Two nonzero vectors u and v are perpendicular (as directed line segments) iff $uv = 0$. (In this case, the vectors u and v are called **orthogonal**.)

VECTOR EQUATIONS FOR LINES AND CIRCLES

Suppose we regard the vector $a = [a_1, a_2]$ as a **direction** in the plane (actually, it represents the position vector OA, where the coordinates of A are the vector components of a). Then from the addition law for vectors (Figure 4.71), a line parallel to a and passing through $B(x_0, y_0)$ would have the vector form for some real

$$OX = OB + BX = OB + ta$$

$$x = ta + x_0 \quad \text{or} \quad \begin{bmatrix} x \\ y \end{bmatrix} = t\begin{bmatrix} a_1 \\ a_2 \end{bmatrix} + \begin{bmatrix} x_0 \\ y_0 \end{bmatrix} = \begin{bmatrix} ta_1 + x_0 \\ ta_2 + y_0 \end{bmatrix}$$

where $x_0 = [x_0, y_0]$.

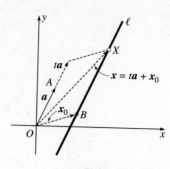

Figure 4.71

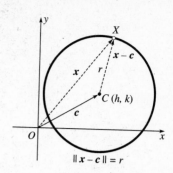

Figure 4.72

Thus, by reading off the components, we have actually found the **parametric equations** for ℓ:

$$\begin{cases} x = ta_1 + x_0 \\ y = ta_2 + y_0 \end{cases} \quad (t \text{ real})$$

It is clear that if $X(x, y)$ is any point on the circle having radius r and centered at $C(c_1, c_2)$, then with x as the vector OX, and $c = OC$, we have $XC = r$, or

$$\|x - c\| = r \text{ , where } x - c = \begin{bmatrix} x \\ y \end{bmatrix} - \begin{bmatrix} c_1 \\ c_2 \end{bmatrix} = \begin{bmatrix} x - c_1 \\ y - c_2 \end{bmatrix}$$

This is the **vector equation** of circle C—totally equivalent to the coordinate form

$$\sqrt{(x - c_1)^2 + (y - c_2)^2} = r$$

or

$$(x - c_1)^2 + (y - c_2)^2 = r^2$$

ELEMENTARY APPLICATIONS

As an example of what we can do with vectors, we are going to prove the Triangle Inequality. This does not sound like much, but you should recall that this was a theorem in Chapter 3 that took practically all the major theorems we had at our disposal to use at the time (in particular, it depended on the Exterior Angle Inequality, which itself involved many key geometric principles, including the SAS Postulate). We think you will be struck with the elegance and efficiency of the vector proof. It definitely demonstrates the power of linear algebra and its usefulness to geometry.

THEOREM 2: TRIANGLE INEQUALITY, VECTOR FORM

For any two vectors u and v, $\|u + v\| \leq \|u\| + \|v\|$, with equality only when u and v represent two collinear directed line segments (Figure 4.73).

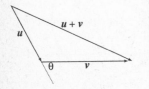

Figure 4.73

THEOREM 2: TRIANGLE INEQUALITY, VECTOR FORM (CONTINUED)

Proof: Since all quantities are nonnegative, let's square the desired inequality.

$$(7) \qquad \|u + v\|^2 \le (\|u\| + \|v\|)^2 = \|u\|^2 + 2\|u\|\,\|v\| + \|v\|^2$$

We also have

$$\|u + v\|^2 = (u + v)^2 = (u + v) \cdot (u + v)$$

$$= u^2 + u \cdot v + v \cdot u + v^2 \le \|u\|^2 + 2uv + \|v\|^2$$

By substitution of this expression into (7) and making the obvious cancellations, we find the desired Triangle Inequality is equivalent to

$$(8) \qquad\qquad uv \le \|u\|\,\|v\|$$

But using the vector Law of Cosines (6), we have, since $\cos\theta < 1$,

$$(9) \qquad\qquad uv = \|u\|\,\|v\|\,\cos\theta \le \|u\|\,\|v\|$$

Thus we have proven the desired inequality. As for equality, this holds only if it is true in both (8) and (9), which can happen only when $\cos\theta = 1$ or $\theta = 0$, or when u and v are collinear.

NOTE: The inequality (8) in the previous proof is the famous **Cauchy-Schwarz Inequality** of vector analysis.

EXAMPLE 2   Use vectors to prove that the medians of a triangle are concurrent, and that the point of concurrency is located on each median at a point that is two-thirds the distance from the vertex to the midpoint of the opposite side.

SOLUTION

It suffices to prove that any two medians, say $\overline{AL}$ and $\overline{BM}$, meet at G such that $AG = \frac{2}{3}AL$ and $BG = \frac{2}{3}BM$ (Figure 4.74). We will prove this by *defining* G to be the point on $\overline{AL}$ such that $AG = \frac{2}{3}AL$, and G' that point on $\overline{BM}$ such that $BG' = \frac{2}{3}BM$, then showing that $G = G'$. Using vectors, we have

$$AL = AB + BL = AB + \tfrac{1}{2}BC \quad \rightarrow \quad AG = \tfrac{2}{3}AL = \tfrac{2}{3}AB + \tfrac{1}{3}BC$$

Similarly,

$$BM = BA + AM = BA + \tfrac{1}{2}AC \quad \rightarrow \quad BG' = \tfrac{2}{3}BM = \tfrac{2}{3}BA + \tfrac{1}{3}AC$$

Then, by substitution,

$$AG' = AB + BG' = AB + \tfrac{2}{3}BA + \tfrac{1}{3}AC$$

$$= AB - \tfrac{2}{3}AB + \tfrac{1}{3}(AB + BC)$$

$$= \tfrac{2}{3}AB + \tfrac{1}{3}BC$$

That is,

$$AG' = AG \quad \text{or} \quad G' = G,$$

as desired. ❏

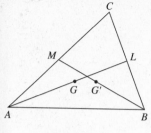

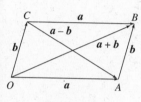

Figure 4.74

Figure 4.75

EXAMPLE 3 Give a vector proof that the diagonals of a parallelogram are perpendicular iff it is a rhombus.

SOLUTION

The sides and diagonals may be labelled as shown in Figure 4.75. Now $OB \perp AC$ iff

$$(\boldsymbol{a} + \boldsymbol{b}) \cdot (\boldsymbol{a} - \boldsymbol{b}) = 0$$

or

$$a^2 - b^2 = 0$$

That is, iff

$$a^2 = b^2 \quad \text{or} \quad \|\boldsymbol{a}\|^2 = \|\boldsymbol{b}\|^2 \quad \leftrightarrow \quad \|\boldsymbol{a}\| = \|\boldsymbol{b}\| \quad \leftrightarrow \quad OA = AB,$$

precisely when $\lozenge OABC$ is a rhombus. ❏

EXAMPLE 4 Use vectors to prove the midpoint formula for the midpoint $M(x_3, y_3)$ of the segment joining $P(x_1, y_1)$ and $Q(x_2, y_2)$:

(10) $$x_3 = \frac{x_1 + x_2}{2}, \quad y_3 = \frac{y_1 + y_2}{2}$$

SOLUTION

With O = origin,

$$\begin{bmatrix} x_3 \\ y_3 \end{bmatrix} = \boldsymbol{OM} = OP + PM = OP + \frac{1}{2} PQ = \begin{bmatrix} x_1 \\ y_1 \end{bmatrix} + \frac{1}{2} \begin{bmatrix} x_2 - x_1 \\ y_2 - y_1 \end{bmatrix}$$

$$= \begin{bmatrix} \frac{1}{2}(x_1 + x_2) \\ \frac{1}{2}(y_1 + y_2) \end{bmatrix}$$ ❏

PROBLEMS ($\S4.6$)

_____ **Group A** _____

All problems in this section are to be worked in the coordinate plane. In most situations, it is best to choose a coordinate system so that a given triangle $\triangle ABC$ has vertices $A(a, 0)$,

$B(b, 0)$, and $C(0, c)$, where $b > 0$, $b > a$, and $c > 0$. ($a < 0$ unless $\angle A \geq 90$). An arbitrary parallelogram can be assumed to have vertices: $A(a, 0)$, $B(b, 0)$, $C(b - a, c)$, and $D(0, c)$.

1. If $A(-2, 5)$, $B(1, 7)$, $C(4, -4)$ are the vertices of a triangle,
    (a) determine the numerical components of $u = AB$, $v = AC$, and $w = BC$. Verify that $u + w = v$.
    (b) Show that $\triangle ABC$ is a right triangle.
    (c) Find $\cos B$ using the Vector Law of Cosines.

2. Give a coordinate proof that the diagonals of a parallelogram bisect each other.

3. Give a coordinate proof that a parallelogram is a rectangle iff its diagonals are congruent.

4. Give a coordinate proof that the segment joining the midpoints of two sides of a triangle is parallel to the third side and has length equal to one-half that of the third side.

5. Give a coordinate proof that the midpoints of the sides of any quadrilateral are the vertices of a parallelogram. Note that three vertices A, B, and D can be taken as above, but C must be taken as a general point, $C(d, e)$.

6. Give a vector proof that the midpoints of the sides of a rhombus are the vertices of a rectangle.

7. State and prove the vector form of the Pythagorean Theorem. Is the converse clear from your proof? (**HINT:** Use $u + v$ for the hypotenuse and u and v for the two legs. In the proof, use $(u + v)^2 = (u + v) \cdot (u + v)$.)

_____ *Group B* _____

8. Two adjacent sides $\overline{OS}$ and $\overline{OT}$ of $\diamond OSRT$ are represented by position vectors $u = [3, 6]$ and $v = [1, 5]$.
    (a) Plot this in the coordinate plane and make a sketch of the parallelogram thereby determined.
    (b) Calculate the area K of $\diamond OSRT$ using either trigonometry or Heron's formula.
    (c) Calculate the number $\sqrt{u^2 v^2 - (uv)^2}$.
    (d) Calculate the determinant (formed by the vector components themselves):
    $$\begin{vmatrix} 3 & 1 \\ 6 & 5 \end{vmatrix}$$
    (e) Did anything happen? (**HINT:** You should have obtained the same answer three times.)

9. **Vector Problem: "Blowin' In the Wind"** A small aircraft with a cruising speed of 145 mph proceeds to a destination 425 miles due north, taking $3\frac{1}{8}$ hours. On the return trip, due south, with the same prevailing wind conditions, the trip took only $2\frac{5}{6}$ hours. In what direction was the wind blowing, and with what velocity? (**HINT:** Let $w = [x, y]$ be the wind vector and $v = [r, s]$ the vector of the aircraft *heading*, not *bearing*. Then by hypothesis, $r^2 + s^2 = 145^2$. Continue.)

10. Using the coordinates for the vertices of $\triangle ABC$ suggested at the head of this problem section, show the following.
    (a) If H is the point having coordinates $(0, -ab/c)$, then $\overline{AH} \perp \overline{BC}$, $\overline{BH} \perp \overline{AC}$, and $\overline{CH} \perp \overline{AB}$. What theorem in geometry does this prove?
    (b) If D, E, and F are the feet of the altitudes from A, B, and C on $\overline{BC}$, $\overline{AC}$, and $\overline{AB}$, respectively (assuming that $\triangle ABC$ is acute-angled), show that $\overleftrightarrow{FD}$ and $\overleftrightarrow{FE}$ have respective equations

$$y = \frac{\pm(bc - ac)x}{(ab + c^2)}$$

What does this prove about the sides of $\angle EFD$ in relation to altitude $\overline{CF}$?

(c) What is the ultimate consequence regarding H and the orthic triangle $\triangle DEF$ for acute-angled triangles $\triangle ABC$?

11. **Using a Hyperbola to Trisect An Angle** The following construction was discovered by Pappus, A.D. 300. Justify and explain each of the steps in the construction, as shown in the figure below:

(1) Let $\angle ABC$ be a given angle. In any xy-coordinate system, consider the graph of the hyperbola $3y^2 - x^2 = 12$, and the line $y = 1$, or $\overleftrightarrow{QW}$.

(2) Construct a line through $F(0, 4)$ (the focus of the hyperbola), cutting the line $\overleftrightarrow{QW}$ at some point D such that $\angle FDW \cong \angle ABC$. (This may be accomplished by first constructing $\angle PQR \cong \angle ABC$ on $\overleftrightarrow{QW}$, then drawing $\overrightarrow{FD} \parallel \overrightarrow{PQ}$.)

(3) With D as center and DF as radius, draw a circle, cutting the hyperbola at $E(x, y)$.

(4) Ray $\overrightarrow{DE}$ is a trisector of $\angle FDW \cong \angle ABC$.

(**HINT**: Drop perpendicular $\overline{EG}$ to line $\overleftrightarrow{QW}$; use coordinate geometry to show that $FE = 2EG$, then consider $DM \perp FE$.)

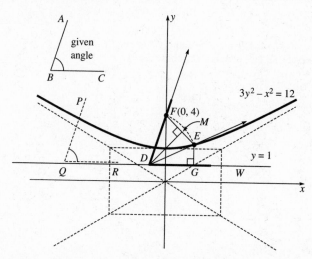

12. Give a coordinate proof of the Secant Theorem proven in Section 3.8. (**HINT**: With no loss of generality, we may assume that the circle has equation $x^2 + y^2 = r^2$ and the interior point A given in the theorem has coordinates $(a, 0)$, where $0 < a < r$. Let ℓ have equation $x = a$ (if ℓ is vertical) or $y = mx + b$ (otherwise), passing through A (thus, $0 = ma + b$ or $b = -ma$ and $y = m(x - a)$). Now solve the system of two equations (for the circle and the line) and prove that there are always two points of intersection.)

13. **The Parallelogram Law For Vectors** A famous law of vectors, known as the **Parallelogram Law,** is useful in topology and real analysis. It is the identity for any two vectors u and v,

$$\|u + v\|^2 + \|u - v\|^2 = 2\|u\|^2 + 2\|v\|^2$$

Interpret this geometrically, and prove in two ways:

(a) by vectors and inner products

(b) by geometry.

(*HINT:* For the vector proof, simply expand $\|u \pm v\|^2 \equiv (u \pm v)^2$ using properties of the inner product, then sum. For the geometric proof, use the formula for the median of a triangle. See Equation (**12**) following Example 5 in §4.3.)

14. **The Euler Line** Let the vertices of $\triangle ABC$ have the coordinates suggested at the beginning of this problem section.

(a) Show that if $d = ab/c$, the orthocenter H, circumcenter O, centroid G, and Nine-Point Center U have the coordinates $H(0, -d)$ (see Problem 10), $O(\frac{1}{2}a + \frac{1}{2}b, \frac{1}{2}c + \frac{1}{2}d)$, $G(\frac{1}{3}a + \frac{1}{3}b, \frac{1}{3}c)$ (see Problem 15), and $U(\frac{1}{4}a + \frac{1}{4}b, \frac{1}{4}c - \frac{1}{4}d)$. (*HINT:* U is the midpoint of $\overline{XL}$, where X and L are the midpoints of $\overline{AH}$ and $\overline{BC}$, respectively.)

(b) Show that the **Euler Line** $\overleftrightarrow{HG}$ has equation $y = -d + (3d + c)x/(a + b)$, and verify that H, O, G, and U all lie on the Euler line of $\triangle ABC$.

15. Prove by vector methods that the centroid $G(x, y)$ of a triangle whose vertices are $A(a_1, a_2)$, $B(b_1, b_2)$ and $C(c_1, c_2)$ has coordinates

$$x = \tfrac{1}{3}(a_1 + b_1 + c_1), \qquad y = \tfrac{1}{3}(a_2 + b_2 + c_2)$$

(*HINT:* Use the fact that if L is the midpoint of $\overline{BC}$, G occupies the point of $\overline{AL}$ two-thirds the distance from A to L, as shown in Example 2.)

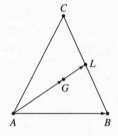

––––––––– *Group C* ––––––––

16. Show that if the endpoints of a segment are $A(p, q)$ and $B(r, s)$, the coordinates x and y of the vertices C_1 and C_2 of the two equilateral triangles having $\overline{AB}$ as base are given by

$$2x = p + r \pm \sqrt{3}\ (q - s)$$
$$2y = q + s \pm \sqrt{3}\ (r - p)$$

where the $+$ or $-$ signs are taken together in the two equations.

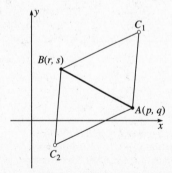

17. Using the formula of Problem 16, prove that if equilateral triangles $\triangle A'BC$, $\triangle AB'C$ and $\triangle ABC'$ be constructed externally on the sides of $\triangle ABC$, then $AA' = BB' = CC'$. (There is a simple synthetic proof of this same proposition.)

18. **NAPOLEON'S THEOREM** The centroids of the equilateral triangles constructed externally on the sides of any triangle ABC are the vertices of another equilateral triangle. (Use the coordinates of A', B', and C' from Problem 17.)

19. **(a)** Squares $\square ABEF$, $\square BCGH$, $\square CDJK$, and $\square DALM$ are constructed externally on the sides of a convex quadrilateral $\lozenge ABCD$, with X, Y, Z, and W their centers. Prove that $XZ = YW$ and $\overline{XZ} \perp \overline{YW}$.

(b) Prove that if $\lozenge ABCD$ is a parallelogram, $\lozenge XYZW$ is a square.

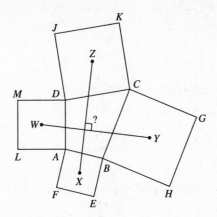

---

*4.7 FAMILIES OF ORTHOGONAL CIRCLES, CIRCULAR INVERSION

The concepts of orthogonal circles and inversion in a circle are closely related topics, as we shall see. They will be used in Chapter 6 in constructing a model for hyperbolic geometry, so we now explore these two topics briefly.

CONDITION FOR ORTHOGONALITY OF CIRCLES

Two circles are said to be **orthogonal** if they meet at right angles. Thus, at either point of intersection, the tangents to the circles are perpendicular, and therefore contain the radii of the two circles, which must also be perpendicular (Figure 4.76). By the Pythagorean Theorem and its converse, we conclude that two circles whose centers are O_1 and O_2, having respective radii r_1 and r_2, are orthogonal iff

(1) $$r_1^2 + r_2^2 = O_1O_2^2$$

We now introduce coordinate geometry to enhance the synthetic method. Recall that a circle with radius r and center $C(h, k)$ has the equation

(2) $$(x - h)^2 + (y - k)^2 = r^2$$

The general equation of a circle, obtained by squaring out the binomials in **(2)**, is

(3) $$x^2 + y^2 + ax + by + c = 0$$

where

$$a = -2h, \quad b = -2k, \quad \text{and} \quad c = h^2 + k^2 - r^2$$

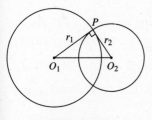

Figure 4.76

Thus, by substitution of $-\frac{1}{2}a$ and $-\frac{1}{2}b$ for h and k in the third equation, we find the formulas

(4) $$h = -\frac{a}{2}, \quad k = -\frac{b}{2}, \quad \text{and} \quad r = \sqrt{\frac{1}{4}a^2 + \frac{1}{4}b^2 - c}$$

which provide the coordinates h and k of the center and the radius r of a circle in terms of the parameters of its general equation **(3)**. In order for the circle not to degenerate to a point or be imaginary, we must obviously require that $r > 0$, or from **(4)**, $\frac{1}{4}a^2 + \frac{1}{4}b^2 - c > 0$, or

(5) $$a^2 + b^2 > 4c$$

LEMMA A: ANALYTIC CRITERION FOR ORTHOGONALITY

The circles $x^2 + y^2 + a_1 x + b_1 y + c_1 = 0$ and $x^2 + y^2 + a_2 x + b_2 y + c_2 = 0$ are orthogonal iff

(6) $$a_1 a_2 + b_1 b_2 = 2c_1 + 2c_2$$

Proof: From **(4)** we conclude that the centers and radii of the two circles are, respectively,

$$O_k(-\tfrac{1}{2}a_k, -\tfrac{1}{2}b_k) \quad \text{and} \quad r_k = \sqrt{\frac{1}{4}a_k^2 + \frac{1}{4}b_k^2 - c_k}, \; k = 1, 2,$$

Hence by **(1)** and the distance formula, assuming the circles are orthogonal,

$$r_1^2 + r_2^2 = O_1 O_2^2 = (-\tfrac{1}{2}a_1 + \tfrac{1}{2}a_2)^2 + (-\tfrac{1}{2}b_1 + \tfrac{1}{2}b_2)^2$$

or, by substituting in the formulas for r_1 and r_2,

$$(\tfrac{1}{4}a_1^2 + \tfrac{1}{4}b_1^2 - c_1) + (\tfrac{1}{4}a_2^2 + \tfrac{1}{4}b_2^2 - c_2) = (\tfrac{1}{4}a_1^2 - \tfrac{1}{2}a_1 a_2 + \tfrac{1}{4}a_2^2) + (\tfrac{1}{4}b_1^2 - \tfrac{1}{2}b_1 b_2 + \tfrac{1}{4}b_2^2)$$

and

$$-c_1 - c_2 = -\tfrac{1}{2}a_1 a_2 - \tfrac{1}{2}b_1 b_2$$

which, by algebra, reduces to the desired result. The steps are reversible, so the theorem is proven in both directions.

COSINE OF ANGLE BETWEEN TWO CIRCLES

Refer back to Figure 4.76. Applying the law of cosines to $\triangle PO_1 O_2$, we find that if θ is the measure of an arbitrary angle between the radii $\overline{O_1 P}$ and $\overline{O_2 P}$, which is also the angle between the two circles (the angle between their tangents at P), its cosine is given by

(7) $$\cos\theta = \frac{r_1^2 + r_2^2 - (O_1 O_2)^2}{2 r_1 r_2}$$

Note that in the proof of Lemma A, had we taken the difference of the two sides of the equation involving $(O_1 O_2)^2$, r_1^2, and r_2^2, the expression would have reduced to

$$r_1^2 + r_2^2 - (O_1 O_2)^2 = \tfrac{1}{2}a_1 a_2 + \tfrac{1}{2}b_1 b_2 - c_1 - c_2 = \tfrac{1}{2}(a_1 a_2 + b_1 b_2 - 2c_1 - 2c_2)$$

This gives us the numerator of the expression in **(7)**. The denominator can be rewritten by use of **(4)** (for r_1 only):

$$r_1^2 = h_1^2 + k_1^2 - c_1 = \tfrac{1}{4}a_1^2 + \tfrac{1}{4}b_1^2 - c_1 \quad \text{or} \quad r_1 = \tfrac{1}{2}\sqrt{a_1^2 + b_1^2 - 4c_1}$$

Hence, making substitutions into (7) and simplifying, we obtain the following interesting formula.

MEASURE OF THE ANGLE BETWEEN TWO CIRCLES

(8)
$$\cos \theta = \frac{a_1 a_2 + b_1 b_2 - 2c_1 - 2c_2}{\sqrt{a_1^2 + b_1^2 - 4c_1}\,\sqrt{a_2^2 + b_2^2 - 4c_2}}$$

This formula depends only on the coefficients a, b, and c of the equations of the two circles written in general form.

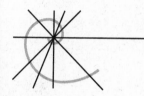

MOMENT FOR DISCOVERY

The Angle Between Two Circles

Consider the circles

$$C_1\colon x^2 + y^2 = 1 \qquad C_2\colon (x - h)^2 + y^2 = 4$$

The first is the unit circle centered at the origin, and the other has its center at $(h, 0)$ on the x-axis, having constant radius 2. Writing these two equations in general form, we have

$$C_1\colon \underbrace{x^2 + y^2 - 1}_{c_1} = 0 \qquad\qquad (a_1 = b_1 = 0)$$

$$C_2\colon x^2 + y^2 - \underbrace{2h}_{a_2}x + \underbrace{h^2 - 4}_{c_2} = 0 \qquad (b_2 = 0)$$

(In the following steps, $\cos \theta$ will be used merely as a symbol for the right side of (8), for convenience. Naturally, when $|\cos \theta| > 1$, a real value for θ does not exist.)

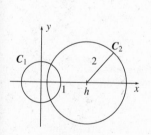

Figure 4.77

1. Compute $\cos \theta$ in terms of h and simplify, to obtain a general formula we can use in subsequent steps. (Did you get $5 - h^2$ for the numerator? The denominator should be a constant.)

2. Calculate $\cos \theta$ for $h = 0$ (when C_1 is inside C_2).

3. Calculate $\cos \theta$ for $h = 1$, $\sqrt{3}$, and $\sqrt{5}$.

4. Calculate $\cos \theta$ when $h = 3$ and 4. What did you discover?

5. Have you discovered a rule that predicts from the general equations of two circles when one is inside or outside the other, whether they intersect, or are tangent?

ORTHOGONAL FAMILIES
OF CIRCLES

We can put Lemma A to work for us in order to quickly write down the equations of circles orthogonal to the unit circle C,

$$x^2 + y^2 = 1$$

(that is, $x^2 + y^2 + 0 \cdot x + 0 \cdot y - 1 = 0$, where $a_1 = b_1 = 0$ and $c_1 = -1$ in the notation of Lemma A).

LEMMA B

A circle is orthogonal to the unit circle C centered at the origin iff its general equation is of the form

$$x^2 + y^2 + ax + by + 1 = 0$$

where a and b are arbitrary reals such that $a^2 + b^2 > 4$.

Proof: By (6), $0 \cdot a_2 + 0 \cdot b_2 = 2(-1 + c_2)$, or $0 = -1 + c_2$. Hence $c_2 = 1$ and $a_2 = a$ and $b_2 = b$ are arbitrary, except for the requirement (5), $a^2 + b^2 > 4c_2 = 4$.

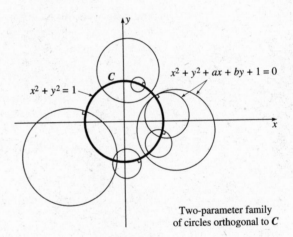

Two-parameter family
of circles orthogonal to C

Figure 4.78

The family of all circles orthogonal to C is thus a **two-parameter** family, because its members depend on both a and b. (See Figure 4.78.) It is interesting to examine a **one-parameter** family, say, by setting $b = 0$. Doing this, of course, restricts all the members of the family to have centers at $(-\frac{1}{2}a, 0)$, that is, on the x-axis, and the condition for nondegeneracy $a^2 + b^2 > 4$ reduces to $|a| > 2$. Thus we find that the one-parameter family of circles given by

$$\mathcal{F}: x^2 + y^2 + ax + 1 = 0 \qquad (a < -2 \text{ or } a > 2)$$

are each orthogonal to C and have their centers on the x-axis. Each member of this family has its center at $(-\frac{1}{2}a, 0)$ which, because of $|a| > 2$, lies *outside* C. Figure 4.79 illustrates this. We note that as the radii of these circles get smaller and smaller, their centers converge to one of the points $A(-1, 0)$ or $B(1, 0)$, and the circles all cluster around those two points.

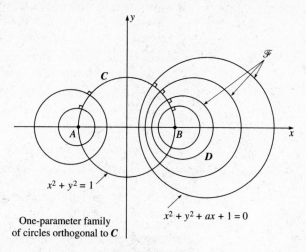

$$x^2 + y^2 = 1$$

One-parameter family
of circles orthogonal to C

$$x^2 + y^2 + ax + 1 = 0$$

Figure 4.79

The next result we derive may seem rather amazing. The Discovery Unit which follows will let you discover a special case of it before we give a formal proof.

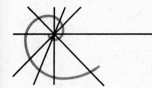

MOMENT FOR DISCOVERY

Orthogonality and Circles With a Common Chord

Consider two particular circles D_1 and D_2 from the family $\mathscr{F}$ mentioned previously, corresponding to $a = -4, -8$:

$$D_1: x^2 + y^2 \underbrace{- 4x}_{a_1} \underbrace{+ 1}_{c_1} = 0$$

$$D_2: x^2 + y^2 \underbrace{- 8x}_{a_2} \underbrace{+ 1}_{c_2} = 0$$

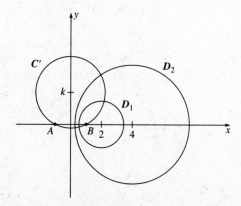

Figure 4.80

1. Write the equation of a circle C' centered at $(0, k)$ on the y-axis and passing through $A(-1, 0)$ and $B(1, 0)$. (***HINT:*** If the center is $C(0, k)$, the radius would be $\sqrt{k^2 + 1}$ by the Pythagorean Theorem.)

2. Expand and simplify, and write this equation in general form $x^2 + y^2 + a'x + b'y + c' = 0$. Identify a', b', c' in terms of k for later use.

3. Using formula (**8**), calculate $\cos \theta_1$ for the angle between C' and D_1, using a', b', c' for C' and a_1, b_1, c_1 for D_1.

4. Likewise, calculate $\cos \theta_2$ for C' and D_2.

5. Did you discover anything?

QUESTION: Do you have a conjecture about circles through A and B and the circles in the family $\mathcal{F}$?

THEOREM 1

Given the unit circle C and any other circle D orthogonal to it, there exist two one-parameter families $\mathcal{F}$ and $\mathcal{G}$ containing the two given circles as particular members, with C in $\mathcal{G}$ and D in $\mathcal{F}$, such that each circle in $\mathcal{F}$ is orthogonal to each circle in $\mathcal{G}$. Moreover, under an appropriate coordinate system, the two families may be characterized by the equations

$$\mathcal{F}: x^2 + y^2 + ax + 1 = 0 \qquad |a| > 2,$$
$$\mathcal{G}: x^2 + y^2 + by - 1 = 0 \qquad b \text{ real}$$

where each member of $\mathcal{F}$ has its center on the x-axis, and each member of $\mathcal{G}$ passes through the points $(\pm 1, 0)$ and is centered on the y-axis. (See Figure 4.81).

Proof: Applying (**6**) to the equations of the circles in each family, we have $a \cdot 0 + 0 \cdot b = 0 = 2(1 - 1)$, hence the orthogonality condition holds and each circle in $\mathcal{F}$ is orthogonal to each circle in $\mathcal{G}$.

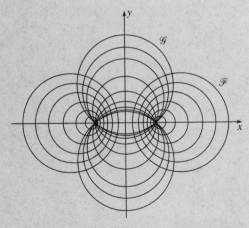

Figure 4.81

Another interesting theorem yields a *synthetic* characterization of orthogonality between two circles, and, in fact, can be used to develop the two systems $\mathcal{F}$ and $\mathcal{G}$ without using coordinates. For that reason, we choose to give a noncoordinate proof.

THEOREM 2

Let **D** be a circle with center D and radius r, and suppose two points A and B are located on a ray from the center D such that the distances DA and DB satisfy the relation

(9)
$$DA \cdot DB = r^2$$

Then every circle **C'** passing through A and B is orthogonal to the given circle **D**.

Proof: If C is the center of circle **C'** (hence C lies on the perpendicular bisector of segment $\overline{AB}$) and s its radius, as in Figure 4.82, then **C'** is orthogonal to **D** iff $r^2 + s^2 = CD^2$. This relation is what we must establish. Working backwards, observe:

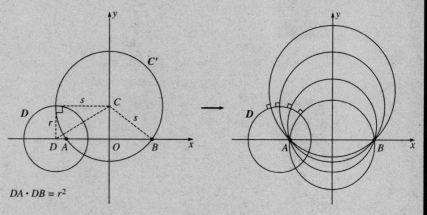

$DA \cdot DB = r^2$

Figure 4.82

(1)	$r^2 = CD^2 - s^2 = CD^2 - CB^2$	(algebra)
(2)	$= (CO^2 + OD^2) - (CO^2 + OB^2)$	(Pythagorean Theorem)
(3)	$= OD^2 - OB^2$	(algebra)
(4)	$= (OD - OB)(OD + OB)$	(algebra)
(5)	$= (OD - OA) \cdot DB$	(betweenness relation)
(6)	$= DA \cdot DB$	(betweenness relation)

Since the steps are reversible, $DA \cdot DB = r^2$ is equivalent to requiring $CD^2 = r^2 + s^2$, or the orthogonality of **C'** and **D**.

The previous theorem automatically yields the entire family $\mathcal{G}$ of circles passing through A and B which are each orthogonal to **D**; to obtain the rest of the family $\mathcal{F}$

we have only to allow point D to vary on line $\overleftrightarrow{AB}$ so that $AD \cdot BD = r^2$ is always satisfied. There will obviously be an infinite number of such points on both sides of O, and these are the centers of the circles belonging to $\mathcal{F}$.

EXAMPLE 1 Find the equation of any circle of radius one unit that is orthogonal to both $x^2 + y^2 = 4$ and $x^2 + y^2 + 2y = 4$. (See Figure 4.83.)

SOLUTION

The given circles intersect at $A(-2, 0)$ and $B(2, 0)$, so we are seeking one of the circles belonging to the family $\mathcal{F}$. We know its equation is of the form

$$x^2 + y^2 + ax + c = 0$$

for some c, since its center lies on the x-axis. Since its radius is 1, we must have (by **(4)**)

$$r^2 = \tfrac{1}{4}a^2 + \tfrac{1}{4}b^2 - c = 1 \qquad \text{or} \qquad c = \tfrac{1}{4}a^2 - 1$$

But also, the orthogonality condition **(6)** applied to both pairs of circles yields

$$0 \cdot a + 0 \cdot 0 = 2 \cdot 4 - 2c \qquad \text{and} \qquad 0 \cdot a + 2 \cdot 0 = 2 \cdot 4 - 2c$$

or $c = 4$. Hence $\tfrac{1}{4}a^2 = 5$ or $a = \pm \sqrt{20}$. Thus, the desired circles are given by

$$x^2 + y^2 \pm \sqrt{20}\, x + 4 = 0$$

centered at $(\pm \sqrt{5}, 0)$.□

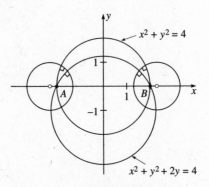

Figure 4.83

CIRCULAR INVERSION   A concept that is very useful in geometry is inversion. In vector geometry, we essentially know what it means to "multiply" two vectors (by taking their dot product). Inversion is the closest we can come to the concept of "dividing by a vector" or finding its "reciprocal." We can think of the **reciprocal** of a nonzero vector OP with respect to the origin O as the vector OP' where P' lies on ray $\overrightarrow{OP}$, and

$$\|OP'\| = \frac{1}{\|OP\|}$$

That is,

$$OP' = \frac{1}{OP}$$

Thus, P' is called the **inverted image** or **inverse** of P, and the relationship, an **inversion**.

It will be convenient to express the above mapping in terms of vectors and coordinates. The two ideas

(1) $\qquad\qquad\qquad\qquad\qquad P'$ lies on ray $\overrightarrow{OP}$

(2) $\qquad\qquad\qquad\qquad\qquad OP' = \frac{1}{OP}$

can be expressed by a single vector equation. Let $v = OP$ and $v' = OP'$. Then $P' \in \overrightarrow{OP}$ implies that $v' = av$ for some positive scalar a; $OP' = 1/OP$ means

$$\|v'\| = \frac{1}{\|v\|}$$

or $\|av\| = a\|v\| = \frac{1}{\|v\|}$, which implies that $a = \frac{1}{\|v\|^2} = \frac{1}{v^2}$. Thus, the above mapping is completely defined by

$$v' = \frac{v}{v^2}$$

(**IMPORTANT NOTE:** The expression $\frac{v}{v^2}$ does not reduce to $\frac{1}{v}$, which would be meaningless.)

If $v = [x, y]$ and $v' = [x', y']$, the coordinate form of this is clearly

$$\begin{bmatrix} x \\ y \end{bmatrix}' = \frac{1}{v^2}\begin{bmatrix} x \\ y \end{bmatrix} = \frac{1}{x^2 + y^2}\begin{bmatrix} x \\ y \end{bmatrix}$$

or

$$x' = \frac{x}{x^2 + y^2}, \quad y' = \frac{y}{x^2 + y^2}$$

This motivates our definition for inversion.

DEFINITION

The mapping of each point $P(x, y) \neq (0, 0)$ in the plane to the unique *image point* $P'(x', y')$ for which

(10) $\qquad\qquad x' = \frac{r^2 x}{x^2 + y^2} \quad$ and $\quad y' = \frac{r^2 y}{x^2 + y^2}$

is called a **circular inversion**. Point P' is called the **inverse** of P, the circle $x^2 + y^2 = r^2$ is called the **circle of inversion**, and O and r are, respectively, the **center** and **radius** of inversion. (See Figure 4.84.)

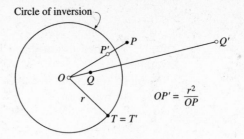

Circle of inversion

$$OP' = \frac{r^2}{OP}$$

Figure 4.84

It is clear by its definition that if P' is the image of P under an inversion in a circle of radius r then $OP' = \sqrt{x'^2 + y'^2} = r^2/\sqrt{x^2 + y^2} = r^2/OP$, or

(11) $OP' \cdot OP = r^2$

Thus it is easy to see that *every inversion is self inverse;* that is, if we apply the inversion to P', we arrive back at point P. This may also be proven directly from equation **(10)** by applying this mapping twice, substituting **(10)** into the equation

$$x'' = \frac{r^2 x'}{x'^2 + y'^2} \qquad y'' = \frac{r^2 y'}{x'^2 + y'^2}$$

and showing that $x'' = x$ and $y'' = y$. We should also point out that the domain of an inversion is the entire plane minus the center of inversion. Such a domain is appropriately called the **punctured plane**.

The particular form of **(11)** has occurred before, in Theorem 2. It was proven (in different notation) that any circle through P and P' is orthogonal to a circle centered at O having radius r iff $OP \cdot OP' = r^2$. (See Figure 4.85.) This proves the third property of inversion we list.

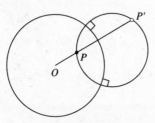

Figure 4.85

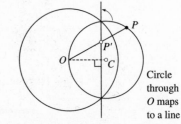

Circle through O maps to a line

Figure 4.86

PROPERTIES OF CIRCULAR INVERSION

1. Points inside C map to points outside of C, points outside map to points inside, and each point on C is self inverse (maps to itself).

2. A line through O is invariant (maps to itself); the individual points of the line need not be self inverse.

PROPERTIES OF CIRCULAR INVERSION (CONTIMUED)

3. A circle orthogonal to the circle of inversion C is invariant.

4. A circle maps to either another circle or a line. A circle maps to a line iff the circle passes through O, and in that case the line is perpendicular to the line passing through O and the center C of that circle. (See Figure 4.86.)

5. A circular inversion preserves the angle measure between any two curves in the plane (is **conformal**).

6. A circular inversion preserves the generalized **cross-ratio** $(AB, CD) \equiv (AC/AD)/(BC/BD)$ of any four distinct points A, B, C, and D in the plane.

Except for Properties 5 and 6, each of these is easy to verify by observation. We illustrate by proving Property 4, and leave properties 1, 2, and 3 for you. Property 5 will be addressed in our third major theorem of this section, and Property 6 will be relegated to Problems 7 and 13.

We want to show that, under inversion, the circle

(12) $$x'^2 + y'^2 + ax' + by' + c = 0$$

corresponds to either a line or a circle. Merely substitute the inversion equations **(10)** into the equation of the circle **(12)**:

$$\frac{r^4 x^2}{\left(x^2 + y^2\right)^2} + \frac{r^4 y^2}{\left(x^2 + y^2\right)^2} + a \cdot \frac{r^2 x}{x^2 + y^2} + b \cdot \frac{r^2 y}{x^2 + y^2} + c = 0$$

and, by multiplying both sides by $(x^2 + y^2)^2$, simplify to

$$r^4 x^2 + r^4 y^2 + ar^2 x(x^2 + y^2) + br^2 y(x^2 + y^2) + c(x^2 + y^2)^2 = 0$$

We find that every term in the above expression has the factor $(x^2 + y^2)$, which is nonzero, so we may divide it out and reduce the expression to

(13) $$c(x^2 + y^2) + ar^2 x + br^2 y + r^4 = 0$$

This is the equation of another circle, provided $c \neq 0$—exactly when the given circle does not pass through O. If $c = 0$, then **(13)** reduces to

(14) $$ax + by + r^2 = 0$$

which is the equation of a line whose slope is $-a/b$, perpendicular to the line through O and $(-\frac{1}{2}a, -\frac{1}{2}b)$, the center of the given circle **(12)**. (Note that the preceding analysis proves the desired property for the *inverse* of a circular inversion, which is also a circular inversion, and is thus sufficient for the purpose.)

THEOREM 3

A circular inversion in the plane is conformal.

Proof: We must show that if C_1 and C_2 are any two curves meeting at point

THEOREM 3 (CONTINUED)

P, and t_1 and t_2 are their respective tangents at P, the angle θ between those tangents is equal to that of the corresponding curves $\mathbf{C}'_1$ and $\mathbf{C}'_2$ (Figure 4.87). Assuming $P \neq O$ (origin), the inversion maps t_1 and t_2 to circles t'_1 and t'_2 through O, and the angle between those two circles at P' equals the angle ϕ at O, and the angle between the circles is also the angle θ' between the tangents to $\mathbf{C}'_1$ and $\mathbf{C}'_2$. But the circles t'_1 and t'_2 are each perpendicular to the lines through O which are perpendicular to t_1 and t_2, respectively, by Property 4. That is, the tangents to the circles at O are parallel to t_1 and t_2. Hence $\theta' = \phi = \theta$, as desired.

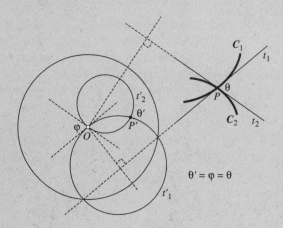

$$\theta' = \phi = \theta$$

Figure 4.87

PROBLEMS   (§4.7)

_____ *Group A* _____

1. **(a)** Determine which of the following pairs of circles with their centers and radii given, are orthogonal pairs, where C_k is the center and r_k is the radius for $k = 1, 2$.

   (1) $C_1 = (4, -3)$, $r_1 = 2$ and $C_2 = (2, 1)$, $r_2 = 4$

   (2) $C_1 = (3, 0)$, $r_1 = 5$ and $C_2 = (6, 1)$, $r_2 = 6$

   (b) Convert the given data to a_k, b_k, c_k form and find the cosine of the measure of the angle between the given circles. (See figures below for actual graphs.)

(a)

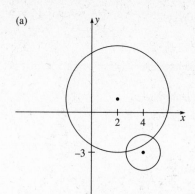

(b)

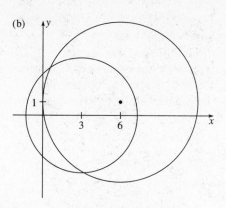

2.   Show that if A lies inside circle C, then no circle centered at A can be orthogonal to C.

3.   (a)   What value must a have if the two circles C_1 and C_2, whose equations are given as follows, are orthogonal:

$$C_1: x^2 + y^2 + ax + 3y = 0$$
$$C_2: x^2 + y^2 - 2x + 8y + 9 = 0$$

   (b)   Locate their centers and carefully graph these two circles.

4.   Precisely how large must a circle centered at $(3, -2)$ be in order to be orthogonal to the circle $x^2 + y^2 - x + y - 1 = 0$?

5.   A family $\mathcal{G}$ of circles pass through $(3, 0)$ and $(-1, 0)$, as shown in the figure. We want to place a circle of radius 6 in the plane so that, if possible, that circle will be orthogonal to every member of $\mathcal{G}$. Where must its center be located? (Give its coordinates.)

6.   A family $\mathcal{F}$ of circles has a common tangent line ℓ at point A. (See the figure below.) Does there exist a family $\mathcal{G}$ of circles all orthogonal to the members of $\mathcal{F}$?

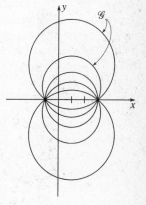

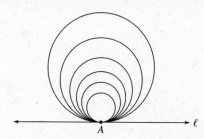

7.   Using the geometric form **(11)** for inversion, show that if A' and B' are the inverse images of A and B, respectively, then $\triangle OAB \sim \triangle OB'A'$. (**HINT:** Apply the SAS Criterion.)

8.   The circle of inversion for an inversion mapping is $x^2 + y^2 = 4$. Find the preimages of the following circles or lines under this inversion and graph all curves involved.

   (a) $x'^2 + y'^2 + 3x' + 4 = 0$

   (b) $x'^2 + y'^2 + 3x' - y' = 0$

   (c) $x' + y' = 1$

   (**HINT:** Write down the transformation equations **(10)** for this inversion. ($r = ?$))

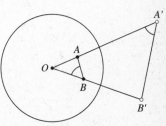

9. Two circles each having unit radius are drawn so that the center of each lies on the other. Find the angle between them in two ways:

    (a) by geometry (properties of circles and their tangents)

    (b) by use of the formula **(8)** in this section (for this part, let the two circles have their centers on the *x*-axis, and find their general equations).

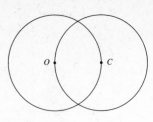

10. Determine whether the following pairs of circles intersect and if so, calculate the angle θ between them.

$$C_1: x^2 + y^2 + 2x - 4y - 11 = 0$$
$$C_2: x^2 + y^2 - 4x + 8y - 101 = 0$$

11. Repeat Problem 10 for the circles

$$C_1: x^2 + y^2 - 6x - 6y - 7 = 0$$
$$C_2: x^2 + y^2 + 14x + 4y + 3 = 0$$

———— *Group B* ————

12. Euclidean Construction For Inverse Points

    (a) Justify the following construction for the inverse of a point *P* lying inside the circle of inversion:

        (1) Let *C* be the circle of inversion (radius *r*, center *O*). Let *P* be a given point inside *C*. Draw the line of center $\overleftrightarrow{OP}$ and erect the perpendicular to $\overleftrightarrow{OP}$ at *P*, cutting the circle at *A*.

        (2) Construct the perpendicular bisector to segment $\overline{OA}$ at *M*, and let this intersect $\overleftrightarrow{OP}$ at *D*.

        (3) With *D* as center and $\overline{DO}$ as radius, inscribe a circular arc through *O* and *A*, cutting line $\overleftrightarrow{OP}$ at *P'*. *P'* is the desired inverse of *P*.

        (*HINT:* The inverse of line $\overleftrightarrow{AP}$ is circle *D*; prove this if you have not already done so.)

    (b) Devise a similar construction when *P* is exterior to the circle of inversion.

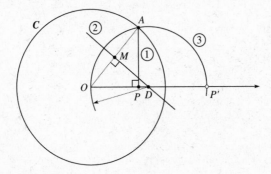

13. Property (6) of Inversion

    (a) Using the result of Problem 7, verify these calculations for the ratios *AC/AD* and *A'C'/A'D'*:

$$\frac{AC}{A'C'} = \frac{OA}{OC'} \quad \text{and} \quad \frac{AD}{A'D'} = \frac{OA}{OD'} \quad \rightarrow \quad \frac{AC}{AD} = \frac{OD'}{OC'} \cdot \frac{A'C'}{A'D'}$$

    (b) Complete the proof that $(AB, CD) = (A'B', C'D')$ under circular inversion.

14. Two nonintersecting circles D_1 and D_2 centered at D_1 and D_2 have radii $r_1 = 9$ and $r_2 = 16$, respectively, with $D_1 D_2 = 35$. Can a circle having radius 15 be orthogonal to both D_1 and D_2? If so, find where such a circle should be located relative to D_1 and D_2 (using coordinates).

15. Formula for Power Prove the following formula for the power of $P(x_0, y_0)$ with respect to the circle $x^2 + y^2 + ax + by + c = 0$:

$$\text{Power } (x_0, y_0) = x_0{}^2 + y_0{}^2 + ax_0 + by_0 + c$$

(This is the result of substituting the coordinates of P into the left member of the general equation of the given circle.) See Problem 16, §4.5 for the definition of power.

16. Radical Axis of Two Circles and Inversion Define the **radical axis** of two circles as the set of all points P such that the powers of P with respect to the two circles are equal. Suppose that under an inversion with respect to the circle C_1, the circle C_2 maps to the line ℓ. Show that line ℓ is the radical axes of the two circles. (Use the result in Problem 15 for a coordinate proof of this; take C_1 as the circle $x^2 + y^2 = r^2$.)

17. **(a)** The equation of the radical axis of two circles is the result of subtracting the equations of the two circles. Illustrate this with the two circles $x^2 + y^2 - 6x + 8y + 24 = 0$ and $x^2 + y^2 - 8x + 7y + 24 = 0$; when you find the equation of the desired line, show that the power of any point $P(x, y)$ on that line with respect to the two circles is equal. (Use the result of Problem 15.) Prove the following general result in coordinate geometry:

The equation of the radical axis of the two circles $x^2 + y^2 + a_k x + b_k y + c_k = 0$ *($k = 1, 2$) is* $(a_1 - a_2)x + (b_1 - b_2)y + (c_1 - c_2) = 0$.

(b) Prove: If two circles intersect, the radical axis is the line containing the common chord of the two circles.

(c) Prove: The common chords formed by the pairwise intersection of three circles are concurrent.

(*HINT:* Use power.)

_____ *Group C* _____

18. Radical Axis and Orthogonal Families of Circles Prove that in any complementary pair $\mathscr{F}$ and $\mathscr{G}$ of mutually orthogonal circles, the radical axis of one family is the line of centers for the other.

19. Referring to Figure 4.81, it is observed that the circles of $\mathscr{F}$ cluster about the endpoints of the common chord of the family $\mathscr{G}$. To generalize, consider any circle $x^2 + y^2 + ax + by + c = 0$ for which a and b are not both zero and $c \neq 0$. For t real, consider the family of circles

$$x^2 + y^2 + t(ax + by + c) = 0, \qquad (a, b) \neq (0, 0), c \neq 0$$

Show that this family is of **type $\mathscr{F}$**, that is, any two members of this family are nonintersecting and there exist two points A and B such that as the radius of a circle in the family tends to zero, its center converges to either A or B. Find the coordinates of A and B in terms of a, b, and c, and verify these facts. What happens if $c = 0$? If $a = b = 0$?

20. Consider the complementary family to that of Problem 19, provided both $a \neq 0$, $b \neq 0$, given by

$$x^2 + y^2 + (bs + \frac{c}{a})x + (-as + \frac{c}{b})y = 0, \qquad s \text{ real}$$

Show that this family is **type $\mathscr{G}$**, that is, any two members intersect at a fixed pair of points A and B. Show that every circle in this family is orthogonal to every circle in the family $\mathscr{F}$ of Problem 19 by using the test for orthogonality **(6)**.

21. **Coordinate Characterization of Orthogonal Systems of Circles**   Obviously, the most general one-parameter family of circles has the form

(15) $$x^2 + y^2 + a(t)x + b(t)y + c(t) = 0$$

where $a(t)$, $b(t)$, and $c(t)$ are functions of a single real variable t (the parameter).

(a) Using the results of Problems 17 and 18, show that a one-parameter family of circles is of type $\mathscr{F}$ or $\mathscr{G}$ iff $a(t)$, $b(t)$, and $c(t)$ are essentially *linear* functions; that is, (15) reduces to

(16) $$x^2 + y^2 + (at + a')x + (bt + b')y + (ct + c') = 0$$

for certain constants a, a', b, b', c, and c' (a change of parameter t from some function $k(t)$ may be necessary).

(b) Define the **discriminant** for a one-parameter family (16):

(17) $$\Delta = (aa' + bb' - 2c)^2 - (a^2 + b^2)(a'^2 + b'^2 - 4c')$$

Prove that the family (16) is of type $\mathscr{F}$, $\mathscr{G}$, or degenerate (all circles have a common tangent and point of intersection) as $\Delta > 0$, $\Delta < 0$, or $\Delta = 0$, respectively.

22. **Euclidean Construction of Radical Axis**   We aim to give a compass/straight-edge construction of the radical axis of two nonintersecting circles C_2 and C_2.

(1) With a compass, draw any circle C cutting the given circles at A_1, B_1 and A_2, B_2, respectively.

(2) Draw lines $\overleftrightarrow{A_1B_1}$ and $\overleftrightarrow{A_2B_2}$, meeting at point P.

(3) From P, drop the perpendicular (standard construction) to the line of centers of the given circles. That perpendicular is the desired radical axis. Prove.

23. Determine (describe, with proof) the locus of all points in the plane the difference of whose powers with respect to two circles is a constant.

24. **Casey's Power Theorem**   Prove that the *difference* of the powers of a point with respect to two circles is a constant multiple of the perpendicular distance from the point to the radical axis.

25. **UNDERGRADUATE RESEARCH PROJECT**   A circle with its center P and radius r is denoted conveniently by $[P, r]$. Find necessary and sufficient conditions that there exists a circle $[P, r]$ orthogonal to each of three given circles $[A, a]$, $[B, b]$, and $[C, c]$. Is the circle $[P, r]$ unique? Explore all possibilities. You will naturally discover your own theorems as you go.

---

*4.8   SOME MODERN GEOMETRY OF THE TRIANGLE

An area known as *modern geometry*, which has developed over approximately the past 300 years, includes material that extends Euclidean geometry far beyond the self-imposed bounds of the ancient Greek geometers. This literature is extensive involving hundreds of articles on scores of different topics, usually named after the geometers who discovered them (like the Gergonne Point and Simson Line, introduced in Chapter 1).

Since many of these results involve collinear points or concurrent lines obtained in a variety of ways, we discuss here two major tools for such results, the theorems of Ceva and Menelaus. Dual to each other, these are two of the most intriguing theorems outside the mainstream of the body of material surrounding Euclid's *Elements*. Curiously, their discoveries were separated by more than a thousand years. The theorem bearing the name Ceva (from which the term "cevian" originates) was discovered by an Italian mathematician Giovanni Ceva (1647–1736), who noted that it complemented the theorem proved approximately 1600 years earlier by the Greek Astronomer Menelaus of Alexandria (around A.D. 100).

AREA OF A CEVIAN TRIANGLE

Let's begin our discussion with a classical problem in geometry—to find the area of a **cevian triangle**, a triangle determined by three cevians of a triangle ($\triangle PQR$ shown in Figure 4.88). The numbers indicated in parentheses along the sides of the triangle represent, not the lengths of the segments so labelled, but rather certain *ratios* of lengths. For instance,

$$p_1 = \frac{BD}{BC} = \frac{BD}{a}, \quad q_1 = \frac{DC}{BC} = \frac{DC}{a}, \quad p_2 = \frac{CE}{b}, \quad \text{etc.}$$

The answer is given by a rather intricate formula. If K is the area of $\triangle ABC$, then it can be shown that

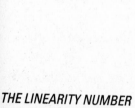

(1) $$\text{Area } \triangle PQR = \frac{\left(p_1 p_2 p_3 - q_1 q_2 q_3\right)^2 K}{\left(p_1 + q_1 q_2\right)\left(p_2 + q_2 q_3\right)\left(p_3 + q_3 q_1\right)}$$

You might find it interesting to test this formula in a few simple cases. The interesting corollary to this is:

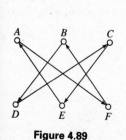

Cevians $\overline{AD}$, $\overline{BE}$, $\overline{CF}$ are concurrent iff Area $\triangle PQR = 0$

Figure 4.88

that is, iff

$$p_1 p_2 p_3 - q_1 q_2 q_3 = 0.$$

Thus, the value of the expression

$$\frac{p_1 p_2 p_3}{q_1 q_2 q_3} = \frac{AF}{FB} \cdot \frac{BD}{DC} \cdot \frac{CE}{EA}$$

determines whether the cevians are concurrent.

THE LINEARITY NUMBER AND DIRECTED DISTANCE

For convenience, we shall define the *linearity number* of D, E, F with respect to $\triangle ABC$ (where D, E, and F lie on the respective sides of the triangle), as the number

(2) $$\begin{bmatrix} ABC \\ DEF \end{bmatrix} = \frac{AF}{FB} \cdot \frac{BD}{DC} \cdot \frac{CE}{EA}$$

A mnemonic device for the defining relation of the linearity number is shown in Figure 4.89

Figure 4.89

NOTE: It is obviously necessary in making this definition that none of the points D, E, and F coincide with the vertices of the triangle A, B, and C. This we assume throughout. If because of special cases in any applications of these ideas it is necessary to assume $D = B$, for example, then the **method of continuity** in geometry is used: Prove the desired property for all cases when D is arbitrarily close to, but distinct from B, then examine what happens in the limit as $D \to B$. Since in all applications (at least in this text) the assertion remains true in the limiting position, the technical requirement is not a hindrance.

Since it is important to deal with cevians when the points opposite the vertices can be anywhere on the *lines* containing the sides of the triangle (as in Figure 4.90) it will be helpful to assume the concept of a *directed* distance formula along lines $\overleftrightarrow{AB}$, $\overleftrightarrow{BC}$, and $\overleftrightarrow{AC}$. This concept was actually introduced in a previous problem, but we repeat it here for your convenience.

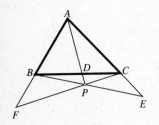

Figure 4.90

DEFINITION OF DIRECTED DISTANCE

For any line ℓ, the Ruler Postulate guarantees a coordinate system for the points of ℓ. Let $P[x_1]$ and $Q[x_2]$ be any two points of ℓ, with their coordinates. The **directed distance** from $P[x_1]$ to $Q[x_2]$ is the number

(3) $$PQ = x_2 - x_1$$

(Note the reversal of the coordinates x_1 and x_2 in this distance formula.) Among several interesting and useful phenomena, we have the following basic rules, for arbitrary points P, Q, R, and X on ℓ:

(a) $PQ = -QP$ (or $PQ + QP = 0$)

(b) $PQ = PR + RQ$ (or $PQ + QR + RP = 0$)

(c) $PX = PQ \to X = Q$ ($X \in \ell$)

(d) $PQ/QR > 0$ iff P-Q-R holds.

To help you get used to this concept, the following example will provide food for thought.

EXAMPLE 1   In Figure 4.91 are illustrated three points on a line, as indicated.

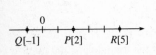

Figure 4.91

(a) Calculate by formula **(1)** the directed distances PQ, PR, and RQ, then verify Property **(b)**.

(b) Show that even though R is not between P and Q in the ordinary sense, $PR + RQ = PQ$.

(c) Calculate QP/PR and verify Property **(d)**.

SOLUTION

(a) We have: $PQ = -1 - 2 = -3$, $PR = 5 - 2 = 3$, $RQ = -1 - 5 = -6$.

(b) Then $PR + RQ = 3 + (-6) = -3 = PQ$.

(c) $QP/PR = \tfrac{3}{3} = 1 > 0$, and we observe that Q-P-R holds, in agreement with property **(d)**.☐

PROPERTIES OF THE LINEARITY NUMBER

Permutations in the columns in the symbol for linearity number will, for odd permutations, transform it into the reciprocal of the original value, as in

$$\begin{bmatrix} ABC \\ DEF \end{bmatrix} = \frac{1}{\begin{bmatrix} ACB \\ DFE \end{bmatrix}}$$

but the *even* permutations in the columns do not affect the value. These details will be left for you to work out. (See Problem 1.)

Another useful property is the following: For any point X on $\overleftrightarrow{BC}$,

(4)
$$\begin{bmatrix} ABC \\ DEF \end{bmatrix} = \begin{bmatrix} ABC \\ XEF \end{bmatrix} \rightarrow D = X$$

Proof: By definition of the linearity numbers involved,

$$\frac{AF}{FB} \cdot \frac{BD}{DC} \cdot \frac{CE}{EA} = \frac{AF}{FB} \cdot \frac{BX}{XC} \cdot \frac{CE}{EA} \rightarrow \frac{BD}{DC} = \frac{BX}{XC}$$

The desired conclusion is reached by "inserting" point C between B and D on the left and between B and X on the right, and by Property **(b)**, using Property **(a)**:

$$\frac{BC + CD}{DC} = \frac{BC + CX}{XC} \rightarrow \frac{BC}{DC} - 1 = \frac{BC}{XC} - 1$$

which yields

$$\frac{BC}{DC} = \frac{BC}{XC} \qquad \text{or} \qquad DC = XC.$$

By Property **(c)** of directed distance, $D = X$.

THE THEOREMS OF CEVA AND MENELAUS

We now turn our attention to the actual statements and proofs of the two theorems mentioned earlier.

THEOREM 1: CEVA'S THEOREM

The cevians $\overline{AD}$, $\overline{BE}$, $\overline{CF}$ of $\triangle ABC$ are concurrent iff $\begin{bmatrix} ABC \\ DEF \end{bmatrix} = 1$.

Proof:

(1) First observe that if P is a point of concurrency of the cevians (Figure 4.92) and all three points D, E, and F are on the sides of the triangle, then the linearity number is positive.

THEOREM 1: CEVA'S THEOREM (CONTINUED)

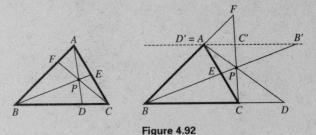

Figure 4.92

(2) If *B-C-D* (other cases similar), then looking at the three possible cases *A-P-D*, *P-A-D*, and *A-D-P*, we find that *exactly one* other cevian falls outside the triangle: For example, if *A-P-D*, then ray $\overrightarrow{CP}$ is interior to $\angle ACD$, which puts $F \in \overrightarrow{CP}$ on the same side of $\overleftrightarrow{AC}$ as D, hence on the opposite side of A as B, or *B-A-F*, and further, ray $\overrightarrow{BP}$ falls interior to $\angle ABC$, so *A-E-C* follows.

(3) Thus, by Property (d), *if there is a point of concurrency* the linearity number is *positive*. (The next part of the proof is to show it equals 1.)

(4) Construct line $\overleftrightarrow{B'C'} \parallel \overleftrightarrow{BC}$ passing through *A*. By central projection through *P* (and using similar triangles), we obtain, in magnitude only,

$$\frac{BD}{DC} = \frac{B'D'}{D'C'} = \frac{B'A}{AC'}$$

(5) By similar triangles,

$$\frac{CE}{EA} = \frac{BC}{B'A} \quad \text{and} \quad \frac{AF}{FB} = \frac{AC'}{BC}$$

(6) Hence

$$\left| \frac{AF}{FB} \cdot \frac{BD}{DC} \cdot \frac{CE}{EA} \right| = \left| \frac{AC'}{BC} \cdot \frac{B'A}{AC'} \cdot \frac{BC}{B'A} \right| = 1$$

and since the linearity number is positive, it must equal 1.

(7) Conversely, suppose the linearity number is 1. If the first two cevian lines meet at *P*, let *X* be the point where the third cevian through *P* meets the remaining side of $\triangle ABC$, say $\overleftrightarrow{AP} \cap \overleftrightarrow{BC} = X$. Then by the first case, and by hypothesis,

$$\begin{bmatrix} ABC \\ XEF \end{bmatrix} = 1 = \begin{bmatrix} ABC \\ DEF \end{bmatrix}$$

By (2), $X = D$ and $\overleftrightarrow{AD}$ coincides with $\overleftrightarrow{AX}$. Therefore, the given cevians are concurrent.

THEOREM 2: MENELAUS' THEOERM

If points D, E, and F lie on the sides of $\triangle ABC$ opposite A, B, and C, respectively, then D, E, and F are collinear iff

$$\begin{bmatrix} ABC \\ DEF \end{bmatrix} = -1$$

Proof: (See Figure 4.93.)

(1) First assume that D, E, and F are collinear, lying on line ℓ. Pasch's postulate implies that if one of the points D, E, and F lies on a side of $\triangle ABC$, then a second point lies on another side and the remaining point lies exterior to the third side, or else all three points lie exterior to the sides of the triangle. Since the ratio AF/FB is positive iff A-F-B, and similarly for the other ratios, this proves that

$$\begin{bmatrix} ABC \\ DEF \end{bmatrix} < 0$$

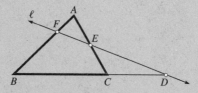

Figure 4.93

(2) For the algebraic part of the proof, let A', B', and C' be the feet of the perpendiculars from A, B, and C on ℓ (Figure 4.94). Then from similar triangles (in magnitude only):

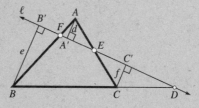

Figure 4.94

$$\frac{AF}{FB} = \frac{d}{e}, \frac{BD}{DC} = \frac{e}{f}, \frac{CE}{EA} = \frac{f}{d}$$

Hence, since the linearity number is negative, we have

$$\begin{bmatrix} ABC \\ DEF \end{bmatrix} = \frac{AF}{FB} \cdot \frac{BD}{DC} \cdot \frac{CE}{EA} = -\left(\frac{d}{e} \cdot \frac{e}{f} \cdot \frac{f}{d} \right) = -1$$

THEOREM 2: MENELAUS' THEOERM (CONTINUED)

(3) Conversely, suppose the linearity number is –1. We shall prove that $\overleftrightarrow{DE}$ meets $\overleftrightarrow{AB}$.

(4) Suppose that $\overleftrightarrow{DE} \parallel \overleftrightarrow{AB}$ (Figure 4.95). Since B-D-C iff A-E-C, BD/DC is positive iff CE/EA is positive, and hence, these two ratios have like signs. Therefore, by the theorem on parallel projections,

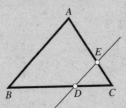 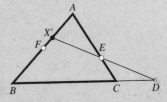

Figure 4.95

$$\frac{BD}{DC} \cdot \frac{CE}{EA} = 1$$

The hypothesis then implies that

$$\frac{AF}{FB} = -1$$

or $AF = -FB = BF$. By Property **(c)** of directed distance, $A = B$ →←

(5) Therefore, $\overleftrightarrow{DE}$ meets $\overleftrightarrow{AB}$ at some point X, and by hypothesis and the first part of the proof,

$$\begin{bmatrix} ABC \\ DEX \end{bmatrix} = -1 = \begin{bmatrix} ABC \\ DEF \end{bmatrix}$$

By permutations and **(2)**, $X = F$, so that D, E, and F are collinear.

The theorems of Menelaus and Ceva provide an excellent example of the duality between collinearity and concurrency in elementary geometry. They can be used to prove a variety of incidence theorems, such as Desargues' and Pappus' Theorems.

EXAMPLE 2   Prove that the medians of a triangle are concurrent using the linearity number.

SOLUTION

In Figure 4.96, let L, M, and N be the midpoints of $\overline{BC}$, $\overline{AC}$, and $\overline{AB}$, respectively, and consider the linearity number of L, M, and N with respect to $\triangle ABC$:

$$\begin{bmatrix} ABC \\ LMN \end{bmatrix} = \frac{AN}{NB} \cdot \frac{BL}{LC} \cdot \frac{CM}{MA} = 1 \cdot 1 \cdot 1 = 1$$

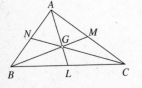

Figure 4.96

Therefore, by Ceva's Theorem, the medians $\overline{AL}$, $\overline{BM}$, and $\overline{CN}$ are concurrent.∎

E X A M P L E   3   Prove the Theorem of Desargues (Problem 12, Section 1.1) by using the linearity number.

SOLUTION

Let the pairs (A, A'), (B, B'), and (C, C') determine three lines concurrent in P, and suppose L, M, and N are the intersections of corresponding sides of $\triangle ABC$ and $\triangle A'B'C'$ (Figure 4.97). We must prove that L, M, and N are collinear. By the Theorem of Menelaus,

$$\begin{bmatrix} PAB \\ NB'A' \end{bmatrix} = \begin{bmatrix} PBC \\ LC'B' \end{bmatrix} = \begin{bmatrix} PCA \\ MA'C' \end{bmatrix} = -1$$

Take the product of these three,

$$\begin{bmatrix} PAB \\ NB'A' \end{bmatrix} \cdot \begin{bmatrix} PBC \\ LC'B' \end{bmatrix} \cdot \begin{bmatrix} PCA \\ MA'C' \end{bmatrix} = (-1)(-1)(-1) = -1$$

which reduces (by Problem 7) to

$$\begin{bmatrix} ABC \\ LMN \end{bmatrix} = -1$$

Again, by Menelaus' Theorem, L, M, and N are collinear.∎

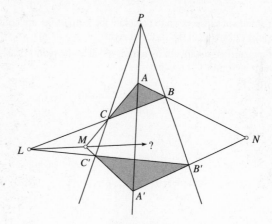

Figure 4.97

PROBLEMS   (§4.8)

_____ *Group A* _____

1. Prove from the definition of linearity number that if $\begin{bmatrix} ABC \\ DEF \end{bmatrix} = x$ then

(a) $\left[\dfrac{ABC}{DEF}\right] = \left[\dfrac{BCA}{EFD}\right] = \left[\dfrac{CAB}{FDE}\right] = x$

(b) $\left[\dfrac{ACB}{DFE}\right] = \left[\dfrac{BAC}{EDF}\right] = \left[\dfrac{CBA}{FED}\right] = \dfrac{1}{x}$

2. Prove from the definition of linearity number that if

$$\left[\dfrac{ABC}{DEF}\right] = \left[\dfrac{ABC}{DEX}\right]$$

where X is some point on line $\vec{AB}$, then $F = X$.

3. Using the property $c/b = a_1/a_2$ for angle bisectors (in the figure, $\vec{AD}$ is the bisector of $\angle CAB$ and $BD = a_1$, $DC = a_2$), use Ceva's Theorem to prove that the angle bisectors of a triangle are concurrent.

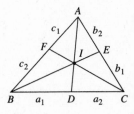

4. Use Ceva's Theorem to prove that the altitudes of a triangle are concurrent. (**HINT:** First establish that $\triangle BEC \sim \triangle ADC$, etc., in the following figure; what does the ratio DC/CE then equal? Continue in like manner around the sides of $\triangle ABC$.)

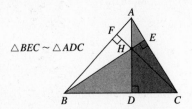

$\triangle BEC \sim \triangle ADC$

5. Use the area formula of this section to calculate the ratio of the area of cevian triangle $\triangle PQR$ to the area of $\triangle ABC$ in the figure at right.

6. If the sides of a quadrilateral are cut by a transversal at the points X, Y, Z, W (as shown in the accompanying figure), show that

$$\dfrac{AX}{XB} \cdot \dfrac{BY}{YC} \cdot \dfrac{CZ}{ZD} \cdot \dfrac{DW}{WA} = 1$$

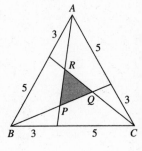

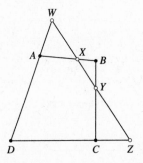

7. Finish the details of Example 3 by providing the algebra to prove

$$\left[\dfrac{PAB}{NB'A'}\right]\left[\dfrac{PBC}{LC'B'}\right]\left[\dfrac{PCA}{MA'C'}\right] = \left[\dfrac{ABC}{LMN}\right]$$

_____ *Group B* _____

8. **Gergonne Point**   The cevians joining the vertices of a triangle with the points of contact of the incircle on the opposite sides are concurrent. Prove. (This point of concurrency is called the **Gergonne Point** of the triangle, after J.D. Gergonne (1771–1859).)

9. **Nagel Point**   Let D, E, and F be the points of contact of the sides opposite A, B, and C
   with the three **excircles** of triangle $\triangle ABC$ (the circles tangent to the three sides which
   lie outside the triangle). The cevians $\overline{AD}$, $\overline{BE}$, and $\overline{CF}$ are concurrent in a point called
   the **Nagel Point** of the triangle. Prove the concurrency of these cevians.

10. **Fermat Point**   Let equilateral triangles $\triangle A'BC$, $\triangle AB'C$, and $\triangle ABC'$ be con-
    structed externally on the respective sides of triangle $\triangle ABC$. Segments $\overline{AA'}$,
    $\overline{BB'}$, and $\overline{CC'}$ are congruent and concurrent, the point of concurrency being
    called the **Fermat Point** of the triangle. Prove this. (**HINT:** Let D, E, and F
    be the intersections of $\overline{AA'}$, $\overline{BB'}$, and $\overline{CC'}$ with $\overline{BC}$, $\overline{AC}$, and $\overline{AB}$, respectively.
    Find all pairs of congruent triangles you can, then apply the following general
    formula, provable from the Law of Sines involving the angle θ shown in the
    figure.)

    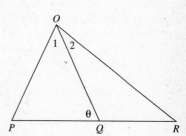

$$\frac{PQ}{QR} = \frac{OP \; \sin m\angle 1}{OR \; \sin m\angle 2}$$

11. A point P is selected at random on the median $\overline{AD}$ of triangle $\triangle ABC$. If E and F are the
    points of intersection of $\overleftrightarrow{BP}$ and $\overleftrightarrow{CP}$ with sides AC and AB, respectively, show that
    $\overleftrightarrow{FE} \parallel \overleftrightarrow{BC}$.

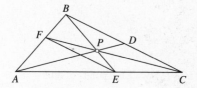

_____ *Group C* _____

12. Prove the existence of the Simson Line using Menelaus' Theorem. (See figure for
    hints.)

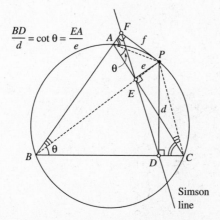

$$\frac{BD}{d} = \cot\theta = \frac{EA}{e}$$

13. **Steiner's Porism**   One area of modern geometry involves the use of circular inversion to obtain results. The following problem is of this type: Suppose we have been given two nonconentric circles C and D (with C inside D), and a third circle C_1 is mutually tangent externally to C and internally to D. Start constructing circles $C_2, C_3, \ldots$ mutually tangent to C and D, and to the previous circle in the sequence. The problem is to determine whether the sequence of circles, called a **Steiner Chain**, closes, with the last one, C_n, tangent to C_1. An inversion transformation immediately reduces the problem to a trivial matter involving a Steiner Chain for two *concentric* circles C' and D', the images of C and D under the inversion. Show this, and thereby establish

Theorem: If one Steiner chain of circles of C and D closes, then any other Steiner chain closes (beginning with an arbitrary circle C_1).

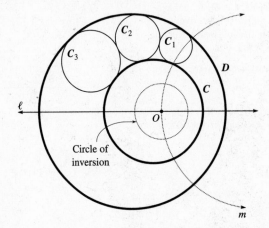

14. Prove Pappus' Theorem (stated in Problem 11, §1.1). (**HINT:** Locate the points of intersection P, Q, and R, calculate the linearity numbers

$$\begin{bmatrix} PQR \\ CBA \end{bmatrix}, \begin{bmatrix} PRQ \\ A'LB \end{bmatrix}, \begin{bmatrix} RQP \\ B'NC \end{bmatrix}, \begin{bmatrix} QPR \\ C'MA \end{bmatrix}, \begin{bmatrix} PQR \\ A'C'B' \end{bmatrix}$$

and multiply.)

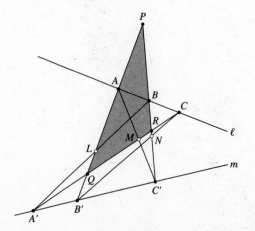

This chapter began with a discussion about various parallel postulates, and the one adopted, the Euclidean parallel postulate, essentially requires a pair of alternate interior angles to be congruent along any transversal cutting two parallel lines. From this logically follows the familiar properties of rectangles, parallelograms and trapezoids. The median of a trapezoid (the line joining midpoints of the two non-parallel sides, or the legs) is always parallel to the two bases, and conversely, a line passing through the midpoint of a leg and parallel to the bases will bisect the other leg. This leads to the important properties of parallel projection and the useful similarity criteria for triangles.

Similar polygons, having corresponding angles congruent and corresponding sides proportional, are pervasive in geometry. This concept leads to the Pythagorean Theorem (not true if the Euclidean Parallel Postulate be denied), the trigonometry of the triangle, the circle theorems, properties of orthogonal circles, circular inversion, and coordinates and vectors. The key concepts for circles were

(1) the measure of an angle inscribed in a circle versus its inscribed arc

(2) the chord-secant relationships (products of lengths of corresponding segments on them are equal)

Regular polygons were introduced briefly, with the main result that an interior angle of a regular polygon having n sides has measure $180(n-2)/n$. A formula was derived from which the cosine of the angle may be calculated between any two (intersecting) circles, which leads to the special case of orthogonality. Orthogonal systems of circles were developed, with the result that each member of a certain entire family of circles is orthogonal to every circle which passes through two fixed points. Circular inversion was then introduced, which maps circles and lines to circles and lines (a line through the center O of inversion maps to itself, a circle through O maps to a line perpendicular to its line of center through O, and a circle not through O maps to a circle not through O). It was then shown how to establish the conformal (angle-measure preserving) property of inversion.

Finally, the last section, an optional section, showed how to derive an important linearity number that is related to the concurrence of cevians of a triangle (lines joining each vertex with a point on the opposite side) and collinearity of points lying on the sides (extended) of a triangle.

Answer each of the following questions True (**T**) or False (**F**).

1. There is only one viable hypothesis about parallel lines, and this was proven to be the case in the 18th century.

2. In this chapter, we developed the concept of similar triangles from properties of parallelograms and trapezoids.

3. The AA Similarity Criterion requires knowledge about the sides of two triangles in order to guarantee they are similar.

4. The SAS Similarity Criterion requires two pairs of proportional sides in two triangles, in addition to a set of congruent angles, in order for the triangles to be similar.

5. A result from vectors allows us to write $AB = AC + BC$.

6. It is not necessarily true that a geometry problem always has its coordinate counterpart.

7. All the circles passing through two fixed points have many other circles orthogonal to them which do not pass through those two points.

8. One of the invariants for circular inversion is curvilinear angle measure.

9. Gauss proved that if n is prime number, then a Euclidean construction can be found for a regular polygon having n sides.

10. The measure of an angle formed by two chords of a circle, whose vertex lies interior to that circle, equals half the difference of the measures of the two arcs inscribed by the angle and its opposite vertical angle.

5

TRANSFORMATIONS IN GEOMETRY

OVERVIEW

The key to the power of modern mathematics is found in the theory of functions, mappings, and transformations. In analysis, the important functions are real-valued functions, vector fields, vector-valued functions, and differential and integral operators. In modern algebra, the operation-preserving mappings between groups, rings and fields (called **homomorphisms**) are important. In topology and geometry, the important functions are the one-to-one mappings between two spaces which are continuous in both directions (called **homeomorphisms**), as well as the distance-preserving mappings (called **isometries**), and in discrete mathematics the important functions are one-to-one structure-preserving correspondences between finite sets or systems. It is appropriate, therefore, to devote some of our attention in a course on geometry to the mappings useful in geometry and some of their applications.

5.1   EUCLID'S SUPERPOSITION PROOF AND PLANE TRANSFORMATIONS

The origin of the concept for transformation theory came from Euclid himself. This is curious, because in his own writings Euclid never referred to it directly, and, in fact, abandoned it entirely after the first several propositions. In the proof of Proposition 4, however, he in effect assumed that one can "move" certain figures around in space without changing their size or shape. He was taking for granted, without saying so, that there exists a certain "rigid motion" which maps one triangle to another.

It is instructive to study the passage in Euclid's *Elements* where the idea of motions is first used to prove a theorem. The following is Euclid's argument for Proposition I.4 (our Axiom C-1, the SAS Postulate), as quoted from Heath's *The Thirteen Books of Euclid's Elements, Book I,* pp. 247–248). We are given $\triangle ABC$ and $\triangle DEF$ with $AB = DE$, $AC = DF$, and $m\angle A = m\angle D$, and we want to prove that $BC = EF$, $m\angle B = m\angle E$, and $m\angle C = m\angle F$. In the following passage, we have italicized Euclid's reference to motions.

> I say that the base *BC* is also equal to the base *EF*, the triangle *ABC* will be equal to the triangle *DEF*, and the remaining angles will be equal to the remaining angles, respectively, namely those which the equal sides subtend, that is, the angles *ABD* to the angle *DEF*, and the angle *ACB* to the angle *DFE*.
>
> For, if the triangle *ABC be applied to the triangle DEF and if the point A be placed on the point D and the straight line AB on DE,* then the point *B* will also coincide with *E*, because *AB* is equal to *DE*.
>
> Again, *AB* coinciding with *DE*, the straight line *AC* will also coincide with *DF*, because the angle *BAC* is equal to the angle *EDF*; hence the point *C* will also coincide with the point *F*, because *AC* is again equal to *DF*.
>
> But *B* also coincided with *E*; hence the base *BC* will coincide with the base EF... and will be equal to it.
>
> Thus the whole triangle *ABC* will coincide with the whole triangle *DEF*...

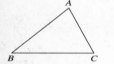

This bit of mathematical poetry contains the germ of the idea for Euclidean motions. We are going to carefully develop this concept from the structure we have built up to this point. First, we need to introduce some key terms in order to talk meaningfully about such mappings.

DEFINITION

A **relation** *f* from the plane ***P*** into itself is a pairing of the points of ***P*** with certain other points of ***P***. If (P, Q) is an ordered pair in this pairing (i.e., point *Q is paired with* point *P*), then *Q* is called an **image** of *P*, and *P* is called a **pre-image** of *Q*, under *f*.

At this point, we can write $P \leftrightarrow Q$ to denote the fact that f pairs P with Q, but we must remember that, so far, there might be many elements paired with P, or none at all, and many with Q, or none at all. That is, a given point P can have several images, and point Q can have several pre-images. (See Figure 5.1 for an example.) However, in order to be of much use, we need to guarantee that a relation has the property that each point has a unique image ("blurred" images are not very useful in geometry!).

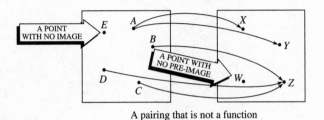

A pairing that is not a function

Figure 5.1

DEFINITION

A **function**, or **mapping**, from the plane P to itself is a relation f for which each point P in P has a unique image Q in P. We write $f: P \rightarrow P$ to denote such a mapping. The **domain** of f is the plane P, and the set of all images under f of the points of P is called the **range** of f. If the range of f is all of P, then f is said to be **onto** P (or, **surjective**). If every point of the range of f has a unique pre-image, then f is called **one-to-one** (or, an **injection**). If f is both one-to-one and onto, then f is called a **bijection**.

See the following table which exhibits all the possible situations for a function involving these concepts; Figure 5.2 gives a pictorial view.

	NOT ONTO	ONTO
NOT ONE-TO-ONE	General Function	Surjection
ONE-TO-ONE	Injection	Bijection

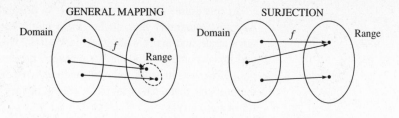

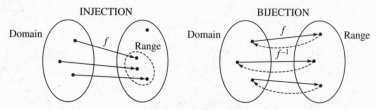

Figure 5.2

For convenience, we let a prime designate the image of a point. Specifically,

$$f(P) = P' \quad \text{or} \quad P \rightarrow P'$$

for all P in $\boldsymbol{P}$. Thus, P' is the unique image of P under f.

DEFINITION

If a mapping $f: \boldsymbol{P} \rightarrow \boldsymbol{P}$ from a plane $\boldsymbol{P}$ to itself is both one-to-one and onto (that is, bijective), then f is called a (**plane**) **transformation**. The **inverse** mapping of a transformation f, denoted by f^{-1}, is the mapping which associates P with Q for each pair of points (P, Q) specified by f. That is, $f^{-1}(Q) = P$ iff $f(P) = Q$. If a transformation maps lines onto lines, it is called a **linear transformation**.

THEOREM 1

Every transformation f in the plane has an inverse mapping f^{-1} such that

$$f[f^{-1}(P)] = f^{-1}[f(P)] = P$$

for every point P in $\boldsymbol{P}$.

Proof: Since f is one-to-one and we have defined $f^{-1}(Q) = P$ iff $f(P) = Q$ for each Q in $\boldsymbol{P}$, then

$$f[f^{-1}(Q)] = f(P) = Q \quad \text{and} \quad f^{-1}[f(P)] = f^{-1}(Q) = P$$

In coordinate geometry, each point P and its image P' under some transformation f has a coordinate pair of reals associated with it, as in $P(x, y)$ and $P'(x', y')$. We shall agree to denote this relationship in the abbreviated form

$$f(x, y) = (x', y') = (f_1(x, y), f_2(x, y))$$

where $x' = f_1(x, y)$ and $y' = f_2(x, y)$. This notation is useful in constructing examples of mappings.

EXAMPLE 1 **A Nonlinear Transformation** Consider the mapping $f: P \to P$ defined in the coordinate plane by

$$f: \begin{cases} x' = x \\ y' = y^3 \end{cases}$$

Find the image of $\triangle ABC$ and its interior under this map if

$$A = (0, 0), \quad B = (2, 2), \quad \text{and} \quad C = (0, 1).$$

SOLUTION

The given triangle maps to a curvilinear triangle, as shown in Figure 5.3, where side $\overline{AC}$ on the y-axis maps to the y'-axis (since $x = 0$ yields $x' = 0$), and side $\overline{AB}$ (on the line $y = x$) maps to the curve $y' = x^3 = x'^3$, $0 \le x' \le 2$. The third side of the triangle, which lies on the line $y = \frac{1}{2}x + 1$, maps to a more complicated cubic curve, namely $y' = \frac{1}{8}(x' + 2)^3$, for $0 \le x' \le 2$. (Do you see why?)❑

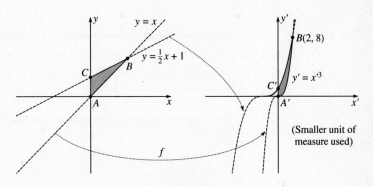

Figure 5.3

EXAMPLE 2 **A Linear Transformation** Show that under the mapping $g : P \to P$ defined by

$$g: \begin{cases} x' = -x + y \\ y' = -2x + y \end{cases}$$

the inverse is given by

$$g^{-1} : \begin{cases} x = x' - y' \\ y = 2x' - y' \end{cases}$$

and use this to show that the images of the two parallel lines $y = 3x$ and $y = 3x + 2$ are parallel lines.

SOLUTION

(a) To reverse the defining equations, write as a 2×2 linear system in x, y and solve for x and y:

$$\begin{cases} -x + y = x' \\ -2x + y = y' \end{cases}$$

Subtract to find $x = x' - y'$, then substitute this into either equation and solve for y:

$$-(x' - y') + y = x' \quad \rightarrow \quad y = 2x' - y'$$

(b) Substitute the equations found in (a) into $y = 3x$ and $y = 3x + 1$:

$$2x' - y' = 3(x' - y') \quad \rightarrow \quad y' = \frac{1}{2}x'$$

$$2x' - y' = 3(x' - y') + 2 \quad \rightarrow \quad y' = \frac{1}{2}x' + 1$$

Thus, the two image lines are also parallel (having slope $\frac{1}{2}$).❏

For later reference, we define two more fundamental terms in transformation theory.

DEFINITION

A transformation f of the plane is said to have A as a **fixed point** iff $f(A) = A$.

DEFINITION

A transformation of the plane is called the the **identity mapping** iff every point of the plane is a fixed point. This transformation is denoted e.

E X A M P L E   3   Find those values of the parameter a for which the linear transformation

$$f : \begin{cases} x' = 3ax + y \\ y' = x - ay \end{cases}$$

has a *nontrivial* fixed point (a point different from the origin $(0, 0)$), and characterize those fixed points (describe geometrically) for each value of a.

SOLUTION

We must have $x' = x$ and $y' = y$ for any fixed point. That is,

$$\begin{cases} x = 3ax + y \\ y = x - ay \end{cases} \quad \text{or} \quad \begin{cases} (3a - 1)x + y = 0 \\ x - (a + 1)y = 0 \end{cases}$$

Since both equations must be satisfied simultaneously for certain $(x, y) \neq (0, 0)$, we must have, by substitution of $x = (a + 1)y$ from the last equation into the preceding equation,

$$(1) \qquad\qquad (3a - 1)(a + 1)y + y = 0$$

If $y = 0$, then these equations force $x = 0$, so we conclude that $y \neq 0$ and, dividing both sides of (1) by y,

$$(3a - 1)(a + 1) + 1 = 0, \quad \text{or} \quad 3a^2 + 2a = 0$$

with roots $a = 0, -\tfrac{2}{3}$. For $a = 0$, we get

$$(3 \cdot 0 - 1)\, x + y = 0, \quad \text{or} \quad y = x,$$

which shows that for $a = 0$, every point of the line $y = x$ is a fixed point. If $a = -\tfrac{2}{3}$, then

$$[3 \cdot (-\tfrac{2}{3}) - 1]x + y = 0, \quad \text{or} \quad y = 3x$$

again a line of fixed points.❏

PROBLEMS   (§5.1)

_____ *Group A* _____

1. Let the point (x, y) in plane **P** be paired with (x', y') iff the conditions

$$x' = x^2 + y^2 \quad \text{and} \quad x' + y'^2 = 1$$

   are satisfied. Find
   (a) all possible images of the points $(x, y) = (\tfrac{3}{10}, \tfrac{2}{5})$, $(\tfrac{5}{13}, \tfrac{12}{13})$, and $(1, -1)$ in **P**.
   (b) all possible pre-images (x, y) in **P** of $(x', y') = (\tfrac{1}{4}, \sqrt{3}/2)$, $(1, 0)$, and $(0, 1)$.
   (c) Is the pairing a function?
   (d) What is the domain and range of the pairing?

2. Let the point (x, y) in plane **P** be paired with (x', y') in plane **P'** iff

$$x' = 2x \quad \text{and} \quad y' = 3x$$

   Find
   (a) all possible images of $(1, -1)$, $(\tfrac{1}{2}, \tfrac{1}{3})$, and $(-3, 4)$
   (b) all possible pre-images of $(2, -3)$ and $(0, 1)$.
   (c) Is the pairing a function? transformation?
   (d) What is the domain of the pairing?

3. **(a)** Show that the image of the parabola $y = x^2$ under the mapping $(x, y) \rightarrow (x', y')$ defined by

$$x' = y \quad \text{and} \quad y' = x^2 + y$$

is a straight line.

(b) Is this mapping a transformation? Show why, or why not.

4. Show that the image of the curve $x^3 + y^2 = 1$ (as shown in the figure) is a broken line under the mapping defined by

$$x' = x^3 \quad \text{and} \quad y' = \delta y^2$$

where $\delta = +1$ if $y \geq 0$ and $\delta = -1$ if $y < 0$. Is this mapping a transformation? Show why or why not.

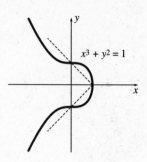

5. Find the image of $\triangle ABC$ and its interior (and sketch) under the transformation defined by

$$x' = 2x \quad \text{and} \quad y' = y$$

if $A = (0, 0)$, $B = (0, 1)$, and $C = (1, 1)$.

6. Find the image of the same triangle $\triangle ABC$ as in Problem 5, under the transformation defined by

$$x' = 2x \quad \text{and} \quad y' = 2y$$

_____ *Group B* _____

7. Find the line into which all points of the plane P map, under the mapping

$$x' = 2x - y \quad \text{and} \quad y' = -6x + 3y$$

Is the mapping a transformation?

8. Show that perpendicularity is preserved under the mapping

$$x' = 2x - 3y \quad \text{and} \quad y' = 3x + 2y,$$

which can be reversed, by algebra, to the equivalent form

$$x = \frac{2x' + 3y'}{13}, \qquad y = \frac{-3x' + 2y'}{13}$$

Follow these steps:

(a) Substitute these last equations into the equations of the lines $y = mx + b$ and $y = (-1/m)x + c$, $m \neq 0$, which are perpendicular lines, and simplify. The result will be the equations of the images of these two perpendiculars.

(b) Find the slopes of the image lines (algebraic expressions involving m) and show that they are also perpendicular, using the rule $m_1 m_2 = -1$.

9. Repeat Problem 8 for the transformation given by

$$x' = 3x + 4y \quad \text{and} \quad y' = -4x + 3y$$

10. If a transformation preserves all distances (that is, $PQ = P'Q'$ for all points P and Q), then it necessarily preserves angle measure as well.

(a) Prove this synthetically.

(b) Give a counterexample to show that the converse of this statement is not true.

11. Characterize geometrically the set of all fixed points of each of the following transformations.

(a) $\begin{cases} x' = x + y - 1 \\ y' = x + 1 \end{cases}$ (b) $\begin{cases} x' = ax + y - a \\ y' = ax + a \end{cases}$

_____ *Group C* _____

12. Show without coordinates that if a linear transformation f preserves perpendicularity, then its inverse f^{-1} also preserves perpendicularity.

13. Show without coordinates that any linear transformation maps parallel lines to parallel lines. (**HINT:** If ℓ' meets m', then where did the point $P' = \ell' \cap m'$ come from?)

14. Using Problem 13, show that a linear transformation maps a parallelogram and its diagonals to another parallelogram and its diagonals. Hence, prove that *midpoints of line segments are preserved* under any linear transformation.

15. Can a transformation of the form

$$\begin{cases} x' = ax + by \\ y' = cx + dy \end{cases}$$

ever have *exactly* one other fixed point besides $(0, 0)$ by choosing the right values for a, b, c, and d? Investigate.

16. The result of Problem 14 leads to a synthetic version of the issue raised in Problem 15. If a linear transformation has *two* fixed points A and B, then the midpoint M_1 of $\overline{AB}$ is a fixed point. Thus the midpoints M_2 of $\overline{AM_1}$ and M_3 of $\overline{M_2B}$ are fixed points, and so on. Show, by continuity of linear transformations (a property you may assume without proof), that every point of segment $\overline{AB}$ is a fixed point.

17. Use Problem 16 to show that if a linear transformation fixes three noncollinear points A, B, and C, then it fixes every point of the plane, and hence is the identity.

5.2 REFLECTIONS: BUILDING BLOCKS FOR ISOMETRIES

We are concerned with two basic types of reflections: reflections in lines, and reflections in points.

DEFINITION OF REFLECTIONS IN THE EUCLIDIAN PLANE

We first formulate our definition for a reflection synthetically.

DEFINITION

Suppose ℓ is a given line and C a given point in some plane P. If the pair of points (P, P') in P is such that ℓ is the perpendicular bisector of segment PP', then P and P' are said to be **reflections** of each other **in line** ℓ. The transformation $P \leftrightarrow P'$ is called a **reflection** in line ℓ, which is called the **axis of reflection**, to be denoted by s_ℓ. (See Figure 5.4.) If, on the other hand, C is always the midpoint of segment $\overline{PP'}$, then P and P' are said to be **reflections** of each other **in point** C (called the **center**), and the transformation $P \leftrightarrow P'$ is called a **reflection** in C, or **central reflection**, denoted by s_C.

REFLECTION IN A LINE   REFLECTION IN A POINT

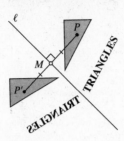

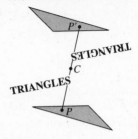

Figure 5.4

To construct the image of a given point P under a reflection in line ℓ, first define $P' = P$ if P lies on ℓ; otherwise, construct the perpendicular $\overline{PM}$ from P to line ℓ, then double the segment $\overline{PM}$ to point P', making M the midpoint of $\overline{PP'}$. Similarly, if C is a given point and P any other, define $P' = P$ if $P = C$; otherwise, double segment $\overline{PC}$ to P', making C the midpoint of $\overline{PP'}$. This proves the *existence* of the mappings s_ℓ and s_C.

REFLECTIONS AS ISOMETRIES AND THE ABCD PROPERTY

The fundamental properties of reflections are all collected into one result. It will establish the invariance of angle measure, betweenness, collinearity, and distance. These same invariant properties are shared by a more general kind of mapping.

> ### DEFINITION
>
> Any mapping which preserves distances (and is therefore one-to-one and a transformation) is called an **isometry** (also **motion**, **rigid motion**, or **Euclidean motion**). Thus, f is an isometry iff for each P and Q, with $P' = f(P)$ and $Q' = f(Q)$,
>
> $$PQ = P'Q'$$

LEMMA

An isometry preserves collinearity, betweenness, and angle measure.

Proof: Let A', B', and C' be the images of points A, B, and C, respectively, under a given distance-preserving transformation, and suppose that A-B-C. Then $AB + BC = AC$. Since the mapping is distance-preserving, $A'B' + B'C' = A'C'$. Therefore, A'-B'-C' (definition of betweenness and the Triangle Inequality). Hence, betweenness and collinearity of any three points is preserved. Finally consider $\angle ABC$ and $\triangle ABC$. By SSS, $\triangle ABC \cong \triangle A'B'C'$. Therefore, $\angle ABC \cong \angle A'B'C'$, and angle measure is preserved.

THEOREM 1: ABCD PROPERTY

Reflections are angle-measure preserving, betweenness preserving, collinearity preserving, and distance preserving.

Proof: In view of the lemma, it suffices to prove that a reflection s_ℓ or s_C is distance-preserving. Suppose that A and B are any two points not on line ℓ (or distinct from C). Consider first the case when A and B are on the same side of ℓ (Figure 5.5). Since ℓ is the perpendicular bisector of $\overline{AA'}$ and $\overline{BB'}$ at midpoints M and N, respectively, then $\Diamond AMNB$ and $\Diamond A'MN'B$ are convex, and by SASAS, $\Diamond AMNB \cong \Diamond A'MNB'$. Hence, $AB = A'B'$. (Note that the quadrilateral $\Diamond ABB'A'$ is an isosceles trapezoid for this case.) The case when A or B lies on ℓ involves congruent right triangles and will be omitted. Finally, when A and B lie on opposite sides of ℓ (Figure 5.6), then A and B' lie on the same side, so that, with $s_\ell(A) = A'$ and $s_\ell(B') = B$, by the first case just proven we have $AB' = A'B$. Hence $\Diamond AB'BA$ is an isosceles trapezoid with congruent diagonals $\overline{AB}$ and $\overline{A'B'}$ (see Problem 20, §4.2.) Therefore, $AB = A'B'$ in all cases. (The proof for point reflections will be left as an exercise.)

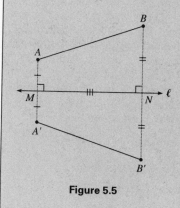

Figure 5.5

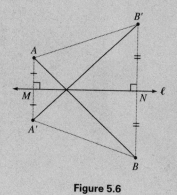

Figure 5.6

APPLICATIONS OF REFLECTIONS, ORIENTATION

A common everyday experience illustrates the geometric principle of a line reflection. A mirror placed along line ℓ will reflect images in the manner of the transformation s_ℓ. A point reflection can be illustrated by the occurrence of light rays passing through a small opening.

It may seem surprising, however, that the two types of reflections have altogether different orientation-preserving properties. When we observe printed letters in a mirror, they are reversed, and a right hand will appear as a left hand in the mirror (Figure 5.7). But for a camera, the image is merely inverted, not reversed. In order to deal with this concept mathematically, we make a couple definitions.

Right hand is reflected as left hand

Figure 5.7

DEFINITION

A linear transformation of the plane is called **direct** iff it preserves the orientation of any triangle, and **opposite** iff it reverses the orientation of each triangle.

OUR GEOMETRIC WORLD

When we look across a lake, the principle of a line reflection is at work. The reflected image of the scenery in the water is the image of the objects on land under a reflection s_ℓ, where ℓ is a line running along the surface of the lake. The image we see is always upside down (inverted). In a similar manner, the pin-hole effect of a camera illustrates the mapping s_C, where C is the lens opening. The image on the film is also inverted.

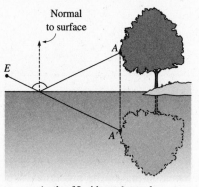

Angle of Incidence θ_i equals
Angle of Reflection θ_r

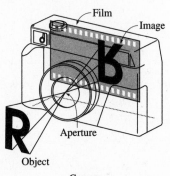

Camera
(Pinhole Effect)

Figure 5.8

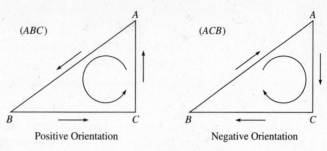

Figure 5.9

> **DEFINITION**
>
> Given a triangle $\triangle ABC$ in the plane, the counterclockwise direction (the path (ABC) in Figure 5.9) is called the **positive orientation** of its vertices, while the clockwise direction (the path (ACB)) is the **negative orientation**.

The idea of orientation can be easily extended to polygons and simple, closed curves. The formal definitions will be omitted.

We encourage you to experiment with orientations relative to line and point reflections in the following Discovery Unit before proceeding further.

MOMENT FOR DISCOVERY

Do Reflections Preserve Orientation?

Suppose $\lozenge ABCD$ is a quadrilateral that has been positively oriented. Consider the image quadrilateral $\lozenge A'B'C'D'$ under a line reflection with axis ℓ, then under a point reflection with center L.

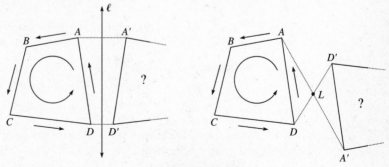

Figure 5.10

1. On your own paper, finish the construction of the image quadrilateral $\lozenge A'B'C'D'$ as carefully as you can in each case.

2. Look at the order of the image vertices, and decide whether the orientation of the image is clockwise or counterclockwise in each case.

3. Did you detect any difference in the orientation-preserving properties of the two types of reflections?

PRODUCTS OF TWO OR MORE REFLECTIONS

An obvious theorem about direct and opposite mappings will be stated after we introduce the necessary idea of the **composition** or **product** of two transformations. If f and g are any two transformations, their product, denoted by gf, is just the ordinary functional composition $g \circ f$ of f and g. That is, for each point P,

$$gf(P) = g \circ f(P) = g[f(P)].$$

This means that we must first apply the mapping f, and then follow this with the mapping g, going from any point P to the point $P' = f(P)$, then on to $P'' = g(P')$; the product gf is the mapping that takes P directly to P''. This can be easily illustrated in an example using coordinates.

EXAMPLE 1 Consider the two linear transformations f and g given in terms of coordinates, as follows.

$$f: \begin{cases} x' = 2x - 3y \\ y' = 3x + 4y \end{cases} \qquad g: \begin{cases} x' = x - y \\ y' = 2x + y \end{cases}$$

Find the coordinate form of the product transformation fg, then repeat for the product gf, and compare the two results.

SOLUTION

For convenience, change the form of g to

$$g: \begin{cases} x'' = x' - y' \\ y'' = 2x' + y' \end{cases}$$

To find gf, merely substitute the equations for f (above) into those just given for g.

$$x'' = (2x - 3y) - (3x + 4y) = 2x - 3y - 3x - 4y = -x - 7y$$
$$y'' = 2(2x - 3y) + (3x + 4y) = 4x - 6y + 3x + 4y = 7x - 2y$$

Hence,

$$gf: \begin{cases} x' = -x - 7y \\ y' = 7x - 2y \end{cases}$$

where we have changed the double primes back to primes. Similarly, if we reverse the mappings we have

$$x'' = 2(x - y) - 3(2x + y) = 2x - 2y - 6x - 3y = -4x - 5y$$
$$y'' = 3(x - y) + 4(2x + y) = 3x - 3y + 8x + 4y = 11x + y$$

or

$$fg: \begin{cases} x' = -4x - 5y \\ y' = 11x + y \end{cases}$$

Thus, $gf \ne fg$, and we would say that f and g **do not commute.**☐

The next two theorems yield further invariance properties in certain situations.

THEOREM 2

The product of an even number of opposite linear transformations is direct, and the product of an odd number is an opposite transformation.

THEOREM 3

The product of two line reflections s_ℓ and s_m, where ℓ and m are parallel lines, is slope-preserving and maps a given line to one that is parallel to it.

Proof:

(1) If line n is parallel to ℓ and m, then obviously $n' = s_\ell(n)$ is parallel to ℓ, hence to m, and $n'' = s_m(n')$ is parallel to m, hence to m and n.

(2) Otherwise, n meets both ℓ and m (Figure 5.11). Let $n' = s_\ell(n)$ and $n'' = s_m(n') \equiv s_m s_\ell(n)$, as before. By the *ABCD* Property, $\angle ABC \cong \angle A'B'C' \equiv \angle A'BC$ and therefore $\angle 1 \cong \angle 2$.

(3) Similarly, $\angle 4 \cong \angle 5$ $(s_m(n') = n'')$.

(4) By the F- and Z-Properties of Parallelism, respectively, $\angle 1 \cong \angle 3$ and $\angle 2 \cong \angle 4$.

(5) Therefore, $\angle 3 \cong \angle 1 \cong \angle 2 \cong \angle 4 \cong \angle 5$, and $\angle 3 \cong \angle 5$.

(6) By the Z-Property of Parallelism, $n \parallel n''$.

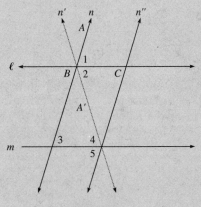

Figure 5.11

QUESTION: Is Theorem 3 true if lines ℓ and m are not parallel?

AN APPLICATION IN PHYSICS

A law of optics which governs the effect of light rays reflecting from a surface, such as a plane mirror (Figure 5.12), is the following: The light ray travels a broken straight line path *OABCE* lying in a plane that is perpendicular to the reflecting surface, such that the **angle of incidence** (the measure of the angle which the incoming ray $\overrightarrow{AB}$ makes with the normal to the reflecting surface) equals the **angle of reflection** (the measure of the angle which the outgoing ray $\overrightarrow{BC}$ makes with the normal).

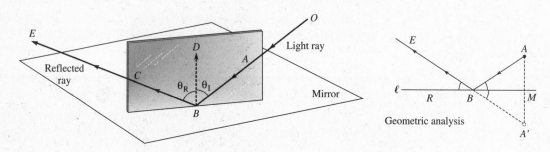

Figure 5.12

Thus, in Figure 5.12, $m\angle ABD = m\angle DBC$ ($\theta_R = \theta_I$). For example, if the light ray comes in at an angle of 55° with the normal (or 35° with the surface of the mirror), it leaves the mirror at an angle of 55° with the normal (or 35° with the mirror).

A geometric analysis of this shows that if the eye of an observer is at *E*, which sees an object at *A* through the mirror along line ℓ, the position of the object appears to the observer to be at the point *A′*, the image of *A* under the geometric line reflection s_ℓ, where $\ell = \overrightarrow{RM}$.

Proof: The previously mentioned law of optics implies that $\angle ABM \cong \angle EBR \cong \angle MBA'$ (Vertical Angle Theorem). Assuming that the depth of the image from *B* as seen in the mirror equals the distance from *B* to the object at *A*, that is, $BA' = BA$, then $\triangle ABM \cong \triangle A'BM$ by SAS, and $\overrightarrow{MB}$ is the perpendicular bisector of segment $\overline{AA'}$. Hence, $A' = s_\ell(A)$.

A mathematical reason for this law has been proposed: The ray of light follows the path of *shortest distance* from *A* to *E*, by way of a point on the mirror (that is, from *A* to *B* to *E*). In other words, the ray of light takes a path so as to minimize the total distance of travel $AP + PE$ from *A* to *E*, where *P* is any point on line ℓ (Figure 5.13). Can you prove that $EB + BA < EP + PA$? Does this remind you of a calculus problem you once had? (See Problem 11 for guidance in proving this fact geometrically, rather than by calculus.)

The law of optics previously mentioned also governs the motion of a projectile when it strikes a resilient surface and bounces off it, as in billiards, or **caroms**, as

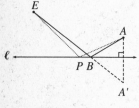

Figure 5.13

the phenomenon is sometimes called. Again, the angle of incidence equals the angle of reflection, as illustrated in Figure 5.14. Further ideas on this are taken up in the problem section which follows.

THE GAME OF BILLARDS

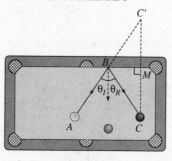

Figure 5.14

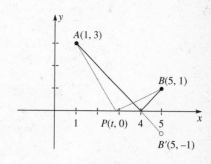

Figure 5.15

EXAMPLE 2 Show that the answer obtained by calculus for minimizing the total distance $AP + PB$, where $A = (1, 3)$, $B = (5, 1)$, and $P = (t, 0)$ for some real t, $1 \leq t \leq 5$, is the same as that obtained by geometry, taking P as the intersection of line $\overleftrightarrow{AB'}$ with the x-axis, where $B' = (5, -1)$, the reflection of B in the x-axis. (See Figure 5.15.)

SOLUTION

(a) By the distance formula,

$$AP = \sqrt{(t - 1)^2 + (0 - 3)^2} = \sqrt{t^2 - 2t + 10}$$
$$BP = \sqrt{(t - 5)^2 + (0 - 1)^2} = \sqrt{t^2 - 10t + 26}$$

We want to minimize the function

$$F(t) = AP + BP = \sqrt{t^2 - 2t + 10} + \sqrt{t^2 - 10t + 26}$$

so we take its derivative. After simplifying, we find

$$F'(t) = \frac{t - 1}{\sqrt{t^2 - 2t + 10}} + \frac{t - 5}{\sqrt{t^2 - 10t + 26}}$$

Setting $F'(t) = 0$ and squaring, we obtain, again after a little algebraic manipulation,

$$\frac{t^2 - 2t + 1}{t^2 - 2t + 10} = \frac{t^2 - 10t + 25}{t^2 - 10t + 26}$$

$$1 - \frac{9}{t^2 - 2t + 10} = 1 - \frac{1}{t^2 - 10t + 26}$$

or, after canceling the ones and cross-multiplying,

$$t^2 - 11t + 28 = 0 \quad \text{or} \quad (t-4)(t-7) = 0$$

The only root lying on [1, 5] is $t = 4$, which produces the answer $P(4, 0)$.

(b) The equation of $\overleftrightarrow{AB'}$ is

$$y - 3 = -(x - 1) = -x + 1 \quad \text{or} \quad y = 4 - x$$

This line cuts the x-axis where $y = 0$, or $x = 4$, in agreement with **(a)**. (*Geometry is better!*)❑

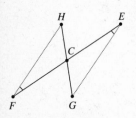

EXAMPLE 3 In Figure 5.16, $s_C(E) = F$ and $s_C(G) = H$. Prove that $\angle GEC \cong \angle HFC$ without using any part of Theorem 1 (*ABCD* Property).

SOLUTION

By definition of s_C, C is the midpoint of segments $\overline{EF}$ and $\overline{GH}$, and $\angle GCE \cong \angle HCF$ by the Vertical Angle Theorem. By SAS, $\triangle GCE \cong \triangle HCF$ and by CPCF, $\angle GEC \cong \angle HFC$.❑

Figure 5.16

Problems (§5.2)

_____ Group A _____

1. Sketch each of the diagrams (a) and (b) in the figure on your paper and draw the image of the segment $\overline{AB}$ in (a), and the image of rectangle $\square XYZW$ in (b) under the reflection s_ℓ in each case.

(a)

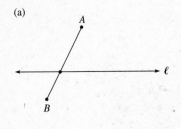

(b)

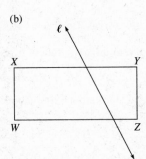

2. The letter "A" of the alphabet has a vertical line of symmetry.

   (a) Which of the letters K, M, L, and X have vertical lines of symmetry?

   (b) Which letters of the alphabet have some line of symmetry?

3. In the accompanying figure we are given that $s_\ell(C) = D$. Prove that $CQ = DQ$ without using any part of Theorem 1. [UCSMP, p. 184]

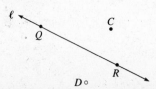

4. Use the information in the figure and prove that $VC = CX$, making use of reflections. [UCSMP, p. 184]

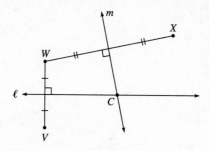

5. Using definitions of s_ℓ and s_C, prove that for any point P,

$$s_\ell^2(P) = s_\ell[s_\ell(P)] = P \quad \text{and} \quad s_C^2(P) = P$$

(Thus, s_ℓ^2 and s_C^2 are each the identity mapping.)

6. Prove the $ABCD$ property for point reflections.

7. Prove that under either a point-reflection or line reflection, a circle with center O maps to a circle with center O', where O' is the image of O under that reflection. What will be the radius of the image circle?

8. Find the coordinate equation for each of the composition mappings gf and fg if

$$f : \begin{cases} x' = y \\ y' = 2x \end{cases} \quad g : \begin{cases} x'' = 5x' + 3y' \\ y'' = 6x' + 5y' \end{cases}$$

(**HINT:** For fg you will need to switch primes and double primes before making substitutions.)

9. The transformation f represents the reflection in the line $y = x$, while g is the reflection in the parallel line $y = x + 2$. (Verify this by trying the mapping out on a few points.) Find, by substitution, the coordinate forms for gf and fg.

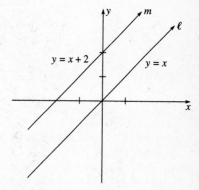

$$f : \begin{cases} x' = y \\ y' = x \end{cases} \quad g : \begin{cases} x'' = y' - 2 \\ y'' = x' + 2 \end{cases}$$

10. (a) Decipher the message in the figure below.

(b) Which letter is written incorrectly? [UCSMP, p. 161]

Help! I'm trapped
inside this page!

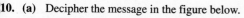

_____ *Group B* _____

11. Use your own sketch of Figure 5.13 and draw $\overline{PA'}$. Use the equidistant locus from A and A' to prove that for all points $P \in \ell$,

$$EB + BA \le EP + PA$$

with equality only when $P = B$. (**HINT:** Use the Triangle Inequality for $\triangle EPA'$.)

12. Taking the x-axis as a straight river bank, at what point $P(t, 0)$ along the x-axis should pipe be laid in order to join two towns $A(1, 5)$ and $B(8, 3)$ to a pumping station at P so as to minimize the amount of pipe used? (Units are in miles.) Work this in two ways:

    (a) by calculus
    (b) by geometry, using reflections.

13. With the coordinate system shown in the following figure, the 7-ball is blocking the cue ball at $A(3, 7)$ from the 4-ball at $B(10, 5)$. In the game of snooker, the balls must be pocketed in numerical order. At what point

    (a) along the x-axis

    (b) along the y-axis

    (c) on the opposite side of the table $y = 48$

    should the cue ball be aimed in order to squarely strike the 4-ball after one carom off that particular side, or sides, of the table? (Units are in inches.)

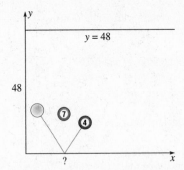

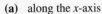

14. Darlene is playing miniature golf. Her golf ball ended up at point A and a barrier blocks a direct shot into the hole at B. By making a geometric construction, show how to locate precisely the desired point P (as illustrated) so the golf ball will bounce to point Q, and then into the hole at B.

15. Lines ℓ and n meet at point A, and n is mapped to line n' by the reflection s_ℓ. If $m \parallel \ell$ and m meets n and n' at points B and C, respectively, prove that $\triangle ABC$ is isosceles.

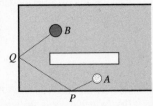

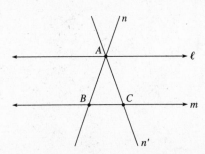

16. Prove that any isometry with a line of fixed points is either the identity or a reflection in that line of fixed points. Is it true that an isometry with exactly one fixed point must be a reflection in that point?

17. Where should a cue ball be located on an empty, triangular pool table with no pockets in order that a shot in some direction will cause the ball to return to its original position after three caroms off the three sides, and then continue on this path indefinitely, assuming no friction?

18. Work Problem 17 assuming the fictitious pool table is a rectangle, allowing four caroms instead of three.

_____ *Group C* _____

19. **Kaleidoscope Effect, 90 Degrees**   Using two mirrors along perpendicular lines ℓ and m, the eye at E will see four copies of triangle T, symmetrically distributed about O, as shown. Explain this phenomenon.

20. **Kaleidoscope Effect, 60 Degrees**   Using two mirrors along lines ℓ and m forming an angle of measure 60, the eye at E will see six copies of the triangle T, symmetrically distributed about O, as shown. Explain this phenomenon.

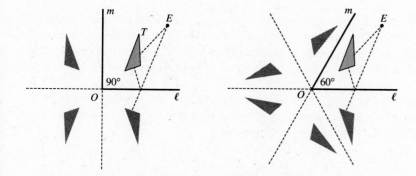

21. The given transformation f has a line ℓ of fixed points and is a reflection in that line. Find the coordinate equation of ℓ, and then verify that ℓ is actually the line of reflection (verify that ℓ is the perpendicular bisector of $\overline{PP'}$ for any point $P(x, y)$ and its image $P'(x', y')$). (**HINT:** Set $x' = x$ and $y' = y$, then solve for x and y to find line ℓ.)

$$f: \begin{cases} x' = \dfrac{3}{5}x + \dfrac{4}{5}y + 2 \\[2mm] y' = \dfrac{4}{5}x - \dfrac{3}{5}y - 4 \end{cases}$$

5.3   TRANSLATIONS, ROTATIONS, AND OTHER EUCLIDEAN MOTIONS

We now take a brief look at the two basic Euclidean motions. We are going to give synthetic definitions of these two types of mappings, then later convert to coordinates. In this section, a *reflection will always be with respect to some line.* Point reflections will not be used here.

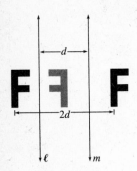

Figure 5.17

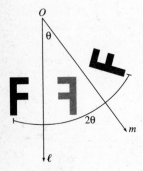

Figure 5.18

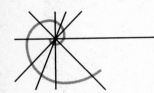

Figure 5.17 is an example of the product of two reflections over parallel lines. The image of the original configuration (the block letter "F") looks as though it could be obtained by directly sliding the figure to the new position a distance 2*d*. In the example of Figure 5.18, the same configuration ("F") has been reflected through two *intersecting lines*. Here it appears as though the letter "F" has been *rotated about point O*.

DEFINITION

A **translation** in the plane is the product of two reflections $s_\ell s_m$, where ℓ and m are parallel lines. A **rotation** is the product of two reflections $s_\ell s_m$, where ℓ and m are lines that meet, and the point of intersection is called the **center** of the rotation.

Since each individual reflection defining a translation or rotation has the *ABCD* property (angle measure, betweenness, collinearity, and distance are invariant), clearly the translation or rotation itself has the *ABCD* property. But a translation or rotation will have the additional property of *preserving orientation* as well, according to Theorem 2 of Section 5.2.

In order to work effectively with translations, it is important to know how to construct the correct parallel lines ℓ and m so that $s_\ell s_m$ will translate one figure to another. (The same is true about rotations.)

MOMENT FOR DISCOVERY

Translations

In Figure 5.19, a triangle $\triangle ABC$ is shown, and we want to find a translation $s_m s$ which will map $\triangle ABC$ to a new *given* position, $\triangle DEF$. It is pretty clear that since the direction of the desired translation is the directed line segment $\overrightarrow{AD}$, the two lines ℓ and m need to be *perpendicular* to line $\overrightarrow{AD}$. The next job is to find out where to locate these two parallel lines. Try the following experiment.

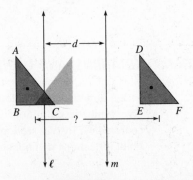

Figure 5.19

1. Draw any line ℓ perpendicular to $\overrightarrow{AD}$, such as that shown here. (In *your* figure, use a different position for ℓ.)

2. Using your best estimate, or by construction, draw line m parallel to line ℓ at the distance $d = \frac{1}{2}AD$ from ℓ.

3. Carefully locate, according to your best estimates, the points A', B', and C', which are the reflected images of A, B, and C, respectively, in line ℓ.

4. Carefully locate, according to your best estimates, the points A'', B'', and C'', the reflected images of A', B', and C', respectively, in line m.

5. What seems to be the result? Can you prove it?

A result similar to that which you may have discovered in the preceding Discovery Unit holds for rotations: If you want to define a rotation about point A, just draw any two lines ℓ and m through point A making an angle of measure $\frac{1}{2}\theta$; the product $r = s_\ell s_m$ will then rotate any point in the plane about A through the given angle θ, regardless of the position of line ℓ.

NOTE: One useful observation about translations is that a translation t is completely determined by specifying one point A and its corresponding image $A' = B$. Thus an appropriate notation for such a translation would be t_{AB}. A rotation r is completely determined by specifying its center O and angle of rotation θ. Hence, we can denote this by $r_\theta[O]$.

E X A M P L E   1 Find the rotation $s_m s_\ell = r_{64}[C]$ that maps the letter "G" to its new position, as shown in Figure 5.20, if line ℓ has already been located, as shown. That is, locate line m. Show how the two line reflections actually work with respect to one of the points being rotated.

SOLUTION

Line m must make an angle of measure $\frac{1}{2} \cdot 64 = 32$ with line ℓ, so we draw line m through point C accordingly. To see how this works, let point P be the end of the "hook" in the letter "G", as shown, and reflect P to P', then to P'', as shown.☐

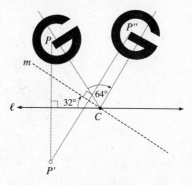

Figure 5.20

*ISOMETRIES AS
PRODUCTS OF
REFLECTIONS,
TRANSIVITY PROPERTY*

An interesting sequence of results ultimately shows that any isometry is the product of at most three reflections. The translation and rotation are special cases, requiring two reflections each to characterize them. The first step is to show that an isometry exists which maps a given triangle into one that is congruent to it.

> ### THEOREM 1
> Given any two congruent triangles $\triangle ABC$ and $\triangle PQR$, there exists a unique isometry which maps one triangle onto the other.

The proof of this theorem has two major parts:

(1) prove that the isometry *exists*

(2) prove that it is *unique*.

We tackle the uniqueness property first.

LEMMA

The only isometry which has three noncollinear fixed points is the identity mapping e (which fixes all points in the plane).

Proof:

(1) First we show that if T is an isometry and $A' = T(A) = A$, $B' = T(B) = B$ for $A \neq B$, then every point P on segment $\overline{AB}$ obeys the property $P' = T(P) = P$, and is a fixed point. (See Figure 5.21.) Suppose that A-P-B. Then A'-P'-B' or A-P'-B, since an isometry preserves betweenness (the lemma of Section 5.2). But $AP' = AP$, and by the Segment Construction Theorem, there is only one such point P, hence $P' = P$.

(2) Thus, all the points on the three sides of $\triangle ABC$ are fixed points. Let two lines ℓ and m pass through a given point P in the plane so that both lines cut two of the sides of $\triangle ABC$ at certain points D, E and F, G, respectively. Since $D' = D$, $E' = E$, $F' = F$, and $G' = G$, T fixes the two lines, hence T fixes P, their point of intersection. Since P was arbitrary, T fixes every point of the plane and is the identity.

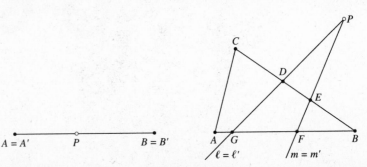

Figure 5.21

COROLLARY

If f and g are two isometries and they both map $\triangle ABC$ to $\triangle PQR$, then $f = g$.

Proof: Consider the product $h = g^{-1}f$. By the preceding Lemma, $g^{-1}f$ is the identity mapping since it takes $\triangle ABC$ to itself (A goes to P, then P goes back to A, etc.) Hence if X is any point,

$$X = g^{-1}[f(X)]$$
$$g(X) = g\{g^{-1}[f(X)]\} = f(X)$$

That is, $g(X) = f(X)$ for all X, hence $f = g$.

It remains to be shown how to construct the unique isometry which maps a given triangle $\triangle ABC$ onto another given triangle $\triangle PQR$ congruent to it ($\triangle ABC \cong \triangle PQR$). If $A \neq P$, construct the perpendicular bisector ℓ of segment $\overline{AP}$ and reflect A to $A_1 = P$, B to B_1, and C to C_1 via ℓ. (If $A = P$, then just apply this argument to any pair of distinct, corresponding vertices; if none exist, then the identity map does the job trivially.) If $B_1 \neq Q$, construct line m, the perpendicular bisector of $\overline{B_1 Q}$ (which passes through $A_1 = P$ since $A_1 B_1 = AB = PQ$), and reflect A_1 to $A_2 = P$, B_1 to $B_2 = Q$ and C_1 to C_2. If $C_2 = R$, we are finished; otherwise, use one more reflection in line $\overleftrightarrow{PQ}$ to map C_2 to R. (Why is $\overleftrightarrow{PQ}$ the perpendicular bisector of segment $\overline{RC_2}$?)

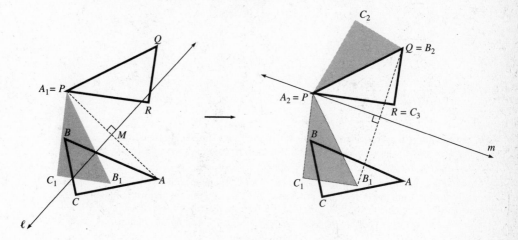

Figure 5.22

Figure 5.22 shows the successive reflections we have constructed in the proof; in the example illustrated, three reflections are actually required. Thus, we have completed the proof of Theorem 1. Indeed, using uniqueness, we have also proven

THEOREM 2

Fundamental Theorem on Isometries Every isometry in the plane is the product of at most three reflections, exactly two if the isometry is direct and not the identity.

COROLLARY

A nontrivial direct isometry is either a translation or rotation.

PROBLEMS ($5.3)

——— Group A ———

1. On your paper, sketch the diagram in the accompanying figure showing a translation of the letter "R." Draw a pair of lines ℓ and m such that "R" maps to its translated image under $s_m s_\ell$.

2. On your paper, sketch the following figure showing a rotation of the letter "R." Draw a pair of lines ℓ and m such that "R" maps onto its rotated image. Estimate, by protractor, the angle of rotation. (**HINT:** Begin by choosing corresponding points and drawing the perpendicular bisector of the segment joining them, following some of the steps of the proof of Theorem 2.)

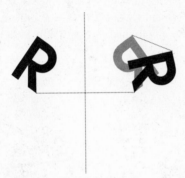

3. When the letter "S" is reflected over line ℓ, and then some line m, its final image is that shown in the accompanying figure. But someone erased line m. Trace the figure and put line m back in. [UCSMP, p. 264]

4. Carla has just had her hair cut. With her back to a mirror on the wall, she holds a hand mirror in front of her face. If Carla's eyes are 1 ft. from the hand mirror, her head is 8 in. thick, and the back of her head is 3 ft. from the wall mirror, about how far from her eyes will the image of the back of Carla's head appear in her mirror? [UCSMP, p. 264]

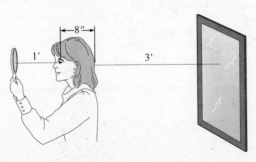

5. The transformation with the rule $T(x, y) = (x + 2, y + 6)$ is a translation.

   (a) Graph $(7, 3)$ and $T(7, 3)$.

   (b) Find the slope of the line through $(7, 3)$ and $T(7, 3)$, its image. (The slope helps to indicate the direction of the translation.)

   (c) Describe in words the effect of T on a figure. [UCSMP, p. 264]

6. If one translation is followed by another, the result is a translation. Verify this by finding $t_2 t_1$ if

$$t_1 : \begin{cases} x' = x + 2 \\ y' = y + 6 \end{cases} \quad t_2 : \begin{cases} x'' = x' + 2 \\ y'' = y' + 2 \end{cases}$$

7. Reflect the point $(4, 2)$ over the y-axis. Reflect its image over the line $y = x$ (interchange coordinates).

   (a) What is the final image?

   (b) What rotation has taken place?

8. Point B is the image of A under a rotation in the figure to the right.

   (a) Identify a point which could be the center of the rotation.

   (b) Identify another point which could be the center.

   (c) Identify a third point which could be the center.

   (d) Generalize parts **(a)**, **(b)**, and **(c)**. [UCSMP, p. 272]

9. Point B is the image of A under a clockwise rotation through an angle of measure 60. Show that this information uniquely identifies the center of rotation C, and give its exact location.

10. Suppose that $t = s_m s_\ell$ is a translation and that $r = s_\ell s_n$ is a rotation. Consider the composite (product) mapping tr. Start with a point A (as shown) and trace its movement under the reflections involved to reach your conclusion. Determine and describe the exact location of the final image of A, then describe the composition mapping tr in general.

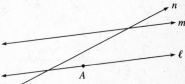

_____ *Group B* _____

11. Recall that the identity transformation e is the mapping which leaves all points unchanged ($e(P) = P$ for all P). Use $s^2 = e$ for any reflection s (Problem 5, § 5.2).

   (a) In Problem 10, prove that

$$tr = s_m s_n$$

   (Hence, tr is a rotation.)

   (b) If $t_1 = s_\ell s_m$ and $t_2 = s_m s_\ell$, then prove that

$$t_1 t_2 = e = t_2 t_1$$

   (Hence, $t_2 = t_1^{-1}$.)

12. Prove that if $t = s_m s_\ell$ is a translation in the direction AB (where $t(A) = B$), then given a line $n \perp \overrightarrow{AB}$, there exists a unique line $k \perp \overrightarrow{AB}$ such that $t = s_k s_n$.

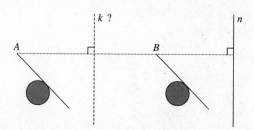

13. Repeat Problem 12 for a rotation $r = s_k s_n$.

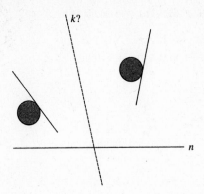

14. In the figure, $\triangle PQR \cong \triangle UVW$. Sketch this figure on your paper, then draw the lines of reflection needed to map $\triangle PQR$ onto $\triangle UVW$ in each case.

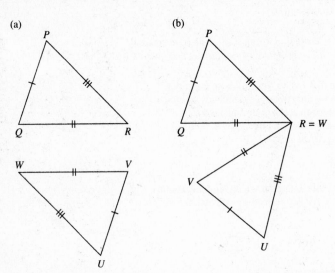

———— *Group C* ————

15. Prove the Corollary of Theorem 2: A nontrivial direct isometry is either a translation or rotation.

16. Conjecture: The only isometry with exactly one fixed point is a rotation. If you can, settle this conjecture (prove, if true, or give a counterexample, if false).

17. What can be said, in general, about the mappings

    (a) $f = s_m s_\ell s_m$, if $\ell \perp m$?

    (b) rt or tr, where r and t are, respectively, a rotation and a translation?

5.4 OTHER TRANSFORMATIONS

We will give a brief account of other important mappings used in geometry, looking at them from the synthetic point of view.

GLIDE REFLECTIONS

The simple act of walking is itself a demonstration of the mapping to be defined next.

> **DEFINITION**
>
> A **glide reflection** is the product of a reflection s_ℓ and a translation t_{AB} in a direction parallel to the axis of reflection (that is, $\overleftrightarrow{AB} \parallel \ell$).

Thus, any glide reflection $g = t_{AB}s_\ell$ is the product of *three* line reflections, since a translation is itself the product of two reflections in parallel lines, which are, in this case, each perpendicular to ℓ. The analogy of walking is made clear by observing that each step is the image of the previous step under a glide reflection (Figure 5.23). Since it is immaterial whether the translation or the reflection is performed first, we easily find that

$$s_\ell t_{AB} = t_{AB}s_\ell = g$$

GLIDE REFLECTION

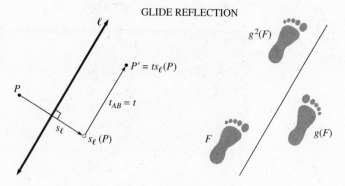

Figure 5.23

DILATIONS AND SIMILITUDES

We next consider linear transformations that preserve angle measure, but not necessarily distance.

> **DEFINITION**
>
> A **similitude**, or **similarity transformation**, is a linear transformation in the plane that preserves angle measure (is **conformal**). That is, if A', B', and C' denote the images of A, B, and C, respectively, then
>
> $$m\angle A'B'C' = m\angle ABC$$
>
> for any three noncollinear points A, B, and C.

EXAMPLE 1 Show that a similitude f preserves equidistant loci. Specifically, show that if line ℓ is the equidistant locus of points A and B, then f maps ℓ to the equidistant locus ℓ' of the images $A' = f(A)$ and $B' = f(B)$.

SOLUTION

Since ℓ is the perpendicular bisector of segment $\overline{AB}$ at M (Figure 5.24) and f is angle preserving, ℓ' will be perpendicular to segment $\overline{A'B'}$ at some point N. Since a linear transformation preserves midpoints of segments, M maps to N, the midpoint of $\overline{A'B'}$. Hence ℓ' is the perpendicular bisector of $\overline{A'B'}$, or equidistant locus.☐

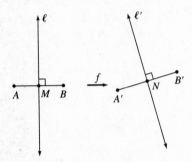

Figure 5.24

A special kind of similitude that is useful in geometry is slope preserving.

DEFINITION

A **dilation**[1] with **center** O and **dilation constant** $k > 0$ is that transformation d_k which leaves O fixed and which maps any other point P to the point P' on ray $\overrightarrow{OP}$ such that $OP' = k \cdot OP$. (If $k < 0$, we agree to locate P' on the ray $\overrightarrow{OQ}$ opposite $\overrightarrow{OP}$, with $OP' = |k|OP$.) Two configurations are said to be **homothetic** iff one is the image of the other under some dilation map.

DILATIONS

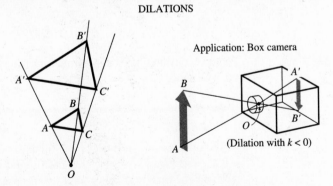

Application: Box camera

(Dilation with $k < 0$)

Figure 5.25

[1]Here the term "dilation" will designate either shrinking *or* stretching. Another term for this which frequently appears is "dilitation," which we will not use here.

It is not difficult to show that instead of having the *ABCD* property, a similitude in general has just the ABC property (having angle measure, betweenness and collinearity as invariants). In addition, it can be proven that a similitude enlarges or shrinks all distances uniformly by a factor *k* for some constant *k*. That is, if *A′* and *B′* are the images of *A* and *B* under the similitude, then $A'B' = kAB$. The factor *k* is called the **dilation constant** of the transformation. See Problem 9 if you would enjoy the challenge of proving this.

In the coordinate plane, a dilation is a slope-preserving similitude, just as a translation is a slope-preserving isometry. The dilations also belong to the class of building block transformations, along with reflections, because a similitude can be shown to be *the product of a dilation and three, or fewer, reflections.*

It is clear that the more general concept of a *conformal* transformation in the plane need not be a similitude, because we have seen that a circular inversion is conformal, is not a linear transformation, and does not have the uniform enlargement/shrinking property (that is, distances are not multiplied uniformly by a constant factor *k*).

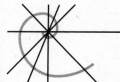

MOMENT FOR DISCOVERY

Dilations and Similitudes

1. Consider the mapping given in coordinate form $f(x, y) = (3x, 3y)$. By plotting points, discover what geometric properties this mapping seems to have. (Is it distance preserving? Is it slope preserving? Is it a similitude or dilation?)

2. If we follow the mapping *f* defined in Step 1 by a rotation or reflection in some line, what would be the over all effect? What type of mapping would you have now?

3. The converse of the result of Step 2 is interesting, and will be pursued in Problem 11. State this converse proposition.

4. What do you think the relationship is between similitudes and dilations in terms of transformation products?

OUR GEOMETRIC WORLD

The operation of a movie projector, slide projector, or TV picture tube are illustrations of dilation mappings. The image in the film, slide, or cathode tube is enlarged by the action of light rays or electrons from some point projected towards the screen, after passing through the image in the film, slide, or electronic signal. An overhead projector and a model train or car randomly placed are illustrations of the more general similitude mapping.

THEOREM 1

A dilation has the ABC Property in Euclidean geometry, and, accordingly, is a similitude.

Proof:

(1) From the definition and SAS Criterion for similar triangles, it follows that $\triangle POQ \sim \triangle P'OQ'$ and $P'Q' = |k| PQ$ for any two points P and Q.

(2) Let A, B, and C be any three collinear points, with A-B-C. Then if A', B', and C' are the images of A, B, and C,

$$A'B' + B'C' = |k| AB + |k| BC = |k| (AB + BC) = |k| AC = A'C'.$$

so A'-B'-C' follows. Thus, collinearity and betweenness are preserved.

(3) Suppose $\triangle ABC$ is any triangle, and $\triangle A'B'C'$ its image. Since $A'B' = |k| AB$, $B'C' = |k| BC$, and $A'C' = |k| AC$, then $\triangle A'B'C' \sim \triangle ABC$ (by SSS). Hence $m\angle A' = m\angle A$, and angle measure is preserved.

THEOREM 2

Given any two similar triangles $\triangle ABC$ and $\triangle A'B'C'$, there exists a unique similitude which maps the first triangle onto the second.

Proof. See Problem 15.

COROLLARY

Any similitude is the product of a dilation and at most three reflections.

Proof: See Problems 9 and 11.

A few examples will demonstrate some applications of similitudes in geometry.

EXAMPLE 2 Problem: Given a triangle, the object is to find a construction which will inscribe a square inside it such that one side of the square lies on the base of the triangle and the other two vertices of the square lie on the other two sides of the triangle.

SOLUTION

We can start with *any* square having its base on $\overrightarrow{BC}$ and opposite vertex on $\overrightarrow{BA}$, the "pull it down" to the desired position by means of a dilation, as shown in Figur

5.26. *Justification:* The construction defines a dilation (how?) and dilations preserve angle measure and ratios of distances, hence a square maps to a square.❑

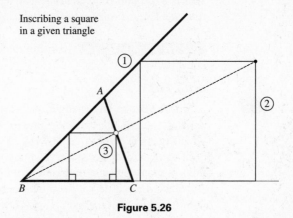

Figure 5.26

A more challenging problem is illustrated next.

EXAMPLE 3 Solve the following, using similitudes: Given a triangle $\triangle ABC$ and three circles **C**, **D**, and **E**, the problem is to find points P, Q, and R on the three circles, respectively, such that $\triangle PQR \sim \triangle ABC$.

SOLUTION

Locate P on circle **C** (P can sometimes be chosen arbitrarily for the purpose), as illustrated in Figure 5.27. Rotate circle **D** to **D'** about P as center through an angle congruent to $\angle A$. Use the dilation d_k, where $k = AB/AC$, to map **D'** to **D''**. If **D''** meets **E** at Q, there will be a solution with the original choice of point P; otherwise, one must change the location of P. (Sometimes no such location for P exists, in which case there is no solution). Let line $\overrightarrow{PQ}$ meet **D'** at R', the image of a unique point R on **D** under the preceding rotation. Then $\triangle PQR$ is the desired triangle. *Justification:* We must show that $\triangle PQR \sim \triangle ABC$. We have $PQ/PR = PQ/PR' = k = AB/AC$, and $m\angle QPR = m\angle R'PR = m\angle A$ by our choice of angle of rotation. Hence, by the SAS Similarity Criterion, $\triangle PQR \sim \triangle ABC$.❑

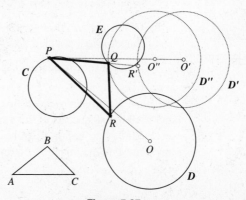

Figure 5.27

NOTE: The famous *Problem of Appolonius* also begins with three given circles, and the object is to find (construct) a circle mutually tangent to those three circles. This is a difficult problem, requiring ten cases for its solution. (For further information, see Eves, *A Survey of Geometry, Volume I*, p. 192.)

EXAMPLE 4 Use similitudes to show that the medians of a triangle are concurrent.

SOLUTION

Let $\triangle ABC$ be any triangle, with medians $\overline{AL}$, $\overline{BM}$ and $\overline{CN}$ (Figure 5.28). Since $\triangle LMN \sim \triangle ABC$ by the SSS Similarity Criterion, there exists a unique similitude f which maps $\triangle ABC$ to $\triangle LMN$ (by Theorem 2). Now consider what f must do to each median. Since S preserves midpoints (in fact, any linear transformation does this), it maps L to the midpoint of the image of $\overline{BC}$, or the midpoint of $\overline{B'C'} = MN$. But since $\overline{AL}$ cuts $\overline{MN}$ at the midpoint of $\overline{MN}$ (properties of parallel lines), that midpoint must lie on $\overline{AL}$, and f maps $\overleftrightarrow{AL}$ to itself. Then any two medians map to themselves, and their point of intersection, G, is a fixed point of f. If G does not lie on the third median, then that third median cuts the other two in distinct points G_1 and G_2, which are also fixed points of f. Then f has three noncollinear fixed points, forcing it to be the identity. But this is impossible. Hence G lies on all three medians. □

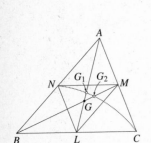

Figure 5.28

LINEAR TRANSFORMATIONS

A more general transformation preserves only betweenness and collinearity (the BC property). It is, accordingly, called a **linear transformation,** meaning, *line preserving,* which was defined previously. A convenient way to construct such a transformation without using coordinates is to use three-dimensional geometry—simply project orthogonally from plane P to plane Q (Figure 5.29), then orthogonally from Q back to P. The resulting composite mapping from P to P is a one-to-one mapping that clearly preserves collinearity and betweenness, but it does not map circles into circles, which, as we shall see, is a property that characterizes similitudes among all linear transformations.

We shall state three interesting facts about linear transformations without proof. Although they *can* be proved synthetically, coordinate proofs are much simpler, and we will defer such proofs until the next section.

- A linear transformation preserves parallelism, betweenness, ratios of lengths of parallel or collinear segments, and cross ratio.

- Given any two triangles $\triangle ABC$ and $\triangle A'B'C'$, there exists a unique linear transformation which takes $\triangle ABC$ onto $\triangle A'B'C'$.

- A linear transformation is a similitude iff it maps circles onto circles.

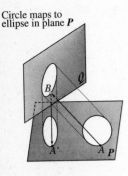

Circle maps to ellipse in plane P

Figure 5.29

INVERSIONS

We have already introduced inversions in Section 4.7 (the reader should go back and review that section at this time). Recall that an inversion in the circle centered at O and having radius r can be defined synthetically as the mapping which takes any point P to the image P' on ray $\overrightarrow{OP}$ such that

$$OP \cdot OP' = r^2$$

As we have see before, inversions have the following important invariant properties.

- Circles or lines map to circles or lines.

- The angle measure between curves is preserved.

- Cross ratio of four points is invariant.

E X A M P L E 5   Use inversion to find a circle that is orthogonal to all the circles in the family G of circles passing through the two points A and B. (See Figure 5.30.)

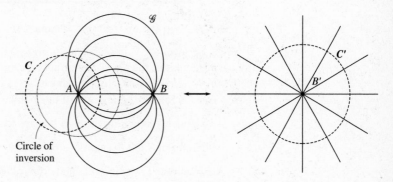

Figure 5.30

SOLUTION

Use point A as the center of an inversion, with any circle of inversion (say, the unit circle centered at A). This will map all the circles through A, including the family $\mathcal{G}$, to a family of lines. Since the circles in $\mathcal{G}$ also pass through point B, these circles map to lines concentric in the image of B under the inversion, B'. Thus, any circle centered at B' will be a circle orthogonal to all the lines through B', such as circle C' as shown, and the pre-image, C, will be a solution. (The conformal property means that C will also be orthogonal to the pre-images of the lines through B', that is, the circles in $\mathcal{G}$.)❑

STEREOGRAPHIC TRANSFORMATION

We mention one last mapping that is useful in mathematics, with many applications in complex variables and elsewhere, the **stereographic projection**. It maps the "punctured" sphere onto the plane. To define it, we again need three-dimensional geometry. Let a sphere S be tangent to a plane P at some point (Figure 5.31), and let N be its opposite pole. For each point P of S, $P \neq N$, the line $\overleftrightarrow{NP}$ will intersect the plane at a unique point P', and we define this point to be the image of P under the mapping.

STEROGRAPHIC PROJECTION

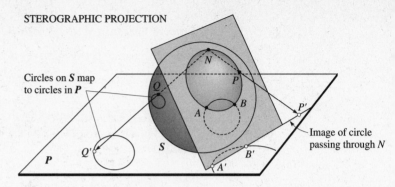

Figure 5.31

Since the mapping is not defined for the single point $P = N$, we must remove it from consideration; thus the stereographic map is defined for the sphere S *minus the point N*, called the **punctured sphere**.

Some of the interesting features of this mapping are

- Any circle of S passing through N maps to a line in P.

- All circles of S not through N are mapped to circles in P.

- Angle measure is preserved, and cross ratio of four points is an invariant.

Problems   (§5.4)

_____ *Group A* _____

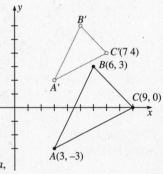

1. Prove that a similitude f maps a circle C with center O to a circle C' with center $O' = f(O)$.

2. Prove that under a dilation d_k a circle C with center O and radius r maps to a circle C' whose center is $O' = d_k(O)$ and whose radius is $|k|r$.

3. Consider the transformation $f(x, y) = (\frac{2}{3}x + 1, \frac{2}{3}y + 4)$.
   (a) Find the images under f of the points $A(3, -3)$, $B(6, 3)$, and $C(9, 0)$.
   (b) Carefully plot these points on your paper and observe the effect of f on the given points. Did $\triangle ABC$ map to one similar to it?
   (c) Verify that a dilation $d_{2/3}$ exists mapping $\triangle ABC$ to $\triangle A'B'C'$, and find its center.

4. Consider the transformation f given by $x' = ax + by$, $y' = cx + dy$ for certain constants a, b, c, and d. Suppose you know that f is a similitude and that $(1, 0)$ is a fixed point. Show that f is either the identity mapping, or a reflection in the x-axis ($x' = x$ and $y' = -y$).
   (**HINT:** After you determine a and c from the given information, by substitution, consider the image of a right triangle $\triangle ABC$ where $A = (0, 0)$, $B = (t, 0)$, and $C = (t, s)$ under f, by substitution.)

5. If you know that a certain mapping is a direct similitude and it has exactly one fixed point, and a line through that fixed point is also fixed by the mapping, what else can definitely be said about this mapping?

6. Analyze the image of the line $\ell: x + y = 1$ under the inversion transformation defined by

$$x' = \frac{x}{x^2 + y^2}, \quad y' = \frac{y}{x^2 + y^2}$$

Show this image is a circle whose center lies on a line through O that is perpendicular to line ℓ. (Find the center and radius of this image circle.) Since an inversion is self-inverse, you may substitute the inversion equations into $x' + y' = 1$ to solve the problem.

7. Use a dilation to inscribe a golden rectangle in a given acute-angled triangle $\triangle ABC$ (with one side of the rectangle lying on the base $\overline{AB}$). (Recall that a Golden Rectangle is one in which nonequal sides are in the ratio of $1/\tau$, where $\tau = \frac{1}{2}(1 + \sqrt{5}\,)$.)

8. Show that under a stereographic transformation, a circle of S passing through N maps to a line in P (refer to Figure 5.31).

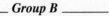

_____ *Group B* _____

9. **Dilation Property of Similitudes** Any similitude has the **Dilation Property:** There exists a positive constant $k > 0$ such that for any two points A and B, with A' and B' their images,

$$A'B' = k \cdot AB$$

Prove this by first considering $\overline{AB}$ and $\overline{BC}$ and their images $\overline{A'B'}$ and $\overline{B'C'}$, with A and B fixed in position, in order to define k, then prove that $C'D' = k \cdot CD$ for arbitrary C and D, using similar triangles.

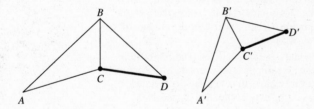

10. Prove the following corollary to the result of Problem 9: If a similitude has two fixed points, it is an isometry.

11. By Problem 9, it follows that a similitude is an isometry followed by a dilation d_k for $k > 0$ from any convenient center C. Prove this explicitly. (**HINT:** Start with point C and define $d_{1/k} = d_k^{-1}$ with center C. If f is the given similitude and it has dilation factor k, what must be true of the product $d_{1/k} f$?)

12. Given three parallel lines ℓ, m, and n, find a construction that will

  (a) locate the vertices of a 45°-right triangle $\triangle PQR$ on those three lines, as indicated in the figure.

  (b) Find a similar solution if the given triangle is equilateral.

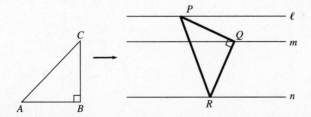

13. Work Problem 12 when the given lines meet at some point.

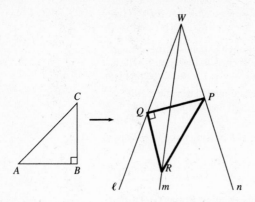

_____ *Group C* _____

14. Prove that a linear transformation is a similitude iff it maps circles to circles. (You may use the fact that a linear transformation preserves ratios of distances between collinear points.)

15. Prove that there exists a unique similitude which maps a given triangle to one that is similar to it.

16. Show that if a direct similitude f maps just one line into one that is parallel to it, then it is slope-preserving, and maps every line to one that is parallel to it. Hence, there exists a point C such that the similitude is either a dilation d_k with center C or the product of d_k and a translation. (In this problem, k may be negative.)

17. Consider the $\mathcal{F}$-$\mathcal{G}$ configuration of two orthogonal systems of circles. The circles in $\mathcal{F}$ which cluster about A are labelled $\mathcal{F}_1$ and those about B, $\mathcal{F}_2$. An inversion is performed with center at B. Use the conformal property of inversions to determine the inverted image of $\mathcal{F}$ and $\mathcal{G}$. In particular, what becomes of the two subfamilies $\mathcal{F}_1$ and $\mathcal{F}_2$?

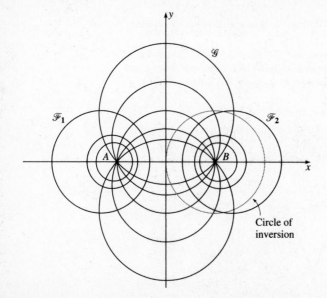

*18. Relation for Radius of Inverted Circle Under Inversion
The purpose of this problem is to find how inversion affects the radius of a circle under circular inversion. Let a circle with center at A and radius a be inverted to the circle centered at B with radius b, where the circle of inversion has radius r and center O, with $c = OA$. Prove that

$$b = \frac{ar^2}{\left| a^2 - c^2 \right|}$$

(**HINT:** Use the fact that $CD = 2a$ and $C'D' = 2b$ to get started, where $OC' = r^2/OC$

and $OD' = r^2/OD$. **WARNING:** $A' \neq B$.)

*This result is used in Problem 22, § 5.7.

Inversion of Circle

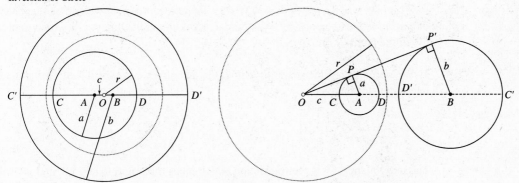

19. PROJECT FOR UNDERGRADUATE RESEARCH   Given a similitude f with dilation factor $k < 1$ and some point A, form the sequence $A_1, A_2, A_3, \ldots$ by defining $A_1 = f(A)$, $A_2 = f(A_1)$, $A_3 = f(A_2)$, and so on. Show that $\{A_n\}$ is a **Cauchy sequence**, that is, given $\varepsilon > 0$, there exists n_0 such that for all $m, n > n_0$,

$$A_m A_n < \varepsilon$$

As such, the limit

$$\lim_{n \to \infty} A_m = B\,(b, c)$$

exists. Find a formula for b and c in terms of the parameters of the mapping f, if possible. Illustrate in the coordinate plane when f is defined explicitly by

$$x' = \frac{2}{3}x + 9, \quad \text{or} \quad y' = \frac{2}{3}y + 27$$

and $A = (0, 0)$. Find the coordinates, by computer if necessary, of the limiting point $B(b, c)$ for this case. Explore whether a general formula for this limit exists in terms of the coordinate formula for f (as given in the next section).

5.5 COORDINATE CHARACTERIZATION OF LINEAR TRANSFORMATIONS

It can be shown that any transformation that maps lines onto lines in the coordinate and vector plane must preserve parallelograms, thus vector addition, and ultimately scalar multiplication of vectors as well (this latter result is a basic theorem of linear algebra whose proof can be found in some texts on the subject). Accordingly, if $f(x, y) = (x', y')$ is any linear transformation, then for certain constants a, b, c, d, x_0, and y_0, we must have

COORDINATE FORM OF A GENERAL LINEAR TRANSFORMATION

(1) $$f: \begin{cases} x' = ax + by + x_0 \\ y' = cx + dy + y_0 \end{cases} \quad (ad \neq bc)$$

NOTE: The condition $ad \neq bc$ characterizes the one-to-one property of f. If f were, for example, $x' = 2x + 2y$ and $y' = 3x + 3y$, then the *entire plane* would be mapped into a *single line*, namely, $3x' = 2y'$, and every point on the line $y = -x$ would map to the single point $(0, 0)$.

We observe that the mapping f in (1) is actually a composition of two simpler linear transformations:

$$f^*: \begin{cases} x' = ax + by \\ y' = cx + dy \end{cases} \qquad t: \begin{cases} x'' = x' + x_0 \\ y'' = y' + y_0 \end{cases}$$

The latter mapping t is the translation which takes the origin to (x_0, y_0), and we call it the **translational component** of f. Since the geometric properties of f^* and f will be virtually the same (but for a translation), we can concentrate our efforts on f^*. Accordingly, we look at the **matrix** defining f^*

$$A = \begin{bmatrix} a & b \\ c & d \end{bmatrix}$$

and its **determinant**

$$\det A \equiv |A| \equiv \begin{vmatrix} a & b \\ c & d \end{vmatrix} = ad - bc$$

(which we require to be nonzero). The geometric behavior of f is thereby completely determined by the four numbers in this 2×2 matrix.

E X A M P L E   1 By merely substituting algebraic expressions and simplifying, find what distortions the linear transformation

$$f: \begin{cases} x' = 4x + 2y \\ y' = -x + 2y \end{cases} \qquad \text{matrix} = \begin{bmatrix} 4 & 2 \\ -1 & 2 \end{bmatrix}$$

imposes on

(a) a square having vertices $A(1, 1)$, $B(2, 1)$, $C(2, 2)$, $D(1, 2)$

(b) the unit circle $x^2 + y^2 = 1$, parameterized as $x = \cos t$, $y = \sin t$, t real (Figure 5.32).

SOLUTION

(a) We know $\Box ABCD$ maps to another parallelogram. To find its vertices, merely substitute the coordinates of A, B, C, and D into the above equations.

$$A(1, 1): \quad x' = 4 \cdot 1 + 2 \cdot 1 = 6, \quad y' = -1 + 2 \cdot 1 = 1$$
$$B(2, 1): \quad x' = 4 \cdot 2 + 2 \cdot 1 = 10, \quad y' = -2 + 2 \cdot 1 = 0$$
$$C(2, 2): \quad x' = 4 \cdot 2 + 2 \cdot 2 = 12, \quad y' = -2 + 2 \cdot 2 = 2$$
$$D(1, 2): \quad x' = 4 \cdot 1 + 2 \cdot 2 = 8, \quad y' = -1 + 2 \cdot 2 = 3$$

The results are shown in Figure 5.32. The image of $\Box ABCD$ is then an oblique-angled parallelogram.

(b) Substitute $x = \cos t$ and $y = \sin t$ into the transformation equations, then eliminate t by squaring and summing.

$$x' = 4 \cos t + 2 \sin t$$
$$x'^2 = 16 \cos^2 t + 16 \sin t \cos t + 4 \sin^2 t$$
$$y' = -\cos t + 2 \sin t$$
$$y'^2 = \cos^2 t - 4 \sin t \cos t + 4 \sin^2 t$$
$$4y'^2 = 4 \cos^2 t - 16 \sin t \cos t + 16 \sin^2 t$$

Sum the resulting squared equations to obtain

$$x'^2 + 4y'^2 = 20 \cos^2 t + 20 \sin^2 t = 20$$

or

$$\frac{(x')^2}{20} + \frac{(y')^2}{5} = 1$$

which is an ellipse. (See graph in Figure 5.32.)❏

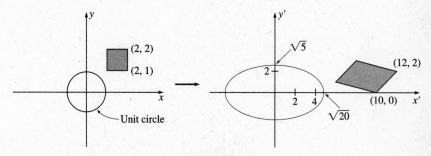

Figure 5.32

It is interesting, and instructional, to see what geometric patterns are effected by the numbers in the matrix defining a transformation. Here are a few examples, where a uniform test pattern is used throughout (unit square plus cross hairs).

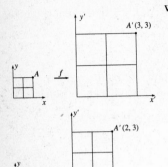

$$f : \begin{bmatrix} 3 & 0 \\ 0 & 3 \end{bmatrix}$$

uniform scale change
for x and y (dilation, $k = 3$)

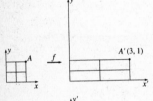

$$f : \begin{bmatrix} 2 & 0 \\ 0 & 3 \end{bmatrix}$$

Different scale changes for x and y—
(distortion along both axes)

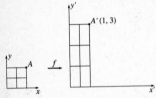

$$f : \begin{bmatrix} 3 & 0 \\ 0 & 1 \end{bmatrix}$$

scale change for x only (distortion is along x-axis)

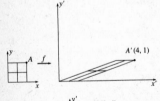

$$f : \begin{bmatrix} 1 & 0 \\ 0 & 3 \end{bmatrix}$$

scale change for y only (distortion is along y-axis)

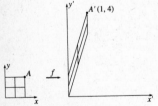

$$f : \begin{bmatrix} 1 & 3 \\ 0 & 1 \end{bmatrix}$$

shear with respect to y-axis (fixes points on x-axis)

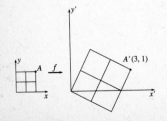

$$f : \begin{bmatrix} 1 & 0 \\ 3 & 1 \end{bmatrix}$$

shear with respect to x-axis (fixes points on y-axis)

$$f : \begin{bmatrix} 2 & 1 \\ -1 & 2 \end{bmatrix}$$

rotation/dilation

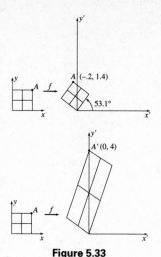

$$f : \begin{bmatrix} 0.6 & -0.8 \\ 0.8 & 0.6 \end{bmatrix} \text{ rotation } (\theta \approx 53.1°)$$

$$f : \begin{bmatrix} -1 & 1 \\ 1 & 3 \end{bmatrix} \quad \text{general linear transformation (random distortions)}$$

Figure 5.33

In working with transformations, it is helpful to have one more tool at our disposal besides coordinates and vectors. It is the theory of matrix products. You may already be familiar with matrix multiplication. In case you are not, the product of two 2×2 matrices is given by the rule

$$\begin{bmatrix} a & b \\ c & d \end{bmatrix} \cdot \begin{bmatrix} x & z \\ y & w \end{bmatrix} = \begin{bmatrix} ax + by & az + bw \\ cx + dy & cz + dw \end{bmatrix}$$

(The rule is to regard each *row* of A and each *column* of B as vectors, then calculate their scalar products for each entry of the product.)

The more general product of a 2×2 matrix and a $2 \times n$ (for $n = 4$) matrix is

$$\begin{bmatrix} a & b \\ c & d \end{bmatrix} \cdot \begin{bmatrix} x_1 & x_2 & x_3 & x_4 \\ y_1 & y_2 & y_3 & y_4 \end{bmatrix} = \begin{bmatrix} ax_1 + by_1 & ax_2 + bx_2 & ax_3 + bx_3 & ax_4 + by_4 \\ cx_1 + dy_1 & cx_2 + dy_2 & cx_3 + dy_3 & cx_4 + dy_4 \end{bmatrix}$$

It should be clear from the general form that the matrix product represents a labor-saving device in dealing with linear transformations. To find the images of two or more points given in coordinate form under a linear transformation, one can use the matrix product as a schematic to organize the various computations. (There is also computer software available which will perform these matrix products automatically.)

E X A M P L E   2 Calculate the images of the four points $P(1, 2)$, $Q(1, -2)$, $R(-1, -2)$, and $S(-1, 2)$ in Figure 5.34 under the transformation

(b) $\quad f: \begin{cases} x' = 2x + 3y \\ y' = x + 2y \end{cases} \quad \text{matrix} : \begin{bmatrix} 2 & 3 \\ 1 & 2 \end{bmatrix}$

SOLUTION

$$\begin{array}{cccc} P & Q & R & S \end{array} \qquad \begin{array}{cccc} P' & Q' & R' & S' \end{array}$$
$$\begin{bmatrix} 2 & 3 \\ 1 & 2 \end{bmatrix} \begin{bmatrix} 1 & 1 & -1 & -1 \\ 2 & -2 & -2 & 2 \end{bmatrix} = \begin{bmatrix} 8 & -4 & -8 & 4 \\ 5 & -3 & -5 & 3 \end{bmatrix}$$

$\therefore P' = (8, 5)$, $Q' = (-4, -3)$, $R' = (-8, -5)$, and $S' = (4, 3)$.

Figure 5.34

In general, the image of $P(x, y)$ under the transformation

$$f: \begin{cases} x' = ax + by + x_0 \\ y' = cx + dy + y_0 \end{cases} \qquad \text{matrix}: A = \begin{bmatrix} c & b \\ c & d \end{bmatrix}$$

may be written in the matrix/vector form

(2)
$$f: \mathbf{x}' = A\mathbf{x} + \mathbf{x}_0, \quad |A| \neq 0$$

where $\mathbf{x}_0$ is the vector $[x_0, y_0]$, $\mathbf{x} = [x, y]$, and $\mathbf{x}' = [x', y']$. A more explicit formulation of the same thing is

(2′)
$$\begin{bmatrix} x' \\ y' \end{bmatrix} = \begin{bmatrix} a & b \\ c & d \end{bmatrix} \begin{bmatrix} x \\ y \end{bmatrix} + \begin{bmatrix} x_0 \\ y_0 \end{bmatrix}, \qquad \begin{vmatrix} a & b \\ c & d \end{vmatrix} \neq 0$$

We are now ready to make a systematic study of the geometric properties of linear transformations as characterized by the form of the matrices defining them. The first step is to obtain a general formula to determine the way a linear transformation, in terms of its matrix, alters the slope of a line under **(2)**. ❑

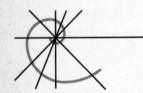

MOMENT FOR DISCOVERY

Slope and Linear Transformations

Consider a line ℓ whose slope is m,

(a)
$$y = mx + k$$

and a general linear transformation

(b)
$$f: \begin{cases} x' = ax + by \\ y' = cx + dy \end{cases} \qquad \text{matrix}: \begin{bmatrix} a & b \\ c & d \end{bmatrix}$$

The transformed line ℓ' under f will generally have the equation

$$y' = m'x' + k'$$

with a different slope m'. To find m', perform the following algebraic calculations.

1. Let $x = 0$ and $x = 1$ in (a). This will produce two points A and B on ℓ. (Find their coordinates.)

2. Find the coordinates of the transformed points A', B', using (b): $x' = ax + by$ and $y' = cx + dy$. (Did you get $B' = (a + bm + bk, c + dm + dk)$?)

3. Using the formula for the slope (formula (2) in § 4.6), the slope of the transformed line $\overleftrightarrow{A'B'}$ is

$$m' = \frac{y_2' - y_1'}{x_2' - x_1'}$$

Find an expression for m', simplifying. You should get a formula involving a, b, c, d, and m.

If you were successful in the preceding Discovery Unit, then you established the slope transformation formula in terms of the matrix for the transformation (1):

EFFECT OF LINEAR TRANSFORMATIONS ON SLOPE

(3) $$m' = \frac{c + dm}{a + bm}$$

An obvious immediate deduction from this formula is a key property of linear transformations in the plane, mentioned earlier.

THEOREM 1

A linear transformation maps parallel lines to parallel lines.

Proof: If two parallel lines have slopes m_1 and m_2, then $m_1 = m_2$ and the above formula obviously yields $m_1' = m_2'$. If the lines are both vertical, then either $b \neq 0$ and the image lines are nonvertical, with slope $m_1' = d/b = m_2'$, or $b = 0$ and both image lines are vertical.

We already know that linear transformations do not always preserve perpendicularity (Example 2). If we impose this property on a transformation, a very surprising thing takes place, and we get more than we might expect. First, we will need to prove a special algebraic lemma.

LEMMA

Suppose the following conditions hold for real numbers a, b, c, and d:

(a) $ad \neq bc$

(b) $a^2 + c^2 = b^2 + d^2$

(c) $ab = -cd$.

Then either $b = -c$ and $d = a$, or $b = c$ and $d = -a$.

Proof:

(1) First, suppose that $a = 0$. Then by **(a)**, $bc \neq 0$ and both $b \neq 0$ and $c \neq 0$. Apply **(b)** and **(c)**: $0 + c^2 = b^2 + d^2$, $0 = -cd$. Since $c \neq 0$, $d = 0$, and we obtain $b^2 = c^2$ or $b = \pm c$. Then either $b = -c$ and $d = a = 0$, or $b = c$ and $d = -a = 0$.

(2) Next, suppose $c = 0$. Then in this case, $a \neq 0$ and **(c)** implies $b = 0 \rightarrow a^2 = d^2$ or $d = \pm a$, and again the desired conclusion results.

(3) The last case is $a \neq 0$ and $c \neq 0$. Let $x = -b/c$, $y = d/a$. That is, $b = -cx$ and $d = ay$. Substitute into **(b)** and **(c)**.

$$a^2 + c^2 = c^2 x^2 + a^2 y^2,$$
$$-acx = -acy$$
$$\therefore x = y$$

$$a^2 + c^2 = (c^2 + a^2)x^2$$
$$x^2 = 1$$
$$\therefore x = y = \pm 1)$$

If $x = 1$, then $b = -c$ and $d = a$, and if $x = -1$, $b = c$ and $d = -a$, as desired.

NOTE: The matrix form of the lemma is: If $ad \neq bc$, $a^2 + c^2 = b^2 + d^2$, and $ab = -cd$, then either

$$\begin{bmatrix} a & b \\ c & d \end{bmatrix} = \begin{bmatrix} a & -c \\ c & a \end{bmatrix} \quad \text{or} \quad \begin{bmatrix} a & b \\ c & d \end{bmatrix} = \begin{bmatrix} a & c \\ c & -a \end{bmatrix}$$

To summarize the two cases, we can simply write

$$\begin{bmatrix} a & b \\ c & d \end{bmatrix} = \begin{bmatrix} a & -\delta c \\ c & \delta a \end{bmatrix}, \quad \text{where } \delta = \pm 1$$

If a linear transformation $x' = Ax$ preserves perpendicularity, then for all slopes such that $m_1 m_2 = -1$, we must have $m'_1 m'_2 = -1$. We shall call such a transformation an **orthomap**. Making use of **(3)**, we have, for all $m_1 m_2 = -1 = m'_1 m'_2$,

$$-1 = \left(\frac{c + m_1 d}{a + m_1 b} \right) \cdot \left(\frac{c + m_2 d}{a + m_2 b} \right) = \frac{c^2 + (m_1 + m_2)\, cd - d^2}{a^2 + (m_1 + m_2)\, ab - b^2}$$

You can show from this (substituting, in turn, $m_1 = 1$, $m_2 = -1$ and $m_1 = 2$, $m_2 = -\frac{1}{2}$) that

$$-1 = \frac{c^2 - d^2}{a^2 - b^2} \quad \text{and} \quad -1 = \frac{c^2 + \dfrac{3}{2}cd - d^2}{a^2 + \dfrac{3}{2}ab - b^2}$$

which ultimately implies that $a^2 + c^2 = b^2 + d^2$, and $ab = -cd$. Since the conditions of the lemma are therefore satisfied, it follows that

$$A = \begin{bmatrix} a & -\delta c \\ c & \delta a \end{bmatrix}$$

Thus, we have (with b replacing c):

COORDINATE FORM OF AN ORTHOMAP

(4) $f : \begin{cases} x' = ax - \delta by \\ y' = bx + \delta ay \end{cases}$ matrix: $\begin{bmatrix} a & -\delta b \\ b & \delta a \end{bmatrix}$

$$(\delta = \pm 1)$$

EXAMPLE 3 Show that, by its form, the following mapping is an orthomap, and determine the sign of δ:

$$f : \begin{cases} x' = 2x - 3y \\ y' = -3x - 2y \end{cases} \quad \text{matrix:} \begin{bmatrix} 2 & -3 \\ -3 & -2 \end{bmatrix}$$

SOLUTION

By direct comparison,

$$\begin{bmatrix} 2 & -3 \\ -3 & -2 \end{bmatrix} \leftrightarrow \begin{bmatrix} a & -\delta b \\ b & \delta a \end{bmatrix}$$

and we find that $a = 2$, $b = -3$, and $\delta a = -2$, or $\delta = -1$. ❏

From algebra and a formula from trigonometry, a surprising generalization follows from **(4)**, which seems to be exceedingly difficult to prove by the synthetic method.

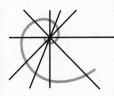

MOMENT FOR DISCOVERY

Experimenting With An Orthomap

Recall the following formula from trigonometry, which gives the angle measure between two lines having slopes m_1 and m_2.

$$\tan \theta = \frac{m_2 - m_1}{1 + m_1 m_2}$$

As an experiment, consider the orthomap whose matrix is

$$\begin{bmatrix} 4 & -3 \\ 3 & 4 \end{bmatrix}$$

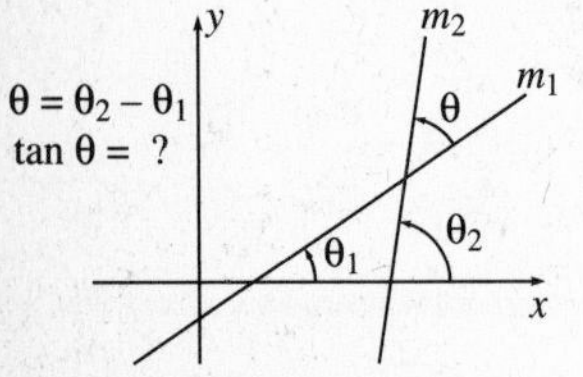

Figure 5.35

and consider lines ℓ_1 and ℓ_2 having slopes $m_1 = 1$ and $m_2 = 3$, respectively.

1. Find $\tan \theta$ for lines ℓ_1, ℓ_2. (You should get $\frac{1}{2}$ for this.)
2. Use **(3)** to find the slopes and m_2' of the images of ℓ_1 and ℓ_2 under this transformation. (Did you get $m_1' = 7$?)
3. Calculate $\tan \theta'$ for ℓ_1' and ℓ_2'.
4. Did anything happen?
5. Repeat the experiment for $m_1 = 0$ and $m_2 = 4$.
6. Have you discovered a theorem? Try proving it in general, using algebra techniques for simplifying complex fractions.

As a result of what you should have found in the Discovery Unit, *any orthomap is a similitude*, which preserves the measure of all angles, not just right angles. This means also that we have found the general coordinate form for a similitude.

> **MATRIX/COORDINATE FORM OF A SIMILITUDE**
> (Dilation factor $\sqrt{a^2 + b^2}$)
>
> **(5)** $f: \begin{cases} x' = ax - \delta by + x_0 \\ y' = bx + \delta ay + y_0 \end{cases}$ matrix: $\begin{bmatrix} a & -\delta b \\ b & \delta a \end{bmatrix}$
>
> $(\delta = \pm 1)$

An important special case of this occurs when $b = 0$ and $\delta = 1$.

> **MATRIX/COORDINATE FORM FOR DILATION CENTERED AT (h, k)**
> (Dilation factor a)
>
> **(6)** $f: \begin{cases} x' = ax + x_0 \\ y' = ay + y_0 \end{cases}$ matrix : $\begin{bmatrix} a & 0 \\ 0 & a \end{bmatrix}$
>
> where $x_0 = h - ah, y_0 = k - ak$

Now, let's work with the distance-preserving property. If f is an isometry, then as we have seen, f preserves angle measure and hence is a similitude, with the form in

(5). Substitute the coordinates $O(0, 0)$ and $P(1, 0)$ into (5); since distance is preserved, $O'U' = OU = 1$, and we obtain the additional requirement

$$a^2 + b^2 = 1$$

Thus

MATRIX/COORDINATE FORM OF AN ISOMETRY

(7)
$$f: \begin{cases} x' = ax - \delta by + x_0 \\ y' = bx + \delta ay + y_0 \end{cases} \quad \text{matrix:} \begin{bmatrix} a & -\delta b \\ b & \delta a \end{bmatrix}$$

where $a^2 + b^2 = 1$ and $\delta = \pm 1$

NOTE: It is interesting that the sign of δ in the above formulas characterizes the direct/opposite nature of the mappings: If $\delta = 1$, the mapping is direct, and if $\delta = -1$, opposite. (See Problem 12.)

The remaining special forms of transformations we introduced synthetically can now be easily found. In particular, if $a = \cos\theta$, $b = \sin\theta$, and $\delta = 1$, we obtain the *rotation about the origin through an angle of measure* θ.

MATRIX/COORDINATE FORM OF ROTATION ABOUT ORIGIN

(8)
$$f: \begin{cases} x' = x\,\cos\theta - y\,\sin\theta \\ y' = x\,\sin\theta + y\,\cos\theta \end{cases} \quad \text{matrix}: \begin{bmatrix} \cos\theta & -\sin\theta \\ \sin\theta & \cos\theta \end{bmatrix}$$

(θ = measure of angle of rotation)

The form in (8) can be easily generalized to give the rotation about an arbitrary point $C(h, k)$, which will be important later. First, translate the center $C(h, k)$ to the origin O under

(9)
$$f: \begin{cases} x' = x - h \\ y' = y - k \end{cases}$$

(That is, the new coordinates of $C(h, k)$ are $(0, 0)$. Now perform the rotation about $(0, 0)$ in the primed coordinate system.

(10)
$$g: \begin{cases} x'' = x'\cos\theta - y'\sin\theta \\ y'' = x'\sin\theta + y'\sin\theta \end{cases}$$

Now translate back to the original center $C(h, k)$ under

$$h: \begin{cases} x''' = x'' + h \\ y''' = y'' + k \end{cases}$$

(11)

Using (9), (10), and (11), the overall effect is the composition hgf, after changing the triple primes to ordinary single primes and by substitution,

---

COORDINATE FORM OF ROTATION ABOUT C(h, k)

(12)
$$f: \begin{cases} (x' - h) = (x - h)\cos\theta - (y - k)\sin\theta \\ (y' - k) = (x - h)\sin\theta + (y - k)\cos\theta \end{cases}$$

(θ = measure of angle of rotation)

We come full circle back to the line reflection, whose matrix representation, curiously enough, is the most complicated of all the mappings considered so far. But it is interesting, and we shall include it, leaving the proof as a problem. (All you have to do is use the definition for s_ℓ in Section 5.2 and apply it directly to coordinate geometry; Problem 18 reveals a different, more elegant, method.)

MATRIX/COORDINATE FORM OF REFLECTION IN A LINE

(line $= \ell : ax + by = c$ where $a^2 + b^2 = 1$)

s_ℓ (coordinate form):

(13)
$$f: \begin{cases} x' = (b^2 - a^2)x - 2aby + 2ac \\ y' = -2abx - (b^2 - a^2)y + 2bc \end{cases}$$

s_ℓ (matrix form):

$$\begin{bmatrix} b^2 - a^2 & -2ab \\ -2ab & a^2 - b^2 \end{bmatrix}$$

EXAMPLE 4 The lines $\ell : y = 3x - 3$ and $m : y = -2x + 7$ meet at $C(2, 3)$, as shown in Figure 5.36.

(a) Calculate the matrices for the reflections s_ℓ and s_m using the form (13).

(b) Verify by the appropriate substitutions that the product $s_m s_\ell$ is a rotation about C, and find the angle of rotation.

(c) Verify that this angle is twice the angle between ℓ and m.

SOLUTION

(a) Let ℓ and m be rearranged in the general form

$$Ax + By = C$$

and normalize the equations by division by $A^2 + B^2$.

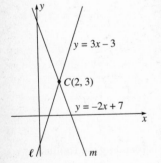

Figure 5.36

$$\ell: \quad 3x - y = 3 \qquad \rightarrow \qquad \frac{3x - y}{\sqrt{10}} = \frac{3}{\sqrt{10}}$$

$$m: \quad 2x + y = 7 \qquad \rightarrow \qquad \frac{2x + y}{\sqrt{5}} = \frac{7}{\sqrt{5}}$$

Hence, for ℓ, a, b, and c are $\frac{3}{\sqrt{10}}$, $\frac{-1}{\sqrt{10}}$, and $\frac{3}{\sqrt{10}}$, respectively. Using **(13)**, $b^2 - a^2 = \frac{1}{10} - \frac{9}{10} = -\frac{4}{5}$, $-2ab = -2(\frac{3}{\sqrt{10}})(\frac{-1}{\sqrt{10}}) = \frac{3}{5}$, $2ac = 2(\frac{3}{\sqrt{10}})(\frac{3}{\sqrt{10}}) = \frac{9}{5}$ and $2bc = 2(\frac{-1}{\sqrt{10}})(\frac{3}{\sqrt{10}}) = -\frac{3}{5}$. Thus, s_ℓ is given by

$$s_\ell : \begin{cases} x' = -\dfrac{4}{5}x + \dfrac{3}{5}y + \dfrac{9}{5} \\ y' = \dfrac{3}{5}x + \dfrac{4}{5}y - \dfrac{3}{5} \end{cases}$$

Similar calculations for s_m yield

$$s_m : \begin{cases} x'' = -\dfrac{3}{5}x' - \dfrac{4}{5}y' + \dfrac{28}{5} \\ y'' = -\dfrac{4}{5}x' + \dfrac{3}{5}y' + \dfrac{14}{5} \end{cases}$$

(b) By substitution,

$$s_m s_\ell : \begin{cases} x'' = -\dfrac{3}{5}(-\dfrac{4}{5}x + \dfrac{3}{5}y + \dfrac{9}{5}) - \dfrac{4}{5}(\dfrac{3}{5}x + \dfrac{4}{5}y - \dfrac{3}{5}) + \dfrac{28}{5} = -y + 5 \\ y'' = -\dfrac{4}{5}(-\dfrac{4}{5}x + \dfrac{3}{5}y + \dfrac{9}{5}) + \dfrac{3}{5}(\dfrac{3}{5}x + \dfrac{4}{5}y - \dfrac{3}{5}) + \dfrac{14}{5} = x + 1 \end{cases}$$

Comparing to **(12)**, we find that this can be put in the form

$$x'' - 2 = -(y - 3) = (\cos 90)(x - 2) - (\sin 90)(y - 3)$$

$$y'' - 3 = x - 2 = (\sin 90)(x - 2) + (\cos 90)(y - 3)$$

This is the form of a rotation about $C(2, 3)$ through the angle $\theta = 90$ (degrees).

(c) The angle between ℓ (slope 3) and m (slope -2) is found by the formula

$$\tan \phi = \frac{m_2 - m_1}{1 + m_1 m_2} = \frac{-2 - 3}{1 - 6} = 1$$

Thus, $\phi = 45$ and $\theta = 90$ (degrees). ❑

PROBLEMS   (§5.5)

_____ *Group A* _____

1. Identify each of the following linear transformations by their coordinate forms.

    (a) $\begin{cases} x' = 2x + y \\ y' = -x + 2y \end{cases}$ (b) $\begin{cases} x' = 3x \\ y' = 3y \end{cases}$ (c) $\begin{cases} x' = x - 3 \\ y' = x - 5 \end{cases}$

2. Identify each of the following linear transformations by their coordinate forms.

    (a) $\begin{cases} x' = -x + 1 \\ y' = -y + 2 \end{cases}$ (b) $\begin{cases} x' = 2x/\sqrt{5} + y/\sqrt{5} \\ y' = x/\sqrt{5} + 2y/\sqrt{5} \end{cases}$ (c) $\begin{cases} x' = y \\ y' = x \end{cases}$

3. The following matrices define certain linear transformations:

    (a) $\begin{bmatrix} 3 & -2 \\ 2 & -3 \end{bmatrix}$ (b) $\begin{bmatrix} 1 & 2 \\ -1 & 1 \end{bmatrix}$

    Find the images of the points $P(3, 0)$, $Q(5, -1)$, $R(-2, 3)$, and $S(0, 2)$ by finding the matrix products

    (a) $\begin{bmatrix} 3 & -2 \\ 2 & -3 \end{bmatrix} \begin{bmatrix} 3 & 5 & -2 & 0 \\ 0 & -1 & 3 & 2 \end{bmatrix}$ (b) $\begin{bmatrix} 1 & 2 \\ -1 & 1 \end{bmatrix} \begin{bmatrix} 3 & 5 & -2 & 0 \\ 0 & -1 & 3 & 2 \end{bmatrix}$

4. If T_1 and T_2 are the linear transformations defined by

    $$T_1 : \begin{cases} x' = 3x - 2y \\ y' = 2x - 3y \end{cases} \quad T_2 : \begin{cases} x'' = 2x' + 2y' \\ y'' = -x' - 2y' \end{cases}$$

    and we want to find the image of $\triangle ABC$, where $A = (-1, 2)$, $B = (3, -2)$, and $C = (2, 0)$ under $T_2 T_1$, what is the best way to proceed?

    (a) Find A'', B'', C'' by direct substitution into the given coordinate equations defining T_1 and T_2, and then do this by use of matrix products.
    (b) Compare your answers. Which way is more efficient?
    (c) What 2×2 matrix characterizes $T_2 T_1$?

5. Verify that

    $$\begin{bmatrix} 3 & 2 \\ 2 & -3 \end{bmatrix}$$

    is the matrix of a similitude by graphing $\triangle ABC$ and its altitude $\overline{BE}$, and the image triangle and altitude, if $A = (0, 0)$, $B = (2, 2)$, $C = (3, 0)$, and $E = (2, 0)$.

6. (a) Using the form (13) for a reflection in a line, find the matrix which represents a reflection in the line ℓ: $3x + 2y = 13$. (You must first "normalize" the equation by dividing by $\sqrt{3^2 + 2^2}$)

    (b) Test your answer by finding the image A' of the point $A(6, 4)$ and that of B, the foot of the perpendicular $\overline{AB}$ to ℓ (find the coordinates of B and verify that $B = B'$), then use the midpoint formula to verify that $B = $ midpoint of segment $\overline{AA'}$. (See figure to the right.)

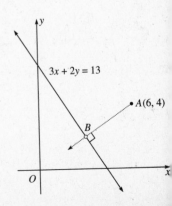

7. (a) Using **(13)** write down the equations for the reflections through each of the parallel lines $x - 2y = 1$ and $x - 2y = 3$. (Don't forget to normalize.)

(b) Show, by coordinates, that the product of the two reflections in **(a)** is a translation.

8. (a) Repeat Problem 7(a) for the lines $x - 2y = 0$ and $3x - y = 0$

(b) Show, by coordinates, that the product of the two reflections in **(a)** is a rotation about the origin, and verify that the angle of rotation is twice that between the given lines.

9. Reflection in a Point $C(h, k)$ Verify that the transformation

$$f: \begin{cases} x' = -x + 2h \\ y' = -y + 2k \end{cases} \qquad \text{matrix:} \begin{bmatrix} -1 & 0 \\ 0 & -1 \end{bmatrix}$$

is the reflection s_C in the point $C(h, k)$.

10. Three of the following mappings, given in matrix form, map squares into squares, but have decidedly different ways of doing it. Without plotting points, find which of the three mappings do so, and tell what the different ways are.

$$\begin{bmatrix} 5 & 2 \\ -2 & -5 \end{bmatrix}, \begin{bmatrix} -4 & 3 \\ 3 & 4 \end{bmatrix}, \begin{bmatrix} 1/3 & -1/4 \\ 1/4 & 1/3 \end{bmatrix}, \begin{bmatrix} -4/3 & 3/5 \\ 3/5 & 4/5 \end{bmatrix}$$

11. Show that the transformation given by

$$f: \begin{cases} x' = x + 3 \\ y' = -y \end{cases} \qquad \text{matrix:} \begin{bmatrix} 1 & 0 \\ 0 & -1 \end{bmatrix}$$

is a glide reflection in the direction of the x-axis.

_____ *Group B* _____

12. (a) Verify, by finding the images of $A(0, 0)$, $B(1, 0)$, and $C(0, 1)$ given in counterclockwise order, that under the linear transformation f whose matrix is

$$A = \begin{bmatrix} a & b \\ c & d \end{bmatrix}$$

the image points A', B', and C' will also be oriented counterclockwise iff $|A| = ad - bc > 0$.

(b) Show that, therefore, an isometry **(7)** is direct iff $\delta = 1$.

(c) Show that the determinant of the 2×2 matrix defining a line reflection **(13)** (deleting the translational component) equals -1, and thus verify that this transformation is opposite, as it should be.

13. (a) Prove that if a linear transformation f preserves the slope of any line and maps $(0, 0)$ to $(0, 0)$, its matrix form must be

$$A = \begin{bmatrix} a & 0 \\ 0 & a \end{bmatrix}$$

for some real a. (**HINT:** Use the formula **(3)** for m' and the fact that m is an arbitrary.)

(b) Show that this is a dilation d_a centered at the origin.

14. Illustrate the result of the fundamental theorem which states that precisely one isometry exists mapping one triangle onto another congruent to it, by solving for the unique parameters a, b, δ, x_0, and y_0 for the isometry which maps $A(0, 0)$, $B(3, 4)$, and $C(4, -3)$ to $A'(-1, 1)$, $B'(4, 1)$, and $C'(-1, 6)$, respectively. (**HINT:** Plug these coordinates into the equations of the transformation, then solve for the parameters.

15. A transformation $x' = Ax + x_0$ is area-preserving iff $|A| = \pm 1$. Verify this directly when $x_0 = [0, 0]$ and

$$A = \begin{bmatrix} 3 & 5 \\ 1 & 2 \end{bmatrix}$$

by finding the image of the triangle having vertices $P(0, 0)$, $Q(2, -1)$, and $R(2, 4)$. Graph, and calculate the areas of both triangles directly.

_____ *Group C* _____

16. **Fundamental Theorem for Linear Transformations** Using coordinates, prove that there exists a unique linear transformation which maps a given triad of noncollinear points onto another given triad. (**HINT:** To get organized, prove first that there exists a unique transformation f which maps $A(0, 0)$, $B(1, 0)$, and $C(0, 1)$ to $P(p, q)$, $Q(r, s)$, and $R(u, v)$ and another, g, which maps A, B, C to $P'(p', q')$, $Q'(r', s')$, and $R'(u', v')$, respectively. Then the desired mapping is gf^{-1}.)

17. A linear transformation is a similitude iff it maps some circle onto another. Prove, using coordinates.

18. **Derivation of Coordinate Form for Line Reflections** Suppose ℓ: $ax + by = c$, with $a^2 + b^2 = 1$; we may then assume that $a = \sin\theta$, $b = \cos\theta$, where θ is the angle between ℓ and the x-axis (line m). Complete the following steps. (See figure at the right.)

(a) Lines ℓ and m meet at $P(c/a, 0)$ at an angle θ, so a rotation through 2θ with center $(c/a, 0)$ is given by $s_m s_\ell$ as follows.

$$s_m s_\ell: \begin{cases} x' - \dfrac{c}{a} = \left(x - \dfrac{c}{a}\right)\cos 2\theta - y \sin 2\theta \\ y' = \left(x - \dfrac{c}{a}\right)\sin 2\theta + y \cos 2\theta \end{cases}$$

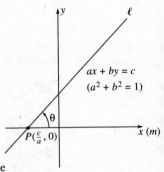

$ax + by = c$
$(a^2 + b^2 = 1)$

Verify.

(b) Verify $s_m^{-1} = s_m$ has the form $x'' = x$, $y'' = -y'$, so substitution of (a) into these two equations gives us the coordinate equations for $s_m^{-1}(s_m s_\ell) = s_\ell$, the desired reflection in line ℓ. Prove this.

(c) Now eliminate the expressions $\cos 2\theta$ and $\sin 2\theta$ by using $b^2 = 1 - a^2$, $\cos 2\theta = \cos^2\theta - \sin^2\theta = b^2 - a^2$, and $\sin 2\theta = 2ab$.

19. **PROJECT FOR UNDERGRADUATE RESEARCH** A linear transformation can be represented as a product of rotations and a **simple affine map** represented in matrix form by

$$A = \begin{bmatrix} a & 0 \\ 0 & b \end{bmatrix}$$

For example,

$$\begin{bmatrix} 2 & -2 \\ 2 & 1 \end{bmatrix} = \begin{bmatrix} 2/\sqrt{5} & -1/\sqrt{5} \\ 1/\sqrt{5} & 2/\sqrt{5} \end{bmatrix}\begin{bmatrix} 3 & 0 \\ 0 & 2 \end{bmatrix}\begin{bmatrix} 2/\sqrt{5} & -1/\sqrt{5} \\ 1/\sqrt{5} & 2/\sqrt{5} \end{bmatrix}$$

Prove that any nonsingular matrix (or transformation) may be so decomposed, and that the given transformation completely determines the parameters a and b of A. (Is there a way to compute a and b directly from the given transformation?)

CONJECTURE: Any linear transformation f may be represented as the product $f = r_\theta A r_\phi$. (Recall that r_θ designates a rotation through an angle of measure θ; this part of the problem requires the consideration of translational components.)

*5.6 TRANSFORMATION GROUPS

The study of geometric invariants over groups of transformations provides important insight for a deeper understanding of geometry. The structure of geometric transformation groups is rich and elegant, as we shall see. We begin with the definition.

DEFINITION

A set of transformations is called a **group**, or **transformation group**, iff

(1) the product fg of any two members f and g belongs to the set (**product closure**), and

(2) the inverse f^{-1} of each member f also belongs to the set (**inverse closure**).

The number of distinct elements in a transformation group is its **order** if it is finite; if there are an infinite number of elements, the group is said to have **infinite order**.

NOTE: The above definition requires that for any f belonging to the group, $f^{-1}f$, which is the identity map, must also belong to the group. Thus, by its definition, every transformation group must contain the identity e.

EXAMPLE 1 Show that the rotations $r_0 = e$, r_{90}, and $r_{180}{}^2$ about some common point do *not* form a transformation group, and find what must be added in order to make it one.

SOLUTION

The set $G' = \{r_0, r_{90}, r_{180}\}$ fails to be a group on both counts: The set is not closed under products (because $r_{180}r_{90} = r_{270} \notin G'$), and $r_{90}{}^{-1} = r_{-90} \notin G'$ (it does not have inverse closure). The remedy is simply to add the rotation r_{270} to the set (which is the same as $r_{-90} = r_{90}{}^{-1}$). The result is a *group of order 4*:

$$G = \{e_0, r_{90}, r_{180}, r_{270}\}$$

To check that this set has product and inverse closure, we will construct a multiplication table, which will allow us to systematically observe all possible products. Each entry in the table is read xy, where x is directly opposite, in the column to the left, and y lies in the top row, directly above.

---

[2]Technically, such rotations are undefined; in general, r_θ is not defined if $\theta \le 0$ or $\theta \ge 180$. But it is customary to extend geometric rotations to arbitrary θ using circular arc measure, as in trigonometry, in the obvious manner.

	e	r_{90}	r_{180}	r_{270}
e	e	r_{90}	r_{180}	r_{270}
r_{90}	r_{90}	r_{180}	r_{270}	e
r_{180}	r_{180}	r_{270}	e	r_{90}
r_{270}	r_{270}	e	r_{90}	r_{180} ❑

CYCLIC GROUP OF ORDER n

Because of the arrangement of the entries of the table in Example 1, where each succeeding row of products is merely a cyclic shift by one position of the row preceding it, the group in Example 1 is known as a **cyclic group of order 4.** By analogy, there is a **cyclic group of order n** for each positive integer n, denoted C_n, and an **infinite cyclic group**, C_∞. Another way to describe cyclic behavior is to observe that all the elements in the group may be generated by taking the *integer powers of a single element* of the group, called a **group generator**. For C_4 in Example 1, one of its generators is r_{90} since

$$(r_{90})^1 = r_{90}, \quad (r_{90})^2 = r_{180}, \quad (r_{90})^3 = r_{270}, \quad \text{and} \quad (r_{90})^4 = r_{360} = e.$$

The element r_{180} is *not* a generator, since its powers do not yield all the elements of the group.

$$(r_{180})^1 = r_{180}, \quad (r_{180})^2 = r_{360} = e, \quad (r_{180})^3 = r_{180}, \quad ....$$

EXAMPLE 2   A rectangle $\square ABCD$ that is not a square has two lines of symmetry ℓ and m, as shown in Figure 5.37, and a *point* of symmetry, E, such that the isometries s_ℓ, s_m, and $r_E \equiv r_{180}[E]$ map the rectangle back into itself. These are the only ones which do, and therefore we call them the **symmetries of the rectangle**. Counting the identity e, there are, again, four transformations which form a group of order 4. Another way to visualize this group is to think of the rectangle as an ordinary sheet of paper, which can be turned over sideways (as if turning a page in a book) and placed on top of its original position, or turned upside down from top to bottom, or both operations. (See Figure 5.38.) Thus, if these operations are denoted, s, u, and b, respectively, we obtain the following table of products, which exhibits product closure.

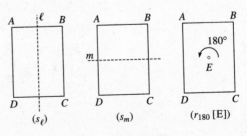

Figure 5.37

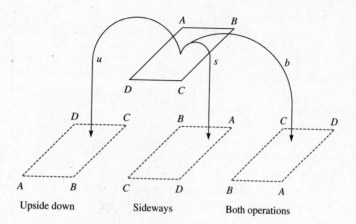

Figure 5.38

	e	s	u	b
e	e	s	u	b
s	s	e	b	u
u	u	b	e	s
b	b	u	s	e

It remains to check inverse closure. However, this, too may be observed from the table: Since the identity appears in each row of products, each element has an inverse (in fact, *is its own inverse,* in this particular case).□

The two groups of order 4 presented in the two preceding examples are distinctly different. The first one was cyclic, generated by a single element, but the second has the property

$$x^2 = e$$

for each element x. Thus, the integer powers of any element will get us no further than itself and e, so it is noncyclic.

The first notable study of such groups and their relation to geometry was carried out by Felix Klein (1849–1925). The significance of the group displayed in Example 2 to geometry was discovered by him, and is sometimes called the **Klein 4-group** or simply the **Fours Group.**

$\mathcal{H}$ISTORICAL $\mathcal{N}$OTE

When he was only 23 years of age, Felix Klein became professor of mathematics at the University of Erlangen. His inaugural lecture there made mathematical history. Known as the **Erlangen Program**, it was a bold proposal to use the group concept to classify and unify the many diverse and seemingly unrelated geometries which had developed in the 1800s. By the time he was 36, when Hilbert met him, he was already a legendary figure, and a year later he went to Göttingen where he became the leading research mathematician (and chairman of the department). He helped to restore the earlier fame enjoyed by the university in the days of Gauss and raised them to new heights. Later, both Hilbert and Minkowski joined his department, and the three of them, although having totally different personalities, worked together in rare harmony. The stimulus toward research and learning which Klein provided can be seen from the fact that he directed a total of 48 doctoral students during his lifetime. His "favorite pupil" in those days was the Englishwoman Grace Chisholm Young, the first female candidate to be granted a German doctoral degree (in 1895) in any subject through the regular examination process.

THE DIHEDRAL GROUPS, SUBGROUPS

In the preceding example, we studied the symmetry groups of a rectangle with unequal sides. Here we look at the symmetry groups of a square. The square obviously has more self-mapping isometries than the rectangle. In Figure 5.39 are shown the lines of symmetry and rotations which lead to all the nonidentity isometries which map the square onto itself. The resulting group of transformations is called the **dihedral group** of the square, denoted D_4, having 8 elements.

SYMMETRIES OF THE SQUARE

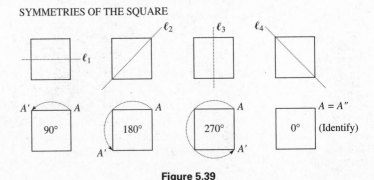

Figure 5.39

> **DEFINITION**
>
> When one transformation group G contains a subset that is itself another transformation group H, then H is called a **subgroup** of G, a relationship that is denoted by
>
> $$H < G$$

In the example, D_4, the set of rotations (which is the cyclic group C_4) is a subset of the larger group D_4, hence one writes

$$C_4 < D_4$$

By analogy, there is a dihedral group D_n of order $2n$ for each positive integer n, and the cyclic group C_n of order n will always be a subgroup of D_n. This is the result of considering the symmetries of a regular n-gon. Because of the geometric properties of a regular polygon, the rotation $r_\theta = r$ (where $\theta = 360/n$ and the center of rotation is the center of the polygon) always maps the n-gon to itself. Thus, the symmetries of the regular n-gon always include as a subgroup the cyclic group

$$C_n = \{r, r^2, r^3, \dots, r^{n-1}, r^n = e\}$$

In order for you to see for yourself what one of these dihedral groups and cyclic subgroups looks like, the following Discovery Unit is recommended.

MOMENT FOR DISCOVERY

Symmetry Group for an Equilateral Triangle

The various members of the dihedral group D_3 are illustrated in Figure 5.40.

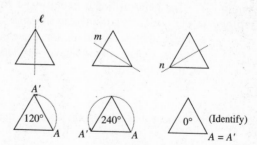

Figure 5.40

1. How many elements are there in this group?
2. What is $s_\ell s_\ell = (s_\ell)^2$? $(s_m)^2$? $(s_n)^2$?
3. What is $s_\ell s_m$? $s_\ell s_n$? $s_m s_n$? Does $s_\ell s_m = s_m s_\ell$?
4. What elements give s_ℓ^{-1}, s_m^{-1}, and s_n^{-1}?

5. How do the rotations behave? What is $(r_{120})^2$?

6. Write out the complete multiplication table for D_3.

7. Can you spot C_3 as a subgroup of D_3?

8. Find all the subgroups you can, and write their individual multiplication tables.

9. Can you imagine how D_2 should be defined, and what this is group is like? (See Problem 7.)

GROUP DIAGRAMS

One way to study group structure is to determine the subgroups and then indicate in a diagram the relationship which these subgroups have to one another. For example, since $C_2 < C_4 < D_4$, and since there are two other cyclic subgroups C_2' and C_2'' of C_4 (what are they?), we have the diagram shown in Figure 5.41. To indicate subgroup relationships, an arrow is directed from any group represented in the diagram to each of its subgroups. Can you figure out the group diagram for D_3 from the previous Discovery Unit?

GROUP DIAGRAM

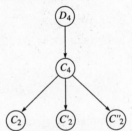

Figure 5.41

THE GENERAL LINEAR GROUP AND MATRICES

The **general linear group**—the group of all linear transformations from the plane to itself—is a tremendously large group, usually denoted by $GL(2)$. The number in parentheses represents the dimension of the underlying space, which is, in this case, a plane. As we have seen, each element of this group may be represented in the matrix form

$$f: x' = Ax + c$$

where A is a nonsingular 2×2 matrix. The product of this mapping with another

$$g: x'' = Bx' + d,$$

may be written, at least symbolically,

$$gf: x'' = B(Ax + c) + d$$

or, by the associative law for matrix multiplication,

$$gf: x' = BAx + Bc + d,$$

where BA denotes the matrix product, as defined previously. This formula for gf can be verified by using the coordinate form introduced earlier, or by observing that Bx' *means* matrix multiplication of B by any column matrix x', and the fact that $x' = Ax + c$ is a column matrix. The latter assumes the **associative** and **distributive laws** for matrix multiplication,

$$(AB)C = A(BC) \quad \text{and} \quad A(B + C) = AB + AC$$

for all matrices A, B, and C.

Another basic concept for matrices is the **identity matrix**, denoted by I and defined by

$$I = \begin{bmatrix} 1 & 0 \\ 0 & 1 \end{bmatrix}$$

It is obvious that the identity matrix represents the identity transformation $x' = Ix = x$. It behaves just like a unit or group identity for matrix multiplication: $AI = IA = A$ for all 2×2 matrices A.

Finally, it is useful to know the formula for the inverse of a nonsingular 2×2 matrix.

$$(1) \qquad A^{-1} = \begin{bmatrix} a & b \\ c & d \end{bmatrix}^{-1} = \frac{1}{k} \begin{bmatrix} d & -b \\ -c & a \end{bmatrix} = \begin{bmatrix} d/k & -b/k \\ -c/k & a/k \end{bmatrix}, \quad k = |A|$$

Matrix theory can be used effectively to verify group structure within $GL(2)$, as the following example shows. First, we give a numerical example to show how one can use (1) in connection with transformations.

EXAMPLE 3 Using matrices, find the coordinate form of the inverse of the linear transformation

$$f: \begin{cases} x' = 3x + 4y + 2 \\ y' = 4x + 5y - 1 \end{cases}$$

SOLUTION

First we transform to matrices.

$$f: \begin{bmatrix} x' \\ y' \end{bmatrix} = \begin{bmatrix} 3 & 4 \\ 4 & 5 \end{bmatrix} \begin{bmatrix} x \\ y \end{bmatrix} + \begin{bmatrix} 2 \\ -1 \end{bmatrix}$$

$$\begin{bmatrix} 3 & 4 \\ 4 & 5 \end{bmatrix}^{-1} \begin{bmatrix} x' \\ y' \end{bmatrix} = \begin{bmatrix} 3 & 4 \\ 4 & 5 \end{bmatrix}^{-1} \left(\begin{bmatrix} 3 & 4 \\ 4 & 5 \end{bmatrix} \begin{bmatrix} x \\ y \end{bmatrix} + \begin{bmatrix} 2 \\ -1 \end{bmatrix} \right)$$

$$\begin{bmatrix} 3 & 4 \\ 4 & 5 \end{bmatrix}^{-1} = \frac{1}{15 - 16} \begin{bmatrix} 5 & -4 \\ -4 & 3 \end{bmatrix} = \begin{bmatrix} -5 & 4 \\ 4 & -3 \end{bmatrix}$$

$$\begin{bmatrix} -5 & 4 \\ 4 & -3 \end{bmatrix} \begin{bmatrix} x' \\ y' \end{bmatrix} = I \begin{bmatrix} x \\ y \end{bmatrix} + \begin{bmatrix} -5 & 4 \\ 4 & -3 \end{bmatrix} \begin{bmatrix} 2 \\ -1 \end{bmatrix} = \begin{bmatrix} x \\ y \end{bmatrix} + \begin{bmatrix} -14 \\ 11 \end{bmatrix}$$

or

$$\begin{bmatrix} x \\ y \end{bmatrix} = \begin{bmatrix} -5 & 4 \\ 4 & -3 \end{bmatrix} \begin{bmatrix} x' \\ y' \end{bmatrix} - \begin{bmatrix} -14 \\ 11 \end{bmatrix}$$

Therefore,

$$f^{-1}: \begin{cases} x = -5x' + 4y' + 14 \\ y = 4x' - 3y' - 11 \end{cases} \quad \square$$

EXAMPLE 4   Show that all transformations having the matrix form

$$\begin{bmatrix} 1 & a \\ 0 & 1 \end{bmatrix}$$

is a subgroup H of $GL(2)$. (Note that this set of matrices represents all shears with respect to the y-axis introduced earlier. It is assumed here that $(0,0)$ is a fixed point.)

SOLUTION

We must prove product and inverse closure. Take any two representative elements of H (these have to be arbitrary and different),

$$\begin{bmatrix} 1 & a \\ 0 & 1 \end{bmatrix} \quad \text{and} \quad \begin{bmatrix} 1 & b \\ 0 & 1 \end{bmatrix}$$

Their product is

$$\begin{bmatrix} 1 & a \\ 0 & 1 \end{bmatrix} \begin{bmatrix} 1 & b \\ 0 & 1 \end{bmatrix} = \begin{bmatrix} 1 \cdot 1 + a \cdot 0 & 1 \cdot b + a \cdot 1 \\ 0 & 1 \end{bmatrix} = \begin{bmatrix} 1 & a + b \\ 0 & 1 \end{bmatrix}$$

which has the desired form, hence belongs to H. To calculate the inverse of any member of H, write

$$\begin{bmatrix} 1 & a \\ 0 & 1 \end{bmatrix}^{-1} = \frac{1}{1 - 0} \begin{bmatrix} 1 & -a \\ -0 & 1 \end{bmatrix} = \begin{bmatrix} 1 & -a \\ 0 & 1 \end{bmatrix}$$

which also has the desired form, and belongs to H. $\square$

The important subgroups of $GL(2)$ include mappings we have studied previously:

Similitudes	Dihedral groups
Dilations	Translations
Isometries	Rotations about a single point
Direct isometries	

THE ROTATION-TRANSLATION GROUP

The tools developed in this section will be particularly useful in one of our applications of group theory to geometry. It may seem surprising to learn that the set of all rotations in the plane (about different points) is *not* a subgroup, while the set of all translations is. More surprising is that, *taken together*, the family of all rotations and translations *is* a subgroup. In order to understand these ideas, let's go back to the basic elements of rotations and translations, the line reflections.

Recall that a translation in the plane is the product of two line reflections s_ℓ and s_m, where ℓ and m are parallel lines, each perpendicular to the direction of the translation. These perpendiculars may be chosen arbitrarily, as long as their distance apart is $\frac{1}{2}$ that of the amount of translation. The rotation $r_\theta[A]$ of θ degrees about center A is the product of two reflections s_ℓ and s_m, where lines ℓ and m intersect at A. These lines may be chosen arbitrarily, as long as they pass through A and form an angle of measure $\frac{1}{2}\theta$.

Now let's consider the product of two rotations having different centers, $r_{2\alpha}[A]$ preceded by $r_{2\beta}[B]$. In Figure 5.42 is shown the case when $\alpha + \beta < 180$, which determine the base angles of A and B of $\triangle ABC$. Suppose we *choose* $\ell = \overleftrightarrow{AB}$, $m = \overleftrightarrow{AC}$, and $n = \overleftrightarrow{BC}$ in representing the two rotations:

$$r_{2\alpha}[A] = s_m s_\ell, \quad r_{2\beta}[B] = s_\ell s_n.$$

Since the product of a line reflection with itself is the identity, we have

$$r_{2\alpha}[A] \cdot r_{2\beta}[B] = (s_m s_\ell) \cdot (s_\ell s_n) = s_m s_\ell^2 s_n = s_m e s_n = s_m s_n$$

Thus, the product of any two rotations is either a translation (if $n \parallel m$), or another rotation (if n meets m). In the latter case, the center is $C = n \cap m$ and the angle of rotation is 2θ, where $\theta = -m\angle ACB = \alpha + \beta - 180$ (the sign denoting the clockwise direction). This proves

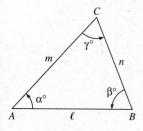

Figure 5.42

THEOREM 1

The product of two rotations $r_{2\alpha}[A]$ and $r_{2\beta}[B]$ is either a translation or rotation. It is a translation iff $\alpha + \beta = 180$, and a rotation otherwise. If $\alpha + \beta < 180$, then

$$r_{2\alpha}[A]r_{2\beta}[B] = r_{-2\gamma}[C]$$

where C is the third vertex of the triangle having base $\overline{AB}$, whose vertices A, B, and C are oriented positively and whose angles are α, β and γ, respectively, each oriented in the positive (counterclockwise) direction.

COROLLARY A

If $\triangle ABC$ is any triangle in the plane, with vertices oriented in the positive (counterclockwise) direction, and α, β, and γ the three angles at A, B, and C, respectively, oriented positively, then the product

$$r_{2\alpha}[A]r_{2\beta}[B]r_{2\gamma}[C]$$

is the identity transformation.

Proof: Use the fact that $r_{2\gamma}[C]$ is the inverse of $r_{-2\gamma}[C]$.

The converse of this corollary is also true, and it is a very interesting, but manageable problem. (See Problem 12 for hints.) We state it for later reference.

COROLLARY B

Suppose that $r_{2\alpha}[A]r_{2\beta}[B]r_{2\gamma}[C] = e$ for any triangle $\triangle ABC$ whose vertices are oriented positively, and such that α, β, and γ are each positive and less than 180, corresponding to counterclockwise rotations in each case. Then α, β, γ are the respective measures of the angles at A, B, and C.

We include one last result that is interesting and also makes a nice problem.

COROLLARY C

If $2\alpha + 2\beta + 2\gamma = 360$ and α, β, $\gamma > 0$, then for any three noncollinear points A, B, and C the product

$$r_{2\alpha}[A]r_{2\beta}[B]r_{2\gamma}[C]$$

is either a translation or the identity.

PROBLEMS   (§5.6)

_____ *Group A* _____

1. If f and g are linear transformations as given by their matrix forms

$$f: \begin{bmatrix} 4 & 1 \\ 3 & 1 \end{bmatrix} \qquad g: \begin{bmatrix} 2 & 2 \\ 6 & -4 \end{bmatrix}$$

find the matrix representation for fg, gf, and f^{-1} in two ways:
   (a) By matrix products and the formula for inverse, (1).
   (b) By coordinates, writing f and g in coordinate form and using primes and double primes for making substitutions, as in previous examples. Compare your results in (a) and (b).

2. Show that the set of matrices of all *direct* similitudes ($\delta = 1$ in (5), §5.5) is closed under multiplication and the taking of inverses, proving that the direct similitudes is a subgroup of the group of all similitudes. That is, show that

   (a) $\begin{bmatrix} a & -b \\ b & a \end{bmatrix} \begin{bmatrix} c & -d \\ d & c \end{bmatrix}$ (b) $\begin{bmatrix} a & -b \\ b & a \end{bmatrix}^{-1}$

   are also matrices of direct similitudes with $\delta = 1$.

3. Show that the set of all dilations, having matrices of the form

$$\begin{bmatrix} a & 0 \\ 0 & a \end{bmatrix}, \quad a \neq 0$$

   is a subgroup of the similitudes, having matrices of the form

$$\begin{bmatrix} a & -\delta b \\ b & \delta a \end{bmatrix}, \quad a^2 + b^2 \neq 0, \quad \delta = \pm 1$$

4. The four matrices

$$E = \begin{bmatrix} 1 & 0 \\ 0 & 1 \end{bmatrix}, \quad S = \begin{bmatrix} -1 & 0 \\ 0 & 1 \end{bmatrix}, \quad U = \begin{bmatrix} 1 & 0 \\ 0 & -1 \end{bmatrix}, \quad B = \begin{bmatrix} -1 & 0 \\ 0 & -1 \end{bmatrix}$$

form a 4-element group, which represent, respectively, the identity, the reflections in the coordinate axes, and reflection in the origin. Verify that this is a group, and show that it has identically the same table of operations as the Fours Group discussed earlier. (When this happens, the two groups are said to be **isomorphic**.)

5. In the complex numbers (where $i^2 = -1$), consider the 4-element set $S = \{1, -1, i, -i\}$. Show that this is a group under ordinary multiplication of complex numbers. Is it the Fours Group, as in Problem 4, or the cyclic group C_4?

6. In the example in the figure to the right, verify that both products

$$r_{260}[B]r_{100}[A] \quad \text{and} \quad r_{100}[A]r_{260}[B]$$

are translations.

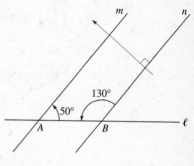

7. The dihedral group D_2 containing C_2 may be generated as the **symmetries of a line segment**. Let $\overline{AB}$ be any line segment, with M the midpoint, and ℓ the perpendicular bisector through M. With $m = \overleftrightarrow{AB}$, analyze the reflections s_ℓ and s_m in order to find the elements of D_2.

8. Find the inverse of each of the following matrices, which represent isometries, the last one being the general case.

$$\begin{bmatrix} \sqrt{3}/2 & -1/2 \\ 1/2 & \sqrt{3}/2 \end{bmatrix}, \quad \begin{bmatrix} 8/17 & 15/17 \\ -15/17 & -8/17 \end{bmatrix}, \quad \begin{bmatrix} a & -\delta b \\ b & \delta a \end{bmatrix} \quad (a^2 + b^2 = 1)$$

9. Give two proofs that the product of any two isometries is an isometry.

_____ *Group B* _____

10. If you have not already done so, analyze the subgroup structure of the dihedral group D_3 explored in the Discovery Unit. Make a group diagram.

11. Making careful sketches or using compass and straight-edge in order to verify Theorem 1, start with the example shown at right and track the mapping of $\triangle UVW$ under the rotations $r_{180}[B]$ and $r_{120}[A]$, comparing this with the result of applying $r_{-60}[C]$ on the same triangle $\triangle UVW$.

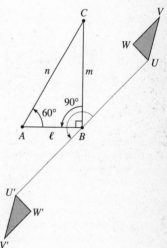

12. (a) Using the fact that $r_{2\gamma}[C]$ maps its center C to itself, write the identity of Corollary B in the form

$$(r_{2\alpha}[A])^{-1} = r_{2\beta}[B]r_{2\gamma}[C]$$

and analyze what the left side must do to point C (locate the point precisely). (See the accompnaying figure.)

(b) What must the right side of the above equation do to point C?

(c) Since the two sides are the same by hypothesis, then the results in (a) and (b) lead to what geometric configuration? Prove that $m\angle CAB = \alpha$, etc.

(d) Complete the proof of Corollary B.

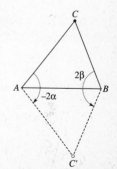

_____ *Group C* _____

13. Prove Corollary C of Theorem 1.

14. A **Generalized Lorentz Transformation** in the plane is defined as follows (a and b are arbitrary real numbers not both zero and $\delta = \pm 1$):

Coordinate Form   Matrix Form

$$f: \begin{cases} x' = ax + \delta by + x_0 \\ y' = bx + \delta ay + y_0 \end{cases} \qquad \begin{bmatrix} a & \delta b \\ b & \delta a \end{bmatrix}$$

(a) Show that the set of matrices of the form

$$\begin{bmatrix} a & \delta b \\ b & \delta a \end{bmatrix}, \quad a^2 \neq b^2, \ \delta = \pm 1$$

is closed under products and the taking of inverses, thus proving that the set of Generalized Lorentz Transformations is another subgroup of **GL[2]**. (Note the close resemblance of these matrices with those of similitudes.)

(b) Show that instead of the invariance of perpendicularity of lines, pairs of lines having *reciprocal slopes* is an invariant. (That is, $m_1 m_2 = 1 \rightarrow m_1' m_2' = 1$; see the following figure.)

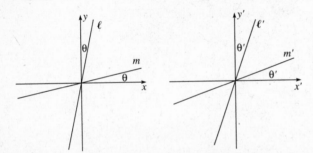

15. Lorentz Group of Special Relativity   If the subgroup in Problem 14 is specialized to the case $a^2 - b^2 = 1$, the result is the famous **Lorentz Group**. If $\delta = 1$, this yields the **Direct Lorentz Group**, prominent in Special Relativity. Show that the Direct Lorentz Group is, indeed, a subgroup of the General Lorentz Group.

16. Project. Complete the following diagram on the next page by adjoining the appropriate directed lines which exhibit the inter-relationships of the various subgroups of **GL[2]** we have considered previously. (Avoid redundancies in the diagram—if a chain of subgroups is indicated, then a line from the first to the last is unnecessary.)

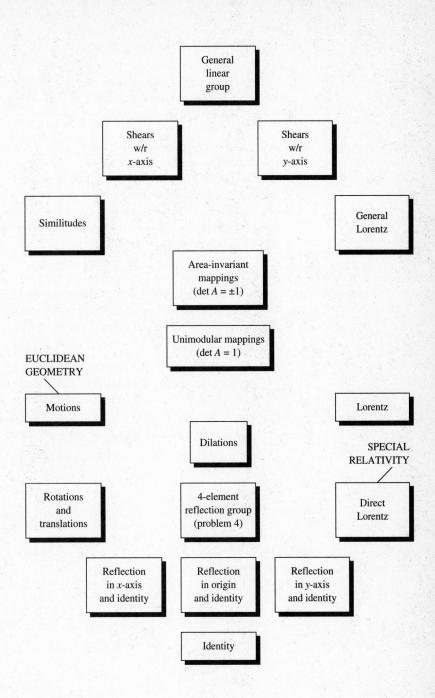

*5.7   Using Transformation Theory in Proofs

Many theorems in geometry can be made simpler and easier to understand if the theory of transformations is brought to bear. We have chosen a few examples to show you for illustration.

NINE-POINT CIRCLE

A dilation can be used not only to derive the existence of the Nine-Point Circle, but to prove other related properties as well. The center U of this famous circle lies on the **Euler Line** of the triangle—the line of collinearity of the orthocenter, centroid, and circumcenter of the triangle (Problem 3 will provide some guidance to enable you to give an elegant transformation proof for the existence of this line). To set things up properly, we will need two results about the relationship of the orthocenter of a triangle and its circumcircle.

LEMMA A

If H is the orthocenter of $\triangle ABC$, $\overline{AD}$ the altitude to side $\overline{BC}$, O the circumcenter, and L' and D' the points of intersection of $\overleftrightarrow{AO}$ and $\overleftrightarrow{AH}$ with the circumcircle, then L is the midpoint of $\overline{HL'}$ and D is the midpoint of $\overline{HD'}$ (Figure 5.43).

Proof:

(1) Since $\angle ABL'$ and $\angle ACL'$ are inscribed angles of semicircles (which, together, make up the entire circumcircle of $\triangle ABC$), these angles are right angles, and hence $\overleftrightarrow{BL'} \parallel \overleftrightarrow{HC}$ (since both are perpendicular line $\overleftrightarrow{AB}$), and similarly, $\overleftrightarrow{L'C} \parallel \overleftrightarrow{BH}$.

(2) Then $\Diamond HCL'B$ is a parallelogram, with diagonals $\overline{HL'}$ and $\overline{BC}$. Hence L is the midpoint of both diagonals, and, in particular, of $\overline{HL'}$.

(3) Since $\angle AD'L'$ is also inscribed in a semicircle, $\overline{AD'} \perp \overline{D'L'}$ and $\overleftrightarrow{DL} \parallel \overleftrightarrow{D'L'}$. By

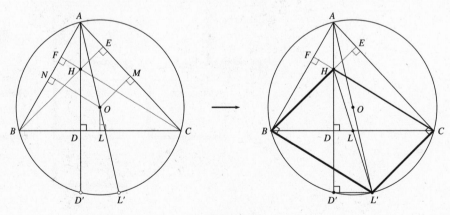

Figure 5.43

the Side-Splitting Theorem, D is the midpoint of $\overline{HD'}$.

Now consider the dilation $d_{1/2}$ with center H and scaling factor $\frac{1}{2}$. Since $d_{1/2}$

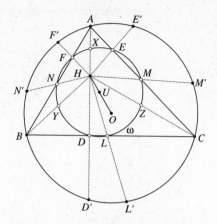

Figure 5.44

maps circles onto circles and is midpoint preserving, the circumcircle of $\triangle ABC$ maps to a circle ω whose center U is the image of O, the circumcenter of $\triangle ABC$ (Figure 5.44). Hence, by Lemma A, the points D', E', F' on the circumcircle map to D, E, and F on ω, and L', M', N' on the circumcircle map to L, M, N on ω. Finally, by definition of X, Y, Z as the **Euler points** of the triangle (midpoints of $\overline{AH}$, $\overline{BH}$, and $\overline{CH}$), the vertices of $\triangle ABC$ map to these latter three points. Hence, the circle ω contains the nine points L, M, N, D, E, F, and X, Y, Z, and is the Nine-Point Circle of $\triangle ABC$. Moreover its center U lies on line $\overleftrightarrow{HO}$, the Euler Line of the triangle, and $HU = \frac{1}{2}HO$.

THEOREM 1

Nine-Point Circle Theorem   The midpoints of the sides of $\triangle ABC$, the feet of the altitudes on sides, and the Euler Points all lie on a circle, the Nine-Point Circle of the triangle. The radius of the Nine-Point Circle is one-half the circumradius of $\triangle ABC$, and the center U lies on the Euler Line. Moreover, if H, U, G, and O are, respectively, the orthocenter, Nine-Point Center, centroid, and circumcenter, the cross ratio $(HG, UO) = 1$. (If directed distances are used, then $(HG, UO) = -1$.)

(The proof of the last part will be left as Problem 2; use the definition of the Euler Line and the fact that $GO = \frac{1}{2}GH$. See Figure 5.45.)

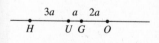

Figure 5.45

FAGNANO'S THEOREM

The second example involves a singularly interesting application of reflections. The property was originally discovered by J.F. Fagnano in 1775; the reflection proof given here is due to L. Fejr.

THEOREM 2

In any acute-angled triangle an inscribed triangle has minimum perimeter precisely when it is the **orthic triangle,** which is the triangle joining the feet of the altitudes on the sides of the given triangle.

Proof: For convenience we denote the perimeter of $\triangle PQR$ by $p(\triangle PQR)$. We want to show that if D, E, and F are the vertices of the orthic triangle on sides $\overline{BC}$, $\overline{AC}$, and $\overline{AB}$, then $p(\triangle DEF) \leq p(\triangle PQR)$.

A reflection in the lines $\overleftrightarrow{AB} = \ell$ and $\overleftrightarrow{AC} = m$ will be used to analyze the perimeter of $\triangle PQR$ (Figure 5.46). Since a reflection preserves distance and angle measure, if $S = s_\ell(\text{P})$ and $T = s_m(\text{P})$, then $\triangle ASR \cong \triangle APR$ and $\triangle APQ \cong \triangle ATQ$. Hence, $SR = RP$ and $PQ = QT$ so that

$$p(\triangle PQR) = PQ + QR + RP = SR + RQ + QT \geq ST$$

Thus, a smaller perimeter is obtained by replacing R and Q by the points of intersection, R' and Q' of $\overline{ST}$ with $\overline{AB}$ and $\overline{AC}$, respectively. This produces a new triangle $\triangle PQ'R'$ such that

$$p(\triangle PQ'R') = ST$$

Note that $AS = AP = AT$, so that $\triangle AST$ is isosceles; the vertex angle at A has measure equal to twice the sum of $m\angle 1$ and $m\angle 2$. Thus, $m\angle SAT = 2m\angle BAC$. Therefore, as P, Q', and R' vary, $\triangle SAT$ varies, but always as an isosceles triangle with a vertex angle congruent to a fixed angle. The base of this triangle is minimal when the legs are minimal, that is, when $SA = AP = AT$ is minimal, i.e., when $\overline{AP} \perp \overline{BC}$ or $P = D$. This new location P' of P will cause Q', R', S, and T to change to new positions, Q', R', S', and T'.

We have so far shown that for all inscribed triangles in $\triangle ABC$, $p(\triangle PQR) \geq p(\triangle DQ'R')$, where $\overline{AD} \perp \overline{BC}$ and Q', R' are collinear with S', T'. If we can prove that $Q' = E$ and $R' = F$, we will be finished since the above proof then gives us

$$p(\triangle PQR) \geq p(\triangle DEF) \text{ (orthic triangle)}$$

Using a simple angle analysis, we can calculate $m\angle BDR'$ and $m\angle Q'DC$.

$$m\angle 3 = 90 - m\angle 4 = 90 - m\angle 5 = \tfrac{1}{2}m\angle S'AT' = m\angle A = m\angle Q'DC$$

By the result of Problem 13, §4.3, $\overline{BQ'}$ and $\overline{CR'}$ are altitudes, and hence $Q' = E$, $R' = F$, as desired.

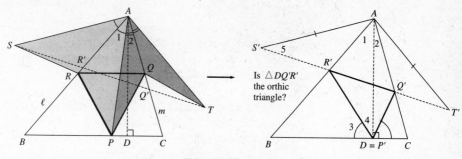

Figure 5.46

Is △ DQ'R' the orthic triangle?

OUR GEOMETRIC WORLD

An important application of line reflections, and *n*-fold product of such reflections, is the technology of fiber optics. Instead of wires conducting electricity, fibers seemingly conduct light along a curved path. Actually, light rays entering one end are merely reflected back and forth along the sides, until they emerge at the other end. (See Problem 20 for the case $n = 3$.)

FIBER OPTICS

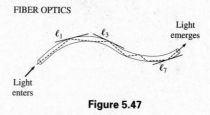

Figure 5.47

NAPOLEON-LIKE THEOREMS

A well-known theorem attributed to an even more famous French general[3] states that if you construct equilateral triangles externally on the sides of any triangle, the centroids of those equilateral triangles also form an equilateral triangle (Figure 5.48). The proof we have in mind is an apt illustration of the power of the method of transformation theory because a coordinate proof is quite messy, and a synthetic proof is all but unmanageable. If you simply apply the theorem on the product of three rotations about the vertices of a triangle (see previous section), the proof falls right out. We are going to let you prove this classic theorem for your own enjoyment. (See Problem 6.)

Several other theorems having the same flavor as Napoleon's Theorem, which apply more generally to convex polygons, can be established using transformation theory. One such result follows. Before you actually read and study through the result, there is an interesting related discovery you can make for yourself.

NAPOLEON'S THEOREM

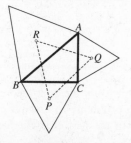

Figure 5.48

---

[3]See J. Wetzel, "Converses of Napoleon's Theorem", *American Mathematical Monthly*, Vol. 99, No. 4 (1992) p. 339–351, for an interesting discussion. According to the article, there is some doubt whether Napoleon actually discovered this theorem.

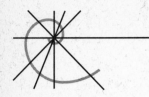

MOMENT FOR DISCOVERY

An Amazing Geometric Effect

1. Start with any triangle $\triangle ABC$. On the sides of this triangle, construct external isosceles triangles $\triangle BPC$, $\triangle AQC$, and $\triangle ABR$ whose vertex angles total 360. (For this example, we have chosen isosceles triangles with vertex angles 90, 120 and 150, respectively, as shown in Figure 5.49

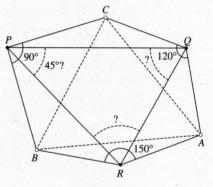

Figure 5.49

2. Join the vertices P, Q, and R, as shown, forming a fifth triangle, $\triangle PQR$.

3. Using a protractor, measure carefully the angles of $\triangle PQR$. Do you notice anything? It might be imagined that $\triangle PQR$ depends, in some way, on the original "mother" triangle $\triangle ABC$—if you change the shape of $\triangle ABC$ then the shape of $\triangle PQR$ will be affected.

4. Draw a different triangle $\triangle ABC$, making its angles radically different (say, by taking one angle to be an obtuse or right angle).

5. Construct isosceles triangles externally on the sides of your triangle *similar* to those of your original diagram (in our example, isosceles triangles having vertex angles of measure 90, 120, and 150 would be constructed).

6. Now measure the angles of the new triangle $\triangle PQR$. Did you discover anything?

7. Write down what seems like a reasonable conjecture. (You get a prize if you discover more than one property here.)

Your actual discovery in the preceding Discovery Unit is *not* the next theorem for a change. Rather, we pursue a related **reconstruction problem:** To find a triangle $\triangle ABC$ given the final triangle $\triangle PQR$ resulting from the isosceles triangles.

THEOREM 3

Let $\triangle PQR$ be any given acute-angled triangle in the plane, oriented positively. There exist points A, B, and C such that $\triangle RAB$, $\triangle QCA$, and $\triangle PBC$ are each isosceles triangles, whose vertex angles are, respectively, twice the measure of the angles of $\triangle PQR$.

Proof: (See Figure 5.50.) Let p, q, and r denote the measures of $\angle P$, $\angle Q$, and $\angle R$ of $\triangle PQR$. By Corollary A of §5.6,

$$r_{2p}[P]r_{2q}[Q]r_{2r}[R] = e \text{ (identity)}$$

For less cumbersome notation, let $r_1 = r_{2r}[R]$, $r_2 = r_{2q}[Q]$, and $r_3 = r_{2p}[P]$. We can take any convenient point $B \neq P, Q, R$, and define

$$A = r_1(B), C = r_2(A)$$

Since $r_3r_2r_1(B) = B$, then $B = r_3(C)$. Thus, points A, B, and C satisfy the requirements stated in the theorem (do you see why?)

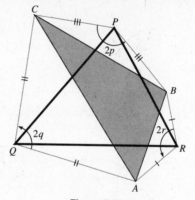

Figure 5.50

NOTE: The solution for points A, B, and C is clearly not unique, since A can be chosen almost at random. It would be interesting to find natural conditions for Theorem 3 that would result in a unique solution.

EXAMPLE 1 Use transformation theory to locate the vertices of the unique triangle which has a given set of three noncollinear points L, M, and N as the midpoints of its sides (Figure 5.51).

SOLUTION

We use the fact that the product of an odd number of point reflections having different centers is another point reflection (there is an easy coordinate proof of this; see Problem 8). Consider the product of point reflections

$$s_M s_L s_N = s_A$$

(the center A is *defined* by the product). Take $B = s_N(A)$ and $C = s_L(B)$, and this will determine the desired $\triangle ABC$. (To find A constructively, simply choose some point P in the plane, and find its image P' under the above product; the midpoint of $\overline{PP'}$ will then be the desired center A, and this is the vertex required in the problem. Proof of the validity of this construction, which can now be converted to an elementary Euclidean construction, will be left as a problem.)❏

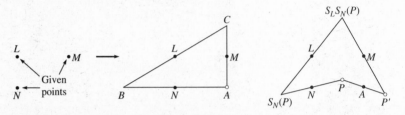

Figure 5.51

THEOREM 4

Yaglom's Theorem Let $\Diamond ABCD$ be a parallelogram, and suppose that squares are constructed externally on the four sides of the parallelogram. Then the centers of these squares also form a square (Figure 5.52).

Proof: Consider the rotation $r = r_{90}[P]$. The square centered at P (having $\overline{AB}$ as side) will rotate under r onto its original position, with $r(B) = A = B'$. The square centered at Q on side $\overline{BC}$ revolves 90 degrees onto the square on side $\overline{DA}$, hence center Q maps to center S (because r preserves distance and angle measure). Thus, $r(Q) = S = Q'$. But this tells us that segment $\overline{PQ}$ rotates 90 degrees onto segment $\overline{PS}$, and therefore $PQ = PS$ and $m\angle QPS = 90$. Since this is true at each of the other vertices Q, R, and S, $\Diamond PQRS$ is a square.

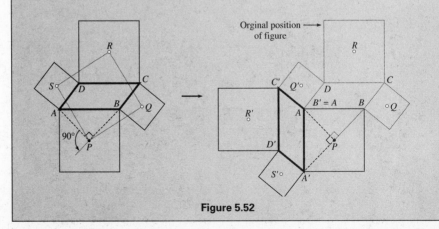

Figure 5.52

The previous proof becomes more dramatic by tiling the plane with copies of the original configuration, as shown in Figure 5.53, with copies of the points P, Q, R, and S generating a lattice. Under the same rotation we considered in the previous proof, $r_{90}[P] = r$, the entire tiling quite clearly maps onto itself (it is invariant under r), and, accordingly, so is the corresponding lattice.

FERMAT'S POINT

A problem that often appears in calculus texts (but for which calculus is one of the more inappropriate methods of solution) goes back to Fermat. The solution presented here was given by J. E. Hofmann in 1929.

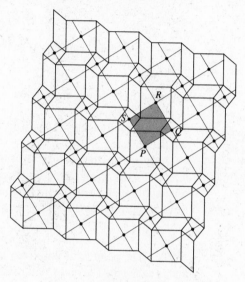

Figure 5.53

The problem is to locate a point inside a triangle so that the sum of the distances from that point to the vertices is a minimum. The solution obviously has all sorts of real-world applications. It turns out that there is a neat *geometric* solution not often realized by students of the calculus. Moreover, once we find out where the point is located, it may be constructed by the ancient tools of Euclidean construction, the compass and straight-edge. The typical engineering student who struggles to work this problem by methods of caclulus will no doubt completely miss the beautiful geometric aspects of this problem.

The pleasant thing about our solution is that its development is "top down," requiring no intermediate lemmas or artificial starting points. We just start with any point P inside a triangle $\triangle ABC$ (which we assume has angle measures < 120), as

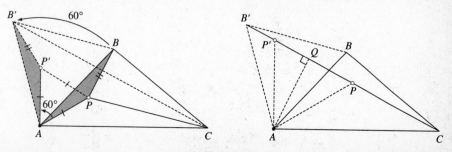

Figure 5.54

shown in Figure 5.54. The only clever idea required is the following: Perform the rotation about vertex A through an angle of measure 60, mapping $\triangle APB$ to $\triangle AP'B'$. Since $AP = AP'$ and $m\angle P'AP = 60$, $\triangle AP'P$ is equilateral. Note that $PB = P'B'$. Can you see why $\triangle ABB'$ is equilateral also? Hence the sum whose minimum we seek,

$$x = PA + PB + PC$$

equals the sum

$$B'P' + P'P + PC$$

and by the triangle inequality (extended to polygons)

$$x \geq B'C = c.$$

Thus, we have found an absolute minimum for all sums x where P is any point inside the triangle. ($B'C$ is a fixed quantity as P varies.) If we can now prove that this sum actually takes on the value c for some choice of P inside the triangle, we shall be finished.

Drop the perpendicular $\overline{AQ}$ to $\overline{B'C}$, and locate point P on $\overline{QC}$ so that $m\angle QAP = 30$, as shown in Figure 5.54. Now perform the same rotation as before, rotating through an angle of 60 about A; again the sum x will equal the above quantity, but this time both P and P' lie on segment $B'C$, with B'-P'-P-C. Hence

$$x = B'P' + P'P + PC = B'C = c$$

NOTE: The point P of minimum sum $PA + PB + PC$ can be shown to be unique and is called the **Fermat Point** of $\triangle ABC$.

There is one thing about the above argument that might cause us concern. There is nothing special about our using vertex A as a center of rotation, which ultimately led to an equilateral triangle $\triangle ABB'$ on side $\overline{AB}$ and diagonal $\overline{B'C}$ of $\triangle AB'BC$. If we consider the other vertices of $\triangle ABC$, the same argument leads to equilateral triangles on the other two sides (constructed externally), and the configuration for Napoleon's Theorem reappears. (See Figure 5.55). The argument seems to tell us that P could lie on any of the three diagonals $\overline{AD}$, $\overline{BE}$, and $\overline{CF}$. How could this be?

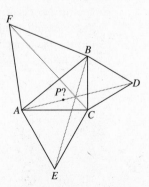

Figure 5.55

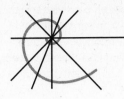

MOMENT FOR DISCOVERY

Locating Fermat's Point

Follow these steps carefully to see what you can deduce for yourself.

1. The preceding argument showed that a minimum sum

$$x = PA + PB + PC$$

occurs when P lies on diagonal $\overline{FC}$, when $x = FC$ (Figure 5.56). Is P the only point on this diagonal for which x is minimal? (Consider its construction; you should try proving that $XA + XB + XC > FC$ if $X \neq P$ but lies on $\overline{FC}$.)

2. Use a similar argument to locate a (unique?) point R on diagonal $\overline{AD}$, and S on $\overline{BE}$, such that $RA + RB + RC = AD$ and $SA + SB + SC = BE$ are minimal values for x. (Give the major steps of that argument for point R only, as indicated in Figure 5.56.)

3. What do you have to know in order to show that R lies on diagonal $\overline{FC}$ and $R = P$? If you can find an argument for this, then this proves $S = P$ also. (**HINT:** Study the angles at P and R; find the measures of $\angle APC$, $\angle APB$,

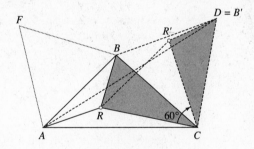

Figure 5.56

and $\angle BPC$.)

4. Is the preceding an airtight argument showing that P lies on all three diagonals $\overline{AD}$, $\overline{BE}$, and $\overline{CF}$?

5. Is it true that $FC \leq AD \leq BE \leq FC$? What does this prove?

6. **Theorem** Given $\triangle ABC$ whose angles are each of measure < 120. Let D, E, and F be those points outside $\triangle ABC$ which form equilateral triangles with the sides. Then ... ?

USING TRANSFORMATION THEORY TO PROVE SAS POSTULATE

One last application provides the foundation for textbooks which use the transformation approach in geometry. If we want to *prove* the SAS Postulate as a theorem, we must go back to the material preceding Chapter 3, and introduce transformations in a manner that is meaningful there. An extra bonus for us is that this same work will save us the trouble of having to prove the SAS Property for non-Euclidean models for which line reflections exist.

The danger in working with these ideas is the tendency to use properties we have been taking for granted in Euclidean geometry, such as the SSS theorem. One cannot assume in this context that an isometry (distance-preserving map) is automatically angle-measure preserving. We shall adopt a term used only infrequently up to now as a synonym for "isometry."

> **DEFINITION**
>
> A **motion** is a transformation in the plane that preserves both distance and angle measure. If a motion has a line of fixed points, that line is called its **axis**. A motion is called **nontrivial** if it is not the identity.

LEMMA B

A nontrivial motion with axis ℓ has only the points of ℓ as fixed points.

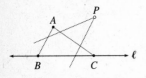

Figure 5.57

Proof: Suppose otherwise—that some point A not on ℓ is left unchanged, along with any two points B and C on ℓ (Figure 5.57). Then, since betweenness is preserved by any distance-preserving map (as argued previously), every point P on segments $\overline{AB}$ and $\overline{AC}$ are fixed. It now follows that every point in the plane is fixed, a contradiction.

LEMMA C

A nontrivial motion with axis ℓ maps every point not on ℓ to a point on the opposite side of ℓ.

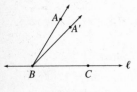

Figure 5.58

Proof: Suppose A is any point not on ℓ, and let A' be the image of A under the given motion (Figure 5.58). Since any two points B and C on ℓ are fixed, then by definition of motions, $m\angle A'BC = m\angle ABC$. If A' and A were on the same side of ℓ, then rays $\overrightarrow{BA}$ and $\overrightarrow{BA'}$ would coincide by the Angle Construction Theorem, and then since $BA = BA'$, by the Segment Construction Theorem, $A = A'$ (since either B-A-A' or B-A'-A must hold if $A \neq A'$). By Lemma B, the motion is the identity, contrary to hypothesis. Hence, A and A' are on opposite sides of ℓ.

The axiom we must assume in place of the SAS Postulate is the following.

AXIOM M

There exists a nontrivial motion having any given line as axis.

LEMMA D

Given any two angles $\angle BAC$ and $\angle YXZ$ having equal measures, there exists a motion which maps point A to point X, ray $\overrightarrow{AB}$ to ray $\overrightarrow{XY}$, and ray $\overrightarrow{AC}$ to ray $\overrightarrow{XZ}$.

(Proof left as a problem—see Problem 1.)

THEOREM 6: SAS CONGRUENCE CRITERION

Suppose that triangles $\triangle ABC$ and $\triangle XYZ$ satisfy the SAS Hypothesis under the correspondence $ABC \leftrightarrow XYZ$. Then $\triangle ABC \cong \triangle XYZ$.

Proof: Suppose that $\overline{AB} \cong \overline{XY}$, $\overline{AC} \cong \overline{XZ}$, and $\angle A \cong \angle X$. By Lemma D, there exists a motion M mapping point A to point X, ray $\overrightarrow{AB}$ to ray $\overrightarrow{XY}$, and ray $\overrightarrow{AC}$ to ray $\overrightarrow{XZ}$. This motion must then map B to Y and C to Z by the Segment Construction Theorem. Hence, $\triangle XYZ$ is the image of $\triangle ABC$ under M, and since M preserves both distance and angle-measure, the corresponding sides and angles of the two triangles are congruent. $\therefore$ $\triangle ABC \cong \triangle XYZ$.

PROBLEMS   (§5.7)

_____ Group A _____

1. (*NOTE: This is the only problem in this section which will not assume the Euclidean axioms.*) Prove Lemma D in the context of a geometry satisfying the axioms of Chapter 2 and Axiom M. (*HINT:* Let ℓ be the perpendicular bisector of line segment $\overline{AX}$ and consider the nontrivial motion M_1 with axis ℓ. Show that A maps to X, and if ray $\overrightarrow{AB}$ is not mapped to ray $\overrightarrow{XY}$, let m bisect $\angle YXB'$ and apply the motion M_2 with axis m. The composite motion M_2M_1 then maps ray $\overrightarrow{AB}$ to ray $\overrightarrow{XY}$ by the Angle Construction Theorem, etc. (details needed))

2. Referring to Figure 5.45, use the facts $GO = \frac{1}{2}GH$ and $HU = \frac{1}{2}HO$ to show that the segments in the figure are correctly labelled, then prove that $(HG, UO) = 1$. (If you are familiar with directed distance, show that this cross ratio is -1.)

3. **Euler Line of a Triangle** A dilation which maps one triangle to another must map the altitudes and orthocenter of the first triangle to those of the second, since perpendicularity and incidence are preserved by dilations. Recall that if G is the centroid of $\triangle ABC$, then $AG = \frac{2}{3}AL = 2GL$, $BG = 2GM$, and $CG = 2GM$. Finally, note that O is the *ortho-*

center of the triangle joining the midpoints of the sides of △*ABC, that is, of* △*LMN.*
(Prove this.) Now define the dilation $d_{-1/2}[G]$ with center G and negative dilation factor
(hence points A, B, and C are reflected through G one-half their distance to G and must
therefore map to L, M, and N). Use this to prove that O, G, and H are collinear and that
$HG = 2GO$.

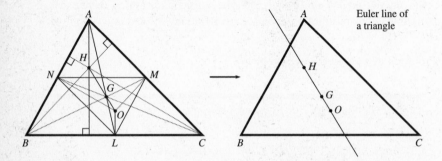

Euler line of
a triangle

4. A map of Alaska and its photocopy are placed on a table at random, face up. Show that
   as point P traces the features of the original map and its correspondent P' traces corresponding features in the copy, the midpoint M of segment $\overline{PP'}$ traces a map that is *similar to the original map.* (Use a coordinate representation of an isometry for this endeav-
   or.) Under what circumstances will the new map be the same size as the original?

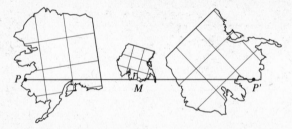

5. **Scheiner's Pantograph**   An instrument invented in 1630 by Christolph Scheiner called
   the **pantograph** may be used to increase or decrease the scale of any drawing. The
   mechanism is fastened at a point of pivot O with the remaining rods allowed to move in
   the manner illustrated. The main ingredient is parallelogram $\diamond ABCA'$. Discuss the operation of this device in connection with ratios and a dilation mapping from center O.

SCHEINER'S PANTOGRAPH

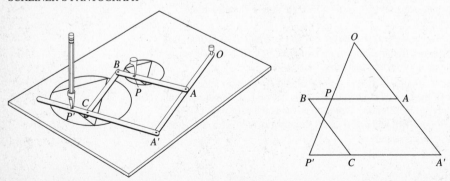

_____ **Group B** _____

6. **Napoleon's Theorem** In Figure 5.48, note that the angles $\angle BPC$, $\angle AQC$, and $\angle ARB$ are each of measure 120. Thus, applying the product

$$r_{120}[P]r_{120}[Q]r_{120}[R]$$

to point B, show that B returns to itself. Hence, the preceding mapping, which must either be a translation or the identity (by Corollary C of 5.6), is the identity. Using the other material in Section 5.6, finish the proof of Napoleon's Theorem.

7. Can the plane be tiled by copies of the diagram for Napoleon's theorem in the manner of the tiling of Theorem 4 in Figures 5.52 and 5.53? If so, find it. (Make a drawing, by hand or via the computer.)

8. For the purpose of finishing the solution of Example 1, consider any three point reflections in coordinate form, s_N, s_L, and s_M, where $N = (x_1, y_1)$, $L = (x_2, y_2)$, and $M = (x_3, y_3)$.

$$s_N : \begin{cases} x' = -x + 2x_1 \\ y' = -y + 2y_1 \end{cases} \quad s_L : \begin{cases} x' = -x + 2x_2 \\ y' = -y + 2y_2 \end{cases} \quad s_M : \begin{cases} x' = -x + 2x_3 \\ y' = -y + 2y_3 \end{cases}$$

Show that the product $s_M s_L s_N$ is a point reflection (of the form $x' = -x + 2a$, $y' = -y + 2b$).

9. **Generalization of Napoleon's Theorem** The following result is what you might have discovered in the Discovery Unit (*Amazing Geometric Effect*). Prove it as the following corollary to Theorem 3.

   Corollary Let $\triangle XYZ$ be a given acute-angled triangle, having angles of measure p, q, and r, respectively. Given any other triangle $\triangle ABC$, suppose that isosceles triangles $\triangle PBC$, $\triangle QCA$, and $\triangle RAB$ are constructed externally on the sides of the given triangle as bases, such that the vertex angles at P, Q, and R have measures $2p$, $2q$, and $2r$, respectively. Then $\triangle PQR$ has angles of measure p, q, and r, respectively, and $\triangle PQR \sim \triangle XYZ$, regardless of the choice of $\triangle ABC$.

10. Reconstruct a triangle $\triangle ABC$ given the feet D, E, and F of the altitudes on the sides. (Transformation theory is not needed for this.)

11. Reconstruct a triangle $\triangle ABC$ given the points of contact P, Q, and R of the sides with the incircle. (Transformation theory is not needed for this.)

12. Given two circles C and D and line ℓ, as shown in the following figure, construct a line $m \parallel \ell$ which cuts off equal chords in the two circles. (**HINT:** Think translations.)

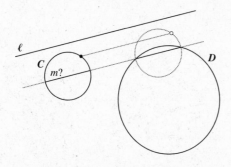

13. Use a similitude to effect the following construction (accompanying figure) : Construct a circle which is tangent to the sides of an angle and passing through some given point P in its interior.

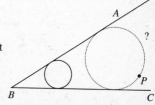

14. Given two circles *C* and *D* intersecting at *A* (as in figure at the right), find a line through *A* which cuts the circles in chords of equal lengths. (***HINT:*** Think central reflections.)

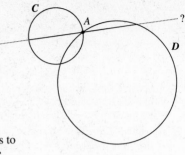

15. (a) In Figure 5.55, use a rotation of △*ABE* through an angle of measure 60 about vertex *A* as center to show that *BE = CF = AD*.

   (b) Show that Fermat's Point (Discovery Unit) is also the point of concurrency of the circumcircles of the three equilateral triangles in Figure 5.55. (***HINT:*** Just show that *P* lies on any one of those circles by analyzing *m*∠*APB* and using the inscribed angle theorems.)

16. How can one determine where to place the vertices of an equilateral triangle so as to inscribed it in the square shown, with one vertex of the triangle on a vertex of the square and the other two vertices lying on the sides of the square opposite that vertex?

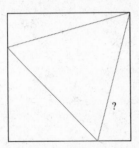

_____ *Group C* _____

17. Given a point *P* in the interior of ∠*ABC*, construct a line through *P* which cuts the sides of the angle in two points *Q* and *R* equidistant from *P*. (***HINT:*** Think central reflections.)

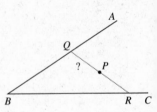

18. In a given convex quadrilateral, inscribe a parallelogram having a given point inside the quadrilateral as center. (See Problem 17.)

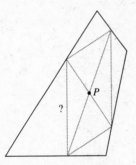

19. Consider three concentric circles. If it exists, construct an equilateral triangle having its three vertices *A*, *B*, and *C* on the given circles, respectively. Determine when the problem does not have a solution.

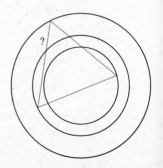

20. **A Mirror Problem**   Three mirrors perpendicular to the plane of this page (accompanying figure) lie along three sides of a convex polygon (double lines). A laser beam is to be shot from point P and aimed in such a way toward the mirrors so as to strike the objective at point Q and stay within a narrow corridor, as shown. How can the direction of the beam be determined? (*HINT:* First solve the problem when there are only two mirrors.)

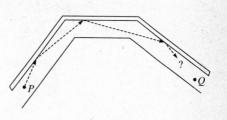

21. **Peucellier's Cell, or How To Draw a Straight Line**   Using our ability to draw perfect circles (as with a well-constructed compass), we can use inversion theory to create a perfect straight line by using a device invented by a French army officer A. Peaucellier (1832–1913). Point O is fixed and $OA = OB$, with $\diamondsuit APBP'$ a rhombus. Show that P and P' are inverse pairs with respect to a circle of inversion centered at O and radius

$$r = \sqrt{OA^2 - AP^2}.$$

As P describes a circle through O, what will the locus of P' be, and why?

PEUCELLIER'S CELL

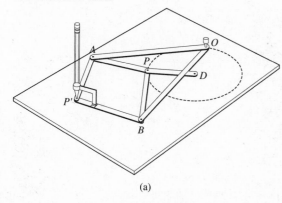

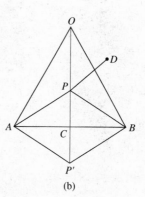

(a)   (b)

22. **Euler's Formula Via Inversion Theory**   Routine methods for proving Euler's Formula $d^2 = R^2 - 2rR$, which relates the distance d between the circumcenter and incenter of any triangle to the circumradius R and inradius r can lead to a cumbersome argument. Complete the steps in the following argument, using inversion: Let X, Y, and Z be the points of contact of the incircle with the sides of $\triangle ABC$ in the figure at the right.

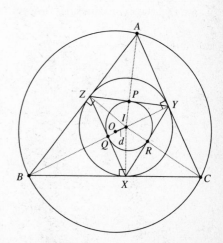

(1)   Show that A and P (midpoint of side $\overline{YZ}$) are inverse points relative to the incenter as circle of inversion by establishing the relation $IP \cdot IA = IY^2 = r^2$. Similarly, B and Q, and C and R, are inverse pairs.

(2)   Hence, the Nine-Point Circle of $\triangle XYZ$ is the inversion of the circumcircle of $\angle ABC$ relative to the incircle, and by Problem 18 of §5.4, $b = Rr^2/(R^2 - d^2)$, where b is the radius of the Nine-Point Circle of $\triangle XYZ$ (= $\frac{1}{2}r$).

(3)   The formula now falls right out. (Show this.)

23. **Ptolemaic Inequality Via Inversion**   The **Ptolemaic Inequality** states that for any four points A, B, C, D in the plane, the inequality $AB \cdot CD + AC \cdot BD \geq AD \cdot BC$ holds, with equality only when the four points lie on a circle. Inversion in one of the points, say D, using *any* circle of inversion (let the radius be 1 for convenience), results in a simple proof. Note that under such an inversion, we have $A'D = \dfrac{1}{AD}$, $B'D = \dfrac{1}{BD}$, and $C'D = \dfrac{1}{CD}$. Use similar triangles to show that $\dfrac{A'B'}{AB} = \dfrac{A'D}{BD} = \dfrac{1}{AD \cdot BD}$. In like manner, $\dfrac{A'C'}{AC} = \dfrac{1}{AD \cdot CD}$ and $\dfrac{B'C'}{BC} = \dfrac{1}{BD \cdot CD}$. But in $\triangle A'B'C'$ we have $A'B' + A'C' \geq B'C'$, hence

$$\frac{AB}{AD \cdot BD} + \frac{AC}{AD \cdot CD} \geq \frac{BC}{BD \cdot CD}$$

Equality holds iff A', B', and C' are collinear (what does this mean regarding A, B, and C?) By multiplying both sides by the quantity $AD \cdot BD \cdot CD$, we obtain the desired inequality. Note that Ptolemy's Theorem (§1.3) is a corollary.

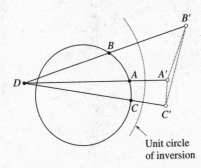

Unit circle
of inversion

24. **PROJECT FOR UNDERGRADUATE RESEARCH**   Explore various extensions of the section on *Napolean-like Theorems* to find either further examples, conditions yielding uniqueness of certain constructions given, or extensions of these results to quadrilaterals and polygons. (For the latter, the development of the last part of §5.6 would have to be revamped.) For example: Does a new theory of products of four rotations exist which yields results similar to Theorem 3 and its corollaries?

The concept of a transformation in geometry was developed, starting with the most basic type—the reflection. The reflection in a line—whose simple model is the common mirror—was used to define all the plane isometries. In particular, a rotation is the product of two reflections in lines which intersect, and a translation is the product of two reflections in lines which are parallel. The fundamental theorem states that any isometry in the plane is the product of at most three reflections. Equally important are the similitudes and dilations. A similitude is any linear transformation which preserves angle measure. A dilation also preserves angle measure, but, in addition, preserves the slopes of all lines and has a central point of projection (i.e., the image of each point lies on a line through the given point and some fixed central point). Every similitude is the product of an isometry and a dilation, making the relationship between the two clear. Finally, glide reflections, inversions, and the stereographic projection were discussed briefly.

The chapter ended with a discussion of groups of transformations and its various subgroups (e.g., the isometries are a subgroup of the similitudes) and several examples showed how transformations can be used to solve problems in geometry.

Answer each of the following questions True (T) or False (F).

1. A line reflection is an example of an opposite transformation.

2. The set of rotations about different centers does not yield a group because the product of two such rotations can be a translation.

3. Only a dilation can map a triangle into one that is similar and noncongruent to it.

4. The most general line-preserving transformation from the plane onto itself would not necessarily preserve parallel lines.

5. Circles and ratios of distances along a line or along parallel lines are invariants of a linear transformation.

6. Circles and ratios of distances along a line or along parallel lines are invariants of a similitude.

7. The essence of any linear transformation in the plane expressed in terms of a coordinate system can be captured by a 2×2 matrix.

8. A linear transformation in the plane could not be represented by a 2×2 matrix having three zeroes in it.

9. Generally speaking, the smaller the group of transformations, the larger its number of invariants.

10. A movie projector, as it projects pictures on the screen, is a prime example of a linear transformation in action.

6

ALTERNATIVE CONCEPTS FOR PARALLELISM: NON-EUCLIDEAN GEOMETRY

OVERVIEW

We will develop here the other two classical geometries possible within the axiomatic framework of absolute geometry begun in Chapter 2 (some modification being required for axiomatic spherical geometry). First, the historical origin of non-Euclidean geometry, which accompanied a geometric revolution in the 1800s, is presented. Then we will take up the mathematical side of this revolution. In the process, possibilities for a wide variety of non-Euclidean geometries (besides the taxicab geometry we studied in Chapter 3) will become apparent, touching on some important areas of geometry of current interest.

*6.1   HISTORICAL BACKGROUND OF NON-EUCLIDEAN GEOMETRY

The first definitive study of parallelism and its effect on geometry was Euclid's, and because of the way he handled this topic, he has been called *the world's first non-Euclidean geometer*. The development of rectangles, parallelograms, area, the Pythagorean Theorem, and volume was based on one suspiciously complex postulate—the famous **Fifth Postulate of Parallels** (Figure 6.1):

> If a straight line falling on two straight lines makes the interior angles on the same side less than two right angles, the two straight lines, if produced indefinitely, meet on that side on which are the angles less than the two right angles.

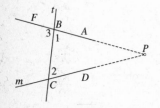

Figure 6.1

That is, in Figure 6.1, if $m\angle 1 + m\angle 2 < 180$, then lines ℓ and m meet on the A-side of line t. Euclid's earliest critic was Proclus (410–485), who wrote detailed commentaries on the works of early Greek geometers. Proclus refused to accept Euclid's postulate because he felt it was too complicated to be a postulate, and thought it could be proven from the other axioms. He constructed his own argument, which he thought settled the issue once and for all (we will take a look at this proof later). His proof was among the first of scores of pseudoproofs of the postulate which were to emerge over the next 1400 years.

The very organization of Euclid's material invites speculation. The use of the parallel postulate is postponed until the last possible moment. Fully 28 propositions appear before that, substantially the same results we obtained in Chapter 3 for absolute geometry. The next-to-last of these results (Proposition 27) reads:

> If a straight line falling on two straight lines make the alternate angles equal to one another, the straight lines will be parallel to one another.

This will be recognized as the lemma we proved in Section 4.1. Euclid's poetic proof, as quoted from Heath's *Thirteen Books of Euclid,* (with his own peculiar illustration, Figure 6.2) reads

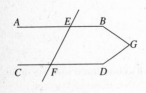

Figure 6.2

> I say that *AB* is parallel to *CD*.
> For, if not, *AB, CD* when produced will meet either in the direction of *B, D* or towards *A, C*.
> Let them be produced and meet, in the direction of *B, D*, at *G*.
> Then, in the triangle *GEF*, the exterior angle *AEF* is equal to the interior and opposite angle *EFG*: which is impossible.
> Therefore *AB, CD* when produced will not meet in the direction of *B, D*.
> Similarly it can be proved that neither will they meet towards *A, C*.
> But straight lines which do not meet in either direction are parallel; [Def. 23] therefore *AB* is parallel to *CD*.
>
> Therefore etc.

Proposition 28 merely uses a different pair of angles along the transversal, so it is a simple corollary to Proposition 27. It is highly fortuitous that the very *next* proposition in the *Elements* is the converse of Propositions 27 and 28, and in proving it, Euclid uses his Parallel Postulate for the first time.

The impressive volume of material and methods associated with proving 28 theorems, the last of which is directly related to parallel lines, has a fatal psychological effect. One begins to wonder why the Parallel Postulate is needed at all. Since one has so much to work with, there emerges the conviction that one should be able to prove the postulate as a theorem, in the style of Proclus. Since geometry abounds with propositions for which a simple logical twist proves their converse, why not try a similar tactic for the converse of Proposition 27, without using the Parallel Postulate?

This was to become the question of the ages, a question whose answer led to a fascinating journey stretching over more than 2000 years, marked with worthwhile developments as well as controversy, and it ultimately revolutionized our thinking about mathematics and axiomatic systems. The problem engaged some of the greatest mathematical minds in history, including Carl Friedrich Gauss and A. Legendre. Even the great analyst Joseph Lagrange worked on the problem, but soon gave it up. (The story is told that on one occasion Lagrange was presenting a paper on parallels to the French Academy, but broke off in the middle of his presentation with the exclamation, "I must meditate further on this!" With that, he put the paper in his pocket, walked out of the meeting, and never spoke of it publicly afterwards.)

During this very long period of time, very few suspected the real source of difficulty. In reality, the postulate is an *undecidable issue* in the context of absolute geometry. (See Section 2.2.) However, one should realize that mathematicians in those days were not well acquainted with models and axiomatic systems. Their postulates were based on the most concrete of observations, and hence they trusted axiomatics more than we do today. With our sophisticated approach, we are much more aware of the logical pitfalls and the use of models in abstract situations. Euclid, Archimedes, and hosts of followers merely postulated what seemed to be self-evident in the world about them—what they thought they could see with their own eyes.

This was very much in line with the teachings of the prominent 18th century philosopher Immanuel Kant, who held that there can be only one valid perception of the universe. As a result, the corresponding belief that there can be only one consistent geometry—the geometry of the world in which we live—came to be firmly rooted in scholarly thinking. Leading experts believed that any other kind of hypothesis would lead to dire consequences, paradoxes, and mathematical foolishness.

The stage was set for a monumental problem to overcome. It defeated G. Saccheri, whom we have already mentioned, who dutifully concluded that his acute angle hypothesis could not be consistent with "the nature of lines." While J.H. Lambert showed great insight in his study of variable angle sums of triangles, he too did not venture far from the Euclidean hypothesis. Although Gauss was the first to recognize the true nature of the problem and to develop a consistent non-Euclidean geometry, even he was to write in a letter to F.W. Bessel in 1829:

> It may take very long before I make public my investigations on this issue; in fact this may not happen in my lifetime for I fear the scream of dullards if I make my views explicit.

So, although Gauss had worked extensively on the problem of parallels, he did not publish his work. How were those who followed to know that he had already constructed rather extensive features of what we now call **hyperbolic geometry**, possibly as early as 1792?

This anti-scholarly environment was responsible for one of the great tragedies and misunderstandings in modern mathematics. It involved a young and talented János Bolyai (1802–1860) of Hungary, who was affected by the incident the rest of his life and was discouraged from further work in mathematics.

When he was only 23, Bolyai discovered a new development of geometry which eventually led to settling the problem of parallels. During the years 1823–1832, he and the Russian mathematician Nicholai Lobachevski (1793–1856) independently, and without knowledge of each other's work, developed an elaborate system of non-Euclidean trigonometry that, from its complete parallel to the well-known formulas of spherical trigonometry, strongly suggested that a denial of the Fifth Postulate will not lead to a contradiction. Instead, one obtains another very bizarre but seemingly consistent geometry (hyperbolic geometry).

After Bolyai made this discovery, it astounded him, and he wrote of his excitement to his father, Wolfgang (Farkas) Bolyai:

> ...I have discovered such magnificent things that I am myself astonished at them...Out of nothing I have created a strange new world.

His work was printed in 1832 as an appendix to a geometry text published by the senior Bolyai, which was a short 26 pages. (Lobachevski's work, which was incredibly similar, was first published in 1829, in Russian.) In that 26 pages, Bolyai develops the bulk of hyperbolic geometry—including the construction of a surface in hyperbolic three-dimensional space on which Euclidean geometry takes place (much like spherical geometry takes place on the surface of the sphere in Euclidean three-dimensional space), which Bolyai called a *parasphere*. (This same surface was constructed by Lobachevski, which he called a *horosphere*.) One starts with an ordinary spherical surface in absolute geometry, deduces formulas on it which are true without the Parallel Postulate, then lets the radius of the sphere become infinite. In Euclidean geometry this produces an ordinary plane, but in hyperbolic geometry a different surface is created. If in the underlying space we assume that the Parallel Postulate is false, then the ultimate result is a two-dimensional Euclidean geometry that lives in a three-dimensional non-Euclidean environment!

The senior Wolfgang Bolyai, who was a friend of Gauss, proudly sent this new work of his son's to him to see what he thought of it. Gauss wrote back, in so many words: "I cannot praise this work, for to do so would be to praise my own work...I myself, long ago came to these same conclusions." This pontifical response was bred of a lifetime of conservative habits by a perfectionist, and the fear of being ridiculed. But his response was viewed by young Bolyai as discouraging, to say the least. Even to imagine that this work of Gauss's may have been locked away in a desk drawer since before he was born, would have been quite bizarre indeed. Bolyai feared Gauss was actually seeking to take credit for his accomplishment, or play down its importance. The incident also turned him against his father for a time, whom he suspected of being in collusion with Gauss.

HISTORICAL NOTE

Although Bolyai and Lobachevski (pictured) were heroes of the geometric revolution of the 1800s, their splendid accomplishments met with total indifference by the mathematical community. János Bolyai was born in 1802, and, taught by his father, he had already mastered the calculus by the time he was 13. Bolyai was a Hungarian, with a flamboyant spirit and quick temper. When his father's request to have Gauss take him as a private student went unanswered, he began his education for the military. He became a skillful fencer, and once accepted the challenge of thirteen officers for a duel on condition that he be allowed to play his violin after each duel; he defeated them all. After his rebuff by Gauss over hyperbolic geometry, he published nothing else, although he left behind some 1,000 pages of manuscript. In his last years, he learned of Lobachevski's work. A posthumous victory was had when, in 1905, the Hungarian Academy of Science established the *Bolyai Prize* in his honor, consisting of 10,000 gold crowns awarded to the mathematician who had most greatly contributed to progress in mathematics. (The first to be awarded this prize was,

appropriately, H. Poincaré; Hilbert was second, in 1910, and Einstein, the third in 1915.)

In contrast to Bolyai's free spirit, Lobachevski was a reserved scholar. He was appointed professor and rector of the University of Kasan (Russia) in 1827. His work in geometry, like Bolyai, courageously challenged the Kantian doctrine of space. It included a complete theory of parallels and the construction of an "imaginary" geometry, which he later called *pangeometry*. Although his development was published three years before Bolyai's, in 1829, he received no recognition for it. The work was largely misunderstood and even stated to be incorrect by one reviewer. If the lack of recognition for his work were not enough, the administration of the University of Kasan decided to revamp its organization in 1846, and Lobachevski was dismissed after 19 years of faithful service. During his final years he suffered blindness and had to dictate his remaining works. But, unlike Bolyai, he never gave up, anticipating the day when his work would finally win approval.

It would take approximately 40 years for mathematicians to realize the significance of the work of Bolyai and Lobachevski. In the 1850s, the concepts of differential geometry, begun by Gauss himself, were developed to the point that surfaces of constant negative curvature (called **pseudospheres**) were more clearly understood, and a variety of models for non-Euclidean geometry were subsequently discovered. After Bernhard Riemann's famous lecture in 1854 on a very general kind of geometry having arbitrary dimension and variable curvature—a development that was to play an important role in the development of the theory of General Relativity—in 1868 Eugenio Beltrami laid out the basic idea for his proof of the relative consistency of non-Euclidean geometry. At first he used the geometry of the pseudosphere as a model, then he developed the upper half-plane and circular disk models. We will study these later in detail. In 1871, Klein gave Beltrami's disk model a new interpretation in the projective plane which made its study more elegant. (This construction is known as the **Beltrami-Klein model**.) Later, in 1882, H. Poincaré reintroduced Beltrami's disk model in connection with transformation groups of complex numbers, known as **Poincaré's Model**.

6.2   AN IMPROBABLE LOGICAL CASE

When we introduced the Euclidean Parallel Postulate in Chapter 4, we observed that there were three logical cases associated with a line ℓ, a point P not on that line, and the lines through P parallel to ℓ.

(1)   There are *no* lines parallel to ℓ. (Postulate for Spherical Geometry)

(2)   There is *exactly one* line parallel to ℓ. (Euclidean Parallel Postulate)

(3)   There are *two or more* lines parallel to ℓ. (Lobachevskian Parallel Postulate)

With a slight modification of our postulates to accommodate a bounded metric, these three hypotheses lead to the three classical geometries, which were given the more esoteric names **elliptic**, **parabolic**, and **hyperbolic geometry** by Felix Klein (1849–1925), in order to place them on an equal footing. Obviously, parabolic geometry is a fancy name for Euclidean geometry; other names commonly used for elliptic and hyperbolic geometry are **spherical** (also **Riemannian**) **geometry**, and **Lobachevskian geometry**.

In absolute geometry, the *existence* of parallel lines is a proven fact (the lemma in Section 4.1). Thus, the set of 16 of axioms we have adopted for absolute geometry eliminates elliptic geometry. In Section 6.8 of this chapter, we show how to properly modify the axioms in order to allow the existence of elliptic geometry, but right now, let's continue to work in absolute geometry. Also, we will not assume Axiom P-1, the Euclidean Parallel Postulate. That is, we take up where we left off at the end of Chapter 3, omitting all the material in Chapters 4 and 5.

PLAUSIBILITY OF THE HYPERBOLIC POSTULATE

The alternative (logical) possibility for Euclidean parallelism in absolute geometry is: Given a point P and a line ℓ not passing through it, there exist *at least two distinct* lines through P, in the same plane as P and ℓ, which are both parallel to line ℓ. Thus we have lines m_1 and m_2 through P parallel to ℓ (which lie in the same plane and do not intersect), as illustrated in Figure 6.3.

No doubt this possibility seems so bizarre that disproving it on purely logical grounds should be a simple matter. But to attempt to do so would be repeating history, and it would lead us directly to that "bottomless night, which extinguished all light and joy," as described by one man who devoted his entire life to the problem without success, Wolfgang Bolyai.

Probably one's first reaction is to attempt to refute this hypothesis from a "practical" standpoint by looking at the apparent behavior of lines in ordinary geometry (a more careful approach might oblige us to ask exactly what the term "ordinary" geometry" means). Nevertheless, if you draw two lines through P and they are both supposed to be parallel to a third line, this would seem to force some lines to be

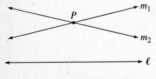

Figure 6.3

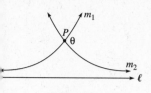

Figure 6.4

curved rather than straight, as illustrated in Figure 6.4. We could reason that the larger the angle θ between m_1 and m_2 is, the more curved the lines have to be. However, we know that drawings and figures do not prove anything from a logical standpoint. Whether lines are "curved" or not depends on the axioms.

Furthermore, the argument is inconclusive because of what could really be taking place *in the world in which we live:* Suppose the angle θ between such parallels were always incredibly small, say less than one-millionth of a degree. No drawing instrument would then be capable of producing precise, distinct lines, and measuring devices would not be accurate enough to detect the small amount of curvature which might be taking place. Even modern physics and the state of the art in technology (the use of lasers, high-tech measuring devices, etc.) can only provide evidence of one hypothesis over another. One might recall, as mentioned earlier, the expedition Gauss undertook to measure the angles between three distant mountain peaks, in which no discrepancy from the Euclidean result was detected. (Had such a discrepancy been found, then it certainly would have *disproven* the validity of the Euclidean Parallel Postulate in the physical world.)

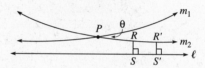

Figure 6.5

A second line of reasoning, which is based more on logic than on physical observation, is to consider the *distances* between m_2 and ℓ. Apparently, we must have $RS > R'S'$ (Figure 6.5). If not detectable by actual measuring devices, this inequality must surely be provable in fact. This would then contradict our idea of parallel lines. Parallel lines are supposedly everywhere equidistant—like the rails of a railroad track. Indeed, this point of view is the basis for several early proofs of the impossibility of hyperbolic geometry, and some early scholars used it to circumvent the problem of dealing with Euclid's Fifth Postulate. But in reality, what they were doing was assuming this property as a substitute for Euclid's Fifth Postulate (which is logically equivalent to it) and proving the postulate from it. So nothing was actually gained. (That is, Euclid's other postulates *by themselves* had not been shown to imply Euclid's Fifth Postulate.)

EQUIVALENT FORMS OF EUCLID'S PARALLEL POSTULATE

Many "natural" geometric properties equivalent to Euclid's Fifth Postulate have been proposed, but each of these, like the equidistant property of parallels, have turned out to be nothing but substitutes for the Fifth Postulate. A table of a few of these follows. Regarding the first property listed, Gauss once wrote:

> If one can prove that there exists a right triangle whose area is greater than any given number, then I am able to establish the entire system of (Euclidean) geometry with complete rigor. I am in possession of several theorems of this sort, but none of them satisfy me.

EQUIVALENT FORMS OF EUCLID'S FIFTH POSTULATE

- The area of a right triangle can be made arbitrarily large.
- Rectangles exist.
- The angle sum of all triangles is constant.
- The angle sum of a single triangle equals 180.
- There exist two similar, noncongruent triangles.
- A circle can be passed through any three noncollinear points.
- A line (transversal) can be drawn through any given interior point of an angle intersecting both sides of that angle.
- Two parallel lines are everywhere equidistant.
- The set of all points lying on one side of a line and equidistant from it (called an **equidistant locus**) is itself a straight line.
- The perpendicular distance from one of two parallel lines to the other is bounded.
- The Pythagorean Theorem holds for all right triangles.

Each of these properties can be readily recognized as results which have either already been proven in Euclidean geometry (Chapter 4), or can be proven without difficulty.

EXAMPLE 1   Show that if a rectangle exists in absolute geometry, then a triangle having angle sum 180 exists.

SOLUTION

In Figure 6.6, we are given a rectangle $\square ABCD$ (all four angles are right angles). We have only to draw a diagonal $\overline{AC}$ and consider the angles of $\triangle ABC$. Since a rectangle is clearly a convex quadrilateral (reason?), diagonal $\overline{AC}$ passes through the interior of $\angle BAD$ and $\angle BCD$. Hence $\overrightarrow{AB}\text{-}\overrightarrow{AC}\text{-}\overrightarrow{AD}$ and $m\angle A = m\angle 1 + m\angle 3$. Similarly, $m\angle C = m\angle 2 + m\angle 4$. Therefore,

(a)
$$(m\angle 1 + m\angle 2 + m\angle B) + (m\angle 3 + m\angle 4 + m\angle D) =$$
$$m\angle A + m\angle B + m\angle C + m\angle D = 360$$

(by definition of a rectangle). But by the Saccheri-Legendre Theorem (Section 3.4),

$$m\angle 1 + m\angle 2 + m\angle B \le 180 \quad \text{and} \quad m\angle 3 + m\angle 4 + m\angle D \le 180$$

If $m\angle 1 + m\angle 2 + m\angle B < 180$, then the above equation **(a)** cannot hold. Therefore,

$$m\angle 1 + m\angle 2 + m\angle B = 180\,\square$$

Another property in the preceding table will be proposed for discovery.

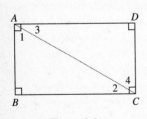

Figure 6.6

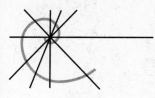

MOMENT FOR DISCOVERY

Are Parallel Lines Everywhere Equidistant?

Suppose we assume that two parallel lines ℓ and m are everywhere equidistant in absolute geometry. This means that if A, C, and E are any three points on line ℓ and B, D, and F are the respective feet of the perpendiculars on m, then

$$AB = CD = EF$$

1. What kind of quadrilateral is $\lozenge ABDC$? $\lozenge CDFE$?

2. What can you deduce conclusively about the angles 1, 2, 3, and 4 in the figure? (Be careful not to assume $\overline{AB} \perp \ell$ from the Z Property of Parallels, which is derived from the Euclidean Parallel Postulate.)

3. In particular, what about $\angle 2$ and $\angle 3$? Can you prove that they are right angles?

4. What have you discovered? Does the assumption "parallel lines are everywhere equidistant" lead to another property listed in the previous table?

Figure 6.7

*TWO FAMOUS "PROOFS"
OF THE EUCLIDEAN
PARALLEL POSTULATE*

One of the earliest proofs of the Parallel Postulate was given in ancient times by Proclus (as mentioned earlier). This argument assumes

(1) that the perpendicular distance from a point on one of two intersecting lines ℓ and m to a point on the other increases without bound as the point varies on that line (that is, in Figure 6.8, $x \to \infty$ as $AP \to \infty$)

(2) that the distance from a point on one of two parallel lines to the other remains bounded ($y < b$ as $BP \to \infty$).

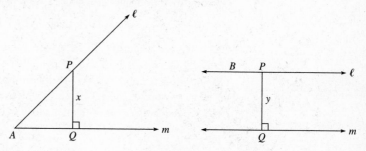

Figure 6.8

PROCLUS

The argument proceeds as follows, relying heavily on properties obtained from the diagram: Let lines ℓ and m be perpendicular to line $\overleftrightarrow{AB}$, as in Figure 6.9. Then by Euclid I.27, the lines are parallel. It is to be proven that ℓ is the only parallel to m

through A. Suppose n is another line through A parallel to m. Since ray $\overrightarrow{AW}$ does not meet line m, $\overrightarrow{AW}$ cuts the perpendicular $\overrightarrow{PQ}$ at some point R, and we let $x = PR$, $y = RQ$. Now let $AP \to \infty$. Then $x \to \infty$ by assumption (1), but $\overrightarrow{PQ}$ is bounded, by assumption (2). Hence, $x > PQ$ for sufficiently large AP, so for some previous position of P, $x = PQ$ and $y = 0$. Thus, n intersects line m, a contradiction, and line ℓ is the unique parallel to m through point A.

PROCLUS' ARGUMENT

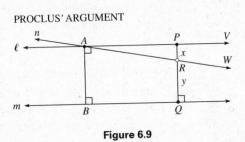

Figure 6.9

This is certainly an interesting argument. Indeed, in Problem 19 you will be asked to prove that Proclus' assumption (1) is valid in absolute geometry. So...

> **QUESTION:** What does Proclus' argument actually prove? Can his argument (as presented here) be made rigorous?

LEGENDRE

A somewhat more sophisticated approach was taken by Legendre (1752–1833). His proof is as follows (as found in R. Bonola's *Non-Euclidean Geometry*, pp. 58-59): Let $\triangle ABC$ have, if possible, angle sum less than 180 (Figure 6.10). Construct $\triangle A'BC$ congruent to $\triangle ABC$, with $A' \in$ Interior $\angle CAB$ and lying on the opposite side of line $\overleftrightarrow{BC}$ as A. (One easy way of accomplishing this construction is to repeat the construction used for the Exterior Angle Inequality, with M the midpoint of $\overline{BC}$ and $\overline{AA'}$.) Through point A', draw a transversal meeting the sides of $\angle CAB$ at B_1 and C_1, respectively. Thus, $\overline{B_1C_1}$ forms a larger triangle $\triangle AB_1C_1$ which contains four subtriangles inside it. Define the **defect** of $\triangle ABC$ to be the number

LEGENDRE'S ARGUMENT

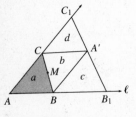

Figure 6.10

$$a = 180 - m\angle A - m\angle B - m\angle C$$

which we are assuming to be positive (where $\angle B = \angle ABC$ and $\angle C = \angle ACB$). Similarly, define the defects of the other three triangles inside $\triangle AB_1C_1$, namely $\triangle BA'C$, $\triangle BA'B_1$, and $\triangle CA'C_1$, as the numbers b, c, and d, respectively. Now, a little later, we prove the principle that *defect is additive*. That is, the defect a_1 of $\triangle AB_1C_1$ equals the *sum* of the defects of each of the four subtriangles:

$$a_1 = a + b + c + d$$

But since the angles of $\triangle A'BC$ are congruent to those of $\triangle ABC$,

$$a = b$$

By the Saccheri-Legendre theorem (Section 3.4),

$$c \geq 0 \quad \text{and} \quad d \geq 0$$

Thus,

$$a_1 \geq a + b = 2a$$

We have thereby constructed a new triangle, $\triangle AB_1C_1$, with defect a_1 at least twice that of $\triangle ABC$. Using an identical construction on $\triangle AB_1C_1$, we can construct $\triangle AB_2C_2$ with defect

$$a_2 \geq 2a_1 \geq 4a.$$

Continuing in this fashion, there exists a triangle with defect

$$a_n \geq 2^n a$$

Since $a > 0$, then $2^n a \to \infty$ and a triangle having defect > 180 exists, which is impossible, by its definition.

> *QUESTION:* What theorem of absolute geometry does Legendre's argument actually prove? Can it be made rigorous?

A USEFUL LEMMA   The following set of problems will explore further interesting arguments which history provides for us, masquerading as proofs of Euclid's Fifth Postulate. Another interesting result will be useful later.

LEMMA[1]

Let $\triangle ABC$ be given, with M the midpoint of side $\overline{AC}$ (Figure 6.11). If the angle sum of $\triangle ABC$ is less than 180, then so is the angle sum of both $\triangle ABM$ and $\triangle BMC$.[1]

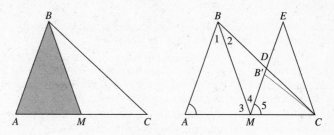

Figure 6.11

Proof: Construct D on side $\overline{BC}$ so that $\angle DMC \cong \angle A$. Now $MD < AB$, for if $MD \geq AB$, we could construct segment $\overline{MB'} \cong \overline{AB}$ on $\overline{MD}$ and by SAS, $\triangle ABM \cong \triangle MB'C$. But this leads to a contradiction of the Exterior Angle Inequality since

$$m\angle BCM \geq m\angle B'CM = m\angle BMA$$

Hence we may locate E on ray $\overrightarrow{MD}$ so that $M\text{-}D\text{-}E$ and $ME = AB$.

(1) $\triangle ABM \cong \triangle MEC.$ (SAS)

---

[1]It is the *converse* of this result that is the familiar consequence of additivity of defect in absolute geometry. (See Problem 16, §6.4.)

(2) Assume that $m\angle A + m\angle 1 + m\angle 3 \geq 180$. Then by the Saccheri-Legendre Theorem, $m\angle A + m\angle 1 + m\angle 3 = 180$.

(3) $m\angle 3 + m\angle 4 + m\angle 5 = 180$. (betweenness properties and Linear Pair Axiom)

(4) By Steps (2) and (3), $m\angle A + m\angle 1 + m\angle 3 = m\angle 3 + m\angle 4 + m\angle 5$, or $m\angle 1 = m\angle 4$.

(5) $BM = EC$ and $m\angle 1 = m\angle E \quad \rightarrow \quad m\angle 4 = m\angle E$. (CPCF)

(6) Thus, since $m\angle BDM = m\angle CDE$, $\triangle BDM \cong \triangle CDE$. (AAS)

(7) $m\angle 2 = m\angle DCE$.

(8) Hence,

$$\text{Angle sum } (\triangle ABC) \quad = m\angle A + m\angle 1 + m\angle 2 + m\angle DCM$$
$$= m\angle 5 + m\angle 4 + m\angle DCE + m\angle DCM$$
$$= m\angle 5 + m\angle 4 + m\angle ECM$$
$$= m\angle 5 + m\angle 4 + m\angle 3 = 180 \qquad \rightarrow\leftarrow$$

$\therefore$ Angle sum $(\triangle ABM) < 180$.   (Assumption in Step (2) false.)

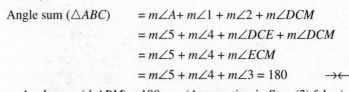

Figure 6.11

The same proof works for $\triangle BMC$ and will be omitted.

PROBLEMS   *(§6.2)*

_____ *Group A* _____

1. If you have not already done so, prove that if the angle sum of all triangles has a constant value k, then $k = 180$. (See Problem 16, §3.5.)

2. Satisfy yourself that the *first six* statements in the previous table (*Equivalent Forms of Euclid's Fifth Postulate*) are actually consequences of our Axiom P-1 (and logically equivalent to Euclid's Fifth Postulate). Find the exact theorem in Chapter 4 which implies each property, providing simple arguments where needed.

3. In the figure, ℓ and m are parallel lines, $\overline{AB} \perp m$, and $\overline{CD} \perp m$. Using Euclid's Fifth Postulate, how would you prove that $AB = CD$?

4. Points A and B are 100 miles apart (figure below). At A and B, lines $\overleftrightarrow{AC}$ and $\overleftrightarrow{BD}$ are constructed, making angles of measure 89.5 and 89.9 (degrees) with line $\overleftrightarrow{AB}$. How can you be sure that $\overleftrightarrow{AC}$ and $\overleftrightarrow{BD}$ will meet if these constructions are taking place in a Euclidean plane?

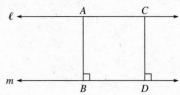

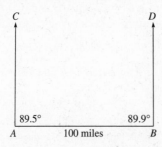

5. Suppose a line ℓ is constructed perpendicular to line $\overleftrightarrow{AB}$ at point A 245 miles away from B, line m is constructed perpendicular to line $\overleftrightarrow{BC}$ at C 658 miles away from B, and that $m\angle ABC = 179.9$ (degrees). If laser beams are shot into outer space from A and C along lines ℓ and m tangent to the earth's surface, will the beams cross at some point in outer space? (See next problem.) If the answer is affirmative, use trigonometry to find how far away the point is from A. (Assume all lines lie in a Euclidean plane.)

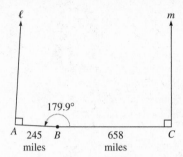

6. In the Euclidean plane, suppose that perpendiculars $\overleftrightarrow{AP}$ and $\overleftrightarrow{CQ}$ to lines $\overleftrightarrow{AB}$ and $\overleftrightarrow{BC}$, respectively, are parallel. Show that A, B, and C are collinear. (See figure at right.)

7. Prove from the Euclidean Parallel Postulate: A circle can be passed through any three noncollinear points. (***HINT:*** See previous problem.)

8. Proclus adopted the following Parallel Postulate: If a line intersects one of two parallel lines, it intersects the other also. Establish from this Euclid's Fifth Postulate of Parallels.

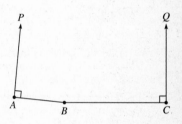

_____ *Group B* _____

9. Prove in Euclidean geometry that if point P lies in the interior of $\angle ABC$, there can be found a line ℓ passing through P which intersects both sides of the angle. (***HINT:*** Draw a line through P parallel to one side of the angle, and choose a point not on that line which lies on the other side of the angle.)

10. Prove in absolute geometry that if the summit angles of a Saccheri Quadrilateral are right angles, the summit and base are congruent. (***HINT:*** Draw a diagonal.)

11. Prove in absolute geometry that the opposite sides of a rectangle are congruent.

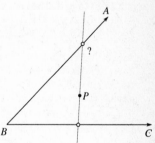

*12. Prove in absolute geometry that if two lines are cut by a transversal such that a pair of alternate interior angles are congruent, the two lines possess a common perpendicular. (***HINT:*** Drop perpendiculars to the lines from the midpoint of the segment joining the points of intersection of the transversal and the lines.)

*Used in the corollary to Lemma C in §6.6. Also occurred as Problem 21, §4.1.

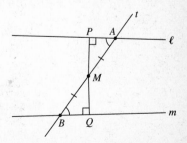

13. Prove in absolute geometry that if a quadrilateral has three right angles and a pair of opposite sides congruent, it is a rectangle.

14. Prove in absolute geometry that if $\Diamond ABCD$ has four right angles (is a rectangle) and $\overline{EF} \perp \overline{AB}$, then both $\Diamond AEFD$ and $\Diamond EBCF$ are rectangles. (See figure at right.)

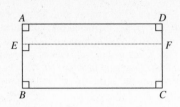

15. Prove in absolute geometry that if a rectangle exists, then a square exists. (See Problems 11 and 14.)

_____ *Group C* _____

16. Prove that if a single rectangle exists in absolute geometry, then one that is arbitrarily large (in terms of its sides) exists. (*HINT:* Double one side of the given rectangle and prove the figure obtained by filling in the remaining sides is a rectangle.)

17. In absolute geometry, prove that if a single rectangle exists, then all Lambert Quadrilaterals are rectangles and hence all Saccheri Quadrilaterals are rectangles. (See Problems 14 and 16.)

18. **Legendre's Second Theorem** From the result of the last problem, using the associated Saccheri Quadrilateral of any triangle, prove Legendre's Second Theorem (the *All-or-None Theorem*): If a single triangle exists in absolute geometry which has angle sum 180, then every triangle has angle sum 180.

19. Prove in absolute geometry that assumption (1) of Proclus' argument is justified, that indeed the distance x from P to side $\overrightarrow{AC}$ of a given angle $\angle BAC$ increases without bound as $AP \to \infty$. (See accompanying figure.)

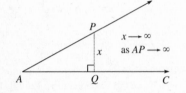

20. **Wallis' Theorem**[2] Wallis introduced the following axiom for Euclidean geometry: Given any triangle and any line segment, a triangle similar to the given triangle may be constructed with the given segment as base. Prove that this axiom implies Axiom P-1 in absolute geometry. (*HINT:* Let A be a point and ℓ a line not on A, and construct $\overline{AB} \perp \ell$ and $m \perp \overline{AB}$ at A. Then $m \parallel \ell$. To show that m is the only parallel to ℓ, consider line n through A. Choose C on n on the B-side of m and drop the perpendicular from C to $\overline{AB}$. Then by Wallis's Axiom, there exists $\triangle ABE \sim \triangle ADC$. Finish the argument.)

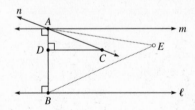

6.3 *THE BELTRAMI-POINCARÉ HALF-PLANE MODEL*

Models of infinite geometries in the Euclidean plane are often derived by deleting part of the plane and taking those remaining points, and the parts of the existing

[2]John Wallis (1616–1703) was famous for products often mentioned in the calculus, called **Wallis products,** such as

$$\frac{\pi}{2} = \frac{2}{1} \cdot \frac{2}{3} \cdot \frac{4}{3} \cdot \frac{4}{5} \cdot \frac{6}{5} \cdot \frac{6}{7} \cdots$$

lines that are left, as the new "points" and "lines." For example, we could take the **punctured plane**, consisting of all points of the plane but for a single point O, which was seen earlier to be the domain of a circular inversion with center O. When we remove part of the plane, like the circular disk in Figure 6.12 centered at the origin, the geometry of what remains consists of all points not lying on the circular disk (shaded portion) and Euclidean lines, with some gaps (where the lines meet the disk). It is not hard to see that in this simple example, two "points" still determine a unique "line." (See Figure 6.12 for a few examples.)

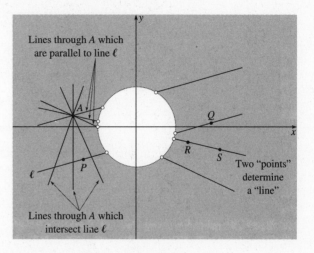

Figure 6.12

The only advantage this has, from our standpoint, is that it causes the Euclidean Parallel Postulate to fail. For example, the figure shows a "point" A and "line" ℓ not passing through it. In this particular case, there are many lines through A which will not meet ℓ at points in this geometry. Thus, according to the original definition of parallel lines, many "lines" through "point" A are parallel to line ℓ in this geometry.

It is this idea that is the prelude to the development of a model for hyperbolic geometry in the Euclidean plane. The Beltrami-Poincaré Half-Plane Model is one of the standard models commonly used to depict hyperbolic geometry and its characteristics, *where Euclidean geometry is used to derive those characteristics!*

The first stage in our presentation here is to consider the ordinary coordinate plane, deleting all points on or below the x-axis, and keeping all points $P(x, y)$ above it ($y > 0$). If the "lines" in this geometry are the ordinary Euclidean lines $ax + by + c = 0$, (a, b not both zero) for which $y > 0$, then clearly the Incidence Postulate "two points determine a line" still holds, as before.

It appears that the Hyperbolic Parallel Postulate holds in every case. In Figure 6.13 is shown a "line" ℓ ($y = x$) and a "point" $A(2, 1)$ not on it. Here, infinitely many lines through A do not meet ℓ. However, our success is short-lived, because horizontal lines will not behave in this manner; instead, they obey the old Euclidean Parallel Postulate (for example: consider m: $y = 2$ and the point $B(-2, 3)$, as shown in Figure 6.13). Here, a unique line, namely $y = 3$, passes through B and is parallel to ℓ.

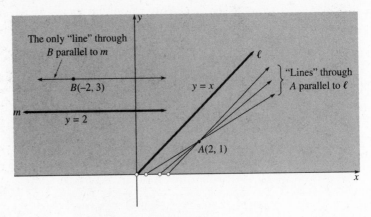

Figure 6.13

In order to obtain a system of "points" and "lines" which work, it is necessary to do something really incredible. Except for all vertical lines, we are going to abandon the Euclidean lines altogether and take, instead, **semicircles** centered on the x-axis lying in the upper half-plane (Figure 6.14).

BELTRAMI-POINCARÉ HALF-PLANE MODEL

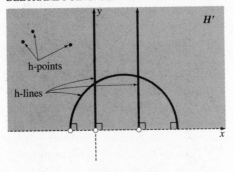

Figure 6.14

Thus, we adopt the following model.

DESCRIPTION OF BELTRAMI-POINCARÉ HALF-PLANE MODEL

POINTS $P(x, y)$ for $y > 0$. (Denote this set of points by H'.)

LINES   Semicircles of the form

$$(x - a)^2 + y^2 = r^2, y > 0 \quad (a, r \text{ constants})$$

Vertical rays of the form

$$x = a \text{ (constant)}, \quad y > 0.$$

NOTE: Instead of continuing to use quotation marks, we shall refer to the "points" and "lines" in H′ as h-points and h-lines. Other special geometric relationships taking place in H′ will be prefixed by the letter "h" as well, such as "h-bisector", "h-perpendicular", etc.

EXAMPLE 1   Show geometrically that, given any two h-points A and B, there is always a unique h-line passing through those points.

SOLUTION

If the ordinary line through A and B is nonvertical, then the perpendicular bisector of segment $\overline{AB}$ will be nonhorizontal, hence will meet the x-axis at a unique point L (Figure 6.15). Take L as center and $\overline{AL}$ as radius, and draw the semicircle through A and B. This will be the unique h-line passing through A and B in this case. If A and B lie on a vertical line $x = c$, the open ray $x = c$, $y > 0$, is the unique h-line passing through A and B.❏

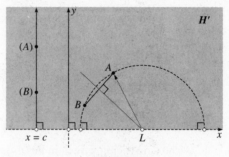

Figure 6.15

The great advantage of the Beltrami-Poincaré model is its **conformal** feature. That is, the angle measure adopted for **H′** can be taken identical to that of the Euclidean plane. Recall that, as in calculus, when dealing with arcs of curves or circles, one always measures the angles between the *tangents* to those curves or circles at their point of intersection. In **H′**, we define angle measure by taking the Euclidean measure of the angle between the h-rays that are the "sides" forming the angle in **H′**.

In order to develop the proper concept for distance in **H′**, which is, after all, dictated by very natural requirements, consider two points A and B on a semicircle h-line ℓ (Figure 6.16), and let M and N be the endpoints of that semicircle on the x-axis. Then if A approaches either M or N on this h-line, we want the distance from A to B to become *unbounded*—like a point approaching the boundary of the universe.

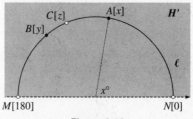

Figure 6.16

Thus, using an asterisk to designate the h-distance we are seeking, $AB^* \to \infty$ as $AM \to 0$ or $AN \to 0$ (Euclidean distance has the same notation as before). We also want $AB^* \to 0$ as A approaches B (or as $AB \to 0$).

Let's temporarily use circular arc coordinates for points on ℓ. Let the points on the semicircle be, with their coordinates, $N[0]$, $A[x]$, $B[y]$, $M[180]$, as shown in Figure 6.16, where $0 < x < y < 180$. Then the h-distance AB^* is just some function of x and y, say

$$AB^* = f(x, y)$$

Repeating what we said before with the new notation, we want the function f to behave as follows:

$$f(x, y) = 0 \text{ if } x = y$$

$$f(x, y) \to \infty \text{ as } x \text{ or } y \to 0$$

$$f(x, y) \to \infty \text{ as } x \text{ or } y \to 180$$

With a little thought, an example of such a function comes to mind:

$$f(x,y) = \left| \frac{1}{x} - \frac{1}{y} - \frac{1}{x - 180} + \frac{1}{y - 180} \right|$$

(The absolute values make $f(x, y) \geq 0$.) Another example is a little less cumbersome, namely

(1) $$f(x, y) = \left| \ln y - \ln x - \ln (180 - y) + \ln (180 - x) \right|$$

(the terms being written in this order since $\ln y > \ln x$ and $\ln (180 - y) < \ln (180 - x)$). Here, we are using the modern notation for the natural logarithm, $\ln x$, rather than the more traditional $\log x$. In the function defined by (1), the desired properties for $f(x, y)$ are due to the basic property of the logarithm, that as $x \to 0$, $\ln x \to -\infty$.

It is interesting that we obtain automatically the desirable property of betweenness with either of these two examples.

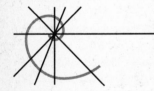

MOMENT FOR DISCOVERY

Exploring Betweenness for the Distance Function $AB^* = f(x, y)$

Suppose that C is a point on the semicircle between A and B. Thus, if z is the coordinate of C, then

$$0 < x < z < y < 180$$

1.   Using (1), write down each of the expressions for $f(x, z)$ and $f(z, y)$, assuming that $x < z$ and $z < y$. (**HINT:** You should get

$$f(x, z) = \ln z - \ln x - \ln (180 - z) + \ln (180 - x)$$

What is the corresponding expression for $f(z, y)$, $z < y$?)

2.   Now write down the expression

$$f(x, z) + f(z, y)$$

and simplify. What did you find?

3.   If $AC^* = f(x, z)$, $CB^* = f(z, y)$, what does $AC^* + CB^*$ equal?

4.   State your discovery.

Using properties of logarithms, note that (1) can be rewritten in the more compact form

$$f(x, y) = \left| \ln \frac{y}{x} + \ln \frac{180 - x}{180 - y} \right|$$

or

$$f(x, y) = \left| \ln \frac{y}{x} \cdot \frac{180 - x}{180 - y} \right|$$

If we now go back to the Euclidean metric, where lengths of chords are used to replace arc length, we obtain still another function obeying the same properties as before (arc $\overgroup{BN} = y \leftrightarrow BN$, arc $\overgroup{AN} = x \leftrightarrow AN$, etc.)

$$AB^* = \left| \ln \frac{BN}{AN} \cdot \frac{AM}{BM} \right| = \left| \ln \frac{AM \cdot BN}{AN \cdot BM} \right|$$

A very useful observation for later use is that the quotient following the logarithm is the *cross ratio*[3] of the points A, B, M, and N

$$AB^* = \left| \ln (AB, MN) \right|$$

Thus we have arrived at what will be defined as the h-metric:

DEFINITION FOR DISTANCE

(3) $AB^* = \left| \ln \dfrac{AM \cdot BN}{BM \cdot AN} \right| \equiv \left| \ln (AB, MN) \right|$ (See figure 6.17)

Along with this, we also have, as discussed earlier:

---

[3]Recall that the cross ratio was extended in Section 4.7 to cover the case when the four points are not collinear.

NOTE: The case when the h-line $\overleftrightarrow{AB}$ lies on a vertical line may be regarded as a special case of **(3)**. The metric becomes simply

$$AB^* = \left| \ln \frac{AM}{BM} \right|$$

which is the limit of **(3)** as the radius of the semicircle through A becomes infinite, and as $AN \to \infty$ and $BN \to \infty$. (Note that as $AN \to \infty$ lim BN/AN = lim $(AB + AN)/AN$ = lim $(1 + AB/AN)/AN = 1 + 0 = 1$.)

DEFINITION FOR ANGLE MEASURE

Given an h-angle $\angle ABC$,

(4) $m\angle ABC^* = \theta = m\angle A'BC'$,

where $\overrightarrow{BA'}$ and $\overrightarrow{BC'}$ are the Euclidean rays tangent to the h-rays $\overrightarrow{BA}$, $\overrightarrow{BC}$, respectively. (See Figure 6.17.)

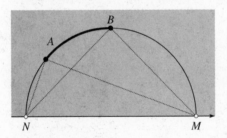
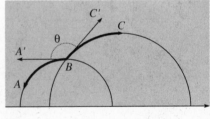

Figure 6.17

It is incredibly remarkable that the geometry of H' now satisfies the entire set of axioms for absolute geometry, in addition to the Hyperbolic Parallel Postulate. This discovery was made by Eugenio Beltrami (1835–1900) in 1868, who began his investigations by studying the geometry on surfaces in Euclidean space.

This model was also studied extensively by Henri Poincaré (1854–1912), who made it popular. Poincaré's work in geometry was so immense that today most mathematicians refer to the circular disk model which we shall introduce in Section 6.7 (as well as the upper half-plane model here) as **Poincaré's Model**.

$\mathcal{H}$ISTORICAL $\mathcal{N}$OTE

When Hilbert met Henri Poincaré for the first time in Paris in 1885, Poincaré had already published more than 100 research papers in mathematics and physics. Poincaré was admitted to the Academy of Paris at the early age of 32. He was the outstanding mathematician of his epoch (ending about 1900). Much of his work was directly connected to geometry, but he also solved some outstanding problems in mathematical physics associated with the so-called *three-body prob-* lem. He founded the field of algebraic topology, and he applied the theory of groups in powerful ways to provide new insight to problems in geometry. His work with complex variables and the fractional linear transformation led to a reformulation of the models for hyperbolic geometry named after him (and which are presented here), but originally discovered by E. Beltrami in 1868.

MOMENT FOR DISCOVERY

Angle Measures and Triangles

Consider the following figure, where h-triangle $\triangle ABC$ has been constructed using the vertical line $x = 3$ and the semicircles $x^2 + y^2 = 25$ and $x^2 + y^2 + 2x = 19$ centered at the origin $(0, 0)$ and the point $(-1, 0)$, respectively.

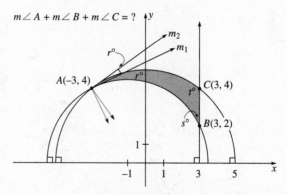

Figure 6.18

1. Show that the points of intersection of these three curves, which are the vertices of the h-triangle, are $A(-3, 4)$, $B(3, 2)$, and $C(3, 4)$.
2. Make a carefully constructed drawing of this figure on your paper, and mea-

sure with a protractor the angles r, s, and t, as shown. (More accuracy results from pursuing the optional Steps 3 and 4 which follow.)

*3. Using properties of slopes for radii and tangents to circles, find the various slopes of the tangents to the semicircles, as indicated.

*4. Using the formula $\tan \theta = \dfrac{m_2 - m_1}{1 + m_1 m_2}$, calculate the degree measures r, s, and t, respectively.

5. Calculate the angle sum $r + s + t$. What did you find?

PROBLEMS ($\S6.3$)

Group A

1. Find an example in the punctured plane where the Euclidean Parallel Postulate fails for some "line" and some "point" not on that "line."

2. Suppose we take as "points" the interior D of some circle in the plane, and as "lines" those portions of ordinary lines cut off by D. In this geometry,

   (a) show that two "points" always determine a unique "line"

   (b) determine the kind of parallel postulate that is valid.

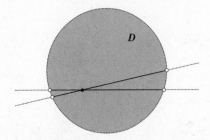

3. Using coordinate geometry, find in H' the equations of the h-lines passing through the h-points $A(0, 2)$ and B, where B has coordinates

   (a) $(0, 5)$

   (b) $(1, \sqrt{3})$

   (c) $(4, 4)$.

4. In H', show that the h-line $x = 5$ ($y > 0$) is h-parallel to the h-line $x^2 + y^2 - 6x + 5 = 0$ ($y > 0$). Sketch the graph of these two h-lines.

5. Find an example of a right h-triangle having one side lying on a vertical line, where it is self-evident that the angle sum of the triangle is less than 180.

_____   **Group B** _____

6.   Using formula **(3)**, calculate the exact h-distance from $C(1, 10)$ to

   (a) $A(7, 8)$

   (**HINT:** Let the semicircle through A and C have equation $(x - h)^2 + y^2 = r^2$, and solve for h and r.

   (b) $B(1, 20)$.

   (c) Show that $\triangle ABC$ is an h-isosceles right triangle.

   (d) Verify that the center of the h-line $\overleftrightarrow{AB}$ (as a Euclidean semicircle) is the point $D(-24, 0)$, and use this to calculate directly each of the h-angles of $\triangle ABC$ at A and B using slope formulas and the formula

$$\tan \theta = \frac{m_2 - m_1}{1 + m_1 m_2} \ .$$

   What do you observe about the angle sum of $\triangle ABC$?

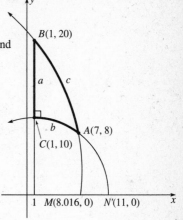

7.   **(a)** Investigate the Pythagorean Theorem for the h-triangle $\triangle ABC$ of Problem 6 by calculating AB^* and using the results for AC^* and BC^* from Problem 6.

   (b) The correct relation for a right $\triangle ABC$ in the Beltrami-Poincaré model is, in standard notation,

$$\cosh c = \cosh a \cosh b$$

   Test this new formula for your answers in **(a)**.

8.   The h-lines $\ell: x^2 + y^2 = 9$ and $m: x^2 + y^2 = 21$ have the y-axis ($y > 0$), as a common h-perpendicular (line $\overleftrightarrow{AB}$ in the figure). Obviously, $\ell \parallel m$. In Euclidean geometry, the two semicircles ℓ and m are concentric, and are everywhere equidistant.

   (a) Show that this is not the case in H' by verifying (by direct calculation) that $CD^* > AB^*$, where $\overleftrightarrow{CD}$ is the semicircle $x^2 + y^2 - 10x + 9 = 0$.

   (b) Verify that the circle $x^2 + y^2 = 9$ is orthogonal to $x^2 + y^2 - 10x + 9 = 0$. What kind of figure is $\diamondsuit ABCD$?

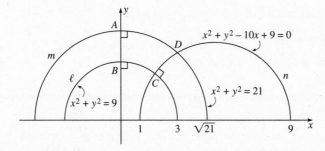

9.   (Refer to Problem 8.) By symmetry, we can infer that since $n': x^2 + y^2 + 10x + 9 = 0$ is the Euclidean reflection of $n: x^2 + y^2 - 10x + 9 = 0$ ($y > 0$) in the y-axis, n' is h-perpendicular to ℓ. Carefully draw a figure of your own to show this and verify that if n' cuts ℓ and m at F and E, respectively, then $CD^* = EF^*$ and $\diamondsuit CDEF$ is a Saccheri Quadrilateral. What kind of angles do the summit angles of $\diamondsuit CDEF$ appear to be?

_____   **Group C** _____

10.   The h-lines $\ell: x^2 + y^2 = 1$ and $m: x^2 + y^2 = 4$ have the h-line $n: x = 0$ ($y > 0$) as a common perpendicular ($\overleftrightarrow{AB}$ in the following figure). Using the theory of orthogonal circles

from § 4.7, show that there exists no circle centered on the x-axis orthogonal to both $x^2 + y^2 = 1$ and $x^2 + y^2 = 4$. What does this information tell you about h-rectangles with base $\overline{AB}$?

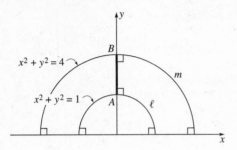

6.4 HYPERBOLIC GEOMETRY 1: ANGLE SUM THEOREM

In this and the next two sections, we will adopt the following axiom, within the context of absolute geometry.

AXIOM P-2: HYPERBOLIC PARALLEL POSTULATE

If ℓ is any line and P any point not on ℓ, there exists more than one line passing through P parallel to ℓ.

ANGLE SUM THEOREM FOR HYPERBOLIC GEOMETRY

A characteristic feature of hyperbolic geometry is that the sum of the measures of the angles of any triangle is always less than 180. This may be proved directly from Axiom P-2 and the lemma established in Section 6.1. The proof we give is an adaptation of one given by Legendre.

THEOREM 1

The sum of the measures of the angles of any right triangle is less than 180.

Proof: Let $\triangle ABC$ be a right triangle with right angle at C (Figure 6.19). Through A, draw the line $\overleftrightarrow{AP} \equiv m$ which makes a right angle with line $\overleftrightarrow{AC}$, with P on the B-side of line $\overleftrightarrow{AC}$. By the lemma of Section 4.1, m is parallel to $\overleftrightarrow{CB}$. By Axiom P-2, there is another line n through A parallel to $\overleftrightarrow{CB}$, and one of the rays $\overrightarrow{AQ}$ on n from A lies on the C-side of m.

THEOREM 1 (CONTINUED)

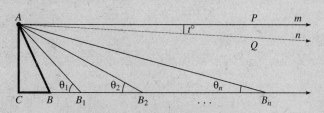

Figure 6.19

(1) We can assume that Q lies on the P-side of $\overleftrightarrow{AC}$, for if not, then Q lies on the P'-side of $\overleftrightarrow{AC}$, where P'-A-P, and we can construct a right triangle $\triangle AB'C$ congruent to $\triangle ABC$ on that side of $\overleftrightarrow{AC}$. The following argument would then apply to $\triangle AB'C$. Set $t = m\angle PAQ$.

(2) $Q \in$ Interior $\angle CAP$, hence $\overrightarrow{AC}$-$\overrightarrow{AQ}$-$\overrightarrow{AP}$.

(3) Construct points $B_1, B_2, \ldots, B_n, \ldots$ on ray $\overrightarrow{CB}$ such that B is the midpoint of $\overline{CB_1}$, B_1 is the midpoint of $\overline{CB_2}$, $\ldots$, B_n is the midpoint of $\overline{CB_{n+1}} \ldots$

(4) Let $\theta_n = m\angle AB_nC$ for each n. You will be asked to prove in Problem 12 that $\theta_1, \theta_2, \ldots, \theta_n, \ldots$ is a decreasing sequence having zero limit (i.e., θ_n becomes arbitrarily small). Thus, for some n, $\theta_n < t$. Consider the corresponding ray $\overrightarrow{AB_n}$.

(5) Now if the order of the rays through A were $\overrightarrow{AC}$-$\overrightarrow{AQ}$-$\overrightarrow{AB_n}$, by the Crossbar Theorem ray $\overrightarrow{AQ}$ would meet segment CB_n. $\rightarrow\leftarrow$ Therefore, the ordering must be $\overrightarrow{AC}$-$\overrightarrow{AB_n}$-$\overrightarrow{AQ}$, and since already $\overrightarrow{AC}$-$\overrightarrow{AQ}$-$\overrightarrow{AP}$, then $\overrightarrow{AC}$-$\overrightarrow{AB_n}$-$\overrightarrow{AQ}$-$\overrightarrow{AP}$.

(6) The sum of the acute angles of $\triangle ACB_n$ may thus be estimated:
$\theta_n + m\angle CAB_n < t + m\angle CAQ = 90$.

(8) Thus, the right triangle $\triangle ACB_n$ has angle sum < 180; by the lemma of Section 6.1, the same is true of $\triangle ACB_{n-1}, \ldots, \triangle ACB_2, \triangle ACB_1$, and $\triangle ACB$.

COROLLARY

The sum of the measures of the angles of any triangle is less than 180.

(To prove this, just drop the perpendicular to the longest side of a triangle from the vertex opposite, forming two right subtriangles, each of whose angle sums is less than 180 by Theorem 1. The given triangle therefore has angle measure less than 180, by algebra.)

AREA IN HYPERBOLIC GEOMETRY: DEFECT OF TRIANGLES AND POLYGONS

One way to use the property just proven, namely, that every triangle has angle sum less than 180, is to define the **defect** of a triangle as the *amount by which the angle sum of a triangle misses the value 180.* That is, we define

$$\delta(\triangle ABC) = 180 - m\angle A - m\angle B - m\angle C$$

The value $\delta \equiv \delta(ABC)$ is called the **defect** of $\triangle ABC$.

Defect may also be defined for polygons, more generally.

DEFINITION

The **defect** of the convex polygon $P_1 P_2 P_3 ... P_n$ is the number

$$\delta(P_1 P_2 P_3 ... P_n) = 180(n-2) - m\angle P_1 - m\angle P_2 - m\angle P_3 - ... - m\angle Pn$$

One of the surprising properties of defect is its additivity. If a convex polygon be subdivided in any manner into convex subpolygons, the sum of the defects of the subpolygons equals that of the original polygon. Thus, all the properties of area outlined in Section 1.3 (except the postulate concerning the area of a unit square, which does not exist in hyperbolic geometry) are valid, as long as we restrict the regions to the family of all convex polygons. Hence, defect defines a perfectly legitimate area function for the hyperbolic plane. We may therefore define, for some constant $k > 0$, the **area** of polygon $P_1 P_2 P_3 ... P_n$ as k *times its defect.* As a special case, the area of $\triangle ABC$ in hyperbolic geometry is given by

$$K = k(180 - m\angle A - m\angle B - m\angle C)$$

The value of k in the previous formula may be determined once a unit for area is agreed upon.

We shall merely prove a special case of the additivity property and leave the rest for problems.

LEMMA

If $\overrightarrow{AD}$ is a cevian of $\triangle ABC$ and δ_1 and δ_2 denote the defects of the subtriangles $\triangle ABD$ and $\triangle ADC$, then

$$\delta(\triangle ABC) = \delta_1 + \delta_2$$

Proof: We have (Figure 6.20)

$$\delta_1 = 180 - m\angle B - m\angle 1 - m\angle 3$$

$$\delta_2 = 180 - m\angle C - m\angle 2 - m\angle 4$$

so that, summing,

$$\delta_1 + \delta_2 = 360 - m\angle B - m\angle 1 - m\angle 3 - m\angle C - m\angle 2 - m\angle 4$$

$$= 360 - (m\angle 1 + m\angle 2) - m\angle B - m\angle C - (m\angle 3 + m\angle 4)$$

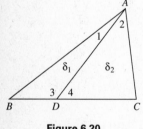

Figure 6.20

$$= 180 - m\angle A - m\angle B - m\angle C$$

$$= \delta(\triangle ABC)$$

EXAMPLE 1   Suppose that the angles shown in Figure 6.21 actually have the values indicated. Calculate the defect of each of the five triangles, and, independent of that, calculate the defect δ of the pentagon using the previous definition. Then test the figure for additivity of defect: Compare δ with $\delta_1 + \delta_2 + \delta_3 + \delta_4 + \delta_5$.

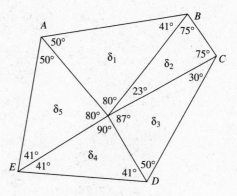

Figure 6.21

SOLUTION

From the figure,

$$\delta_1 = 180 - 50 - 41 - 80 = 9$$

$$\delta_2 = 180 - 75 - 75 - 23 = 7$$

$$\delta_3 = 180 - 87 - 30 - 50 = 13$$

$$\delta_4 = 180 - 90 - 41 - 41 = 8$$

$$\delta_5 = \delta_1 = 9$$

The angles of the pentagon are given by

$$m\angle A = 50 + 50 = 100$$

$$m\angle B = 41 + 75 = 116$$

etc. Hence, the defect δ of the pentagon is

$$180(5 - 2) - 100 - 116 - 105 - 91 - 82 = 540 - 494 = 46$$

The sum of the defects of the five triangles is

$$9 + 7 + 13 + 8 + 9 = 46$$

in agreement. ❑

EXAMPLE 2   In Figure 6.22 is shown a line m parallel to ℓ, with $m\angle ACD = 88$, and right $\triangle ABC$ with acute angles of measures $\frac{1}{2}$ and $89\frac{1}{4}$.

(a) Calculate the area of $\triangle ABC$ using defect.

(b) Assume that point P moves on line ℓ so that $AP \to \infty$, and that $\theta \to 0$ and $\phi \to 0$. Show that the area of $\triangle APC$ is *bounded,* and find a bound for it. (Recall in this connection the comment of Gauss, quoted in Section 6.2.)

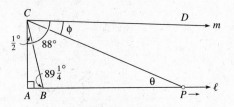

Figure 6.22

SOLUTION

(a) Area $\triangle ABC = k(180 - 90 - 89\frac{1}{4} - \frac{1}{2}) = \frac{1}{4}k$.

(b) Area $\triangle APC = k(180 - 90 - \theta - (88 - \phi)) = k(2 + \phi - \theta)$
$$\leq \lim k(2 + \phi - \theta) = 2k$$

Hence Area $\triangle APC \leq 2k$ (the function Area $\triangle APC$ is increasing as $AP \to \infty$, since A-P-P' implies that Area $\triangle APC <$ Area $\triangle AP'C$).□

NONEXISTENCE OF SIMILAR TRIANGLES IN HYPERBOLIC GEOMETRY

A major consequence of the positive value of defect for all triangles is that it abolishes all possibility of similar triangles or of a concept for similarity transformations in hyperbolic geometry. A dramatic conclusion can be your own discovery if you work out the Discovery Unit which follows.

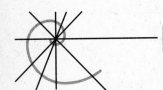

MOMENT FOR DISCOVERY

An Unusual Criterion For Congruence

Suppose that $\triangle ABC$ and $\triangle PQR$ are two triangles having corresponding angles congruent: $\angle A \cong \angle P$, $\angle B \cong \angle Q$, and $\angle C \cong \angle R$. If $AB = PQ$, then the triangles are congruent by ASA. If the triangles are *not* congruent, then one of the triangles has at least two sides which are each of greater length than the corresponding sides in the other, say $AB > PQ$ and $AC > PR$.

1. Construct points D and E on $\overline{AB}$ and $\overline{AC}$ such that $AD = PQ$ and $AE = PR$. Draw $\overline{DE}$ and $\overline{DC}$. Is $\triangle ADE \cong \triangle PQR$?

2. Label the defects of the various triangles as shown in the figure, with δ the defect of $\triangle ABC$. What must be true of δ_1 and δ_4? Of δ_4 and δ?

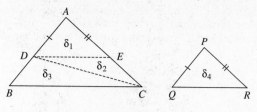

Figure 6.23

3. What is the sum of the defects δ_1 and δ_2? (See lemma in this section.)
4. What equals the sum of the defects δ_1, δ_2, and δ_3?
5. How is the defect of $\triangle ABC$ and δ_1 related? Do you see any contradiction here?
6. What does this prove?

A successful outcome of the preceding Discovery Unit proves the following incredible result in hyperbolic geometry.

> ### THEOREM 2: AAA CONGRUENCE CRITERION FOR HYPERBOLIC GEOMETRY
> If two triangles have the three angles of one congruent, respectively, to the three angles of the other, the triangles are congruent.

COROLLARY

There do not exist any similar, noncongruent triangles in hyperbolic geometry.

PROBLEMS (§6.4)

_____ *Group A* _____

1. A certain equilateral triangle in hyperbolic geometry has angles of measure 55 each. Find its defect.
2. The defect of $\triangle ABC$ is 7.5 and $\triangle ACD \cong \triangle CAB$.
    (a) Find the defect of the convex quadrilateral $\lozenge ABCD$.
    (b) If $\angle B$ is a right angle, find $m\angle BAD$.

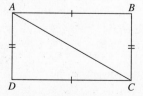

3. Suppose the angles in the figure have the values indicated. Calculate the defects δ, δ_1, δ_2, δ_3, and δ_4 of quadrilateral $\lozenge ABCD$ and the two inner triangles and quadrilaterals shown, respectively, then test the additivity property for defect as in Example 1.

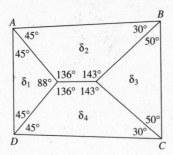

4. Isosceles triangle $\triangle DEF$ is formed from the angle trisectors of the angles of $\triangle ABC$ (accompanying figure). If $m\angle A = m\angle B = 66$, $m\angle C = 27$, and the rest of the angles are as marked, find the defect of $\triangle DEF$ in two different ways.

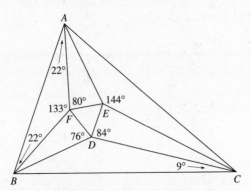

5. Write out the details of the proof of the corollary to Theorem 1.

6. The acute angle of a certain Lambert Quadrilateral has measure 83. Find the defect of the quadrilateral. What is the maximum defect of any Lambert Quadrilateral, and what has to happen if a Lambert Quadrilateral has nearly the maximum defect?

7. The summit angles of a certain Saccheri Quadrilateral each have measure 83. Find the defect of the quadrilateral. (Compare with Problem 6; why should your answer to this problem be precisely twice that of Problem 6?)

8. The defect of a certain regular hexagon in hyperbolic geometry is 12.

   (a) Find the measure of each angle of the hexagon.

   (b) If O is the center of the hexagon, find the measure of each subtriangle making up the hexagon, such as $\triangle ABO$ shown in the figure.

   (c) Are each of these subtriangles equilateral triangles, as they would be if the geometry were Euclidean?

9. The defect of a certain regular dodecagon in hyperbolic geometry is 12. If O is the center, find the measure of

   (a) each angle of the polygon

   (b) each angle of the subtriangle $\triangle AOB$, where O is the center of the polygon (in the figure at the right).

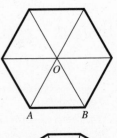

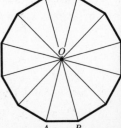

10. In the figure shown at the right, the two nontriangular polygons are regular polygons and the defects of the three polygons are as indicated. Find the measures of the angles of the triangle.

11. In the figure shown below, the two nontriangular polygons are regular polygons, and the various defects are as indicated. Find the measures of the angles of the triangle.

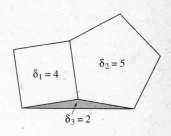

$\delta_1 = 4$ $\delta_2 = 5$ $\delta_3 = 2$

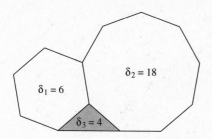

$\delta_2 = 18$
$\delta_1 = 6$
$\delta_3 = 4$

_____ **Group B** _____

12. The proof of Theorem 1 requires that we find a single point Q on ray $\overrightarrow{CB}$ such that $m\angle AQC < t$, since, for large n, B_n will fall beyond Q (i.e., C-Q-B_n) and $m\angle AB_nC < m\angle AQC < t$ by the Exterior Angle Inequality. In the figure, locate points $Q_1, Q_2, Q_3, \dots$ on ray $\overrightarrow{CB}$ such that

$$BQ_1 = BA, \quad Q_1Q_2 = Q_1A, \quad Q_2Q_3 = Q_2A, \dots$$

If $t_n = m\angle AQ_nC$, show that for some n, $t_n < t$, or if $Q = Q_n$, $m\angle AQC < t$, as desired.

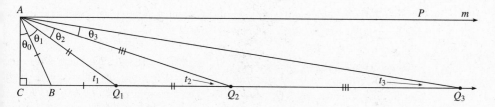

13. Two regular pentagons have equal defects. Show they must be congruent.

14. At point A, three congruent isosceles triangles $\triangle APQ$, $\triangle AQR$, and $\triangle ARP$ are constructed with legs on the fixed lines ℓ, m, and n. Suppose that AP, AQ, and AR become infinite.

   (a) Assuming that $m\angle APQ \to 0$ as $AP \to \infty$, what happens to the defect of $\triangle PQR$? the area of $\triangle PQR$?

   (b) What seems to be the area of the entire hyperbolic plane from this? (See Problems 16 and 17.)

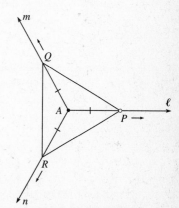

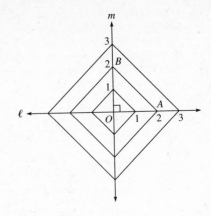

15. Lines ℓ and m are perpendicular at point O, and points at intervals of one unit apart are constructed on each of the four outgoing rays. If we draw in the regular quadrilaterals shown, then determine the answers (with proofs) to these questions:

   (a) Are the angles of any of the resulting quadrilaterals right angles?

   (b) Assuming that the angles of the sequence of quadrilaterals become arbitrarily small, what seems to be the area of the hyperbolic plane from this? (See Problems 16 and 17.)

——— **Group C** ———

16. Prove in Problems 14 and 15 that there are points in the hyperbolic plane not covered by any member of the sequence of triangles or quadrilaterals. (**HINT:** Think about Legendre's proof that $\delta(\triangle ABC) = 0$ for all triangles, which contradicts Axiom P-2.)

17. Prove that the area of the hyperbolic plane is infinite. (**HINT:** Think tiling, *using regular pentagons, not squares or hexagons.* See Problem 14, Section 6.7.)

18. In the figure at the right, triangle $\triangle ABC$ is an equilateral triangle in the hyperbolic plane, with defect $\delta(\triangle ABC) = 180 - 3\theta$, where $m\angle A = \theta$. If L, M, and N are the midpoints of the sides of $\triangle ABC$, and $m\angle LMN = \phi$, show that $\phi > \theta$ and $\delta(\triangle AMN) = \phi - \theta$.

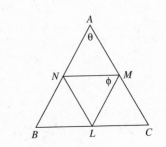

*6.5   HYPERBOLIC GEOMETRY 2: ASYMPTOTIC TRIANGLES

We continue the development of hyperbolic geometry and explore here, and in the next section, the complete theory of parallelism.

ASYMPTOTIC PARALLELISM BETWEEN LINES AND RAYS

A concept which Bolyai used will greatly simplify our task. We start with any angle $\angle ABC$, which can be obtuse, right, or acute (Figure 6.24). Take any point Q on ray $\vec{BC}$ and consider ray $\vec{AQ}$ as Q varies. Thus, various rays $\vec{AC}, \vec{AD}, \vec{AE}, \dots$ are determined, which define (with ray $\vec{AB}$) a variable angle having measure $\theta = m\angle BAQ$. Consider the set of real numbers defined by θ

$$\{\theta\} \equiv \{m\angle BAQ : Q \in \vec{BC}\}$$

Since the sum of the measures of two angles of a triangle is < 180, $\theta < 180 - m\angle B$. Hence this set is bounded and will have a positive least upper bound

$$\gamma = \sup\{\theta\} \leq 180 - m\angle B$$

Notice that the set $\{\theta\}$ has the property that for any given member of this set, there exists another member that is larger. For example, in Figure 6.24, if D is constructed on ray $\vec{BC}$ so that B-Q-D, then $\theta' = m\angle BAD > \theta$. Thus, the least upper bound γ of $\{\theta\}$ cannot belong to $\{\theta\}$. In Euclidean geometry, $\gamma = 180 - m\angle B$, but in hyperbolic geometry, $\gamma < 180 - m\angle B$, as we shall ultimately discover.

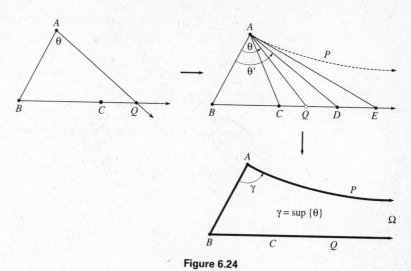

Figure 6.24

Suppose we now construct the unique ray $\overrightarrow{AP}$ on the C-side of $\overleftrightarrow{AB}$ such that

$$m\angle BAP = \gamma$$

Since $\gamma \notin \{\theta\}$, ray $\overrightarrow{AP}$ cannot meet $\overrightarrow{BC}$, and thus we obtain in Figure 6.24 a configuration consisting of two rays $\overrightarrow{AP}$ and $\overrightarrow{BC} = \overrightarrow{BQ}$ lying on parallel lines, and a segment, $\overline{AB}$. This configuration resembles a triangle having one vertex "at infinity," and is called an **asymptotic triangle** (a formal definition will be given momentarily). Normally a Greek character Ω (omega) is used to represent the missing vertex. This is merely for convenience, and it must be remembered that Ω *does not represent an ordinary point.* Some authors refer to it as an "ideal point." To think of it as a "point at infinity" may help your intuition, but care must be used, obviously.

To make the relationship between the rays $\overrightarrow{AP}$ and $\overrightarrow{BQ}$ easier to deal with, we give a descriptive definition that does not depend on the idea of a least upper bound. It can be seen that the rays $\overrightarrow{AP}$ and $\overrightarrow{BQ}$ just constructed will satisfy these properties.

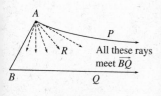

Figure 6.25

DEFINITION

A ray $\overrightarrow{AP}$ is **asymptotically parallel** to ray $\overrightarrow{BQ}$, denoted by $\overrightarrow{AP} \parallel_a \overrightarrow{BQ}$, iff (Figure 6.25)

(1) $\overrightarrow{AP}$ and $\overrightarrow{BQ}$ lie on the same side of line $\overleftrightarrow{AB}$

(2) $\overrightarrow{AP}$ does not meet $\overrightarrow{BQ}$

(3) every ray $\overrightarrow{AR}$ interior to $\angle PAB$ meets ray $\overrightarrow{BQ}$.

Further, line ℓ is **asymptotically parallel** to line m iff ℓ contains a ray $\overrightarrow{AP}$ that is asymptotically parallel to a ray $\overrightarrow{BQ}$ contained by m. (This relationship will similarly be denoted by $\ell \parallel_a m$.)

It turns out that asymptotic parallelism obeys the same fundamental laws that Euclidean parallelism does, like symmetry and transitivity. These properties will be taken up in due course.

PROPERTIES OF ASYMPTOTIC TRIANGLES

We begin with the definition for asymptotic triangles.

> **DEFINITION**
>
> An **asymptotic triangle** is the set of points consisting of an ordinary line segment $\overline{AB}$, called its **base,** and two rays $\overrightarrow{AP}$ and $\overrightarrow{BQ}$, called its **legs,** such that $\overrightarrow{AP} \parallel_a \overrightarrow{BQ}$. We denote such a figure by the symbol $\triangle AB\Omega$. The **base angles** of $\triangle AB\Omega$ are the angles at A and B, and the **angle sum** is just $m\angle A + m\angle B$. A **right asymptotic triangle** is an asymptotic triangle having a right angle, and an **isosceles asymptotic triangle** is one whose base angles are congruent (Figure 6.26).

ASYMPTOTIC TRIANGLES

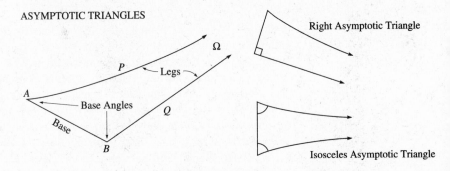

Figure 6.26

As before, we can think of the third vertex of an asymptotic triangle as being a "point at infinity" designated by Ω, and we can take the measure of that third angle as zero. Such imagery can help us remember the correct relationships involved, but again, care must be used.

An asymptotic triangle thus has three "sides," but only two real angles—the *base angles* $\angle A$ and $\angle B$ in Figure 6.26. The only side whose length is defined is the base $\overline{AB}$. Thus, the *measurable parts of an asymptotic triangle* $\triangle AB\Omega$ *are*

$$\overline{AB} \;\; \rightarrow \;\; AB$$

$$\angle A \;\; \rightarrow \;\; m\angle A$$

$$\angle B \;\; \rightarrow \;\; m\angle B$$

Asymptotic triangles can therefore come in all sizes and shapes. We will say that two asymptotic triangles are **congruent** iff under some correspondence the measurable corresponding parts of the two triangles are congruent.

Asymptotic triangles behave much the way ordinary triangles do in Euclidean geometry. In fact, every property of a Euclidean triangle $\triangle ABC$ that does not involve vertex C corresponds to some valid property of asymptotic triangle $\triangle AB\Omega$ in the hyperbolic plane, including the fact

$$m\angle A + m\angle B < 180$$

as we shall see. First, we prove a fundamental property which will be used frequently.

LEMMA A

If $\overrightarrow{AC} \parallel_a \overrightarrow{BD}$, A-K-C, and B-M-D, then $\overrightarrow{KC} \parallel_a \overrightarrow{MD}$.

Proof: (See Figure 6.27.) We must prove that the three conditions in the definition for asymptotic parallelism are fulfilled.

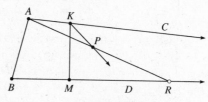

Figure 6.27

(1) Since $\overrightarrow{AC}$ and $\overrightarrow{BD}$ lie on the same side of $\overleftrightarrow{AB}$, so do K and M. One can apply the Theorem of Pasch to show that A and B cannot lie on opposite sides of $\overleftrightarrow{KM}$, hence they lie on the same side. Thus, since A-K-C and B-M-D, both C and D lie on the opposite side of $\overleftrightarrow{KM}$ as A and B, hence $\overrightarrow{KC}$ and $\overrightarrow{MD}$ lie on the same side of line $\overleftrightarrow{KM}$.

(2) Since ray $\overrightarrow{AC}$ is a nonintersector of $\overrightarrow{BD}$ and $\overrightarrow{KC} \subseteq \overrightarrow{AC}$, ray $\overrightarrow{KC}$ does not meet $\overrightarrow{MD}$.

(3) Let ray $\overrightarrow{KP}$ lie in the interior of $\angle CKM$. Then P lies on the B-side of $\overleftrightarrow{AC}$ and on the C-side of $\overleftrightarrow{KM}$, hence also on the C-side of $\overleftrightarrow{AB}$, or $P \in$ Interior $\angle BAC$. Hence ray $\overrightarrow{AP} \subseteq$ Interior $\angle BAC$, and since $\overrightarrow{AC} \parallel_a \overrightarrow{BD}$, $\overrightarrow{AP}$ meets $\overrightarrow{BD}$ at some point R. The Postulate of Pasch applied to $\triangle ABR$ then implies that $\overrightarrow{KP}$ meets segment $\overline{BR}$ and thus ray $\overrightarrow{MD}$.

LEMMA B

If $\overrightarrow{KC} \parallel_a \overrightarrow{MD}$, A-K-C, and B-M-D, then $\overrightarrow{AC} \parallel_a \overrightarrow{BD}$.

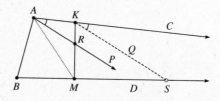

Figure 6.28

Proof: (See Figure 6.28.) As before, we can prove that $\overrightarrow{AC}$ and $\overrightarrow{BD}$ lie on the same side of line $\overleftrightarrow{AB}$, and that $\overrightarrow{AC}$ is a nonintersector of $\overrightarrow{BD}$. It remains to prove that every ray $\overrightarrow{AP} \subseteq$ Interior $\angle BAC$ meets ray $\overrightarrow{BD}$. By the Crossbar Theorem, there is nothing to prove unless we assume that $\overrightarrow{AP} \subseteq$ Interior $\angle MAC$. Therefore, by the Crossbar Theorem, $\overrightarrow{AP}$ meets segment $\overline{KM}$ at some point R. By the Exterior Angle Inequality, $m\angle CKM > m\angle KAM > m\angle CAP$, so we may construct ray $\overrightarrow{KQ}$ interior to $\angle CKM$ such that $m\angle CKQ = m\angle CAP$. Since $\overrightarrow{KC} \parallel_a \overrightarrow{MD}$, $\overrightarrow{KQ}$ meets $\overrightarrow{MD}$ at some point S. Lines $\overleftrightarrow{KQ}$ and $\overleftrightarrow{AP}$ are parallel, so by the Theorem of Pasch, $\overrightarrow{AP}$ must meet side $\overline{MS}$ of $\triangle KMS$ and hence, ray $\overrightarrow{BD}$, finishing the proof.

SYMMETRY OF ASYMPTOTIC PARALLELISM

The next result is a prelude to the Symmetry Law for Asymptotic Parallelism.

LEMMA C

If an asymptotic triangle $\triangle AB\Omega$ has $\overrightarrow{AC} \parallel_a \overrightarrow{BD}$ and is isosceles, then $\overrightarrow{BD} \parallel_a \overrightarrow{AC}$.

Proof: We already know that $\overrightarrow{BD}$ and $\overrightarrow{AC}$ lie on the same side of $\overleftrightarrow{AB}$ and do not meet, so it remains to show that every ray $\overrightarrow{BP} \subseteq$ Interior $\angle ABD$ meets ray $\overrightarrow{AC}$ (Figure 6.29). Construct ray $\overrightarrow{AR}$ in the interior of $\angle BAC$ such that $\angle BAR \cong \angle ABP$. Since $\overrightarrow{AC} \parallel_a \overrightarrow{BD}$, ray $\overrightarrow{AR}$ meets $\overrightarrow{BD}$ at some point Q, and we may construct S on $\overrightarrow{AC}$ such that $AS = BQ$. Since $\triangle AB\Omega$ is isosceles, $\angle BAC \cong \angle ABD$. By SAS, $\triangle SAB \cong \triangle QBA$. But, by CPCF, $m\angle ABS = m\angle BAR = m\angle ABP$, so ray $\overrightarrow{BP}$ coincides with ray $\overrightarrow{BS}$ and $\overrightarrow{BP}$ meets $\overrightarrow{AC}$, as desired.

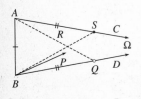

Figure 6.29

THEOREM 1: SYMMETRY LAW FOR ASYMPTOTIC PARALLELISM

In any asymptotic triangle $\triangle AB\Omega$, if $\overrightarrow{AC} \parallel_a \overrightarrow{BD}$, then $\overrightarrow{BD} \parallel_a \overrightarrow{AD}$, and each leg is asymptotically parallel to the other.

Proof: We must reduce the general situation to that of Lemma C. Let $\overrightarrow{AC} \parallel_a \overrightarrow{BD}$ as in Figure 6.30, and construct the angle bisectors of the base angles $\angle A$ and $\angle B$, meeting at a point P (since the angle bisector of $\angle A$ meets ray $\overrightarrow{BD}$).

(1) Now P is easily seen to be equidistant from $\overrightarrow{AC}$, $\overrightarrow{AB}$, and $\overrightarrow{BD}$.[4] Drop perpendiculars $\overline{PQ}$, $\overline{PS}$, and $\overline{PR}$ to lines $\overleftrightarrow{AC}$, $\overleftrightarrow{AB}$, and $\overleftrightarrow{BD}$, respectively. Then $PQ = PS = PR$. (By the Exterior Angle Inequality, Q and R will lie on the rays $\overrightarrow{AC}$ and $\overrightarrow{BD}$ as illustrated, so we may assume that A-Q-C and B-R-D, and we shall assume $\overrightarrow{QP}$-$\overrightarrow{QR}$-$\overrightarrow{QC}$ holds. There are actually two other logical cases besides that illustrated, namely, $\overrightarrow{QR}$-$\overrightarrow{QP}$-$\overrightarrow{QC}$ and $QR = QP$; these details will be left for you to work out.)

(2) Since $\triangle PQR$ is isosceles, $m\angle RQC = 90 - m\angle PQR = 90 - m\angle PRQ = m\angle QRD$.

(3) By Lemma A, $\overrightarrow{QC} \parallel_a \overrightarrow{RD}$, so $\triangle QR\Omega$ is an isosceles asymptotic triangle.

(4) $\overrightarrow{RD} \parallel_a \overrightarrow{QC}$ (Lemma C) and therefore, by Lemma B, $\overrightarrow{BD} \parallel_a \overrightarrow{AC}$.

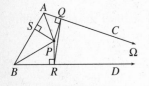

Figure 6.30

[4]A general property of the angle bisector of an angle in absolute geometry (see Problem 11, §3.6)

CONGRUENCE CRITERION FOR ASYMPTOTIC TRIANGLES

The first congruence criteria for asymptotic triangles will be presented as a Discovery Unit.

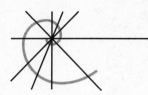

MOMENT FOR DISCOVERY

Asymptotic Triangles and Congruence

Consider two asymptotic triangles, as shown in Figure 6.31. Suppose that $AB \cong CD$ and $\angle B \cong \angle D$, as indicated. Does it seem that $\angle A$ ought to be congruent to $\angle C$? Let's analyze this carefully.

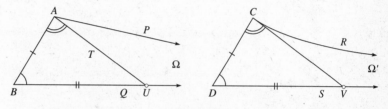

Figure 6.31

1. Assume, for sake of argument, that $m\angle A > m\angle C$.
2. Can you construct $\angle C$ inside $\angle BAP$? (Use ray $\overrightarrow{AT}$ for this purpose.)
3. What must be true of rays $\overrightarrow{AT}$ and $\overrightarrow{BQ}$?
4. Construct $\triangle CDV \cong \triangle ABU$. What must be true of rays $\overrightarrow{CR}$ and $\overrightarrow{CV}$?
5. Can you gain a contradiction? What have you shown?

If you were successful in the reasoning experiment outlined in the preceding Discovery Unit, then you have proven

THEOREM 2 BA CONGRUENCE CRITERION

If two asymptotic triangles have congruent bases and a pair of congruent base angles, the triangles are congruent (that is, the remaining pair of base angles are congruent).

COROLLARY: CONGRUENCE CRITERION FOR RIGHT TRIANGLES

If two right asymptotic triangles have congruent bases, the triangles are congruent.

There is also the AA Congruence Criterion, but it depends on the transitivity property of asymptotic parallelism (proved in the next section), so we merely state it for later reference.

AA CONGRUENCE CRITERION

If two asymptotic triangles have the base angles of one congruent, respectively, to the base angles of the other, then the triangles have congruent bases, and the triangles are congruent.

INEQUALITIES RELATED TO ASYMPTOTIC TRIANGLES

It may surprise you to learn that so far in this section *we have not explicitly used the Postulate of Parallels for hyperbolic geometry!* Everything we have done thus far is valid and meaningful in Euclidean geometry, even though some of the preceding ideas reduce to trivia. However, we are about to invoke the full power of Axiom P-2 for the first time.

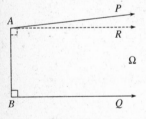

Figure 6.32

THEOREM 3

A right asymptotic triangle can have but one right angle, and the other base angle is acute.

Proof: Suppose in Figure 6.32 that $\triangle AB\Omega$ is an asymptotic triangle with right angle at B.

(1) First suppose that $m\angle A > 90$. In this case, construct ray $\overrightarrow{AR}$ such that $m\angle BAR = 90$ and $\overrightarrow{AB}\text{-}\overrightarrow{AR}\text{-}\overrightarrow{AP}$ by the Angle Construction Theorem. Since $\overrightarrow{AR} \subseteq$ Interior $\angle PAB$, ray $\overrightarrow{AR}$ meets $\overrightarrow{BQ}$.

(2) But lines perpendicular to the same line in absolute geometry are parallel (by the lemma in Section 4.1). $\rightarrow\leftarrow$ (We still have not used Axiom P-2.)

(3) Suppose $m\angle A = 90$. By constructing ray $\overrightarrow{BQ'}$ opposite ray $\overrightarrow{BQ}$ and ray $\overrightarrow{AP'} \parallel_a \overrightarrow{BQ'}$, as shown in Figure 6.33, we obtain another right asymptotic triangle, $\triangle AB\Omega'$.

(4) $\triangle AB\Omega' \cong \triangle AB\Omega$ by the BA Congruence Criterion, so $m\angle P'AB = 90$ and P', A, and P are collinear points, lying on a line $m_1 \parallel_a \overrightarrow{BQ} = \ell$.

(5) By Axiom P-2, there exists a second line m_2 through A parallel to ℓ. Taking points R-A-S on m_2, either R or S (say S) lies on the B-side of $\overrightarrow{AP}$ and on either the Q-side (as shown) or the Q'-side of $\overleftrightarrow{AB}$. Hence ray $\overrightarrow{AS}$ lies either in the interior of $\angle BAP$ or of $\angle BAP'$, and in either case, $\overrightarrow{AS}$ meets $\overrightarrow{BQ}$ or $\overrightarrow{BQ'}$. That is, m_2 meets ℓ. $\rightarrow\leftarrow$

(6) $\therefore$ $m\angle A < 90$ (the only remaining possibility).

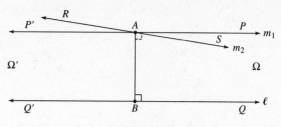

Figure 6.33

The next two results make good problems. However, the Exterior Angle Inequality, analogous to that for ordinary triangles, is a simple consequence of the Discovery Unit which follows.

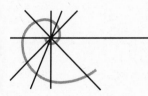

MOMENT FOR DISCOVERY

Angle Sum Theorem for Asymptotic Triangles

In Figure 6.34 are shown two asymptotic triangles, with

(a) $m\angle A + m\angle B > 180$

(b) $m\angle A + m\angle B \equiv m\angle 1 + m\angle 2 = 180$ (if possible)

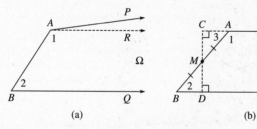

Figure 6.34

In (a), we have shown the construction of a ray $\overrightarrow{AR}$ such that $m\angle 1 + m\angle 2 = 180$ (which makes $m\angle 1 < m\angle BAP$), and in (b), perpendiculars have been constructed from the midpoint M of $\overline{AB}$ to lines $\overrightarrow{AP}$ and $\overrightarrow{BQ}$, respectively.

1. In (a), why is $\overrightarrow{AR} \parallel \overrightarrow{BQ}$? (You do not need the Euclidean Parallel Postulate here.)

2. Since $\overrightarrow{AP} \parallel_a \overrightarrow{BQ}$, what must happen? Did you get a contradiction?

3. In (b), why is $\angle 2 \cong \angle 3$? Can you conclude that $\triangle AMC \cong \triangle BMD$?

4. What theorem guarantees that C, M, and D are collinear? (See Problem 11 in Section 2.8.)

5. Is $\triangle CD\Omega$ an asymptotic triangle? Why? What previous theorem is violated?

6. Have you reached any conclusions?

If you were successful with the preceding Discovery Unit, then you have proven

THEOREM 4

The angle sum of any asymptotic triangle is less than 180.

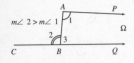

Figure 6.35

COROLLARY

The measure of an exterior angle of an asymptotic triangle is greater than that of the opposite base angle. (See Figure 6.35.)

PROBLEMS   (§6.5)

_____ *Group A* _____

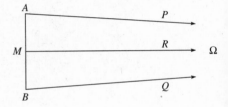

1. Given an isosceles asymptotic triangle $\triangle AB\Omega$ and the midpoint M of $\overline{AB}$, suppose that ray $\overrightarrow{MR}$ is asymptotically parallel to both $\overrightarrow{AP}$ and $\overrightarrow{BQ}$. Prove that $\overrightarrow{MR} \perp \overline{AB}$. (See the figure at the right.)

2. Given an isosceles asymptotic triangle $\triangle AB\Omega$ and M some point on base $\overline{AB}$, suppose that ray $\overrightarrow{MR}$ is asymptotically parallel to both legs of $\triangle AB\Omega$, $\overrightarrow{AP}$ and $\overrightarrow{BQ}$, and that $\overrightarrow{MR} \perp \overline{AB}$. Prove that M is the midpoint of $\overline{AB}$.

3. Prove the Corollary to Theorem 4 (that $m\angle 2 > m\angle 1$ in Figure 6.35).

4. Suppose that A-B-C-D and that rays $\overrightarrow{AP}$, $\overrightarrow{BQ}$, $\overrightarrow{CR}$, and $\overrightarrow{DS}$ are asymptotically parallel to each other. Explain why the measures of the angles at A, B, and C are in increasing order.

_____ *Group B* _____

5. Prove that the perpendicular bisector of the base of an isosceles asymptotic triangle "passes through" Ω (that is, it is asymptotically parallel to the legs).

6. Prove that in asymptotic $\triangle AB\Omega$, as shown in the following figure, if line ℓ intersects ray $\overrightarrow{AP}$ at point D and does not pass through A or B, then either ℓ meets $\overline{AB}$ at an interior point E or ray $\overrightarrow{BQ}$ at an interior point F. (***HINT:*** Draw $\overrightarrow{BD}$.)

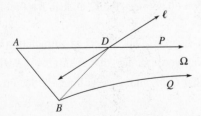

The next three problems, and Problem 10, make use of the Transitive Law for Asymptotic Parallelism to be proven in the next section.

7. **Postulate of Pasch for Asymptotic Triangles** If a line ℓ meets one of the sides of an asymptotic triangle $\triangle AB\Omega$ at an interior point D and does not pass through A, B, or Ω, then ℓ must meet one of the other two sides at an interior point E or F. (**HINT:** See Problem 6 for one case. For the other case, construct $\overrightarrow{DR} \parallel_a \overrightarrow{BQ}$ and use Transitivity of Asymptotic Parallelism.)

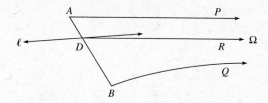

8. Prove the AA Congruence Criterion for Asymptotic Triangles. (**HINT:** In the following figure, assume $AB < CD$ and construct $\overline{DE} \cong \overline{AB}$ on $\overline{CD}$ and $\angle DEF \cong \angle A$.)

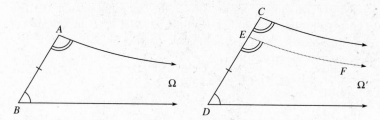

9. Explicitly prove that the angle bisectors of the base angles of an asymptotic triangle meet in a point that is equidistant from the three sides.

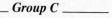 **Group C**

10. Given asymptotic triangle $\triangle AB\Omega$, find a construction for point C on $\overrightarrow{AB}$ and ray $\overrightarrow{CR}$ asymptotically parallel to both $\overrightarrow{AP}$ and $\overrightarrow{BQ}$. (**HINT:** Study carefully the proof of Theorem 1.) (See the accompanying figure.)

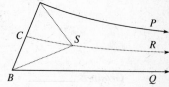

11. In ordinary triangles $\triangle ABC$ and $\triangle DEF$, if $\angle B \cong \angle E$ and $\overline{BC} \cong \overline{EF}$ but $AB < DE$, then $m\angle A > m\angle D$, a result provable in absolute geometry. State and prove the direct analogue for this for asymptotic triangles (where the condition $\overline{BC} \cong \overline{EF}$ is deleted, since $\overrightarrow{BC}$ and $\overrightarrow{EF}$ are rays with infinite lengths).

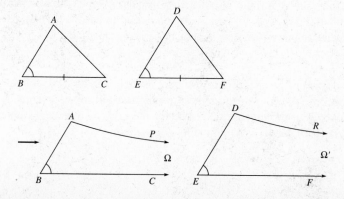

*6.6   HYPERBOLIC GEOMETRY 3: THEORY OF PARALLELS

As mentioned earlier, in hyperbolic geometry there are two distinct types of parallelism, unlike the situation for Euclidean geometry. One, asymptotic parallelism, was introduced and developed in the preceding section. The other will be introduced here shortly. We begin with a point A and line ℓ not passing through A (Figure 6.36).

THREE CLASSES OF LINES THROUGH POINT A

Consider all the lines in the plane which pass through point A. Some of these, like the perpendicular $\overrightarrow{AB}$ to line ℓ, will intersect ℓ (called **intersectors**), and the rest are the **nonintersectors**, the lines which are parallel to ℓ. (The same symbol for parallelism as was used for Euclidean geometry can be used here if we are not concerned about the type of parallelism.) As we saw in the last section, as a line through point A pivots about A in, say, the counterclockwise direction from ray $\overrightarrow{AB}$, we come to a "first" parallel m_1 to line ℓ. As in the preceding section, this line will be said to be **asymptotically parallel**, or **limit parallel**, to ℓ on this side of $\overrightarrow{AB}$; symmetric to this, there is a second asymptotic parallel m_2 to line ℓ on the other side of $\overrightarrow{AB}$. As before, we write

$$m_1 \parallel_a \ell \quad \text{and} \quad m_2 \parallel_a \ell$$

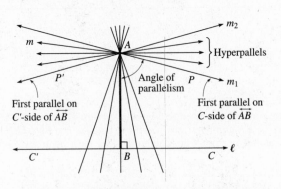

Figure 6.36

However, there are an infinity of other lines through A which are also parallel to ℓ, such as line m in Figure 6.36. Any such line is said to be **divergently parallel**, or **hyperparallel**, to line ℓ (for reasons soon to be made clear), and we write

$$m \parallel_d \ell$$

The two types of parallels are as different as night and day. We contrast the different properties for convenience; these properties will all be eventually proven here or covered in the problems. Except for the Law of Symmetry, note that all other properties of parallelism are in direct opposition.

PROPERTIES OF PARALLELISM IN HYPERBOLIC GEOMETRY

Asymptotic Parallelism	**Divergent Parallelism**
The Law of Symmetry holds (if $\ell \parallel_a m$, then $m \parallel_a \ell$).	The Law of Symmetry holds (if $\ell \parallel_d m$, then $m \parallel_d \ell$).
The Transitive Law holds for asymptotic parallelism in the same direction.	There is no Transitive Law for divergent parallelism.
If $\ell \parallel_a m$, lines ℓ and m have no common perpendicular.	If $\ell \parallel_d m$, lines ℓ and m possess a common perpendicular, and conversely.
If $\ell \parallel_a m$ at A, the perpendicular distance from point P on ℓ to line m decreases to zero as $AP \to \infty$.	If $\ell \parallel_d m$ at A, the perpendicular distance from point P on ℓ to line m increases to ∞ as $AP \to \infty$.

ASYMPTOTIC PARALLELISM, ANGLE OF PARALLELISM

In Figure 6.36, $\angle BAP$ and $\angle BAP'$ are called the **angles of parallelism**. (It is customary also to refer to the *measures* of these angles as "angles of parallelism," the meaning always being clear by context.) Up to congruence, these angles are uniquely determined by the perpendicular distance AB (hence $m\angle BAP = m\angle BAP'$). This is due to the BA Congruence Criterion for Asymptotic Triangles proved in the previous section. In fact, as the distance AB *decreases*, the angle of parallelism *increases*, as shown in Figure 6.37. (See Problem 2.) Thus we can introduce a one-to-one function $p(a)$ which is defined as the measure of the angle of parallelism $\angle BAP$ in terms of the distance $a = AB$. The inverse function $p^{-1}(\theta)$ gives the distance AB corresponding to a given angle of parallelism $\theta = m\angle BAP$.

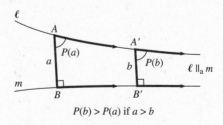

$P(b) > P(a)$ if $a > b$

Figure 6.37

It can be shown that as a varies over all positive reals, $p(a)$ ranges from 90 to zero, exclusive. This brings out a dramatically non-Euclidean feature of hyperbolic geometry: Given a positive real number $\theta < 90$, say $\theta = 1$ (degree) as shown in Figure 6.38, there exists a corresponding distance $AB = p^{-1}(\theta)$ such that the angle of parallelism at A is the given number θ!

Bolyai and Lobachevski independently derived the same formula for $p(a)$. A derivation may be found in Sommerville's *Elements of Non-Euclidean Geometry*, pp. 60–61, along with all the formulas for hyperbolic trigonometry, pp. 66–67.

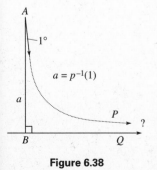

Figure 6.38

> **BOLYAI-LOBACHEVSKI FORMULA FOR ANGLE OF PARALLELISM**
>
> **(1)** $\tan \frac{1}{2}A = e^{-ka}$,
>
> where k is some constant determined by the unit of measure

A ramification of **(1)** is that a natural unit for length exists, unlike the situation in Euclidean geometry where there is no link between angle measure and distance. For example, we could choose as the unit of length the segment corresponding to the angle of parallelism of measure 45—the result of bisecting a right angle. (This means that in **(1)** we set $A = 45$ (degrees) and $a = 1$, and this will determine the constant k.) If, on the other hand, we take $k = 1$ in order to achieve simplicity, then this amounts to choosing as unit of measure ($a = 1$) the segment corresponding to an angle of parallelism having measure $A = 2 \cdot \tan^{-1}(1/e) \approx 20.2$ (degrees).

OUR GEOMETRIC WORLD

In our supposedly Euclidean world, we may change the unit of distance by simply mulitplying all distances by a constant. A triangle three times the size of a given right triangle is still a right triangle. In hyperbolic geometry, however, a change of scale also involves a change in angle measure. A triangle three times the size of a given right triangle is no longer a right triangle. (This follows from the formulas of hyperbolic trigonometry given in the next set of problems. In particular, see Problem 14.)

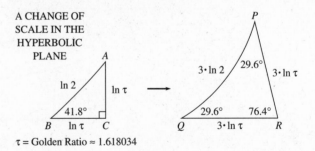

A CHANGE OF
SCALE IN THE
HYPERBOLIC
PLANE

$\tau = $ Golden Ratio ≈ 1.618034

Figure 6.39

E X A M P L E   1   Find the value of k necessary in **(1)** if the angle of parallelism corresponding to the unit of length has measure 80. Then use this value to determine the measure of the angle of parallelism corresponding to a distance of 10 units.

SOLUTION

Set $A = 80°$ and $a = 1$ in **(1)**. Thus

$$\tan 40° = e^{-k \cdot 1}$$

$$\cot 40° = e^k$$

$$k = \ln(\cot 40°)$$

$$\therefore k \approx 0.1754258$$

The formula **(1)** becomes

$$\tan \tfrac{1}{2}A \approx e^{-(0.1754258)a}$$

Setting $a = 10$, we obtain

$$\tan \tfrac{1}{2}A \approx e^{-1.754258} \approx 0.1730356$$

$$\therefore A \approx 19.634°\square$$

TRANSITIVITY OF ASYMPTOTIC PARALLELISM

In Figure 6.40 is illustrated the Transitive Law for Asymptotic Parallelism. We are given three distinct points A, B, and C on some line, and we determine rays r_1, r_2, and r_3 such that

$$r_1 \parallel_a r_2 \quad \text{and} \quad r_2 \parallel_a r_3$$

Then it is to be established that $r_1 \parallel_a r_3$, regardless of the ordering of the points A, B, and C.

LEMMA A

If $A\text{-}C\text{-}B$ and $\overrightarrow{AP} \parallel_a \overrightarrow{BQ}$, $\overrightarrow{BQ} \parallel_a \overrightarrow{CR}$, then $\overrightarrow{AP} \parallel_a \overrightarrow{CR}$.

Proof: In Figure 6.40, we must prove that $\overrightarrow{AX}$ meets $\overrightarrow{CR}$, but that $\overrightarrow{AP}$ does not.

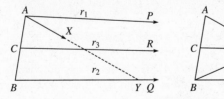

Figure 6.40

(1) Since $\overrightarrow{AP} \parallel_a \overrightarrow{BQ}$, $\overrightarrow{AX}$ meets $\overrightarrow{BQ}$ at some point Y; since $\overrightarrow{BY}$ and $\overrightarrow{CR}$ are non-intersectors, $\overrightarrow{CR}$ does not meet segment $\overline{BY}$, hence it must meet $\overline{AY}$ (Postulate of Pasch). Thus, $\overrightarrow{AX}$ and $\overrightarrow{CR}$ meet.

(2) Suppose $\overrightarrow{AP}$ meets $\overrightarrow{CR}$ at S; we may assume that $A\text{-}S\text{-}P$ and $C\text{-}S\text{-}R$. Then $P \in$ Interior $\angle BSR$ and $\overrightarrow{SB}\text{-}\overrightarrow{SP}\text{-}\overrightarrow{SR}$.

(3) By the law of symmetry, $\overrightarrow{CR} \parallel_a \overrightarrow{BQ}$, and by Lemma A of §6.4, $\overrightarrow{SR} \parallel_a \overrightarrow{BQ}$.

(4) Hence ray $\overrightarrow{SP}$ meets ray $\overrightarrow{BQ}$. That is, ray $\overrightarrow{AP}$ meets $\overrightarrow{BQ}$. →←

(5) The contradiction proves ray $\overrightarrow{AP}$ does not meet $\overrightarrow{CR}$.

COROLLARY

If B-A-C and $\overrightarrow{AP} \parallel_a \overrightarrow{BQ}$, $\overrightarrow{BQ} \parallel_a \overrightarrow{CR}$, then $\overrightarrow{AP} \parallel_a \overrightarrow{CR}$.

(The proof of this case reduces to that of Lemma A via the Symmetry Law proven in Section 6.4.)

LEMMA B

If A-B-C and $\overrightarrow{AP} \parallel_a \overrightarrow{BQ}$, $\overrightarrow{BQ} \parallel_a \overrightarrow{CR}$, then $\overrightarrow{AP} \parallel_a \overrightarrow{CR}$.

(See Problem 1 for the proof.)

Our work then establishes in general

THEOREM 1: TRANSITIVE LAW FOR ASYMPTOTIC PARALLELISM

Let A, B, and C by any three collinear points. In terms of the preceding notation, if $\overrightarrow{AP} \parallel_a \overrightarrow{BQ}$ and $\overrightarrow{BQ} \parallel_a \overrightarrow{CR}$, then $\overrightarrow{AP} \parallel_a \overrightarrow{CR}$.

A PROPERTY OF DIVERGENT PARALLELISM

We encourage you to work through the next Discovery Unit, which will reveal the reason for the term "divergent parallelism."

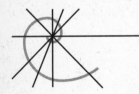

MOMENT FOR DISCOVERY

Exploring Divergent Parallelism

In Figure 6.41, we have m_1 asymptotically parallel to ℓ and m any line divergently parallel to ℓ. Choose P any point on m and let Q be the foot of the perpendicular on ℓ.

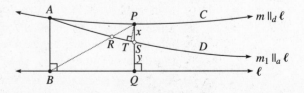

Figure 6.41

1. Why must $\overrightarrow{AC}$-$\overrightarrow{AD}$-$\overrightarrow{AB}$ hold, rather than $\overrightarrow{AD}$-$\overrightarrow{AC}$-$\overrightarrow{AB}$? (Clearly, $\overrightarrow{AC} \neq \overrightarrow{AD}$.)
2. Why must ray $\overrightarrow{AD}$ meet segment $\overline{BP}$ at a point R?
3. Why must ray $\overrightarrow{AD}$ meet segment $\overline{PQ}$ at a point S?
4. Let T be the foot of the perpendicular from P to $\overrightarrow{AD}$.
5. If $z = PT$, why does $z \to \infty$ as $AP \to \infty$? (See Problem 19, §6.2.)
6. What conclusion can you draw concerning $x + y = PQ$ as $AP \to \infty$?

If your work in the Discovery Unit was successful, then you have proven

THEOREM 2

If $m \parallel_d \ell$ at A not on ℓ, P is any point on m, and Q is the foot of the perpendicular to ℓ from P, then $PQ \to \infty$ as $AP \to \infty$.

COROLLARY

If $m \parallel_d \ell$ at A not on ℓ and C is any other point on m, then $m \parallel_d \ell$ at C.

(The property of Theorem 2 is independent of point A on line m, hence it would be true at any other point C on line m, *provided we prove the converse of Theorem 2*. (See Problem 7.)

NOTE: The corollary clears up a potentially troublesome point for divergent parallelism that arises for both types of parallelism (Lemmas A and B of Section 6.5 answer the question for asymptotic parallelism). Thus, divergent parallelism between lines ℓ and m is a universal property of the lines, independent of the choice of point A on line ℓ in Figure 6.36.

A COMMON PERPENDICU-LAR FOR DIVERGENTLY PARALLEL LINES

We begin with two relatively simple results.

LEMMA C

If lines ℓ and m have a common perpendicular, then $\ell \parallel_d m$.

Proof: Since the perpendicular m at A is parallel to ℓ (Figure 6.42) but ℓ cannot be asymptotically parallel to m by a property of asymptotic triangles (which one?), then $\ell \parallel_d m$.

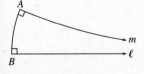

Figure 6.42

COROLLARY

If lines ℓ and m make congruent alternate interior angles with a common transversal, then $\ell \parallel_d m$.

Proof: If two lines make congruent alternate angles with a common transversal, they have a common perpendicular. (See Problem 12, §6.2.)

The converse of the previous lemma is also true. This is a remarkable result in hyperbolic geometry that was proven later in its history by David Hilbert (1862–1943). This theorem seems highly unnatural, for consider the situation in Figure 6.43, where $\angle BAC$ is an angle of parallelism having measure 1. Then a line m through A making an angle of only 2 (degrees) with $\overrightarrow{AB}$ could not meet ℓ and would not be asymptotically parallel to ℓ, hence m must be divergently parallel to ℓ. Then, according to the theorem, the two lines ℓ and m have a *common perpendicular!*

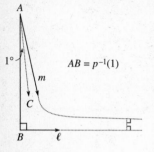

$AB = p^{-1}(1)$

Figure 6.43

THEOREM 3: HILBERT'S THEOREM ON PARALLELS

Two lines are divergently parallel iff they possess a common perpendicular.

Proof: In view of the Corollary to Lemma C, we need only prove this in one direction. Suppose $\ell \parallel_d m$. The idea of the proof is to show that there exist two congruent segments between points on ℓ and m which are both perpendicular to m, which then forms a Saccheri Quadrilateral whose base and summit have a common perpendicular—the line we are seeking. Let A and B be any two points on line ℓ, and drop the perpendiculars $\overline{AC}$ and $\overline{BD}$ to line m (Figure 6.44). If $AC = BD$, we are finished; otherwise, assume that $AC > BD$, and choose P any point on ℓ such that A-B-P.

(1) Construct point E on $\overline{AC}$ such that $EC = BD$.

(2) Construct ray $\overrightarrow{EF}$ on the P-side of $\overleftrightarrow{AC}$ such that $\angle CEF \cong \angle DBP$. Then $\overleftrightarrow{EF} \parallel_d \overleftrightarrow{CK}$ since $\overrightarrow{BP} \parallel_d \overrightarrow{DK}$

(3) It follows that line $\overleftrightarrow{EF}$ meets line ℓ at some point S. (Proof is as follows; see inset of Figure 6.44.) If ray $\overrightarrow{EF}$ meets segment $\overline{AB}$ we are finished. Otherwise, $\overleftrightarrow{EF}$ meets segment $\overline{BD}$ at some point G, since it meets diagonal $\overline{AD}$ by the Postulate of Pasch. Construct ray $\overrightarrow{DQ}$ asymptotically parallel to $\overrightarrow{BP}$, forming asymptotic triangle $\triangle BD\Omega$. Also, construct ray $\overrightarrow{CR}$

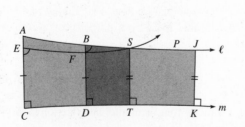

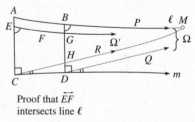

Proof that $\overleftrightarrow{EF}$ intersects line ℓ

Figure 6.44

THEOREM 3: HILBERT'S THEOREM ON PARALLELS (CONTINUED)

asymptotically parallel to $\overrightarrow{EF}$, forming asymptotic triangle $\triangle CE\Omega'$. By BA, $\triangle EC\Omega' \cong \triangle BD\Omega$, and $\angle ECR \cong \angle BDQ$. Hence, the lines $\overleftrightarrow{CR}$ and $\overleftrightarrow{DQ}$ make congruent angles with transversal m and are divergently parallel. Since $\overleftrightarrow{CR}$ cannot meet segment $\overline{EG}$, and it meets segment $\overline{DE}$ by the Crossbar Theorem, $\overleftrightarrow{CR}$ meets $\overrightarrow{GD}$ at some point H by the Postulate of Pasch, hence by the Postulate of Pasch for Asymptotic Triangles, must meet either side $\overrightarrow{DQ}$ or $\overrightarrow{BP}$ of $\triangle BD\Omega$ (or be asymptotically parallel to $\overrightarrow{DQ}$, which is impossible). But line $\overleftrightarrow{CR}$ is parallel to line $\overleftrightarrow{DQ}$, so must meet $\overrightarrow{BP}$ at some point M, forming the ordinary triangle $\triangle ACM$. Hence, by the Postulate of Pasch, ray $\overrightarrow{EF}$ meets either $\overline{CM}$ or $\overline{AM}$. Since $\overrightarrow{EF} \parallel_a \overrightarrow{CR}$ and $\overrightarrow{EF}$ does not meet segment $\overline{CM}$, $\overrightarrow{EF}$ must meet segment $\overline{AM}$ at some point S.

(4) Construct $\overleftrightarrow{ST} \perp m$, construct J on ray $\overrightarrow{AP}$ such that A-B-J and $ES = BJ$, and drop the perpendicular $\overline{JK}$ to line m.

(5) By SASAA, $\Diamond ESTC \cong \Diamond BJKD$, and $ST = JK$, forming Saccheri Quadrilateral $\Diamond SJKT$, as desired.

COROLLARY

Divergent parallelism is symmetric: If $\ell \parallel_d m$, then $m \parallel_d \ell$.

PROBLEMS   (§6.6)

_____ **Group A** _____

1. Prove the final case of the Transitive Law for Asymptotic Parallelism (Lemma B for Theorem 1): If A-B-C and $\overrightarrow{AP} \parallel_a \overrightarrow{BQ}$, $\overrightarrow{BQ} \parallel_a \overrightarrow{CR}$, then $\overrightarrow{AP} \parallel_a \overrightarrow{CR}$. (See the figure below.) (**HINT:** Draw segment $\overline{CS}$ and assume that A-S-T.)

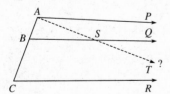

2. If $a < b$, show that $p(a) > p(b)$. Give a synthetic proof.

3. Using the Bolyai-Lobachevski Formula (**1**), show that $\lim p(a) = 90$ as $a \to 0$, and $\lim p(a) = 0$ as $a \to \infty$.

4. (a) Assuming that $k = 1$ in the Bolyai-Lobachevski formula **(1)**, calculate $p^{-1}(1)$ and give an approximation on your pocket calculator accurate to four decimals.

  (b) Compare your answer for **(a)** with the value $p^{-1}(80)$ calculated to four decimals.

  (These answers are the distances corresponding to angles of parallelism of measure 1 and 80 (degrees), respectively.)

5. Determine the value k in the Bolyai-Lobachevski Formula **(1)** which will make the unit of measure $a = 1$ when $p(a) = 45$. Use a pocket calculator to give an approximation accurate to four decimals.

6. Complete the details of the following argument, and *state what property of parallelism it proves* (see accompanying figure): Let $\overrightarrow{AP} \parallel_a \overrightarrow{BQ}$, with $\overline{AB} \perp \overrightarrow{BQ}$, and let $UV = \varepsilon$. Construct the perpendiculars ℓ and m to segment $\overline{UV}$ at U and V.

  (1) $\ell \parallel_d m$?

  (2) There exists C on ℓ such that the perpendicular distance CD ?
     to line m is greater than AB.

  (3) Locate E on $\overline{CD}$ so that $DE = AB$, and construct ray $\overrightarrow{EF}$ on
     the U-side of $\overleftrightarrow{CD}$ such that $\overrightarrow{EF} \parallel_a \overrightarrow{DV}$.   ?

  (4) If $\overrightarrow{EF}$ meets $\overrightarrow{CU}$ at a point T (figure insert), then $\overrightarrow{CU}$ must intersect
     line m (because $\overrightarrow{EF} \parallel_a \overrightarrow{DV}$, then $\overrightarrow{TF} \parallel_a \overrightarrow{DV}$). →←   ?

  (5) Then in $\triangle CDU$, $\overrightarrow{EF}$ meets $\overline{DU}$ at some point R.   ?

  (6) EF cuts segment $\overline{UV}$ at some point W.   ?

  (7) Hence, there exists a point G on ray $\overrightarrow{AP}$ such that the   ?
     perpendicular distance from G to line $\overrightarrow{BQ}$ is less than UV.
     (Show how this construction follows logically.)

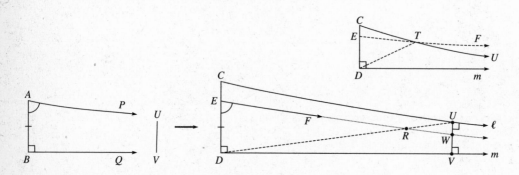

7. Prove the converse of Theorem 2: If $PQ \to \infty$ as $AP \to \infty$ and $\ell \parallel m$, then $\ell \parallel_d m$. (Use the result of Problem 6.)

8. Explicitly prove the corollary to Hilbert's Theorem (Symmetry Law for Divergent Parallelism).

_____ *Group B* _____

9. Prove the Hyperbolic Parallel Postulate from the following weaker form: There exists a line ℓ and a point P not on it such that at least two lines passing through P are parallel to ℓ.

The next group of problems are designed to let you become familiar with the Bolyai-Lobachevski formulas for hyperbolic trigonometry, given in the table below. The calculus formulas for the hyperbolic functions sinh x, cosh x, and tanh x are assumed here.

HYPERBOLIC TRIGONOMETRY OF THE RIGHT TRIANGLE

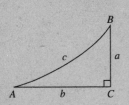

$$\sin A = \frac{\sinh a}{\sinh c} \qquad \sin B = \frac{\sinh b}{\sinh c}$$

$$\cos A = \frac{\tanh b}{\tanh c} \qquad \cos B = \frac{\tanh a}{\tanh c}$$

$$\tan A = \frac{\tanh a}{\sinh b} \qquad \tan B = \frac{\tanh b}{\sinh a}$$

cosh c = cosh a cosh b (Pythagorean Theorem for Hyperbolic Geometry)

NOTE: These formulas are based on the value $k = 1$ in (1).

---

10. (a) In the formula for tan A, find $m\angle A$ to two-decimal accuracy for an isosceles right triangle having $a = b = 2$ units.

   (b) Repeat the calculation for $a = b = \frac{1}{2}$. What seems to be happening as the sides of the triangle become small?

11. Work Problem 10 if $a = b = 4$. What seems to be happening as $a = b$ becomes large without bound? (Try $a = b = 8$.)

12. In a certain hyperbolic right triangle $\triangle ABC$, the legs are of length $a = 3$, $b = 4$. Find c to four-decimal accuracy using the Pythagorean Theorem for Hyperbolic Geometry. (*HINT:* A formula from calculus will be needed here):

$$\cosh^{-1}x = \ln (x + \sqrt{x^2 - 1}) \text{ for } x \geq 1.$$

Some calculators have a function "hyp," which when activated just before "sin," "cos," or "tan," yields the calculations of the three corresponding hyperbolic functions, and their inverses can then be accessed through the usual inverse function key.)

13. In a certain hyperbolic right triangle $\triangle ABC$, $a = .03$ and $b = .04$. Find c (as in Problem 12). Show that $c \approx .05$, approximating a Euclidean right triangle. Solve for A and B and show that $A + B \approx 90$. (This illustrates the principle that sufficiently small hyperbolic triangles behave like their Euclidean counterparts.)

14. Show that if τ is the Golden Ratio (as defined in the problem set for §1.2), then ln τ, ln τ, and ln 2 are the sides of a hyperbolic right triangle. Do not use a calculator. (*HINT:* Verify that cosh c = cosh a cosh b holds for these values; recall that $\tau = \frac{1}{2}(1 + \sqrt{5})$, $\tau^2 = \tau + 1$, $\tau^{-1} = \tau - 1$, etc.)

_____ *Group C* _____

15. According to the AAA Congruence Criterion for Hyperbolic Geometry, there is only one right triangle $\triangle ABC$ for which $A = 30$, $B = 45$ ($C = 90$). Find its three sides, accurate to four decimals, using the appropriate formulas from hyperbolic trigonometry. (The following formulas from the calculus may be found useful:

$$\sinh^{-1}x = \ln (x + \sqrt{x^2 + 1}), \quad \tanh^{-1}x = \frac{1}{2} \ln \frac{1 + x}{1 - x} \text{ for } 0 \leq x < 1.)$$

16. A right triangle has acute angles of measure 20 and 60. Find the three sides using four-place accuracy.

17. Use the formula for tan B in hyperbolic trigonometry to derive the Bolyai-Lobachevski formula **(1)** for the angle of parallelism (with $k = 1$). (Merely take the limit as $b \to \infty$; no calculators allowed for this problem.)

18. Use hyperbolic trigonometry to validate the following construction due to Beltrami for the angle of parallelism (a variation of one given by J. Bolyai). (See figure below.) Let $AB = a$ be given, and construct the perpendicular ℓ to $\overleftrightarrow{AB}$ at B.

   (1) Construct a Lambert Quadrilateral $\Diamond ABCD$ with $\overline{AB}$ as base, C on ℓ, and right angles at A, B, and D.

   (2) With A as center and $BC = c$ as radius, swing an arc, cutting segment $\overline{CD}$ at E. (Why will this circle intersect $\overline{CD}$? Show that $AC > c$ and $AD < c$.)

   (3) Then $\overrightarrow{AE} \parallel_a \overrightarrow{BC}$, and the angle of parallelism is $\angle BAE \equiv p(a)$.

   (**HINT:** You must show that if $\theta = m\angle BAE$, then $\tan \frac{1}{2}\theta = e^{-a}$. First prove that this is equivalent to $\sin \theta = \operatorname{sech} a$, either from Problem 17 or directly from $\tan \frac{1}{2}\theta = e^{-a}$. Then use hyperbolic trigonometry for a, c, x, and y in the accompanying figure.)

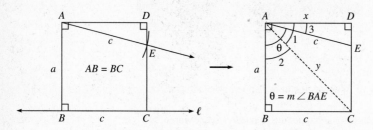

19. Prove that if the perpendicular bisectors of two sides of a triangle are divergently parallel, then that of the third side is divergently parallel to each of the other two. (**HINT:** In the following figure, let m be the common perpendicular of the perpendicular bisectors of sides $\overline{AC}$ and $\overline{BC}$; drop perpendiculars from the three vertices A, B, and C to m.

NOTE: Points A, B, and C lie on what is known as the **equidistant locus E** with respect to line m and distance $AA' = a$—the set of all points whose distances from m are constant and equal to a.

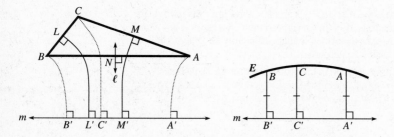

20. Given any angle $\angle ABC$, find a construction (proof) for the unique line lying in its interior that is asymptotically parallel to both its sides. (You may use the function $p(a)$ defined in **(1)** as an aid to construction.)

21. Prove the Pythagorean Theorem for hyperbolic geometry from the relations for $\sin A$, $\cos A$, and $\tan A$.

6.7 MODELS FOR HYPERBOLIC GEOMETRY: RELATIVE CONSISTENCY

The modern philosophy towards axiomatics in mathematics is to avoid an axiom system until, or unless, you can find a model for it. Much of mathematics research today is done within thoroughly tried systems, such as the real numbers or *n*-dimensional geometry. So most of the time we feel pretty safe. But often one must venture far from home, and try out new theories to see where they lead. The demands of modern rigor dictate that these new systems be free of contradictions, and to that end, models for those systems that live in a less "hostile," more familiar, environment must be found.

Note that finding a model for an axiomatic system does not prove absolute consistency, only relative consistency—that is, the system is as consistent as the environment in which the model has been constructed. In modern algebra, for example, rings and fields are normally defined axiomatically, and we study them and prove theorems about them, but this does not imply that they exist or that the definitions are consistent beyond all shadow of doubt. An example of a ring is, of course, reassuring, but it does not prove that the larger family of rings we may have in mind exists. There is the story of a topologist who once published a complicated theorem which had several incredibly fascinating corollaries in topology. However, it was ultimately discovered that the only set to which the theorem applied was the empty set!

This story has a lesson for everyone working in axiomatics, and the same difficulty pursues us in the field of geometry. We might ask the question: Can we be certain that Euclidean geometry is consistent, that is, contradiction-free? We really cannot, but it is well known that the Euclidean plane can be constructed entirely within the set of real numbers and its ordered pairs (that is, the coordinate plane). Thus we can say that *Euclidean geometry is as consistent as the real number field*. The real surprise is that hyperbolic geometry—the one which ought to lead to a contradiction—also has a model in Euclidean geometry, and it too is as consistent as Euclidean geometry. Had those 17th and 18th century mathematicians actually succeeded in proving Euclid's Fifth Postulate by obtaining a contradiction, then this contradiction would have shown up in the model, and ultimately in Euclidean geometry itself. Instead of "repairing" Euclid's geometry, as Saccheri had set out to do, he would have *destroyed* it!

OUR GEOMETRIC WORLD

Where relatively small distances are concerned, Euclidean geometry is certainly the simplest model for our physical world. But it is not the proper model for physicists, astronomers, and astronauts. Even the basic question of what constitutes a straight line in outer space is problematic. Since light rays are bent by large gravitational forces, the visual image of a distant star may be distorted, and the apparent shortest path to it along a light ray can be a curved line!

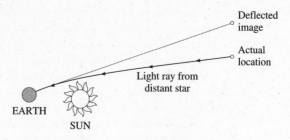

Figure 6.45

We shall concentrate our efforts here in developing one of the famous models for hyperbolic geometry.

*POINCARÉ'S CIRCULAR
DISK MODEL*

In Section 6.2 we briefly explored the Beltrami-Poincaré Half-Plane Model for Hyperbolic Geometry. There, the distance formula was different for the two types of lines that existed. This is unnecessary for Poincaré's Circular Disk Model, as we shall see.

Begin with a unit circle C in the Euclidean plane (Figure 6.46), and let H denote the set of interior points of C. Then the objects and measurements in the model are as follows (on the next page).

POINCARÉ CIRCULAR DISK MODEL

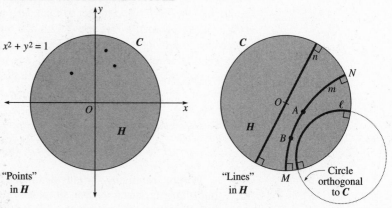

Figure 6.46

DESCRIPTION OF POINCARÉ DISK MODEL

POINTS   Any ordinary point in *H*, called an **h-point**.

LINES   Any arc of a circle in *H* orthogonal to *C* or diameter of *C*,
                called an **h-line**. Define the **endpoints** of an h-line as the
                ordinary endpoints of the arc or diameter on **C**, not belong-
                ing to *H*.

DISTANCE   For any two h-points *A* and *B*, define the **h-distance** from *A*
                to *B* to be the real number

(1) $$AB^* = \left| \ln (AB, MN) \right| = \left| \ln \frac{AM \cdot BN}{AN \cdot BM} \right|$$

                where *M* and *N* are the endpoints of line $\overrightarrow{AB}$.

ANGLE   Any angle in *H* has the measure ascribed to it by Euclidean
MEASURE   geometry, that is, the measure of the angle between the tan-
                gents to its sides:

(2) $m\angle CDE = \theta$ (Figure 6.47)

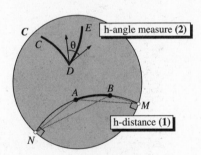

Figure 6.47

REALIZATION OF THE
AXIOMS FOR ABSOLUTE
GEOMETRY IN H

We now take the previous axioms which defined absolute geometry (Chapters 2
and 3) and verify their validity, one by one. Since our model is for a single plane,
the only incidence axiom we need to worry about is "two points determine a line."
Not only is this true, a Euclidean construction can be given. Let *A* and *B* be any
two h-points. If *A* and *B* are collinear with the center of *C* (point *O* in Figure 6.48),
then the Euclidean diameter through *A* and *B* is the unique h-line we seek.
Otherwise, take the inverse point *A′* of *A* with respect to *C* as circle of inversion.
Thus, *OA′* = 1/*OA* (Figure 6.48). Then draw the unique Euclidean circle through *A*,
B, and *A′*. This circle will be orthogonal to *C* because all circles passing through *A*
and *A′* are orthogonal to *C*. (See Theorem 2, Section 4.7.) Hence, in either case, we
have found the unique h-line passing through *A* and *B*.

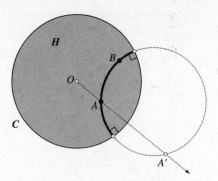

Figure 6.48

Next, we come to the metric axioms, Axioms D-1, D-2, and D-3. Since existence is clear by definition of AB^* in **(1)**, we may combine these axioms into the following two statements:

(a)   For any two h-points A and B, $AB^* \geq 0$, with equality only when $A = B$.

(b) $AB^* = BA^*$

In **(a)**, note that the order in which the endpoints M and N are listed is immaterial, for if M and N were interchanged in **(1)**, we obtain $x' = (AB, NM) = 1/(AB, MN) = 1/x$, so that $\ln x' = \ln (1/x) = -\ln x$. The absolute value takes care of the rest. By its definition, therefore, $AB^* \geq 0$, and if $A = B$, then from **(1)**

$$AB^* = \left| \ln (AA, MN) \right| = \ln 1 = 0$$

Conversely, if $AB^* = 0$, then $(AB, MN) = 1$ and

$$\frac{AM}{AN} \cdot \frac{BN}{BM} = 1 \quad \rightarrow \quad \frac{AM}{AN} = \frac{BM}{BN}$$

If $A \neq B$, then by proper choice of notation for the endpoints,

$$AM > BM \quad \text{and} \quad AN < BN \quad \rightarrow \quad \frac{AM}{AN} > \frac{BM}{BN}$$

a contradiction. Therefore, $A = B$. The assertion in **(b)** will be left as Problem 13.

EXAMPLE 1   If C lies on arc $\overset{\frown}{AB}$, which is a subarc of the h-line $\overset{\leftrightarrow}{AB}$ (Figure 6.49), show that C lies between A and B under the metric given by **(1)**.

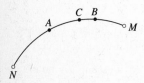

Figure 6.49

SOLUTION

We must show that $AC^* + CB^* = AB^*$.

If the endpoints of the h-line are M and N (as shown), then we have

$$AC^* = \left| \ln \frac{AM}{AN} \cdot \frac{CN}{CM} \right| = \left| \ln \frac{AM}{CM} \cdot \frac{CN}{AN} \right| = \ln \frac{AM}{CM} \cdot \frac{CN}{AN}$$

(since $AM/CM > 1$ and $CN/AN > 1$). Similarly,

$$CB^* = \left| \ln \frac{CM}{CN} \cdot \frac{BN}{BM} \right| = \left| \ln \frac{CM}{BM} \cdot \frac{BN}{CN} \right| = \ln \frac{CM}{BM} \cdot \frac{BN}{CN}$$

Now we use the property of logarithms $\ln x + \ln y = \ln xy$.

From the preceding expressions for AC^* and CB^*,

$$AC^* + CB^* = \ln\frac{AM}{CM} \cdot \frac{CN}{AN} + \ln\frac{CM}{BM} \cdot \frac{BN}{CN}$$

$$= \ln\frac{AM}{CM} \cdot \frac{CN}{AN} \cdot \frac{CM}{BM} \cdot \frac{BN}{CN} = \ln\frac{AM}{AN} \cdot \frac{BN}{BM} = AB^*$$

as desired.◻

The result in this example clearly shows that h-betweenness essentially coincides with Euclidean betweenness on arcs of circles, or diameters, orthogonal to C. Accordingly, the sorts of objects that constitute **h-segments**, **h-rays**, and **h-angles** are quite natural, as shown in Figure 6.50. Thus, Axiom D-4, the next metric axiom, is clear:

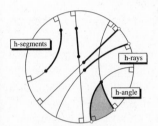

Figure 6.50

Given A-B-C, then $\overrightarrow{BA} \cup \overrightarrow{BC} = \overleftrightarrow{AB}$.

To proceed, we consider next the Plane Separation Postulate, Axiom P-1. Let the h-line ℓ be given. The Plane Separation Postulate requires us to first define the two "sides" of ℓ, that is, **h-half-planes**, to prove they are h-convex, and to show that any h-segment which "straddles" ℓ must actually intersect ℓ. In $\boldsymbol{H}$, it appears that an h-half-plane for line ℓ (Figure 6.51) is the set of h-points either exterior, or interior to the circle determined by ℓ.

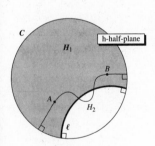

Figure 6.51

It looks like it might be difficult to prove that both such sets are "convex."But we can reason this way. In the case of H_1 shown in Figure 6.51, let A and B be any two points in H_1, and consider the arc $\widehat{AB}$ of the circle orthogonal to C with endpoints A and B (this is the h-segment $\overline{AB}$). Now we know that two circles which are orthogonal to C cannot be tangent to each other (they cannot meet at an angle of zero degrees), and must either cross over each other or have no points in common. Therefore, if $\widehat{AB}$ is not entirely contained by H_1, then there exists a first encounter (intersection) of arc $\widehat{AB}$ with line ℓ as we proceed from A to B, at which point $\widehat{AB}$ crosses over into H_2, then a *second* encounter with ℓ. This gives us two h-points in common between h-lines $\overleftrightarrow{AB}$ and ℓ, which is impossible unless $\overleftrightarrow{AB} = \ell$. Therefore, h-segment $\overline{AB}$ is devoid of points of $\boldsymbol{H}$ not in H_1, so must lie completely in H_1. Finally, it is clear that any arc, circular or otherwise, joining points in opposite h-half-planes H_1 and H_2 must meet ℓ.

Since the model is conformal (coincides with Euclidean angle measure), all the axioms on angle measure (Axioms A-1–A-4) are clearly valid. A potentially difficult task is verifying the Ruler Postulate, Axiom D-5. This is essentially an extension of Example 1, however. If ℓ is a line and A and B are two points on ℓ, the key to coordinatizing points on ℓ is the definition

$$x = \ln\,(AP, MN)$$

for the coordinate of P on ℓ, *without absolute values*. Point A will have coordinate *0* (why?), and B will have a positive coordinate by proper choice of the endpoints M and N. It may thus be shown that there is a one-to-one correspondence between points and real numbers, and that if $P[x]$ and $Q[y]$ are two given points with their coordinates, $PQ* = |x - y|$. These details will be left as Problem 13.

This leaves the SAS Postulate as the last remaining postulate of absolute geometry to be verified.

"REFLECTIONS" IN H

The SAS Postulate will be settled once we show that an h-reflection in an arbitrary h-line ℓ exists. This transformation will consist of a circular inversion with itself as the circle of inversion. Our earlier discussions then make short work of the SAS Postulate since

(a) a circular inversion preserves cross ratio, hence h-distance

(b) a circular inversion is conformal relative to curvilinear angle measure, hence preserves h-angle measure (as shown in Section 4.7)

(c) Euclid's superposition argument, as presented in Section 5.7, proves SAS in any setting that is rich enough to allow reflections in arbitrary lines.

To construct an h-reflection in an h-line ℓ, first, if ℓ is a diameter of C, take just the Euclidean reflection in the Euclidean line containing ℓ; since this is a Euclidean isometry, certainly cross ratios, h-distance, and h-angle-measure are preserved (Figure 6.52). Otherwise, if ℓ is the arc of a circle C_1 orthogonal to C, consider the inversion with C_1 as circle of inversion (Figure 6.52 shows how it works). Thus, again we have found the desired h-reflection, since ℓ maps to itself (because C_1 contains ℓ), the half-planes of ℓ map to each other, and an inversion is h-distance preserving and h-conformal.

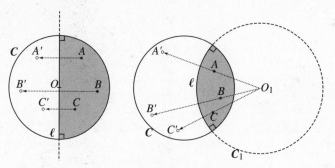

Figure 6.52

PARALLEL POSTULATE IN H

The Postulate of Parallelism for **H** is immediately clear after studying the diagram of Figure 6.53 and its ramifications. This completes the proof.

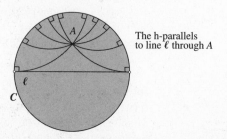

The h-parallels to line ℓ through A

Figure 6.53

OTHER MODELS FOR HYPERBOLIC GEOMETRY

After having found one model, it is not surprising, given the number of mathematicians who worked on this problem, that many others were constructed. We already introduced the half-plane model, widely used for research in hyperbolic geometry.

Another model, due to F. Klein, resembles Poincaré's model, but is not conformal. It is easier to describe, but both distance and angle measure must be defined using cross ratios. Conceptually, it consists of all the points inside the unit circle C, with "lines" consisting of all chords of C, and "segments," "rays," and "angles" all being the intersection of their Euclidean counterparts with the interior of C (Figure 6.54). We shall not define distance and angle measure for this model, but simply provide you with a reference where a complete, readable, development may be found: M. J. Greenberg, *Euclidean and Noneuclidean Geometries*, Chapter 7. The reason this model cannot be conformal is obvious: Any "triangle" in the model is an ordinary Euclidean triangle, which would have angle sum 180.

KLEIN'S MODEL

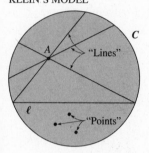

Figure 6.54

A most interesting fact is that Klein's model can be mapped onto the Poincaré Circular Disk Model H using a modified stereographic projection. The latter model H can, in turn, be mapped into the Beltrami-Poincaré Half-Plane Model by simply taking a point T on C as a center of inversion, and inverting in some suitable circle centered at T. Some of these ideas will be pursued in the problems (in particular, Problem 18). As it turns out, all models of hyperbolic geometry are isomorphic, hence are logically equivalent. This bears witness to the fact that the three classical geometries (parabolic, hyperbolic, and elliptic) are **categorical**. (See Section 2.2 for a discussion of this term.)

We mention one other source of models. If you take a right circular cylinder or cone and roll them on a plane, as indicated in Figure 6.55, we can see that they will "pick up" the geometric properties of Euclidean geometry, as if we were inking all the Euclidean configurations onto the cylinder and cone. (The particular feature illustrated is the Euclidean property concerning an exterior angle of a triangle.) In fact, this rolling process suggests a way to define formally a one-to-one mapping from a portion of the Euclidean plane to the cylinder or cone. A straight line, for example, maps to either a circle or helix on the cylinder, and to a kind of spiraling helix on the cone.

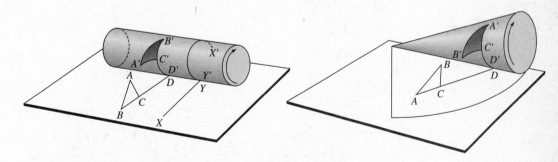

Figure 6.55

What is interesting, however, is that the geometry of the cylinder and cone can be defined independently of this mapping by taking as metric the length of the shortest arc on these surfaces between two given points, and, as angle measure, the Euclidean angle measure between arcs of curves in three-dimensional space. The resulting geometry is called the **intrinsic geometry** of the cylinder or cone. In theory, this procedure can be carried out for any reasonably-behaved surface S, yielding a two-dimensional **intrinsic geometry of S**. It turns out that the intrinsic geometry of both the cone and cylinder are locally isometric to the Euclidean plane, that is, small regions of the cylinder and cone are isometric to the Euclidean plane.

Analogously, the hyperbolic plane can be locally realized on a surface in Euclidean space. A rare find was had in the field of differential geometry dating back to Beltrami (1872). A surface S having *constant negative curvature* (defined in terms of the curvatures of certain curves on S) was discovered. Moreover, its intrinsic geometry was found to be locally isometric to the hyperbolic plane. This remarkable surface, called a **tractroid** or **pseudosphere**, is obtained by revolving the curve

$$ y = \ln\left(\frac{1 + \sqrt{1 - x^2}}{x} \right) - \sqrt{1 - x^2}\,, \quad 0 < x \leq 1, $$

called the **tractrix**, about the y-axis, as shown in Figure 6.56. Thus, with the proper units of measure, right triangle $\triangle ABC$, with standard notation representing the intrinsic lengths of segments, satisfies the relation

$$ \cosh c = \cosh a \cosh b, $$

which is one of the formulas for a right triangle in hyperbolic geometry, as introduced in the problem set for Section 6.6. It should be realized, however, that the pseudosphere is only a *local* realization of hyperbolic geometry; that is, only small portions of the hyperbolic plane can be mapped isometrically onto this surface.

SURFACE MODEL FOR HYPERBOLIC GEOMETRY

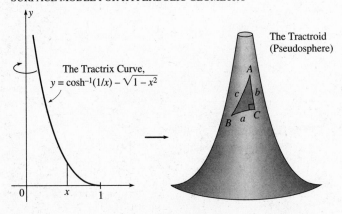

The Tractrix Curve,
$y = \cosh^{-1}(1/x) - \sqrt{1 - x^2}$

The Tractroid
(Pseudosphere)

Figure 6.56

PROBLEMS   (§6.7)

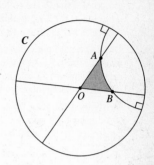

———— *Group A* ————

1. The figure shows a certain h-triangle △*ABO* in the Poincaré Model **H**. What important theorem in hyperbolic geometry is exhibited?

2. What important configuration is exhibited in the Poincaré Model shown in the figure below?

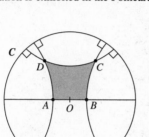

3. In the accompanying figure, what kinds of lines ℓ and *m* are depicted in each case (i.e., intersectors, asymptotic parallels, or divergent parallels)?

(a)

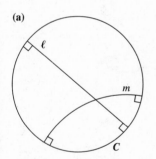

(b)

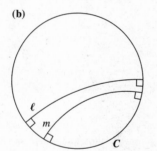

(c)

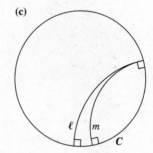

4. Draw two diameters of **C** through center *O*, and at the endpoints *U* and *V* draw Euclidean perpendiculars, meeting at point *W*. Then with *W* as center, draw a circle through *U* and *V*. This circle will be orthogonal to **C** and will represent an h-line not passing through *O*. (See figure to the right.)

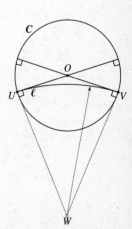

   (a) Identify and sketch all the h-lines through *O* which are parallel to ℓ.

   (b) Which of these are asymptotic parallels? Divergent parallels?

5. Draw an example of an asymptotic triangle △*AB*Ω in **H**.

6. Verify to your own satisfaction that if the h-line ℓ is a diameter of **C** and *A* is any point not on ℓ, the h-perpendicular to ℓ through *A* will also pass through

the point B that is the reflection of A in ℓ. Thus, give an actual Euclidean construction of the h-perpendicular to ℓ through A. (**WARNING:** The h-perpendicular is *not* the Euclidean line $\overleftrightarrow{AB}$.) (See figure at right.)

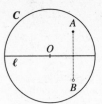

7. Draw two perpendicular diameters ℓ and m of C, and locate A any h-point not on these lines. Making use of your construction in Problem 6, draw the two h-perpendiculars to ℓ and m from A. What kind of figure is formed? What characteristics can be observed from the model?

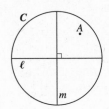

8. Verify the formula $e^{-a} = \tan \frac{1}{2}A$ for the measure of the angle of parallelism $\angle A$ corresponding to the distance a in the Poincaré Model for the example illustrated in the figure below. (**HINT:** Slope $PW = \tan A$; use the identity

$$\tan \tfrac{1}{2}A = \frac{1 - \cos A}{\sin A} = \frac{\sqrt{1 + \tan^2 A} - 1}{\tan A}$$

THE RELATION $e^{-a} = \tan \frac{1}{2}A$

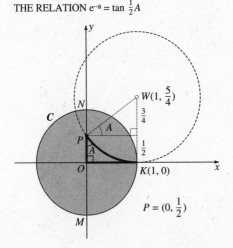

9. Let ℓ be a diameter of C. In the Poincaré Model, what configuration represents

   (a) the family of all the asymptotic parallels to ℓ in the same direction?

   (b) the asymptotic parallels to ℓ in the direction opposite to that of (a)?

   (c) Are symmetry and transitivity of asymptotic parallelism evident from the model? Make a sketch.

_____ *Group B* _____

10. What kind of curve in hyperbolic geometry is represented by a Euclidean circle in the Poincaré Model? (**HINT:** Can an inversion map this circle to one having center O—the center of C?)

11. In the figure below (on next page), h-lines ℓ and m meet at A. As parts of circles orthogonal to C, their extensions must meet at a second point A' in the Euclidean plane, and these extended h-lines are two specific members of the family $\mathcal{G}$ of circles orthogonal to C.

   (a) What do the intersections of these circles with H represent, specifically?

   (b) Verify to your own satisfaction that precisely one of these circles determines the unique h-line which bisects the h-angle $\angle BAC$. Name the one that appears to do so in the figure.

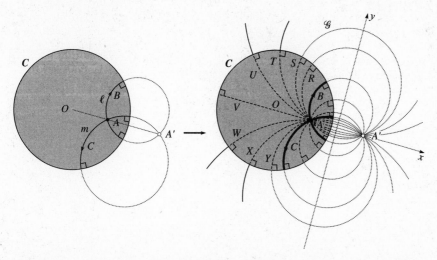

12. (*NOTE:* If you have trouble with this problem, you may find it helpful to work Problem 11 first, if you have not already done so.) The figure shows two h-lines ℓ and m which do not meet.

   (a) What kind of parallels are they?

   (b) What important theorem is true concerning ℓ and m in hyperbolic geometry?

   (c) To see this in the model, recall the orthogonal families of circles $\mathcal{F}$ and $\mathcal{G}$ (§4.7); m and C are two specific members of $\mathcal{G}$ in this case. Verify to your own satisfaction that one of the circles from the family $\mathcal{F}$ is orthogonal to ℓ, as well as to m and C.

HILBERT'S THEOREM DEPICTED
IN THE POINCARÉ MODEL

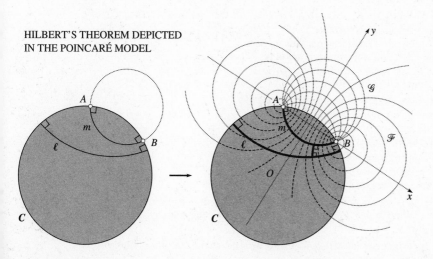

13. Prove that if $P[x]$ and $Q[y]$ are located such that $x = \ln(AP, MN)$ and $y = \ln(AQ, MN)$, then

$$PQ^* = \left| y - x \right|$$

(Since x and y can be interchanged without affecting this equation, it
follows that for all P and Q, $PQ^* = QP^*$.)

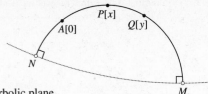

_____ **Group C** _____

14. Show that the smallest regular pentagon with which one could tile the hyperbolic plane
has area $90k = \frac{1}{2}\pi$ (here, k is the conversion factor from degrees to radians). Complete
the drawing started in the following figure which depicts this regular tiling in the
Poincaré Model, by computer or otherwise. Verify to your own satisfaction that the
tiling is legitimate, i.e., that each tile is congruent to the regular pentagon at the center.
(Note that each pentagon in this tiling constitutes a counterexample to the transitivity of
divergent parallelism.)

REGULAR PENTAGONS IN
THE POINCARÉ MODEL

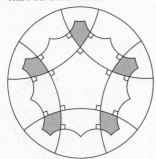

15. Use the Poincaré Model to show that in hyperbolic geometry there exist h-convex sets
which do not cover the entire "plane" (the set **H**), yet contain two intersecting lines.
Hence, uncover the flaw in Legendre's argument proving Euclid's Fifth Postulate in
absolute geometry as presented in Section 6.2.

16. Use h-reflections to show that the construction depicted in the figure at
the right locates the h-midpoint M of h-segment $\overline{AB}$ ($\overline{O_1A}$ and $\overline{O_1B}$ are
tangent to ℓ). (**HINT:** O_1 is the center of a Euclidean circle through A
and B orthogonal to ℓ; find an h-reflection which maps point M to
point O, thus ℓ to a diameter of C, and use the fact that orthogonal
Euclidean circles map to orthogonal Euclidean circles.)

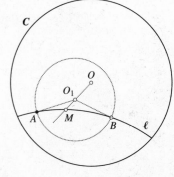

17. Consider the h-line ℓ with endpoints M and N, with common chord $\overline{MN}$
of ℓ and C as shown in the figure below. By the following analysis, it
can be shown that the open Euclidean segment $(\overline{MN})$ *is an equidistant
locus for line ℓ* in the Poincaré Model (complete all details):

(1) Let M and N be limiting points of the centers of the family $\mathscr{F}$ of
circles orthogonal to the family $\mathscr{G}$ of all circles passing through M
and N, of which C and ℓ are specific members.

(2) Since the circles in $\mathscr{F}$ have their centers lying on line $\overleftrightarrow{MN}$, they are each orthogonal
to the common chord $\overline{MN}$, and to C and ℓ. Let $\overline{AD}$ be the h-segment perpendicular
to ℓ and Euclidean segment $\overline{MN}$.

(3) For any point C on $\overline{MN}$, the h-line $\overleftrightarrow{CD}$ makes equal angles with the chord $\overline{MN}$ in
the Euclidean plane, hence with the h-lines $\overleftrightarrow{AD}$ and $\overleftrightarrow{BC}$.

(4) Using a simple argument in synthetic geometry, it may be proven that $\lozenge ABCD$ is a Saccheri Quadrilateral, with $BC = AD$. (Give this argument.)

(5) Therefore, $BC = AD = a$ (constant), and since C was arbitrary, all h-points of chord $\overline{MN}$ are h-equidistant from ℓ.

EQUIDISTANT LOCUS DEPICTED IN
THE POINCARÉ MODEL

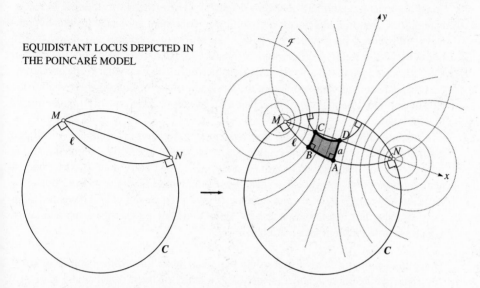

18. Show that the inversion indicated in the accompanying figure maps the Poincaré Circular Disk Model onto the Beltrami-Poincaré Half-Plane Model, with, in particular, h-lines ℓ and m mapping to ℓ' and m', respectively.

MAPPING THE POINCARÉ MODEL
TO HALF-PLANE MODEL

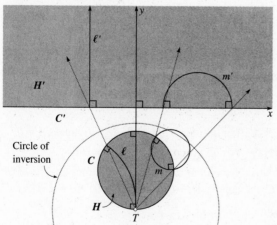

*6.8   Axioms for a Bounded Metric: Elliptic Geometry

It had been assumed by early geometers that the hypothesis of the obtuse angle could not exist in axiomatic geometry. Riemann (1826–1866) was the first to recognize that by assuming distances are bounded, a consistent geometry for this hypothesis could indeed be constructed. This observation of Riemann was partly due to his work in the field of differential geometry, so extensive and revolutionary that Einstein found Riemann's work indispensable for his own epochal work in relativity. One of Riemann's models was a modification of the ordinary sphere, with each pair of antipodal points identified. His other models require advanced mathematics to understand.

Unfortunately, if we look for an *elementary* axiomatic development of elliptic geometry—one that does not require a knowledge of projective geometry or the introduction of unnatural betweenness relations—apparently none is to be had. It actually is possible, however, to make very minor modifications in the axioms we have already adopted for absolute geometry, and provide not only a basis for elliptic geometry, but a foundation for *all three* classical geometries simultaneously, which we will call **neutral geometry**.

AXIOMS FOR NEUTRAL GEOMETRY

Since the original axioms for absolute geometry were three dimensional (in keeping with the desirable feature of including a study of three-dimensional geometry in high school geometry) the new axioms will also be stated for three dimensions, even though we will focus our attention here on the geometry of a single plane.

For reasons which will soon become obvious, the Axioms for Distance must precede the Axioms for Incidence. The definition for the betweenness relation will be taken verbatim from that given in Section 2.5. This leads to the corresponding definitions for segments, rays, and angles.

Undefined Terms: Point, Line, Plane, Space

Axioms:

AXIOM N-0 Space is the set of all points. Lines and planes are nonempty subsets of space.

AXIOM N-1 To each pair of points (A, B) is associated a unique real number $AB \geq 0$.

AXIOM N-2   For all points A and B, $AB > 0$ unless $A = B$.

AXIOM N-3   For all points A and B, $AB = BA$.

AXIOM N-4   Given any four distinct collinear points A, B, C, and D such that A-B-C, then either D-A-B, A-D-B, B-D-C, or B-C-D.

DEFINITION   Let α denote the least upper bound for the set of all distances AB, A and B arbitrary points in space.

AXIOM N-5: Ruler Postulate   The points of each ray may be assigned to the set of real numbers x which lie in the range $0 \leq x \leq \alpha$, called **coordinates,** in such a manner that

(1) every point of the ray is assigned to a unique coordinate

(2) each coordinate $x \geq 0$ is assigned to a unique point

(3) the endpoint of the ray has zero coordinate

(4) if points A and B have coordinates a and b, then $AB = \left| a - b \right|$.

AXIOM N-6   Each two points A and B lie on a line, and if $AB < \alpha$, that line is unique.

AXIOM N-7   Each three noncollinear points determine a plane.

AXIOM N-8   If points A and B lie in a plane and $AB < \alpha$, then the line determined by A and B lies in that plane.

AXIOM N-9   If two planes meet, their intersection is a line.

AXIOM N-10   Space consists of at least four noncoplanar points, and contains three noncollinear points. Each plane contains at least three noncollinear points, and each line contains at least two distinct points A and B such that $AB < \alpha$.

AXIOM N-11: Plane Separation Postulate   Let ℓ be any line lying in any plane P. The set of all points in P not on ℓ consists of the union of two subsets H_1 and H_2 of P such that

(1) H_1 and H_2 are convex sets

(2) H_1 and H_2 have no points in common

(3) if A lies in H_1 and B lies in H_2 such that $AB < \alpha$, line ℓ intersects segment $\overline{AB}$.

AXIOM N-12: Existence of Angle Measure   To every angle there corresponds a unique, positive real number θ, $0 < \theta < 180$.

AXIOM N-13: Angle Addition Postulate   If D lies in the interior of $\angle ABC$, then $m\angle ABC = m\angle ABD + m\angle BDC$, and conversely.

AXIOM N-14: Protractor Postulate   The set of rays lying on one side of a line $\ell = \overrightarrow{OA}$ having the same endpoint O including ray $\overrightarrow{OA}$ may be assigned to the real numbers θ, for which $0 \leq \theta < 180$, called **coordinates**, in such a manner that

(1) each ray is assigned a unique coordinate θ

(2) each coordinate is assigned to a unique ray

(3) the coordinate of ray $\overrightarrow{OA}$ is 0

(4) if rays $\overrightarrow{OP}$ and $\overrightarrow{OQ}$ have coordinates θ and ϕ, then $m\angle POQ = \left| \theta - \phi \right|$.

AXIOM N-15   A linear pair of angles is a supplementary pair.

AXIOM N-16: SAS Postulate   If the SAS Hypothesis holds for two triangles under some correspondence between their vertices, then the triangles are congruent.

NOTE: These axioms are meaningful whether the metric is bounded or unbounded. If distances are unbounded, which case will be designated by the symbol $\alpha = \infty$, since $AB < \alpha$ for all A and B, then the Incidence Axiom N-6, for example, becomes the usual axiom "two points determine a line." Thus, with a moment's reflection, we can see that *the axioms for neutral geometry become the axioms for absolute geometry under the hypothesis of an unbounded metric.*

BETWEENNESS RESULTS FOR A BOUNDED METRIC

We define betweenness exactly as in absolute geometry. The segment $\overline{AB}$ and ray $\overrightarrow{AB}$ are also defined as before, provided $AB < \alpha$. This will guarantee that the line $\overleftrightarrow{AB}$ through A and B is unique in this case. We also note that the material on betweenness in Section 2.7 is valid here by restricting to nonnegative coordinates.

LEMMA A

Given a line ℓ and a point $A \in \ell$, there exist B and C on ℓ such that B-A-C and $BC < \alpha$.

Proof: This is an exercise on interpreting the Ruler Postulate for nonnegative coordinates.

LEMMA B

If A-B-C, then $AB < \alpha$, and $\overrightarrow{AB} = \overrightarrow{BA} \cup \overrightarrow{BC}$.

Proof: Since $AB + BC = AC \le \alpha$, then $AB < \alpha$. The rest follows from Axioms N-4 and N-6.

LEMMA C

Suppose that P-A-Q, $\ell = \overrightarrow{AP} \cup \overrightarrow{AQ}$, and $AB = \alpha$ for some point B lying on ray $\overrightarrow{AP}$. For any two distinct points $C \in \overrightarrow{AP}$ and $D \in \overrightarrow{AQ}$, distinct from A and B and such that $AC + AD > \alpha$, then C-B-D.

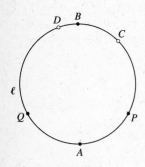

Figure 6.57

Proof: (Figure 6.57)

(1) By the Ruler Postulate, we may take coordinates on ray $\overrightarrow{AP}$ such that $A[0]$, $C[c]$, and $B[\alpha]$ where $0 < c < \alpha$. Hence, A-C-B.

(2) Now by Axiom N-4, either D-A-C, A-D-C, C-D-B, or C-B-D (the desired relation).

(3) If D-A-C, then $DC = DA + AC > \alpha$, which contradicts the definition of α.

(4) If A-D-C, then $D \in \overrightarrow{AC}$ (since $AC = c < \alpha$). By Theorem 2, Section 2.7, which is valid here, $\overrightarrow{AP} = \overrightarrow{AC} = \overrightarrow{AD} = \overrightarrow{AQ}$, and $P \in \overrightarrow{AQ} \rightarrow\leftarrow$ (It was assumed that P-A-Q).

(5) If C-D-B, then since the points B, D, C, and A all belong to ray $\overrightarrow{BC}$, A-C-$B \rightarrow A$-C-D-B or, again, $D \in \overrightarrow{AC}$, $AD < AB = \alpha$, and $\overrightarrow{AQ} = \overrightarrow{AD} = \overrightarrow{AC} = \overrightarrow{AP} \rightarrow\leftarrow$

(6) This leaves the only remaining case, C-B-D.

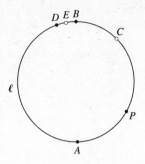

Figure 6.58

THEOREM 1

If A lies on line ℓ, there is a unique point B on ℓ such that $AB = \alpha$.

Proof: Lemma A and the Ruler Postulate guarantee the existence of a ray $\overrightarrow{AP}$ on ℓ ($AP < \alpha$) and $B \in$ ray $\overrightarrow{AP}$ such that $AB = \alpha$. Suppose a second point $D \neq B$ exists on line ℓ such that $AD = \alpha$. Applying Lemma C, where point C is chosen arbitrarily on ray $\overrightarrow{AP}$, we obtain C-B-D. Hence $BD < \alpha$ and there exists E on $\overline{BD}$ such that B-E-D (Figure 6.58). By Axiom N-4, either A-B-E, B-A-E, E-A-D, or E-D-A. But none of these can hold, since they all lead to either $AB < \alpha$ or $AD < \alpha$, a contradiction. Hence, $B = D$.

ELLIPTIC GEOMETRY

The next result generalizes what is a particularly obvious property on the earth's surface: *Each meridian line (great circle) which passes through the north pole also passes through the south pole.*

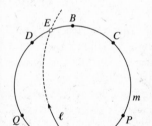

Figure 6.59

THEOREM 2

If $AB = \alpha$, then every line passing through A also passes through B.

Proof: (Figure 6.59) Let ℓ be any line passing through A. If m is a line through A and B, since the two rays from A on m contain points with arbitrary positive coordinates, we may choose the points $P[\frac{1}{2}\alpha]$ and $C[\frac{3}{4}\alpha]$ on one ray, and $Q[\frac{1}{2}\alpha]$ and $D[\frac{3}{4}\alpha]$ on the other, with $A[0]$ as origin, and such that P-A-Q. Then since $AC + AD = \frac{3}{4}\alpha + \frac{3}{4}\alpha > \alpha$, by Lemma C, C-B-D. Now if $\ell \neq m$, P and Q lie on opposite sides of ℓ, and C and D do also (rays $\overrightarrow{AP}$ and $\overrightarrow{AQ}$ lie in opposite half-planes). Hence ℓ cuts segment $\overline{CD}$ at some point E (note that $CD = CB + BD = \frac{1}{2}\alpha < \alpha$), and ℓ and m have both A and E in common. By Axiom N-6, $AE = \alpha$. By Theorem 1, $E = B$ and ℓ passes through B, as desired.

COROLLARY

If $AB = \alpha$ and C is any other point in the plane distinct from A or B, then A-C-B holds.

Proof: This proof is an easy consequence of the previous results, and will be left as Problem 12.

Note the important fact which the preceding result implies:

If A, B, and C are any three noncollinear points, then AB, BC, and AC are each less than α, hence segments $\overline{AB}$, $\overline{BC}$, and $\overline{AC}$ exist and $\triangle ABC$ is uniquely determined.

Another consequence of Theorem 2 is that to each point *A* there corresponds a unique point *A'* in the plane for which $AA' = \alpha$. Hence, we can define point *A'* as the **opposite pole** of point *A*.

We now set up the results needed in order to obtain the Angle Sum Theorem that characterizes elliptic geometry. The Crossbar Theorem (Section 2.8) will be needed for the proof of one of the lemmas which follows, which is seen to be valid in the present setting. (You should convince yourself that the proof given earlier works here.)

LEMMA D

If the sides of a triangle are each of length less than $\frac{1}{2}\alpha$, then the measure of an exterior angle of the triangle is greater than that of either opposite interior angle.

The proof of Lemma D depends on another lemma.

LEMMA E

In $\triangle ABC$, if $AB < \frac{1}{2}\alpha$, $AC < \frac{1}{2}\alpha$, and *M* is the midpoint of segment $\overline{BC}$, then $AM < \frac{1}{2}\alpha$ (Figure 6.60).

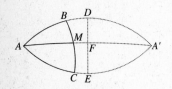

Figure 6.60

Proof: First, $AM < \alpha$, or else line $\overleftrightarrow{AB}$ passes through *M* (Theorem 2) and it would follow that *A*, *B*, and *C* are collinear. The lines $\overrightarrow{AB}$, $\overrightarrow{AM}$, and $\overrightarrow{AC}$ all meet at point *A'* such that $AA' = \alpha$ (Theorem 2), and since $AB < \frac{1}{2}\alpha$ and $AC < \frac{1}{2}\alpha$, then there exist points *D* and *E* on rays $\overrightarrow{AB}$ and $\overrightarrow{AC}$ such that *A-B-D*, *A-C-E*, and $AD = AE = \frac{1}{2}\alpha$. Now $DE < \alpha$, or else line $\overleftrightarrow{BD}$ passes through *E* and again *A*, *B*, and *C* would be collinear. Hence, by the Crossbar Theorem, ray $\overrightarrow{AM}$ meets segment $\overline{DE}$ at some interior point *F*, hence either *A-M-F*, *M = F*, or *A-F-M*. The remainder of the proof breaks down into three major steps.

(1) Show that $AF = \frac{1}{2}\alpha$. (See Problem 18.)

(2) *M* lies on the *A*-side of line $\overleftrightarrow{DE}$. (See Problem 19.)

(3) Therefore, *A-M-F* and $AM < AF$, or $AM < \frac{1}{2}\alpha$, as desired.

The proof of Lemma D now follows exactly the same pattern set in Section 3.4 (Figure 6.61). The only modification is in the use of the Segment Doubling Theorem; here we must be certain that segment $\overline{AM}$ has length less than $\frac{1}{2}\alpha$. But since $AB < \frac{1}{2}\alpha$ and $AC < \frac{1}{2}\alpha$, Lemma E guarantees this. The rest of the argument is exactly the same as Euclid originally constructed it.

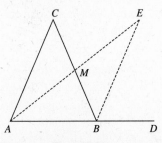

Figure 6.61

COROLLARY

The base angles of an isosceles triangle whose legs are of length less than $\frac{1}{2}\alpha$ are acute.

(Proof left as a problem; see Figure 6.62 for a hint.)

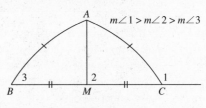

$$m\angle 1 > m\angle 2 > m\angle 3$$

Figure 6.62

ANGLE SUM THEOREM FOR ELLIPTIC GEOMETRY

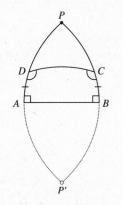

Figure 6.63

THEOREM 3

The angle sum of any triangle exceeds 180.

Proof: It suffices to prove the hypothesis of the obtuse angle for Saccheri Quadrilaterals (see Section 3.7). Thus, let $\diamond ABCD$ be any Saccheri Quadrilateral, with base $\overline{AB}$ (Figure 6.63). Locate point P on ray $\overrightarrow{AD}$ such that $AP = \frac{1}{2}\alpha$, and P' on the opposite ray such that $PP' = \alpha$.

(1) P-A-P' and $AP' = \frac{1}{2}\alpha = AP$ so that $\triangle APB \cong \triangle AP'B$ by SAS.

(2) P-B-P' and $BP = BP' = \frac{1}{2}\alpha$.

(3) $\triangle APB$ is isosceles, with $m\angle PAB = m\angle PBA = 90$.

(4) Since the perpendicular to $\overleftrightarrow{AB}$ at B is unique, $C \in$ line $\overleftrightarrow{BP}$; since C and D lie on the same side of $\overleftrightarrow{AB}$, $C \in$ ray $B\overset{\frown}{P}$.

(5) It follows that A-D-P and B-C-P, for otherwise, since $BC = AD$, either $D = P = C$ or A-P-D and B-P-C, so that segments $\overline{AD}$ and $\overline{BC}$ meet at P. →←

(6) Since $AD = BC$, we have $PD = PA - AD = \frac{1}{2}\alpha - BC = PB - BC = PC$.

(7) $\triangle PCD$ is an isosceles triangle with legs of length less than $\frac{1}{2}\alpha$, hence by the corollary to Lemma D, $m\angle PCD < 90 \rightarrow m\angle BCD = m\angle ADC > 90$. This proves the Hypothesis of the Obtuse Angle.

THE CONCEPT OF EXCESS AND AREA FOR ELLIPTICAL GEOMETRY

Classical spherical geometry involves the concept of **excess**—the amount by which the angle sum of a triangle (or polygon) exceeds its Euclidean counterpart (180 for a triangle and $180(n - 2)$ for an n-gon). We have been able to derive the whole thing axiomatically, however. Continuing with the development, we define excess for a polygon in general.

DEFINITION

The **excess** of a convex, n-sided polygon $P_1P_2P_3...P_n$ is defined to be the number

$$\varepsilon(P_1P_2P_3...P_n) \equiv m\angle P_1 + m\angle P_2 + m\angle P_3 + ... + m\angle P_n - 180(n-2)$$

By Theorem 3, we know that the excess of any triangle is positive. It can be proven that this is also true of any polygon as well, but its proof depends on an inductive argument showing that we may decompose an n-gon into $n-2$ triangles, which we will not pursue. The following theorem also requires an inductive argument, which we shall omit.

THEOREM 4: ADDITIVITY OF EXCESS

Suppose that a convex n-gon P has been decomposed into a finite number m of convex subpolygons $P_1, P_2, ... , P_m$, and the excess of P_k is denoted $\varepsilon(P_k)$. Then

$$\varepsilon(P) = \varepsilon(P_1) + \varepsilon(P_2) + ... + \varepsilon(P_m)$$

We shall be content to prove only the following special case, which nevertheless is precisely the case we need later for another significant result in elliptic geometry

LEMMA F

Suppose that $\overline{AD}$ is any cevian of $\triangle ABC$ to side $\overline{BC}$. Then

$$\varepsilon(\triangle ABC) = \varepsilon(\triangle ABD) + \varepsilon(\triangle ADC)$$

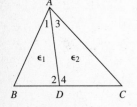

Figure 6.64

Proof: (Figure 6.64.) Let $\varepsilon, \varepsilon_1, \varepsilon_2$ be the respective excesses. Then

$$\varepsilon = m\angle A + m\angle B + m\angle C - 180$$

$$\varepsilon_1 = m\angle B + m\angle 1 + m\angle 2 - 180$$

$$\varepsilon_2 = m\angle 3 + m\angle 4 + m\angle C - 180$$

Summing the last two equations yields

$$\begin{aligned} \varepsilon_1 + \varepsilon_2 &= m\angle B + m\angle 1 + m\angle 2 + m\angle 3 + m\angle 4 + m\angle C - 360 \\ &= m\angle B + (m\angle 1 + m\angle 3) + m\angle C + 180 - 360 \\ &= m\angle B + m\angle A + m\angle C - 180 = \varepsilon \end{aligned}$$

or

$$\varepsilon_1 + \varepsilon_2 = \varepsilon$$

In view of the additivity of excess, and the fact that it is positive for all polygons, excess provides a concept for area, just as defect did for hyperbolic geometry. Accordingly, we define the **area** K of $\triangle ABC$ to be, for some positive constant k,

$$K = k(m\angle A + m\angle B + m\angle C - 180),$$

and similarly for convex polygons. (The constant k. depends on the unit of area chosen.) Again, just as in the case of defect in hyperbolic goemetry, this concept will be found to satisfy the properties for area outlined in Section 2.3. However, in the case of elliptic geometry, it is possible to make this result a little less mysterious by going to the spherical model.

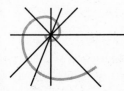

MOMENT FOR DISCOVERY

Area of a Spherical Triangle on a Unit Sphere

The area of a sphere of radius $r = 1$ is 4π. The area of a hemisphere is therefore 2π. A **lune** is the region of a sphere bounded by two great semicircles, meeting at two poles. The **angle** of the lune is just the obvious spherical angle formed by the

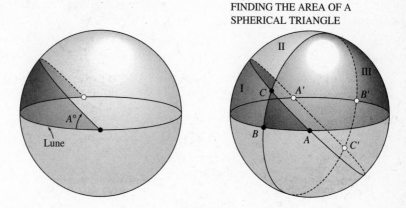

FINDING THE AREA OF A SPHERICAL TRIANGLE

Figure 6.65

two spherical rays emanating from either pole. A lune L with angle of measure A has area equal to the proportionate amount of L to a hemisphere, which is a "lune" with angle measure 180. That proportional amount is $A/180$. Thus,

$$\frac{\text{Area } (L)}{2\pi} = \frac{A}{180} \quad \text{or} \quad \text{Area } (L) = \frac{\pi}{90} \cdot A$$

Use this formula in the following analysis for the area (as ordinarily understood) of any spherical triangle.

In the figure is indicated the extension of the sides of $\triangle ABC$ to obtain the great circles shown, meeting at the opposite poles A', B', and C'. It is clear that

$$\triangle ABC \cong \triangle A'B'C' \;\rightarrow\; K = K'$$

where K and K' are the areas of the triangles. Note that Regions I and III make up the rest of lunes A and B outside $\triangle ABC$, having areas $(\pi/90)A$ and $(\pi/90)B$. Thus,

$$\text{Area (Lune } A) = K + I = (\pi/90)A$$

where I and K also denote the areas of the figures they represent. Region II is the remainder of the lune of $\angle C' \cong \angle C$ after deleting $\triangle A'B'C'$.

1. Obtain formulas for $K + III$ and $K' + II = II + K$.
2. What type of region is $H \equiv K \cup I \cup II \cup III$? What is its area?
3. Deduce a formula for $K = \text{Area } \triangle ABC$. What did you discover? (Use algebra to write the expression you obtained in terms of the quantity $A + B + C - 180$.)

NONEXISTENCE OF SIMILAR TRIANGLES IN ELLIPTIC GEOMETRY

A direct result of this concept for area is the AAA Congruence Criterion for triangles in elliptic geometry (completely analogous to the situation in hyperbolic geometry).

> If two triangles have the three angles of one congruent, respectively, to the three angles of the other, the triangles are congruent.

The proof of this will be left as Problem 5. As in the case of hyperbolic geometry, a corollary is that there can be no similar, noncongruent triangles in elliptic geometry.

The AAA Congruence Criterion, obtained axiomatically, must obviously be a property of the sphere. It is somewhat difficult, however, to visualize this result or to obtain a direct proof (on the sphere) without resorting to rather extensive use of spherical trigonometry. Thus, in this instance, we can appreciate the power of the axiomatic method.

PROBLEMS   (§6.8)

_____ *Group A* _____

1. A triangle in elliptic geometry has angles of measure 35.1, 63.8, and 84.5. Find its excess.
2. A pentagon in elliptic geometry has angles of measure 158.3, 69.1, 93.5, 105.6, and 118.6. Find its excess.
3. In the previous Discovery Unit, you should have found the formula

$$\text{Area } \triangle ABC = \frac{\pi}{180}\,(\text{Excess } \triangle ABC) \equiv \frac{\pi}{180}\,(A + B + C - 180)$$

which gives the actual surface area on the unit sphere covered by $\triangle ABC$. Use this to solve the following: An equilateral triangle covers an area equivalent to one-half a hemisphere on a unit sphere. What size must each of its angles be?

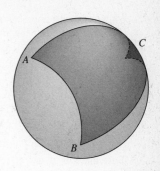

4. If a sphere has radius $100/\pi$, what must the value of α be for the geometry taking place on that sphere?

5. The AAA Congruence Criterion for Elliptic Geometry Suppose that in $\triangle ABC$ and $\triangle PQR$, $\angle A \cong \angle P$, $\angle B \cong \angle Q$, and $\angle C \cong \angle R$. Assume that no two corresponding sides are congruent. Then two of the sides of one of the triangles have the greater lengths, so assume that $AB > PQ$ and $AC > PR$. Locate D on $\overline{AB}$ such that $AD = PQ$ and E on $\overline{AC}$ so that $AE = PR$. Use the excesses as labelled in the figure to gain a contradiction, and complete the proof that $\triangle ABC \cong \triangle PQR$.

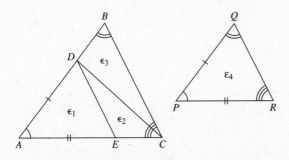

6. Prove that the elliptic plane has bounded area, and that relative to the area constant k introduced earlier, its area is $720k$. (**HINT:** Show that a half-plane is the union of 4 nonoverlapping trirectangular triangles (three right angles) and their interiors.)

7. With the excesses as labelled in the figure, prove that

$$\varepsilon(\triangle ABC) = \varepsilon_1 + \varepsilon_2 + \varepsilon_3 + \varepsilon_4 + \varepsilon_5 + \varepsilon_6$$

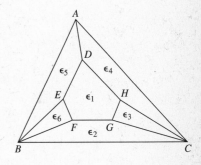

8. Identify, with proof, the set of points on the same side of a line in elliptic geometry and equidistant from that line (See the figure below.). Is it

EQUIDISTANT LOCUS IN
SPHERICAL GEOMETRY

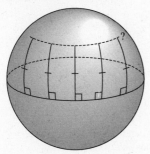

another line as it is in Euclidean geometry?

9. What is the least upper bound of either of the nonright angles of a right triangle on the sphere? Attempt proving your result axiomatically.

10. What is the least upper bound of the area covered on the unit sphere by an equiangular quadrilateral?

_____ *Group B* _____

11. Prove that every pair of distinct lines in elliptic geometry has a common perpendicular. (You may use the result of Problem 23, which is independent of this result.)

12. Write an explicit proof of the corollary of Theorem 2.

13. Prove that if A, B, and C are the measures of the angles of a right triangle in the elliptic plane ($C = 90$), then

$$90 < A + B < 270$$

(It can be shown that there exists a right triangle for each pair of values A, B whose sum lies in this range.)

ELLIPTIC TRIGONOMETRY FOR A RIGHT TRIANGLE

(Based on the value $\alpha = \pi$—sphere of unit radius)

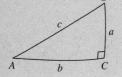

$$\sin A = \frac{\sin a}{\sin c} \qquad \sin B = \frac{\sin b}{\sin c}$$

$$\cos A = \frac{\tan b}{\tan c} \qquad \cos B = \frac{\tan a}{\tan c}$$

$$\tan A = \frac{\tan a}{\sin b} \qquad \tan B = \frac{\tan b}{\sin a}$$

Pythagorean Relation: $\cos c = \cos a \cos b$

(**NOTE:** Radian mode on calculator must be used for all quantities involving spherical distance.)

In the following exercises some formulas from spherical trigonometry will be needed. (See table.)

14. A spherical right triangle $\triangle ABC$ with hypotenuse $\overline{AB}$ has $c = \frac{1}{2}\pi$ and $b = \frac{1}{4}\pi$. Calculate a, A ($= m\angle BAC$), and B ($= m\angle ABC$). Interpret your result on the unit sphere.

15. Find the hypotenuse c and the angles A and B of a right triangle if $a = \frac{1}{2}\pi$ and $b = \frac{1}{3}\pi$. Interpret on the unit sphere.

16. In right triangle $\triangle ABC$ with right angle at C, we are given $A = 45$ and $B = 60$. Show that this information uniquely determines the triangle. (Find all possible solutions for a, b, and c, accurate to four decimals.)

17. If in right triangle $\triangle ABC$ $a = .03$ and $b = .04$, find c, and verify that $c \approx .05$, approximating a Euclidean right triangle. Solve for A and B and show that $A + B \approx 90$.

(**NOTE:** These results illustrate the fact that sufficiently small spherical triangles have Euclidean-like behavior.)

18. In Step 1 of the proof of Lemma E for Theorem 3 (Figure 6.60), prove that $AD = AE = AF = \frac{1}{2}\alpha$ by use of congruence criteria for triangles. (**HINT:** First prove that $m\angle ADE = m\angle EDA' = 90$.

_____ *Group C* _____
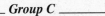

19. In Step 2 of the proof of Lemma E for Theorem 3 (Figure 6.60), prove that M lies on the A-side of line $\overleftrightarrow{DE}$ by using rays, half-planes, and betweenness considerations.

20. Show that the set of all points due north of a line that is uniformly $\frac{1}{10}$th of a unit south of the equator on the unit sphere is not convex (by example, using some suitable line segment).

21. Triangle $\triangle ABC$ is an equilateral triangle in the elliptic plane, with excess $\varepsilon(\triangle ABC) = 3\theta - 180$. If L, M, and N are the midpoints of the sides of $\triangle ABC$, and $m\angle LMN = \phi$, show that $\theta > \phi$ and $\varepsilon(\triangle AMN) = \theta - \phi$.

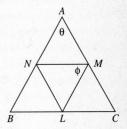

22. Show that if A, B, and C are distinct, collinear points, and if $AB + AC \leq \alpha$, then either A-B-C, B-A-C, or A-C-B holds. Interpret this result if $\alpha = \infty$.

23. **Parallel Postulate for Elliptic Geometry** Prove axiomatically that any pair of lines in the elliptic plane intersect. Hence, if line ℓ and point A not on ℓ be given, no line through A is parallel to line ℓ. (**HINT:** Choose A and B on ℓ such that $AB = \alpha$, and C any point on the other line m.)

The origin of the concept of non-Euclidean geometry was actually Euclid himself, who has been called the first non-Euclidean geometer. His development of parallelism led directly to the question of whether the Fifth Postulate is provable in absolute geometry. The Half-Plane Model and Circular Disk Model discovered by E. Beltrami and H. Poincaré show conclusively that such a proof is not possible, although many very good mathematicians in the past have gone to great effort to find a proof. János Bolyai's development and that of N. Lobachevski were the first extensive treatments of the hypothesis that deny the Parallel Postulate—all without apparent contradiction. Bolyai called this development **absolute science**—that which is *absolutely true* whichever assumption about parallels is made. In his development, Bolyai examined the geometry of a sphere in the ordinary three-dimensional space of absolute geometry, then let the radius of the sphere approach infinity. The resulting surface (called a **horocycle**) has as its intrinsic geometry ordinary Euclidean geometry. The formulas gleaned from this limiting case, however, were the trigonometric formulas for hyperbolic geometry.

The chief property of triangles, when there is more than one parallel to a line from a given point (the **Hyperbolic Parallel Postulate**), is that regarding angle sums: The sum of the measures of the angles of any triangle is less than 180. The study of Bolyai on asymptotic triangles was presented here in some detail, and the resulting study of limit parallels (here called **asymptotic parallelism**) and hyper-parallels (**divergent parallelism**) followed. One of the chief distinctions between these two types of parallelism is that although the property of symmetry holds for both, the Transitivity Law holds only for asymptotic parallelism. Another distinctive property is that the distance between two lines which are asymptotically parallel converges to zero, while that between two divergent parallels tends to infinity. (In Euclidean geometry, the distance between parallel lines remains constant.)

A brief development of the axioms for absolute geometry adapted to allow for a bounded metric (called **neutral geometry**) was undertaken, and the result is an axiomatic basis for spherical or **elliptic geometry**. The chief property of a triangle characterizing elliptic geometry is that the sum of the measures of the angles of any triangle is greater than 180. The two geometries hyperbolic and elliptic thus have certain analogies, such as the AAA Congruence Criterion for both, nonexistence of similar, noncongruent triangles, and most spectacularly, the completely parallel formulas for non-Euclidean trigonometry, which appeared in the problem sections for Sections 6.6 and 6.8.

Answer each of the following questions either True (**T**) or False (**F**).

1. In the Beltrami-Poincaré Half-Plane Model, an ordinary Euclidean ray can sometimes represent a ray in hyperbolic geometry.

2. In Poincaré's Circular Disk Model, an ordinary Euclidean ray can sometimes represent a ray in hyperbolic geometry.

3. Poincaré's Circular Disk Model, like the Half-Plane Model, is conformal, and every aspect of the hyperbolic plane regarding angle measure may be perfectly illustrated.

4. The AAA Congruence Criterion for both hyperbolic and elliptic geometry implies that a triangle is completely determined by specifying its three angles.

5. In hyperbolic geometry, as in Euclidean geometry, a circle may be drawn through any three given points, as long as those points do not lie on a straight line.

6. In hyperbolic geometry, as in Euclidean geometry, if two lines have a common perpendicular, they are parallel.

7. In hyperbolic geometry, as in Euclidean geometry, if two lines are parallel, they have a common perpendicular.

8. All the lines in the hyperbolic plane divergently parallel to the same fixed line possess the same fixed common perpendicular.

9. It can be proven axiomatically that the elliptic plane has bounded area.

10. As late as 1830, it was thought by most scholars that Euclid's Parallel Postulate could be proven within the framework of absolute geometry.

APPENDIX A: BIBLIOGRAPHY

Baravalle, H., *Eighteenth Yearbook.* Washington, D.C.: National Council of Teachers of Mathematics, 1945.

Bonola, R., *Non-Euclidean Geometry.* New York: Dover Publications, Inc., 1955.

Burton, D.M., *History of Mathematics: An Introduction, Second Edition.* Dubuque, Iowa: William C. Brown Publishers, 1991.

Coxeter, H.S.M., *Introduction to Geometry.* New York and London: John Wiley and Sons, Inc. 1961.

Coxeter, H.S.M., and S.L. Grietzer, *Geometry Revisited (The New Mathematics Library, Volume 19).* Washington, D.C.: Mathematical Association of America, 1967.

Dudeney, H.E., *Amusements in Mathematics.* New York: Dover Publications, Inc., 1958.

Eves, H., *A Survey of Geometry, Volume One.* Boston: Allyn and Bacon, Inc., 1963.

_____, *An Introduction to the History of Mathematics, Third Edition.* New York: Holt, Rinehart and Winston, 1969.

Graustein, W.C., *Introduction to Higher Geometry.* New York: Macmillan, 1930.

Greenberg, M.J., *Euclidean and Non-Euclidean Geometries.* San Francisco: W.H. Freeman and Company, 1974.

Grünbaum, B., and G.C. Shephard, *Tilings and Patterns.* New York: W.H. Freeman and Company, Inc., 1987.

Heath, T.L., *The Thirteen Books of Euclid's Elements, Volume One.* New York: Dover Publications, Inc., 1956.

Hirsch, C.R., H.L. Schoen, A.J. Samide, D.O. Coblentz, and M.A. Norton, *Geometry.* Glenview, Illinois: Scott, Foresman and Company, 1990.

Jacobs, H.R., *Geometry.* New York: W.H. Freeman and Company, 1987.

Krause, E.F., *Taxicab Geometry.* New York: Dover Publications, Inc., 1975.

Lockwood, E.H., *A Book of Curves.* New York and London: Cambridge University Press, 1963.

Milnor, J., "Hyperbolic Geometry: The First 150 Years," *The Mathematical Heritage of Henri Poincaré, Proceedings of Symposia in Pure Mathematics of the American Mathematical Society,* **39**, 25–40.

Moise, E., *Elementary Geometry from an Advanced Standpoint.* Reading, Mass.: Addison-Wesley Publishing Company, Inc., 1963.

Coxford, A., Z. Usiskin, and D. Hirschorn, *The University of Chicago Study Mathematics Project, Geometry.* Glenview, Illinois: Scott, Foresman and Company, 1991.

Sommerville, D.M.Y., *The Elements of Non-Euclidean Geometry.* New York: Dover Publications, Inc., 1958.

Yaglom, I.M., *Geometry Transformations.* New York: Random House, Inc., 1962.

Appendix B: Answers to Selected Problems

Section 1.1

1. Angle sum equals 360. **5.** By summing angle measures in isosceles triangles $\triangle AMC$ and $\triangle MCB$, deduce that $m\angle C = 90$. **7. (b)** From (a), $2 = 6x + z$, $4 = 8x + y + z$, and $6 = 6x + 4y + z$ or $x = \frac{1}{2}$, $y = 1$, $z = -1$. $\therefore K = \frac{1}{2}B + I - 1$. **13. (b)** If $m\angle A = 120°$ in isosceles $\triangle ABC$, draw lines AD and AE forming angles of measure 30°, 60°, and 30°, then subdivide equilateral $\triangle ADE$.

Section 1.2

1. Place point P of each of the four pieces at A, B, M, and N. **3. (b)** $I = 450\pi$, $II = 128\pi$. **(c)** $2(I) = \pi r^2 = \pi \cdot \frac{1}{4}a^2$, $2(II) = \pi \cdot \frac{1}{4}b^2$, $2(III) = \pi \cdot \frac{1}{4}c^2$; use $c^2 = a^2 + b^2$. **5. (b)** $\tau = \frac{1}{2}(\sqrt{5} + 1)$. **7.** Let $AD = BC = 2a$; $MC^2 = 4a^2 + a^2 = 5a^2$; $AE = AM + ME = AM + MC = \sqrt{5}a + a \to \therefore AE/AD = (\sqrt{5} + 1)a/2a = \tau$. **9.** Use a series expansion for $\sin(\pi + \varepsilon) = -\sin \varepsilon$, where $\varepsilon = $ error; if $P = \pi + \varepsilon$, what estimate holds for $P + \sin P$? **11.** (1) Measure of angle of $\diamond$ at $C = 270° - C \to$ Area $= ab \sin(270° - C) = -ab \cos C$. (2) From the square, cut off a triangle $\cong \triangle ABC$ at top and right, and fit these into spaces at bottom and left. (3) Area of square $= c^2 = $ square of side $a + $ square of side $b + $ two parallelograms $= a^2 + b^2 - 2ab \cos C$. **13.** In general, $s_n = $ radical expression of formula (starting with s_1, s_2, etc.); polygon has $3 \cdot 2^n$ sides. **15.** $4 \sin^{-1} \frac{1}{2}\sqrt{a}$ (twice the length of arc of unit circle subtended by chord of length a). **17.** By similar triangles, $GD/DE = FD/DB = (DE/DB)/DB \to GD = DE^2/DB^2 = (\frac{1}{2})^2/[(\frac{7}{8})^2 + 1] = 16/113$; $CL = 3 + 16/113$ (error < 0.00000027).

Section 1.3

1. $AC = 10$, $BD = 11$. **3.** $m = 20$. **5. (a)** $K = 138\sqrt{6}$; $m = \sqrt{415}$, $n = 6\sqrt{10}$. **9.** $m\angle CAB = m\angle CDB \to \triangle ABE \sim \triangle DBC \to x/c = a/n$ and $d/y = n/b$; $ac = nx$ and $bd = ny$, or $ac + bd = n(x + y) = nm$. **11.** Let M play role of m in Brahmagupta's Formula, $A = az$, $B = cy$, $C = bz$, $D = cx$; simplify by algebra (similarly for N). **13.** $m\angle CAB = \frac{1}{2}m\overarc{BC} = m\angle BDC \to \triangle ABE \sim \triangle DBC \to h/b = a/\delta$. **15.** Write $m^2 = (mn) \cdot (m/n)$ and use results of Problem 14 (b). **17.** M, N, and $K = \frac{1}{2}MN$ are integers since one of M or N must be even; by results of Problems 11 and 14 (c), and since $K = \frac{1}{2}MN$ the diagonals are $\perp$, $MN\delta^2 = (AB + CD) \cdot (AD + BC) = (az \cdot cx + bz \cdot cy) \cdot (az \cdot cy + cx \cdot bz) = c^2z^2MN \to \delta = cz$, an integer.

Section 1.4

3. By Law of Cosines, $x^2 = 2y^2(1 - \cos 20°) = 4y^2\sin^2\frac{1}{2}(20°)$; $y = AM \cdot \sec 20° = \frac{1}{2}s \sec 20°$. **5. (a)** Cardioid with cusp point at A. **(b)** Limaçon with point of self-intersection at A. **7.** $D = (5.28, 5.04)$ (exact). $DE \approx 5.532$, $EF = \sqrt{13} \approx 3.606$, $DF = (4/5)\sqrt{58} \approx 6.093$, $x \approx 9.220$, $y \approx 7.616$, $z \approx 3.606$, $R = \frac{1}{2}AB = 7.5$ (exact), $ax/2R \approx 9 \cdot (9.220)/15 = 5.532 \approx DE$, etc. **9.** Since triangle is equilateral, incircle passes through midpoints of sides and its center is the orthocenter, so it is the Nine-Point Circle of the triangle and it will bisect segments mentioned. **11.** (2) $\triangle ABD$. (3) Law of Cosines for $\triangle ABC$. (4) Substitution of (3) into (2). (5) $d^2 = p(p - 1)a^2 + (1 - p)c^2 + pb^2 = -pqa^2 + qc^2 + pb^2$. **13.** $ax = by + cz$ (Ptolemy's Theorem) or $ax/2R = by/2R + cz/2R \to EF = DF + DE$.

Section 2.1

1. (c) If it rains, then Mary is unhappy. **3.** *Given:* Right triangle. *Prove:* It has only one right angle. *Converse:* If triangle has but one right angle, it is a right triangle. **5.** *Given:* Medians of a triangle. *Prove:* Medians are concurrent. *Converse:* If three lines passing through the vertices of a triangle are concurrent, they are the medians. **7. (b)** Inflation accompanied by low unemployment. **9.** Conclusions (Justifications): (1) Either John lost key in homeroom, or on way home (all the logical cases). (2) Assume John lost key in homeroom (first logical case). (3) $\to\gets$ (John did not go to homeroom). (4) $\therefore$ John lost key on way home (Rule of Elimination). **11.** Conclusions (Justifications): (1) p (Given). (2) r or s (all the logical cases since $p \to r$ or $p \to s$). (3) r (first logical case). (4) x (Theorem 2). (5) $\therefore q$ (Theorem 4). (6) s (remaining logical case). (7) y (Theorem 3). (8) $\sim p \to\gets$ (Theorem 1 and Given). (9) $\therefore q$ (Rule of Elimination). **13.** There will be precisely 10 points and 5 lines.

Section 2.2

1. A, B, C, D not collinear; area of $\lozenge ACBD$ is one unit square. (See figure.)   **7. (b)** neither categorical nor complete   **9.** Point O has coordinates $(10, -30)$, so certain sums involving angles used in proof do not follow.   **11.** Through "point" PR (representing President), draw three "lines" E, P, and N (representing Executive, Program, and Nominating Committees), and locate "points" VP (Vice President) and ST (Secretary-Treasurer) on E; VP must belong to two further "lines," say A and B, as must ST (C and D), and A and P must have a member M_4 in common. Continue analysis in this fashion; show that more than seven members will ultimately violate By-Laws.   **13.** If points are taken as ordinary points in the plane and a segment joins two points iff they are adjacent, Axiom 5 disallows a pentagon or any triangle. If there is a quadrilateral present, the fifth vertex must join two nonadjacent vertices of that quadrilateral, which yields one model; if not, then one of the points joins the other four, yielding the other model.   **15.** Argument in problem also proves that M is not an interior point of circle O.

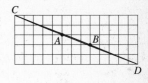

Exercise 1

Section 2.3

1 (a) ≈ 28.6667 **(b)** ≈ 29.4647 **(c)** ≈ 29.8233.   **3.** 24 sq. units   **5.** triangle   **7. (a)** 18 units **(b)** 12 sq. units, when $PF_1 = PF_2$ **(c)** area constant ($= 12$), perimeter variable with minimum 18 units (when $PF_1 = PF_2$)   **9.** $\sqrt{\pi}$ **11. (a)** 112.5 sq. yds. (7.5×15) **(b)** 112.5 sq. yds. **(c)** 143.2 sq. yds.   **13.** By result of Problem 6(b), area of central triangle in polygon equals $\frac{1}{2}as$; each central triangle is congruent, so by area axioms, $K = n \cdot \frac{1}{2}as = \frac{1}{2}a(ns) = \frac{1}{2}ap$.   **15. (a)** Horizontal line at height y cuts region in segment of constant length a, so desired area equals that of a rectangle, ab. **(b)** Use vertical lines; desired area equals that of right triangle with hypotenuse $y = 6x$, $0 \le x \le 4$, or 48 sq. units.   **17.** $C'/2r' = C/2r = $ constant $= \pi$; assuming $\lim\limits_{n\to\infty} p_n = C$, $\lim\limits_{n\to\infty} a_n = r$, and $\lim\limits_{n\to\infty} p'n = C$ (latter two were proven separately in Euclid's *Elements*), then by theorems on limits, $\lim\limits_{n\to\infty} \frac{1}{2}a_n p_n = \lim\limits_{n\to\infty} \frac{1}{2}rp'n = \frac{1}{2}Cr \to K = \frac{1}{2}Cr$. $\therefore$ $K = \frac{1}{2}(2\pi r)r = \pi r^2$.

19. Let $P_n = \text{Vol } \mathbf{P}_n$ for each n; in Step 3, $P_2 = \frac{1}{4}B_2 h_2 + 2P_3$, where $B_2 = \frac{1}{4}B_1 = (1/16)B$ and $h_2 = \frac{1}{2}h_1 = \frac{1}{4}h$ or $P_2 = (1/256)Bh + 2P_3 \to 4P_2 = (1/64)Bh + 8P_3$; final result is geometric series with $r = \frac{1}{4}$.

Section 2.4

1. See figure.   **3.** All by Axiom I-5. **5.** *Lines:* $\{1, 2\}$, $\{1, n\}$, $\{2, n\}$ for each $n \ge 3$, and $\{n : n \ge 3\}$; *Planes:* $\{n : n \ne 2\}$, $\{n : n \ne 1\}$, $\{1, 2, n\}$ for each $n \ge 3$.   **7.** For Axiom I-1, observation of figure for Problem 7 is adequate; the 12 given lines may be divided into 4 families of observable parallel lines: $\{\ell_1, \ell_2, \ell_3\}$, $\{\ell_4, \ell_5, \ell_6\}$, $\{\ell_7, \ell_8, \ell_9\}$, $\{\ell_{10}, \ell_{11}, \ell_{12}\}$; taken together, these lines pass through all points of the plane, hence, the Parallel Postulate.   **9.** *Points:* $S = \{A, B, C, D, E\}$; *Lines:* $\{A, B\}$, $\{A, B, C\}$, $\{A, D\}$, $\{A, E\}$, $\{B, C\}$, $\{B, D\}$, $\{B, E\}$, $\{C, D\}$, $\{C, E\}$, $\{D, E\}$; *Planes:* $\{A, B, C, D\}$, $\{A, B, C, E\}$, $\{A, D, E\}$, $\{B, D, E\}$, $\{C, D, E\}$.   **11. (a)** Use Axiom I-5. **(b)** Consider any line $\overleftrightarrow{AB}$ (Axiom I-5), point C not on $\overleftrightarrow{AB}$ (Axiom I-5), plane P determined by A, B, C (Axiom I-2), and $D \notin P$ (Axiom I-5). The plane P' determined by A, B, D (Axiom I-2) has two points in common with P, so $P \cap P'$ is a line (Axiom I-4), which must be line $\overleftrightarrow{AB}$ (Axiom I-1).   **13. (a)** It suffices to prove the Parallel Postulate; consider $\overleftrightarrow{AB}$ and $C \notin \overleftrightarrow{AB}$ in a given plane P—work with $D \notin P$ (Axiom I-5), and noncoplanar lines $\overleftrightarrow{AB}$ and $\overleftrightarrow{CD}$ Axiom I-6 yields a unique plane P' containing $\overleftrightarrow{CD}$ parallel to $\overleftrightarrow{AB}$. **(b)** Let ℓ and m be any two lines in P, and Q the plane of A and ℓ; by (a), Q contains a line $\ell' \parallel \ell$ passing through A, and by Axiom I-6, there is a plane P' through ℓ' parallel to m, which will be the unique plane through A parallel to P. **(c)** If ℓ has n points $A_1, \ldots, A_n$, and m is any other line containing point B_1, take the parallels to $\overleftrightarrow{A_1 B_1}$ passing through $A_2, \ldots, A_n$ to obtain n points on m, then show there can be no others; there are $n + 1$ lines through each point and n lines in each family of parallel lines, thus $n(n + 1)$ lines altogether. **(d)** To count the points in space, let ℓ and m be any two noncoplanar lines; there are n planes containing ℓ, one for each point on m, plus the plane parallel to m, each containing $n^2 - n$ points not counting those on ℓ, so there are $n(n^2 - n) + n^2 = n^3$ points; to count the number of planes, first prove there are $n^2 + n + 1$ planes passing through a fixed point A and n planes in each family of parallel planes $\to$ there are $n(n^2 + n + 1)$ planes in space.   **15. (a)** $\{A, K, U\}$, $\{V, L, H\}$, $\{S, I, N\}$ **(b)** $\{G, H, I, P, Q, R, Y, Z, \Sigma\}$ and $\{J, E, \Sigma, Y, K, F, D, Z, L\}$ **(c)** $\{F, T, P, U, Q, D, R, E, S\} \cap \{G, W, L, Q, F, S, \Sigma, M, B\} = \{F, Q, S\}$ **(d)** $\{O, A, Z, V, K, I, E, U, P\}$

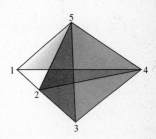

Exercise 1

Section 2.5

1. G-E-D-F **3.** A-B-D, A-C-E **5.** segments $\overline{AB}$, $\overline{BC}$, $\overline{AC}$; rays $\overrightarrow{AB}$, $\overrightarrow{AC}$, $\overrightarrow{BC}$ **7.** segments $\overline{AB}$, $\overline{AC}$; rays $\overrightarrow{AB}$, $\overrightarrow{BA}$, $\overrightarrow{CA}$ **9.** Conclusions: (1) A-B-C and B-A-C (Given) (2) $AB + BC = AC$, $BA + AC = BC$ (3) $AB + (BA + AC) = AC$ (4) $2AB = 0$ or $AB = 0$ (5) $A = B$ $\to\leftarrow$ **11.** Choose C on ℓ such that A-B-C, then use Theorem 1.   **13.** Let $c_1 = AM$, $c_2 = MB$, $b_1 = AN$, $b_2 = NC$, $d = MN$, and $e = MC$.

(1) $d \leq b_1 + c_1, d \leq b_2 + e$, and $e \leq a + c_2$ (2) $d \leq b_2 + a + c_2$ (3) $2d \leq b_1 + c_1 + b_2 + a + c_2 = a + b + c$ or $d \leq \frac{1}{2}(a + b + c)$ **15.** Assume $W \in B_t(R)$, hence $WR < t$. To show $WP < r$, write $WP \leq WR + RP < t + PR \leq (r - PR) + PR = r$; similarly, $WQ < s$. $\therefore W \in B_r(P) \cap B_s(Q)$.

Section 2.6

1. $H(B, \overrightarrow{AD})$; $H(A, \overrightarrow{DC}) \cap H(C, \overrightarrow{DA})$ **3. (a)** $H(E, \overrightarrow{DG})$. **(b)** $H(F, \overrightarrow{DE})$ **5.** No; in Euclidean geometry it can be the region between two parallel lines.   **7.** Of three points, at least two must always lie on same side of given line.   **9.** Either another half-plane, region between two parallel lines, interior of an angle, interior of some polygon, or region between two parallel lines to one side of a polygonal arc. **11.** Let $D \in H(A, \overrightarrow{BC}) \cap H(C, \overrightarrow{BA})$; since $B \in$ line $\overleftrightarrow{BC}$, $(\overrightarrow{BD}] \subseteq H(A, \overrightarrow{BC})$ and $(\overrightarrow{BD}] \subseteq H(C, \overrightarrow{BA}) \rightarrow (\overrightarrow{BD}] \subseteq$ Interior $\angle ABC$.   **13.** a closed or open half-plane, or half-plane plus either an open or closed ray along the boundary.

Section 2.7

1. Since $AO = 3$, $OD = 4$, and $AD = |3 - (-4)| = 7$, A-O-D holds, but not A-O-B, A-O-C, nor B-O-C.   **3. (a)** $3 \leq x \leq 10$ **(b)** $x \leq 5$ **(c)** $3 \leq x \leq 5$ **5.** Locate $T_1[\frac{1}{3}b]$ and $T_2[\frac{2}{3}b]$ by Ruler Postulate.   **7.** $XY = 15, 19, 27,$ or 61 **9.** an open ray   **11.** Conclusions, using the Lemma: (1) $\overline{AB} \subseteq \overline{AC}$. (2) $B \in \overline{AC}$. (3) $\overrightarrow{AC} \subseteq \overrightarrow{AB}$.   **13.** Let $A[a], B[b], C[c]$ be three collinear points on given line: six possible orders of three real numbers are: (1) $a < b < c$ or $c < b < a \rightarrow$ A-B-C (Theorem 1), (2) $a < c < b$ or $b < c < a \rightarrow$ A-C-B, (3) $b < a < c$ or $c < a < b \rightarrow$ B-A-C.

Section 2.8

1. (a) 40 **(b)** 160 **(c)** 160 **(d)** none   **3. (a)** 110 **(b)** 80 **(c)** 149 **(d)** 170   **5.** Conclusions: (1) Let $\overrightarrow{AS'}$ be the opposite ray of $\overrightarrow{AS}$. (2) $m\angle S'AW = 180 - m\angle SAK = 91$ (Linear Pair Axiom) (3) In H_2, $\overrightarrow{AS'}[91]$ and $\overrightarrow{AW}[60]$. (4) $m\angle S'AE = |91 - 60| = 31$ (Protractor Postulate) (5) $\therefore m\angle SAW = 180 - 31 = 149$ (Step (2))   **7.** dual to proof for trisection of a segment (Problem 5, §2.7)   **9.** I. If $\ell \perp m$, with $m\angle CAB = m\angle CAB'$; by Linear Pair Axiom, $m\angle CAB + m\angle CAB' = 180 \rightarrow m\angle CAB = m\angle CAB' = 90$. Let ray $\overrightarrow{AC'}$ be opposite ray $\overrightarrow{AC}$; by the Linear Pair Axiom, $m\angle C'AB = 180 - m\angle CAB = 90 \rightarrow m\angle C'AB' = 90$. II. If ℓ and m form four right angles at $A = \ell \cap m$, then $m\angle B'AC = 90 = m\angle CAB$ and $\ell \perp m$.   **11.** Consider the opposite ray $\overrightarrow{AE'}$ of $\overrightarrow{AD}$ ($\therefore$ D-A-E′) and $m\angle E'AC = m\angle BAD = m\angle EAC$. By the Angle Construction Theorem, there is only one ray $\overrightarrow{AE'}$ on the E-side of $\overrightarrow{AB}$ forming an angle with ray $\overrightarrow{AC}$ whose measure is that of $\angle EAC$; $\therefore \overrightarrow{AE} = \overrightarrow{AE'}$ and $E \in$ line $\overleftrightarrow{AD}$, with E-A-D.

Testing Your Knowledge (Chapter 2)

1-F, 2-T, 3-F, 4-T, 5-F, 6-F, 7-F, 8-F, 9-F, 10-T

Section 3.1

1. (a) $\overline{RS} \cong \overline{VW}, \overline{RT} \cong \overline{UW}$ **(b)** $ST = RT - RS$ (R-S-T), and by U-V-W, $UV = UW - VW = RT - RS = ST$. $\therefore \overline{ST} \cong \overline{UV}$ **3.** There are 3 pairs of distinct congruent segments and 6 pairs of distinct congruent angles.   **5.** $\angle Y \cong \angle Z$ **7.** When $\triangle ABC$ or $\triangle XYZ$ is isosceles, with $AC = BC$ (or $XZ = YZ$).   **9.** $\triangle DEF$ and $\triangle KHG$ under $DEF \leftrightarrow KHG$.   **11. (a)** equality of sets **(b)** reflexive property of $\cong$ **(c)** when $\triangle ABC$ is equilateral and equiangular   **13.** If $A \neq C$ and $A \neq D$, then C-A-D, and $CD = CA \cup AD \neq CA, AD$; $B \in \overline{CD} \rightarrow B \in \overline{CA}$ or $B \in \overline{AD}$. If $B \in \overline{CA}$, then $\overline{AB} \subseteq \overline{CA} \subset \overline{CD}$. $\rightarrow \leftarrow$ If $B \in \overline{AD}$, then $\overline{AB} \subseteq \overline{AD} \subset \overline{CD}$. $\rightarrow \leftarrow \therefore A = C$ or $A = D$. Assume $A = C$; argument similar to preceding shows that $B = D$. (If $A = D$, then $B = C$ follows.)   **15.** When R-Q-S, U-V-W, $\overleftrightarrow{TV} \perp \overleftrightarrow{UW}$, $RQ = UV$ (or VW), $QS = VW$ (or UV), and $PQ = TV$.   **17.** Consider two similar triangles in Euclidean geometry, one having sides of length 8, 12, and 18, the other with sides 12, 18, and 27

Section 3.2

1. $AB = AC = \sqrt{10} \neq BC$, so it is isosceles but not equilateral.   **3.** Let $A = (1, 0), B = (0, 1), C = (0, 0)$, then $\triangle ABC$ is a right triangle with $a = b = 1, c = 2 \neq \sqrt{2}$.   **5. (a)** $R(3, 3)$ **7.** For an example where no taxicab circle exists through A, B, C, take $A(0, 0), B(6, 4)$, and $C(10, 2)$; if it does exist, it need not be unique: start with a taxicab circle and look for three noncollinear points on it that would permit sliding it back and forth.   **11.** an ordinary straight line   **15.** resembles Figure 3.13, Example 3,. either with or without symmetry. Example: Focus $= (0, 1)$ and directrix $y = x - 1$ leads to "parabola" consisting of negative x-axis, segment $y = x$ ($0 \leq x \leq 1$), and ray $x = 1$ ($y \geq 1$).

Section 3.3

1. (a) $\overline{UV} \cong \overline{UY}, \overline{VW} \cong \overline{YX}, \overline{UW} \cong \overline{UX}, \overline{VX} \cong \overline{WY}$ **(c)** $\overline{AC} \cong \overline{CB}$ and $\overline{DE} \cong \overline{DF}$ **3.** $x = 5, y = 6.7$ by ASA.   **5.** $x = 23, y = z = 90$ by SAS **7. (a)** Reflexive Law for $\cong$ **(b)** ASA **(c)** CPCF **(d)** definition of $\cong$ **9.** S and T are both equidistant from Q and R, so $\overleftrightarrow{ST}$ is the perpendicular bisector of $\overline{QR}$.   **11. (a)** Isosceles Triangle Theorem **(b)** betweenness **(c)** definition of $\cong$ **(d)** $\triangle ABD \cong \triangle AGE$ **(e)** SAS

(f) CPCF **(g)** SAS **(h)** CPCF   **13.** Conclusions: (1) ∠DBC ≅ ∠ACB. (2) m∠ABC = m∠DCB = 90 (3) ∠ABC ≅ ∠DCB (4) $\overline{BC} \cong \overline{BC}$ (5) △ABC ≅ △DCB. (6) ∴ $\overline{AB} \cong \overline{CD}$ and AB = CD.   **15.** *Given:* △ABC with AB = AC, line ℓ is the perpendicular bisector of $\overline{BC}$ at M. *Prove:* A ∈ ℓ. Conclusions: (1) $\overline{AB} \cong \overline{AC}$ (2) $\overline{BM} \cong \overline{MC}$ (3) ∠B ≅ ∠C (4) △ABM ≅ △ACM (5) m∠BMA = m∠CMA (6) m∠BMA + m∠CMA = 180 (7) m∠BMA = 90 (8) $\overline{AM} \perp \overline{BC}$ (9) ∴ ℓ = $\overleftrightarrow{AM}$ and A ∈ ℓ   **17.** Conclusions: (1) Line BF meets $\overleftrightarrow{AC}$ or $\overrightarrow{AD}$ at some interior point E or G, respectively. (2) Assume $\overrightarrow{BF}$ meets $\overrightarrow{AD}$ at G and A-G-D. (3) A-G-D-B (4) G ≠ B and G, B ∈ line $\overrightarrow{AD} \to \overrightarrow{BG} = \overrightarrow{BE} = \overrightarrow{AD}$ (5) F ∈ $\overrightarrow{BE} \to F \in \overrightarrow{AD} \cap \overrightarrow{CD} = D$ or F = D. →← (6) Line $\overrightarrow{BF}$ meets $\overleftrightarrow{AC}$ at E → A-E-C. (7) B and E lie on opposite sides of line $\overleftrightarrow{CD}$. (8) B-F-E (9) ∠DFB ≅ ∠EFC (10) △DFB ≅ △EFC (11) ∠BDF ≅ ∠CEF (12) ∠ADC ≅ ∠AEB (13) ∠C ≅ ∠B, BF = BF (14) BE = BF + FE = CF + FD = CD (15) △ADC ≅ △AEB (ASA) (16) ∴ $\overline{AD} \cong \overline{AE}$ and AD = AE   **21.** Conclusions: (1) ∠ABE ≅ ∠DBC (2) △ABE ≅ △DBC (3) ∴ AE = DC   **23. (a)** Take any point on the perpendicular bisector of segment $\overline{AB}$. **(b)** From any point P take rays $\overrightarrow{PA}$, $\overrightarrow{PB}$, $\overrightarrow{PC}$ such that m∠APB = m∠BPC = m∠CPA = 120, with PA = PB = PC.

Section 3.4

1. (a) 30 < x < 70, 0 < y < 70 **(b)** 70 < x < 180, 0 < y < 110   **3.** ∠A and ∠B are acute (Exterior Angle Inequality)   **5.** If K-T-R, ∠WTK is exterior angle of △WTR → m∠WTK > m∠WRT = 90. →←   **9.** In Figure 3.33, m∠3 > m∠2 → m∠1 + m∠3 > m∠1 + m∠2 → 180 > m∠1 + m∠2 (Linear Pair Axiom).   **11.** Let m∠ACB = 90. Suppose AB = BC; then m∠A = m∠C = 90. →← If AB < BC, construct $\overrightarrow{BD} \cong \overrightarrow{BA}$ on $\overrightarrow{BC}$; m∠BAC = m∠BAD = m∠BDA > m∠DCA = 90. →←   **13.** On segment $\overline{AC}$, construct $\overline{AD} \cong \overline{AB}$; CD = AC − AD = AC − AB = BC. In isosceles triangles △ABD and △BDC, m∠ADB + m∠BDC = m∠ABD + m∠DBC = m∠ABC < 180. →← ∴ B ∈ $\overleftrightarrow{AC}$ → B-A-C, A-C-B, or A-B-C (Problem 13, §2.7). If B-A-C, then BA + AC = BC = AC − AB or BA = −AB → AB = 0. →← Similarly if A-C-B.

Section 3.5

1. $\overline{CD}$ is longest, $\overline{BC}$ is shortest.   **3.** ∠TVW: m∠TVW < m∠T (Scalene Inequality) and m∠TVW < m∠U (SAS Inequality)   **5.** $\overline{OA}$, $\overline{OE}$, $\overline{OB}$, $\overline{OD}$, and $\overline{DC}$ **7. (a)** y = 31 **(b)** By the Triangle Inequality, 2x < 35 + y = 66 → x < 33; 2x + y > 35 → 2x > 4 → x > 2. ∴ 2 < x < 33..   **9.** 2   **11. (a)** Assume AC = DF and BC > EF (figure), and prove AB > DE: On segment $\overline{BC}$, construct $\overline{CG} \cong \overline{EF}$. Conclusions: (1) △AGC ≅ △DEF (2) m∠AGC < 90. (3) m∠AGB > 90 (4) m∠ABG < 90 < m∠AGB (5) ∴ AB > AG = DE. **(b)** No   **13.** AB ≤ AC + CB → AB − BC ≤ AC ≤ AB + BC (Triangle Inequality)   **15.** Use m∠3 + m∠4 = 180.   **17.** If respective sides of △ABC and △XYZ are congruent, but m∠A > m∠X, then by the SAS Inequality, BC > YZ. →←   **19.** Suppose A and B ∈ B_r(O), where B_r(O) is as defined in Problem 15, §2.5. Then OA < r and OB < r. If A-C-B, one angle at C is right or obtuse, say m∠OCA ≥ 90. OC < OA < r → C ∈ B_r(O) → AB ⊆ B_r(O) and B_r(O) is convex.

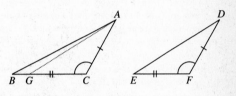

Exercise 11

Section 3.6

3. Assume right triangles △ABC and △XYZ have AB = XY and AC = XZ (figure). On the opposite ray of $\overrightarrow{ZY}$ construct segment $\overline{ZW} \cong \overline{BC}$. Conclusions: (1) △ABC ≅ △XWZ (2) XY = AB = XW (3) △XYW is isosceles, so ⊥ $\overline{XZ}$ bisects base $\overline{WY}$ at Z. (4) YZ = ZW = BC (5) ∴ △ABC ≅ △XYZ by SSS.   **5. (a)** △ABD ≅ △CBD (AAS) **(b)** △BFP ≅ △BFQ (HA) **(c)** inconclusive   **7.** No; location of P is not unique.   **9.** Conclusions: (1) △ABD ≅ △XYZ (2) BD = YZ = BC (3) △BCD is isosceles with base angles at C and D. (4) m∠ADB = 180 − m∠BDC = 90 (5) AB > BD = BC →← (6) ∴ AC ≤ XZ (7) Similarly, XZ ≤ AC. (8) AC = XZ and △ABC ≅ △XYZ by SAS.   **11.** For any point I on the bisector, let lines $\overrightarrow{IA}$, $\overrightarrow{IC}$ be perpendicluar to sides $\overrightarrow{BA}$, $\overrightarrow{BC}$; △IAB ≅ △ICB (HA), hence IA = IC.   **13.** The first two angle bisectors meet at some point I (Crossbar Theorem); I is then equidistant from the sides, hence lies on the third bisector (Problem 12).   **15.** Since m∠ABC > m∠ACB, m∠DBC = ½m∠ABC ≥ ½m∠ACB = m∠ECB → m∠FBC > m∠FCB → BF < CF. Construct segment $\overline{CF'} \cong \overline{BF}$ on $\overline{CF}$ and ∠CF'H ≅ ∠BFD → $\overline{F'H}$ does not cross $\overline{FG}$ and C-H-G by the Postulate of Pasch. By SAS, △FBD ≅ △F'CH → BD = CH < CG < CE. →← ∴ AC ≤ AB. Similarly, AB ≤ AC. ∴ AB = AC.

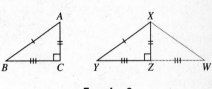

Exercise 3

Section 3.7

1. (a) By SAS, △ABC ≅ △BAD → $\overline{AC} \cong \overline{BD}$.   **3.** If ◇ABCD and ◇XYZW are convex quadrilaterals with ∠A ≅ ∠X, $\overline{AB} \cong \overline{XY}$, ∠B ≅ ∠Y, $\overline{BC} \cong \overline{YZ}$ and ∠C ≅ ∠Z, then ◇ABCD ≅ ◇XYZW.   **5.** Conclusions: (1) △ABC ≅ △XYZ (SAS) (2) ∠1 ≅ ∠3 → ∠2 ≅ ∠4 (3) △ACD ≅ △XZW (4) $\overline{AD} \cong \overline{XW}$, ∠D ≅ ∠W, and m∠BAD = m∠5 + m∠6 = m∠7 + m∠8 = m∠YXZ, or ∠BAD ≅ ∠YXZ.   **7.** By HL, △ABD ≅

$\triangle CBD$, hence $AD = CD$ and $\angle ABD \cong \angle CBD$. Since D lies interior to $\angle ABC$, $\overrightarrow{BD}$ bisects $\angle ABD$.   **9.** Suppose $\Diamond ABCD$ is a rectangle. If $AB < CD$, on $\overline{CD}$ construct $CE \cong \overline{AB}$ forming Saccheri Quadrilateral $\Diamond ABCE$; show that $m\angle CEA > m\angle BAE$. →←   **11.** Conclusions: (1) B and D lie on opposite sides of line $\overleftrightarrow{AC}$ (definition of convex quadrilateral). (2) $\overline{AB}$, $\overline{CD}$ lie on opposite sides of $\overleftrightarrow{AC}$ → E, F lie on opposite sides of $\overleftrightarrow{AC}$. (3) $\overleftrightarrow{AC}$ meets $\overline{EF}$ at a point G. (4) E-G-F → E lies on the G-side of $\overleftrightarrow{AF}$. (5) C lies on the G-side of $\overleftrightarrow{AF}$ → E lies on the C-side of $\overleftrightarrow{AF}$. (6) C-F-D → D lies on the opposite side of $\overleftrightarrow{AF}$ as C. (7) D and E lie on opposite sides of $\overleftrightarrow{AF}$. Similarly, A and F lie on opposite sides of $\overleftrightarrow{DE}$.

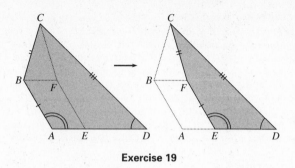

Exercise 19

13. Conclusions: (1) $m\angle 2 + m\angle 3 + m\angle B \le 180$ → $m\angle 2 + m\angle 3 \le 90$ (2) $m\angle 1 + m\angle 2 = m\angle DAB = 90$ (3) $m\angle 2 + m\angle 3 \le m\angle 1 + m\angle 2$ → $m\angle 3 \le m\angle 1$ (4) $AD = CB$ and $AC = CA$ (5) $\therefore AB \le CD$.   **15.** Counterexample in Euclidean geometry: A unit square and a non-square rectangle having one side of unit length.   **17.** In Euclidean geometry, consider square $\Box ABCD$; draw two congruent segments $\overline{AE}$ and $\overline{AF}$ to side $\overline{CD}$ with D the midpoint of $\overline{EF}$, and consider $\Diamond ABCE$ and $\Diamond ABCF$.   **19.** Counterexample for Euclidean geometry shown in figure ($\triangle CFB$ is isosceles, $\Diamond ABFE$ is a parallelogram).

Section 3.8

1. The center of a circle is equidistant from the endpoints of any chord, so must lie on its perpendicular bisector.   **3. (a)** If chords $\overline{AB}$, $\overline{CD}$ are equidistant from center O, and P and Q are the feet of the perpendiculars to $\overline{AB}$ and $\overline{CD}$ from O, $OP = OQ$; $OA = OC$ → $\triangle AOP \cong \triangle COQ$ (HL) → $AP = CQ$. Since P and Q are the midpoints of $\overline{AB}$ and $\overline{CD}$, $AB = 2AP = 2CQ = CD$. Steps are reversible, so converse true.   **5.** First prove *Lemma:* An equilateral polygon inscribed in a circle is equiangular. (Consider O the center of the circumscribed circle, and radii drawn to the vertices; use congruent triangles to show that $m\angle P = m\angle Q = m\angle R = ...$) Now $\angle PQR \cong \angle RST$, $\overline{PQ} \cong \overline{RS}$, $\overline{QR} \cong \overline{ST}$ → $\triangle PQR \cong \triangle RST$ → $\overline{PR} \cong \overline{RT}$. It follows that $\Diamond PRTV$ is equilateral; by the lemma, $\Diamond PRTV$ is equiangular.   **7.** (1) By the Tangent Theorem, $\triangle APO$ and $\triangle BPO$ are right triangles, with $PO = PO$ and $OA = OB$ → $\triangle APO \cong \triangle BPO$ (HL) → $m\angle APO = m\angle OPB$. (2) Points of circle lie entirely on one side of tangent, and O is an interior point of diameter $\overline{AOC}$ → A-O-C and O lies on the C-side of $\overrightarrow{PA}$ → O lies on the B-side of $\overrightarrow{AP}$, and similarly, it lies on the A-side of $\overrightarrow{PB}$ → $O \in$ Interior $\angle APB$. $\therefore$ $\overrightarrow{PO}$ bisects $\angle APB$.   **9.** $AM = MC = MB$ (corollary, Theorem 2)   **11.** If the circle at common vertex O of the n angles meets sides at $P_1, P_2, ... , P_n$, since interior $\angle P_1 OP_n$ is disjoint from that of other angles, the arc $A = \overparen{P_1 P_2 P_n}$ lies in the exterior of $\angle P_1 OP_n$, so is a major arc, with $mA = 360 - m\angle P_1 OP_n$; by Additivity of Arc Length, $mA = \overparen{mP_1 P_2} + \overparen{mP_2 P_3}$ $... + \overparen{mP_{n-1} P_n} = m\angle P_1 OP_2 + m\angle P_2 OP_3 + ... + m\angle P_{n-1} OP_n = 360 - m\angle P_1 OP_n$.   **13.** Line $y = 1$ does not meet circle $x^2 + y^2 = 25$ since $x^2 + 1 = 25$ → $x = \pm 2\sqrt{6}$, irrational.   **15.** (1) Let A-B-F with $AB = $ c and $BF = $ a. (2) Draw circle centered at A and radius $r = b$, and one centered at B, radius $r' = a$. Since $x = c = $ distance between centers, by the condition of Problem 14, $r - r' \le x \le r + r'$ becomes $b - a \le c \le b + a$, which is satisfied since $b \le a + c$. Hence by the Two-Circle Theorem, the circles intersect at some point C, producing desired triangle.

Testing Your Knowledge (Chapter 3)

1-T, 2-F, 3-T, 4-F, 5-T, 6-T, 7-T, 8-T, 9-F, 10-F

Section 4.1

1. $m\angle A = 59 = m\angle ECD$, $m\angle E = 31$ **3.** $x = 47 = z$, $y = 43$ **5.** Top and middle strokes are parallel, hence $\angle 3 \cong \angle 4$ (F-property)   **7.** Shortcut: Vertical Pair Theorem and Linear Pair Axiom reduces F- and U-properties to Property Z. To prove Property Z, use the lemma for one direction, and Theorem 1 for other.   **9.** $m\angle C = 60 = m\angle BAC$; $m\angle EAC = \frac{1}{2}(180 - m\angle BAC) = 60 \rightarrow \overrightarrow{AE} \parallel \overrightarrow{BC}$ (Z-property). Generalizes to any isosceles triangle.   **11.** Prove first that one side of one angle meets a side of other, then use Linear Pair Axiom, Property F.   **13.** Use Corollary D of Theorem 1.   **15.** (1) $\triangle PQR$ and $\triangle PQS$ are isosceles. (2) $m\angle SPQ = 180 - 2m\angle SQP$ and $m\angle QPR = 180 - 2m\angle PQR$. (3) $\therefore$ $m\angle SQP + m\angle PQR = 90$.   **17.** By the Parallel Postulate, the perpendicular t to ℓ at A must cut the parallel line m at a point B; take $C \in \ell$ and $D \in m$ on the same side of t. Then $m\angle CAB + m\angle ABD = 180$ (U-property) → $m\angle ABD = 90$ or $t \perp m$.   **19.** I. (Fifth Postulate → Axiom P-1): Let $A \notin \ell$. By the lemma, there exists $m \parallel \ell$ passing through A. Suppose another line $m' \parallel \ell$ passes through A. Angles which m' makes with perpendicular t from A to ℓ are distinct from those of m, hence angles on same side of t cannot both be supplementary for both lines m and m'. By the Fifth Postulate, either m or m' meets ℓ. →← $\therefore$ parallel through A is unique and Axiom P-1 holds. II. (Axiom P-1 → Fifth Postulate): Let transversal t meet lines ℓ and m at A and B such that $m\angle CAB + m\angle ABD < 180$ (C and D on same side of t). If $\ell' \parallel m$ and $A \in m'$, ℓ' makes supplementary angles with t (Theorem 1, proven from Axiom P-1), and $\ell \ne \ell'$. By Axiom P-1, ℓ meets m at some point P; P lies on C-side of t, for otherwise angle sum of two angles of $\triangle PAB > 180$. $\therefore$ Euclid's Fifth Postulate holds.   **21.** $\triangle AMC \cong \triangle BMD$ (HA) → $\angle AMC \cong \angle BMD$. Since A-M-B and C and D lie on opposite sides of M, by the result of Problem 11, §2.8, C-M-D and line $\overleftrightarrow{CD}$ is a common perpendicular to ℓ and m.

Section 4.2

1. (1) Suppose $\Diamond ABCD$ is a rhombus. (2) $AB = BC$ (definition) (3) $AB = CD$, $BC = CA = AB$ (property of parallelogram, Step 2).   **3.** If $\Diamond ABCD$ is rhombus, $AB = BC$ and $CD = AD \rightarrow B$ and D are both equidistant from A and $C \rightarrow BD$ is perpendicular bisector of segment $\overline{AC}$. **5.** 97   **7.** Square $\Box ABCD$ is a rhombus $\rightarrow$ diagonals perpendicular (Problem 3). $m\angle A = m\angle B = 90 \rightarrow \triangle DAB \cong \triangle CBA$ (SAS) $\rightarrow AC = BD$. Converse: If $\Diamond ABCD$ has $\overline{AC} \perp \overline{BD}$ and $AC = BD$, then $\triangle DAB \cong \triangle CBA \rightarrow \angle A$. $\angle B \rightarrow m\angle A + m\angle B = 2m\angle A = 180$ or $m\angle A = 90 = m\angle B$. $\therefore$ all four angles of $\Diamond ABCD$ are right angles. Since $\overrightarrow{BD}$ is perpendicular bisector of AC, $AB = BC$, and $\Diamond ABCD$ is a square.   **9.** 13 **11.** (1) Since $ED = AC$ and $\overrightarrow{ED} \parallel \overrightarrow{AC}, \Diamond ACDE$ is a parallelogram. (2) $AB = BE$, $m\angle BAE = m\angle AEB$ (3) $m\angle EAC = m\angle AED$ (4) $m\angle EAC = 90$ and $\Diamond ACDE$ is a rectangle. (5) $BC = BD$ (6) $m\angle BCD = 60 = m\angle BDC$ (7) $AC = BC = CD$ and $\triangle BCD$ is equilateral. (8) $\therefore \Diamond ACDE$ = rhombus $\rightarrow \Diamond ACDE$ = square.   **13.** (1) $\overrightarrow{AD} \parallel \overrightarrow{BC}$ and $\overline{AD} \cong \overline{BC}$ (2) $\angle ADE \cong \angle CBF$ (Property Z) (3) $\triangle ADE \cong \triangle CBF$ (HA) $\rightarrow AE = CF$. **15.** Given segment AB, draw line ℓ through A and locate points A-P-Q-R on ℓ such that $AP = PQ = QR$; construct lines through P and Q parallel to $\overrightarrow{RB}$, intersecting $\overline{AB}$ at T_1 and T_2. By Theorem 3, $AT_1 = T_1T_2 = T_2B$.   **17.** Each quadrilateral formed is a rhombus.   **19.** If $\Diamond ABCD$ is an isosceles trapezoid with bases $\overline{AB}$ and $\overline{CD}$, then $AB \neq CD$ (since it is not a parallelogram), and we may assume $AB > CD$. Locate E on $\overline{AB}$ such that $AE = CD \rightarrow \Diamond AECD$ is a parallelogram $\rightarrow \overline{AD} \parallel \overline{CE}$. $CD = AD = CB \rightarrow m\angle B = m\angle CEB = m\angle A$ (F-Property). To reverse, locate E on AB as before; $m\angle B = m\angle A = m\angle CEB \rightarrow BC = CE = AD$.   **21.** (b) False in absolute geometry; to prove rectangle exists, use Problem 21, §4.1. (Do not use Parallel Postulate.)   **23.** Angle sum of convex quadrilateral is $360 \rightarrow$ remaining fourth pair of angles are congruent and we may invoke SASAA (Problem 10, §3.7).

Section 4.3

1. (a) yes (SSS criterion) (b) $m\angle X = 104.5$, etc. (c) $\triangle ZXY$.   **3.** If D and W are feet of altitudes on $\overrightarrow{AB}$ and $\overrightarrow{XY}$, respectively, then $\triangle ACD \sim \triangle XZW$ (AA Criterion) $\rightarrow CD/ZW = AC/XZ = k$.   **5.** (a) $ABC \leftrightarrow YXZ$ (b) 2 (c) $m\angle Y = 60$.   **7.** 111.25 (exact)   **9.** (a) Let $a/b = c/d = k$ or $a = kb$ and $c = kd \rightarrow a + c = kb + kd = k(b + d)$ or $(a + c)/(b + d) = k = a/b$. (c) 14/3   **11.** Let $\overrightarrow{BE} \parallel \overrightarrow{CA}$; then $m\angle E = m\angle DCA = m\angle BCD \rightarrow BC = BE = a$; $\triangle BDE \sim \triangle ADC \rightarrow BD/DA = BE/AC = a/b$. For converse, locate D' on $\overline{AB}$ so that $\overrightarrow{CD'}$ bisects $\angle BCA$; by the first part, $BD/DA = a/b = BD'/D'A \rightarrow BD = BD' \rightarrow D' = D$. (b) With $x = PM = PS = QR$, $r = AP$, $s = QB$, since $\triangle APS \sim \triangle ACB$, $\triangle BQR \sim \triangle BCA \rightarrow r/b = x/a$, $s/a = x/b \rightarrow (r + x)/(b + a) = x/a$, $(s + x)/(a + b) = x/b$. Division yields $(r + x)/(s + x) = b/a \rightarrow AM/MB = b/a$. $\therefore \overrightarrow{CM}$ bisects $\angle ACB$.   **13.** (a) $\triangle ADC \sim \triangle BEC \rightarrow CE/CD = b/a \rightarrow \triangle CED \sim \triangle CBA$ (SAS Criterion) $\rightarrow m\angle EDC = m\angle BAC \equiv A$; similarly, $m\angle BDF \equiv A$, and $\overrightarrow{DA}$ bisects $\angle FDE$. (b) $\overline{AD} \perp \overline{BC}$; since $\triangle ACB$ and $\triangle DCE$ have two pairs of congruent angles, they are similar $\rightarrow CE/CD = b/a \rightarrow \triangle ADC \sim \triangle BEC \rightarrow m\angle BEC = m\angle ADC = 90$. $\therefore \overline{BE} \perp \overline{AC}$.   **15.** Suppose $AB = kXY$, $BC = kYZ$, and $AC = kXZ$; if $k > 1$, locate D on $\overline{AB}$ and E on $\overline{AC}$ such that $AD = XY$ and $AE = XZ$. Prove: (1) $\triangle ADE \cong \triangle XYZ$. (2) $\triangle ADE \sim \triangle ABC$ and $\angle Y \cong \angle ADE \cong \angle B$ (3) $\triangle ABC \sim \triangle XYZ$ by SAS.   **17.** (a) $m\angle BAC = 180 - 2m\angle B = m\angle D$ (both triangles are isosceles) $\rightarrow \triangle ABC \sim \triangle DBA \rightarrow AB/BC = BD/AB = x \rightarrow AB^2 = BC \cdot BD = BC(BC + CD) \rightarrow (AB/BC)^2 = 1 + CD/BC = 1 + AB/BC$, or $x^2 = 1 + x \rightarrow x = \tau$. Steps reversible, so converse is true. (c) Since $m\angle B = 72$, $\cos 72 = \cos B = \frac{1}{2}BC/AB = \frac{1}{2}\tau^{-1}$. To show that $\cos 36 = \frac{1}{2}\tau$, use identity $2\cos^2\theta = 1 + \cos 2\theta$.   **19.** In $\triangle ABC$, let $c^2 = a^2 + b^2$. By Law of Cosines, $\cos C = 0$ or $C = 90$.   **21.** (a) Quantity on left equals $QA^2 \cdot BC + (QA + AB)^2 \cdot CA + (QA + AC)^2 \cdot AB + BC \cdot CA \cdot AB = QA^2 \cdot BC + QA^2 \cdot CA + AB^2 \cdot CA + QA^2 \cdot AB + AC^2 \cdot AB + BC \cdot CA \cdot AB = AB \cdot CA \cdot (AB + CA + BC) = AB \cdot CA \cdot 0 = 0$. (b) Let $h = PQ$. By Pythagorean relation, quantity on left $= (QA^2 + h^2)BC + (QB^2 + h^2)CA + (QC^2 + h^2)AB + BC \cdot CA \cdot AB = 0 + h^2BC + h^2CA + h^2AB = h^2 \cdot (BC + CA + AB) = 0$.   **23.** (c) By general cevian formula, $h_c^2 = xa^2 + (1 - x)b^2 - x(1 - x)c^2 = x(a^2 - b^2) + b^2 - xc^2 + x^2c^2$ or $c^2x^2 + (a^2 - b^2 - c^2)x + (b^2 - h_c^2) = 0$—a quadratic equation in x; there can be only one real root. (b) Discriminant must equal zero: $(a^2 - b^2 - c^2)^2 - 4c^2(b^2 - h_c^2) = 0 \rightarrow h_c^2 = [4b^2c^2 - (a^2 - b^2 - c^2)^2]/4c^2)$. Use algebra to factor the expression in numerator, with $2s = a + b + c$.

Section 4.4

1. All three.   **3.** See figure.   **5.** (b) See figure.   **7.** 45, 67.5, 67.5   **9.** (b) Each point is the common vertex of the same three regular polygons (square, hexagon, dodecagon) having angle measures $90 + 120 + 150 = 360$.   **11.** See figure.   **13.** $AF = AE = \tau$ by construction (Problem 7, §1.2); if M is the midpoint of $\overline{AF}$, then in $\triangle AOM$, $\cos \theta = AM/AO = \frac{1}{2}\tau$.   **15.** If sides of regular pentagon have unit length, by the Law of Cosines, each diagonal (= side of star) has length:

$$\sqrt{1^2 + 1^2 - 2 \cdot 1 \cdot 1 \cdot \cos 108} = \sqrt{2 + 2 \cos 72} = \sqrt{4 \cos^2 36} = \tau.$$

For angle measure θ at each point, $1^2 = \tau^2 + \tau^2 - 2\tau \cdot \tau \cdot \cos \theta \rightarrow (3 + \sqrt{5}) \cos \theta = \sqrt{5} + 2 \rightarrow \cos \theta = \frac{1}{4}(1 + \sqrt{5}) = \frac{1}{2}\tau \rightarrow \theta = 36$.

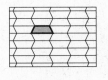

Exercise 3

Exercise 5

Exercise 11

17. $n = 24, 30, 32, 34, 40, 48, 51, 60, 64, 68, 80, 85,$ and 96

Section 4.5

1. 8   **3.** $4\sqrt{3}$ **5. (a)** $\frac{1}{4}\pi$ **(b)** $\frac{1}{2}$ **(c)** $\frac{1}{4}\sqrt{3}$ **(d)** $\frac{3}{8}\sqrt{3}$ **(e)** If r = radius of large circle, Area = $\frac{3}{4}\pi r^2 + \frac{1}{2}\pi(r/\sqrt{2})^2 + \frac{1}{2}r^2 = $
$(2\pi + 1)/(6 + 4\sqrt{3})$.   **7. (a)** $m\angle BAC = \theta + \phi = \frac{1}{2}x + \frac{1}{2}y$ (Euclidean Exterior Angle Theorem)   **(b)** $\theta = m\angle BAC + \phi \to \frac{1}{2}x = $
$m\angle BAC + \frac{1}{2}y$ **9. (c)** 150   **11.** $r = 10\sqrt{2}$ (P lies on AB)   **13.** By Law of Sines, $r = 12 \sin 20/\sin 85 \approx 4.1199$.   **15.** Since
$RM = MA = MB = MC$, A, B, C, and R are concyclic; R bisects semicircle $\overarc{ARB} \to m\angle BCR = m\angle RCA$.   **17. (a)** With $\overline{OP}$ as cevian and $r = $
$OA = OB$ as radius of circle, $OP^2 = (AP/AB)r^2 + (PB/AB)r^2 - (AP/AB)(PB/AB) \cdot AB^2 = r^2 - AP \cdot PB \to$ Power $(P) = PA \cdot PB$. **(b)** $PA \cdot PB = $
Power (P) = constant = $PC \cdot PD$ **19.** $OP > 10$, provided $m\angle APB < 90$.   **21.** $\triangle BOC \cong \triangle COD \to m\angle AOD = 3 \cdot m\angle AOC = 270 - 3\theta$.
$m\angle EDC = \phi$, $m\angle E = m\angle ODE = \theta - \phi$. $\therefore$ $m\angle EOA = \theta + 2\phi - 90$, $m\angle OAE = 270 - 2\theta - \phi$.   **23. (a)** $m\angle BXC = A + (90 - \frac{1}{2}A) = 90 + $
$\frac{1}{2}A$ = constant, so X varies on circle having $\overline{BC}$ as chord. **(b)** $\triangle PXB$ is equilateral for all P; generalization involves similar triangles.   **25.**
If C_1, C_2 are the centers of $\mathbf{C}_1$, $\mathbf{C}_2$, respectively, $CC_1 + CC_2$ = constant (ellipse with foci at C_1, C_2), or $CC_1 - CC_2$ = constant (hyperbola with
foci at C_1, C_2).

Section 4.6

1. (a) $u = [3, 2]$, $v = [6, -9]$, $w = [3, -11]$ **(b)** $uv = 3 \cdot 6 + 2 \cdot (-9) = 0$ **(c)** $\cos B = 1/\sqrt{10}$.   **3.** If $|AC| = \sqrt{(b - 2a)^2 + (c - 0)^2} = $
$|BD| = \sqrt{(b - 0)^2 + (0 - c)^2}$ then $b^2 - 4ab + 4a^2 + c^2 = b^2 + c^2 \to a = 0$ (since $a \neq b$) $\to$ vertices of $\diamond ABCD$ are $A(0, 0)$, $B(b, 0)$,
$C(b, c)$, $D(0, c)$—a rectangle.   **5.** If $A(a, 0)$, $B(b, 0)$, $C(d, e)$, $D(0, c)$ are vertices of $\diamond ABCD$, midpoints of sides are $L(\frac{1}{2}(a + b), 0)$,
$M(\frac{1}{2}(b + d), \frac{1}{2}e)$, $N(\frac{1}{2}d, \frac{1}{2}(c + e))$, $P(\frac{1}{2}a, \frac{1}{2}c) \to$ Slope $\overrightarrow{PL} = -c/b$ = Slope $\overrightarrow{MN}$, Slope $\overrightarrow{PN} = e/(d - a)$ = Slope $\overrightarrow{LM} \to \diamond LMNP$ is a parallel-
ogram.   **7.** For any two orthogonal vectors u and v, $\|u + v\|^2 = \|u\|^2 + \|v\|^2$. $Proof: (u+v)^2 \equiv (u + v) \cdot (u + v) = u \cdot u + u \cdot v + v \cdot u + v$
$\cdot v = u^2 + v^2$.   **9.** 25 mph and $\approx 16.26°$ south of east or south of west (Conditions of problem lead to $s + y = 136$, $s - y = 150$.)   **11.** If
M is midpoint of FE, then from $FE = 2EG$ we obtain $ME = EG \to \triangle DEM \cong \triangle DEG \to \angle FDM \cong \angle MDE \cong \angle EDG$.   **13.** Sum of squares
of lengths of diagonals of a parallelogram equals sum of squares of sides.   **15.** Let $G = (x, y)$. $AG = \frac{2}{3}AL = \frac{2}{3}(AB + BL) = \frac{2}{3}(AB + $
$\frac{1}{2}BC) = \frac{2}{3}AB + \frac{1}{3}BC \to [x - a_1, y - a_2] = \frac{2}{3}[b_1 - a_1, b_2 - a_2] + \frac{1}{3}[c_1 - b_1, c_2 - b_2] \to x - a_1 = \frac{2}{3}(b_1 - a_1) + \frac{1}{3}(c_1 - b_1)$, etc.   **17.** $A' = $
$(\frac{1}{2}b + \frac{1}{2}\sqrt{3}c, \frac{1}{2}c + \frac{1}{2}\sqrt{3}b)$, $B' = (\frac{1}{2}a - \frac{1}{2}\sqrt{3}c, \frac{1}{2}c - \frac{1}{2}\sqrt{3}a)$, $C' = (\frac{1}{2}a + \frac{1}{2}b, -\frac{1}{2}\sqrt{3}b + \frac{1}{2}\sqrt{3}a)$; $AA'^2 = a^2 + b^2 + c^2 - ab + \sqrt{3}bc - $
$\sqrt{3}ac = BB'^2 = CC'^2$.   **19.** Let $\diamond ABCD$ be as in Problem 5; show $X = (\frac{1}{2}(a + b), \frac{1}{2}(a - b))$, $Y = (\frac{1}{2}(b + d + e), \frac{1}{2}(b - d + e))$, $Z = $
$(\frac{1}{2}(c + d - e), \frac{1}{2}(c + d + e))$, $W = (\frac{1}{2}(a - c), \frac{1}{2}(c - a))$. **(a)** $4XZ^2 = 4YW^2$, Slope $\overrightarrow{XZ} = -1/$Slope $\overrightarrow{YW}$ **(b)** Show $M(\frac{1}{2}b, \frac{1}{2}c)$ is common midpoint of both $\overline{XY}$, $\overline{WY}$ when $d = b - a$, $e = c$.

Section 4.7

1. (a) orthogonal.   **3. (a)** $a = 3$ **5.** $C(1 \pm 2\sqrt{10}, 0)$ **7.** $OA \cdot OA' = OB \cdot OB' \to OA/OB = OB/OA' \to \triangle AOB \sim \triangle B'OA'$ (SAS
Criterion)   **9.** $\theta = 60$ **11.** $\theta = 135$ **13. (b)** $AC/A'C' = OA/OC'$ and $AD/A'D' = OA/OD'$ by similar triangles.

$$\frac{AC}{AD} = \frac{OD'}{OC'} \cdot \frac{A'C'}{A'D'} \text{ and similarly, } \frac{BD}{BC} = \frac{OC'}{OD'} \cdot \frac{B'D'}{B'C'} \text{. Therefore, } (AB, CD) = \frac{AC}{AD} \cdot \frac{BD}{BC} = \frac{OD'}{OC'} \cdot \frac{A'C'}{A'D'} \cdot \frac{OC'}{OD'} \cdot \frac{B'D'}{B'C'}$$

$$\frac{A'C'}{A'D'} \cdot \frac{B'D'}{B'C'} = (A'B', C'D')$$ **15.** $C(-\frac{1}{2}a, -\frac{1}{2}b)$ = center and r = radius = $\sqrt{1/4a^2 + 1/4b^2 - c}$; Power $(P) = PC^2 - r^2 = (x_0 + \frac{1}{2}a)^2$

$+ (y + \frac{1}{2}b)^2 - (\frac{1}{4}a^2 + \frac{1}{4}b^2 - c) = x_0^2 + y_0^2 + ax_0 + by_0 + c$ **17. (c)** Let circles be C_1, C_2, C_3 and P = intersection of common chord of C_1,
C_2; Power(P) wr/C_1 = Power (P) wr/C_2 and Power(P) wr/C_2 = Power(P) wr/C_3 $\to$ Power(P) wr/C_1 = Power(P) wr/C_3 $\to P$ lies on common
chord of C_1, C_3.   **19.** Show that the circles $C(t)$ corresponding to $t = t_1$, t_2 do not intersect if $c \neq 0$. (Algebra shows the only possible point
of intersection would be $(0, 0)$.) Radius r of $C(t)$ given by $r^2 = \frac{1}{4}(t^2a^2 + t^2b^2 - tc) = \frac{1}{4}t(ta^2 + tb^2 - c)$, which converges to zero as either $t \to 0$
or $t \to c/(a^2 + b^2)$. Center of $C(t)$ is $C(-\frac{1}{2}ta, -\frac{1}{2}tb)$, so by letting $t \to 0$ or $t \to c/(a^2 + b^2)$, we get the two distinct points $A(0, 0)$ and
$B(-\frac{1}{2}ac/(a^2 + b^2), -\frac{1}{2}bc/(a^2 + b^2))$, and family is Type F. (Complementary family G consists of all circles passing through A and B, or $x^2 + y^2$
$+ (bt + c/4a)x - aty = 0$, t real.) If $c = 0$, given family consists of all circles having a common tangent line at $(0, 0)$; if $a = b = 0$ (and $c < 0$),
family consists of concentric circles centered at the origin.   **21. (a)** Use fact that radical axis between any two members is fixed line. **(b)**
Radius r of any circle of family is given by $4r^2 = (at + a')^2 + (bt + b')^2 - 4(ct + c') \equiv (a^2 + b^2)t^2 + 2(aa' + bb' - 2c)t + a'^2 + b'^2 - 4c'$; two real
zeros of this quadratic in t yields values to which t converges in order that radii of circles in family converge to zero for type F. If there are no
real zeros, radii do not converge to zero and family is type G. Discriminant of quadratic is Δ, which therefore must be positive for type F and
negative for type G.   **23.** a line parallel to the radical axis (For algebraic proof, use result of Problem 15.)

Section 4.8

1. (b) $\begin{bmatrix} ABC \\ DEF \end{bmatrix} = \dfrac{AE}{EC} \cdot \dfrac{CD}{DB} \cdot \dfrac{BF}{FA} = \dfrac{FB}{AF} \cdot \dfrac{BD}{DC} \cdot \dfrac{DC}{CE} = \dfrac{1}{x}$.   **3.** $\begin{bmatrix} ABC \\ DEF \end{bmatrix} = (b/a) \cdot (c/b) \cdot (a/c) = 1$ **5.** $9604/117{,}649 \approx$

.082 sq. units   **9.** Let $a_1 = BD$, $a_2 = DC$, $b_1 = CE$, $b_2 = EA$, $c_1 = AF$, $c_2 = FB$. If D', D'' are other two points of contact of excircle D on sides of triangle, then $BD' = BD = a_1$ and $CD'' = CD = a_2$; $AD' = AD'' \to c + a_1 = b + a_2 \to 2a_1 = a + b - c = 2s - 2c$ or $a_1 = s - c$. Similarly,

$a_2 = s - b$, $b_1 = s - a$, $b_2 = s - c$, $c_1 = s - b$, and $c_2 = s - a$. Hence $\begin{bmatrix} ABC \\ DEF \end{bmatrix} = \dfrac{s - b}{s - a} \cdot \dfrac{s - c}{s - b} \cdot \dfrac{s - a}{s - c} = 1$.   **11.** Since lines are

concurrent, $\begin{bmatrix} ABC \\ DEF \end{bmatrix} = 1$; since $BD = DC$, $AF/FB \cdot CE/EA = 1 \to AE/EC = AF/FB \to \overleftrightarrow{EF} \parallel \overleftrightarrow{BC}$.   **13.** Let m be a circle $\perp C, D$ cutting ℓ at

O, and take O as center of inversion. Configuration $m\,C, D$ maps to a straight line and two concentric circles, with circles in proposed chain tangent to these. Thus, if chain closes, it can be rotated to any desired position to prove that every Steiner Chain closes.

Testing Your Knowledge (Chapter 4)

1-F, 2-T, 3-F, 4-T, 5-F, 6-F, 7-T, 8-T, 9-F, 10-F

Section 5.1

1. (a) $(\frac{1}{4}, \pm\frac{1}{2}\sqrt{3}\,)$; $(1, 0)$; $(1, -1)$ has no image. **(b)** Every point on $x^2 + y^2 = \frac{1}{4}$; every point on $x^2 + y^2 = 1$; $(0, 0)$. **(c)** No **(d)** all $P = (x, y)$ such that $x^2 + y^2 \le 1$ **3. (a)** If $y = x^2$ then $y' = y + y = 2y = 2x'$, or $y' = 2x'$. **(b)** No, $(\pm1, 0)$ both map to $(0, 1)$.   **7.** $y' = -3x'$; No **9.** $x = (3x' - 4y')/25$, $y = (4x' + 3y')/25$. **(a)** By substitution, $y = mx + b$ becomes $(4x' + 3y')/25 = m(3x' - 4y')/25 + b \to y' = (3m - 4)x'/(4m + 3) + 25b/(4m + 3)$ and $y = (-1/m)x + c$ becomes $y' = -(4m + 3)x'/(3m - 4) + 25c/(3m - 4)$. **(b)** Since $m_1'm_2' = [(3m - 4)/(4m + 3)] \cdot [(-4m - 3)/(3m - 4)] = -1$, the image lines are perpendicular.   **11. (a)** only one fixed point $(0, 1)$ **(b)** If $a \ne \frac{1}{2}$, the only fixed point is $(0, a)$; if $a = \frac{1}{2}$, every point on the line $y = \frac{1}{2}x + \frac{1}{2}$ is fixed point.   **13.** If ℓ' meets m' at P', consider $f^{-1}(P') = P$; P must lie on both ℓ and m, but $\ell \parallel m. \to\leftarrow$ **15.** Since $x = ax + by$ and $y = cx + dy$ for some $x, y \ne 0$, $bc = (a - 1)(d - 1)$; if $b \ne 0$, then $y = (1 - a)x/b$ is a line of fixed points, or if $b = 0 \;(\therefore a = 1)$, or $d = 1$, then $cx = 0$ and y-axis is a line of fixed points. Finally, if $a = 1$ and $d \ne 1$, $y = cx/(1 - d)$ is a line of fixed points.   **17.** By the result of Problem 16, $\overline{AB}$ and $\overline{BC}$ are two noncollinear segments of fixed points. Let P be any point and ℓ and m any two lines through P cutting $\overline{AB}$ and $\overline{BC}$ at two pairs of fixed points. Then ℓ and m are fixed by the mapping, and P is a fixed point; since P was arbitrary, the mapping is the identity.

Section 5.2

1. (a) See figure. **(b)** See figure.   **3.** By definition of s_ℓ, ℓ is the perpendicular bisector of $\overline{CD}$, so Q is equidistant from C and D.   **5.** Line ℓ is perpendicular bisector of $\overline{PP'}(\equiv \overline{P'P})$, hence $s_\ell(P') = P$ or $s_\ell[s_\ell(P)] = P$.   **7.** Use isometry property of reflections.   **9.** gf, fg both given by $x'' = x - 2$, $y'' = y + 2$ **11.** $EB + BA = EB + BA' = AA' \le EP + PA' = EP + PA$ (Triangle Inequality)   **13. (a)** $(85/12, 0)$ **(b)** $(0, 55/13)$ **(c)** $(77/12, 48)$ **15.** Let $B' = s_\ell(B)\ (\in n')$ and let ℓ meet BB' at point M (midpoint of BB'). Then (figure) (1) $\triangle B'AM \cong \triangle BAM$ (2) $\angle 1 \cong \angle 2$ (3) $\angle 2 \cong \angle 3$ ($\ell \parallel m$ and Property Z) (4) $\angle 1 \cong \angle 4$ (5) $\angle 4 \cong \angle 3$ (Steps 2 and 3) $\to AB = AC$. **17.** any point on the orthic triangle; see Problem 13, §4.3   **19.** As line of vision from E moves counterclockwise, first image seen is reflection of T in vertical mirror, then T itself, then reflection of T in intersection of mirrors (reflection in both mirrors), and finally reflection of T in horizontal mirror.   **21.** $x - 2y = 5$; $P(x, y)$ has as its image $P'(3x/5 + 4y/5 + 2, 4x/5 - 3y/5 - 4)$; the midpoint of P and P' is given by $M(\frac{1}{2}(x + 3x/5 + 4y/5 + 2), \frac{1}{2}(y + 4x/5 - 3y/5 - 4)) = M(4x/5 + 2y/5 + 1, 2x/5 + y/5 - 2)$; x-coordinate minus twice y-coordinate of $M = 4x/5 + 2y/5 + 1 - 4x/5 - 2y/5 + 4 = 5$, so M lies on ℓ.

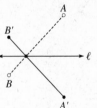

Exercise 1(a)

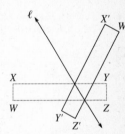

Exercise 1(b)

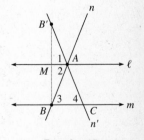

Exercise 15

Section 5.3

1. See figure.   **3.** See figure.   **5. (b)** 3 **(c)** Each point will be moved a distance of $2\sqrt{10}$ along a line having slope 3.
7. $(4, 2) \to (-4, 2) \to (2, -4)$ **(b)** 90° clockwise turn about the origin
9. Center of rotation is vertex C of an equilateral triangle $\triangle ABC$ with base $\overline{AB}$ (lying below the line $\overleftrightarrow{AB}$). Any center O of such a rotation lies on perpendicular bisector of $\overline{AB}$, and $OA = OB$ with $m\angle AOB = 60$ $\to \triangle OAB$ is equilateral, with $O = C$ or C', the reflection of C in line $\overleftrightarrow{AB}$; clockwise orientation requires that $O = C$.   **11. (a)** $tr = (s_m s_n)(s_n s_\ell) = s_m s_n{}^2 s_\ell = s_m e s_\ell = s_m s_\ell$; since $m \parallel \ell$, tr is a translation. **(b)** $t_1 t_2 = (s_\ell s_m)(s_m s_\ell) = s_\ell s_m{}^2 s_\ell = s_\ell e s_\ell = s_\ell{}^2 = e$.   **13.** Let P, P' be a corresponding pair (e.g., center of each disk); drop $\overline{PM} \perp n$ ($M \in n$), double segment $\overline{PM}$ to Q, and draw perpendicular bisector of $\overline{P'Q} \equiv k$—desired line. If rotation were $s_m s_n$ for any other line m, then by construction, $s_m s_n(P) = s_m(Q) = P'$; last equation shows that m is perpendicular bisector of $P'Q$, hence $m = k$.   **15.** By Theorem 2, f must be product of exactly two line reflections: $f = s_\ell s_m$. If $\ell \parallel m$, f is a translation; if ℓ meets m, f is rotation, by definition.   **17. (a)** $f = s_\ell$ **(b)** Suppose $t = s_\ell s_m$, where $\ell \parallel m$. By the result of Problem 13, we can choose line k such that $r = s_k s_\ell$. Then $rt = s_k s_\ell{}^2 s_m = s_k s_m$, so rt is itself a translation or rotation.

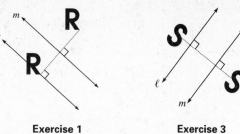

Exercise 1   **Exercise 3**

Section 5.4

1. Let P be any point on a circle centered at O, radius r. With $\overline{AB}$ any fixed diameter, $\angle APB = 90 = m\angle A'OP'$, where, because of the ABC property, O' is the midpoint of $\overline{A'B'}$. $\therefore$ P' lies on circle centered at O', diameter $\overline{A'B'}$.   **3. (a)** $A' = (3, 2)$, $B' = (5, 6)$, $C' = (7, 4)$ **(c)** $O(3, 12)$ **5.** It is a dilation d_k with center at the only fixed point, O (all lines through O are fixed by d_k).   **7.** See Example 1 (the square is replaced by a Golden Rectangle).   **9.** By definition, $k = A'B'/AB$; $\triangle ABC \sim \triangle A'B'C' \to B'C'/BC = A'B'/AB = k$. Let D be arbitrary. $\triangle BCD \sim \triangle B'C'D' \to C'D'/CD = B'C'/BC = k$, or $C'D' = k \cdot CD$ for arbitrary C and D.   **11.** $S = d_k(d_{1/k}S)$, so S is the isometry $d_{1/k}S$ followed by the dilation d_k.   **13. (a)** Locate Q on line ℓ; with $Q =$ center, rotate line m through 90° to new position m'. If $P = m' \cap n$, rotate P 90° in reverse direction to obtain R on m. $\triangle PQR$ will be desired triangle. **(b)** Use rotation of 60°.   **15.** For uniqueness, use method of Theorem 1, §5.3. To prove existence, let k be constant of proportionality between $\triangle ABC$ and $\triangle DEF$ ($DE = k \cdot AB$); with any point as center, $d_k^{-1} = d_{1/k}$ maps $\triangle DEF$ to $\triangle D'E'F'$ congruent to $\triangle ABC$. By Theorem 1, §5.3, there exists isometry f taking $\triangle ABC$ to $\triangle D'E'F'$; $d_k f = S$ is desired similitude.
17. Circles of G map to family of concurrent lines through A' (image of point A), while F_1, F_2 map to family of concentric circles centered at A', inside and outside image ω of y-axis, respectively, as shown in figure.

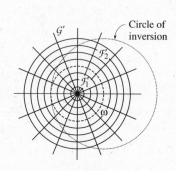

Circle of inversion

Exercise 17

Section 5.5

1. (a) similitude with center $(0, 0)$ **(b)** dilation d_3 with center $(0, 0)$ **(c)** translation t_{AB}, where $A = (0, 0)$ and $B = (-3, -5)$ **3. (a)** $P'(9, 6)$, $Q'(17, 13)$, $R'(-12, -13)$, $S'(-4, -6)$ **(b)** $P'(3, -3)$, $Q'(3, -6)$, $R'(4, 5)$, $S'(4, 2)$ **7. (a)** $x' = 3x/5 + 4y/5 + 2/5$, $y' = 4x/5 - 3y/5 - 4/5$ and $x'' = 3x'/5 + 4y'/5 + 6/5$, $y'' = 4x'/5 - 3y'/5 - 12/5$ **(b)** $x'' = x + 4/5$, $y'' = y - 8/5$ **9.** Midpoint of $P(x, y)$ and $P'(x', y')$ is $(1/2(x - x + 2h), 1/2(y - y + 2k)) = (h, k)$.   **11.** f is the reflection s_ℓ: $x' = x$, $y' = y$, where $\ell =$ x-axis, followed by the translation t_{AB}: $x'' = x' + 3$, $y'' = y'$ in direction $AB = [3, 0]$ (x-axis).   **13. (a)** By (3), $m = m' = (c + md)/(a + mb)$ for all $m \to bm^2 + (a - d)m - c = 0$, or $b = 0$, $a = d$, $c = 0$. **(b)** From **(6)** we see that $x_0 = y_0 = 0$.   **15.** $K = K' = 5$ **17.** It suffices to prove for $x' = ax + by$, $y' = cx + dy$, with inverse $x = kdx' - kby'$, $y = -kcx' + kay'$ ($k = 1/(ad - bc)$). If circle $x^2 + y^2 + Ax + By + C = 0$ maps to another circle, by substitution $k^2(d^2 + c^2) \cdot x'^2 + k^2(b^2 + a^2)y'^2 + 2k^2(ad + bc)x'y' + \{$linear terms in $x', y'\} = 0 \to d^2 + c^2 = b^2 + a^2$ and $ad + bc = 0$; use lemma.

Section 5.6

1. $fg = gf$: $x' = 14x + 4y$, $y' = 12x + 2y$; f^{-1}: $x' = x - y$, $y' = -3x + 4y$ **3.** Both are transformation groups (using geometry); set of dilations is a subset of the set of similitudes given by setting $b = 0$.   **5.** Cyclic.   **7.** $D_2 = \{e, s_\ell, r_{180}, s_\ell r_{180}\}$, where r_{180} has center at M; D_2 is the Fours Group.   **9.** Coordinate proof: If $A = [a, -\delta b, b, \delta a]$, $B = [c, -\varepsilon d, d, \varepsilon c]$, with $a^2 + b^2 = 1 = c^2 + d^2$, then $AB = [ac - \delta bd, -\delta\varepsilon (bc + \delta ad), bc + \delta ad, \delta\varepsilon (ac - \delta bd)] \equiv [r, -\delta's, s, \delta'r]$, where $r = ac - \delta bd$, $s = bc + \delta ad$, and $\delta' = \delta\varepsilon$. Now show, by algebra, that $r^2 + s^2 = 1$.
13. (**WARNING:** Do not assume angles of $\triangle ABC$ are of measure α, β, γ.) Consider $\triangle ABD$, where $m\angle A = \alpha$, $m\angle B = \beta$, and $\therefore$ $m\angle D = \gamma$. By Theorem 1, $r_{2\alpha}[A]r_{2\beta}[B]r_{2\gamma}[C] = r_{-2\gamma}[D]r_{2\gamma}[C]$. If this is not the identity, then $C \neq D$. Let $\ell = CD$, and $m, n \parallel$ lines through C and D, making

alternate interior angles of measure γ with $\ell \to r_{2\gamma}[C] = s_m s_\ell$ and $r_{-2\gamma}[D] = s_\ell s_n \to r_{-2\gamma}[D]r_{2\gamma}[C] = s_\ell s_n s_m s_\ell = s_\ell t s_\ell = t$ (where t is the translation $s_n s_m$, since $m \parallel n$).   **15.** By formulas given, the direct LorentzTransformations $\boldsymbol{L}$ are a subset of the generalized Lorentz Group; must show

$\boldsymbol{L}$ closed under products, inverses: If $A = \begin{bmatrix} a & b \\ b & a \end{bmatrix}$, $B = \begin{bmatrix} c & d \\ d & c \end{bmatrix}$, and $a^2 - b^2 = 1 = c^2 - d^2$, then $AB = \begin{bmatrix} ac + bd & ad + bc \\ bc + ad & bd = ac \end{bmatrix} \equiv \begin{bmatrix} r & s \\ s & r \end{bmatrix}$

and $A^{-1} = \begin{bmatrix} a & -b \\ -b & a \end{bmatrix}$. Show by algebra that $r^2 - s^2 = (ac + bd)^2 + (ad + bc)^2 = 1$.

Section 5.7

3. Since O lies on the perpendicular bisectors of sides of $\triangle ABC$, O lies on altitudes of $\triangle LMN$, hence O = orthocenter of $\triangle LMN$. Since $d_{-1/2}[G]$ maps $\triangle ABC$ to $\triangle LMN$, O maps to orthocenter H, hence O = $H' \to OG = H'G' = \frac{1}{2}HG$ and H-G-O. $\therefore HG = 2GO$.   **5.** $OP'/OP = OA'/OA = k$ (Side-Splitting Theorem) $\to OP' = kOP \to P'$ is image of P under d_k (center O). By adjusting A on OA', k can theoretically assume any value > 1.   **7.** Take copies of $\triangle ABC$ with common vertex A, filling in with equilateral triangles for the fundamental region of a tesselation (figure).   **9.** Starting with $\triangle ABC$ and constructing isosceles triangles $\triangle QCA$, $\triangle RAB$, $\triangle PCB$ having vertex angles of measure $2p$, $2q$, and $2r$, as stated in the problem, Corollary C of Theorem 3, §5.6 implies that the product $r_{2p}[P]r_{2q}[Q]r_{2r}[R]$ maps B back to itself, hence is the identity. By Corollary B of Theorem 3, $\triangle PQR$ has angle measures p, q, and r.   **11.** Circumcenter I of $\triangle PQR$ is incenter of $\triangle ABC$, determined by constructing perpendiculars to $\overline{PI}$, $\overline{QI}$, $\overline{RI}$ at P, Q, R, respectively.   **13.** Draw any circle tangent to sides of $\angle ABC$ and draw $\overrightarrow{BP}$, where P is the given internal point, cutting circle at Q, then map circle to one passing through P using $d_k[B]$, where $k = BP/BQ$.   **15. (b)** Circumcircle of $\triangle ABF$ will cut $\overline{FD}$ at some point Q (if $\angle ACB < 120$); by circle theorems, $\to m\angle AQB = 120 = m\angle APB \to P = Q$.   **17.** Reflection of line $\overleftrightarrow{BC}$ in P will intersect $\overleftrightarrow{BA}$ at Q; draw $\overleftrightarrow{QP}$, cutting $\overleftrightarrow{BC}$ at R.   **19.** Let A be any point on outer circle; rotate innermost circle about A through $60°$, and find B, its intersection with middle circle (no solution if B does not exist). Rotate B about point A $60°$ in reverse direction to find point C; $\triangle ABC$ will be desired triangle.   **21.** $PO \cdot P'O = (PC + CO)(P'C + CO) = (-CP + CO) \cdot (CP + CO) = CO^2 - PC^2 = (OB^2 - BC^2) - (PB^2 - BC^2) = OB^2 - PB^2 = r^2$ (constant since linkages $\overline{OB}$ and $\overline{PB}$ are rigid as P and P' move) $\to P$ and P' are inverse pairs under circular inversion, with circle passing through O traced by P mapping to straight line drawn by P'.   **23.** Complete this as a project.

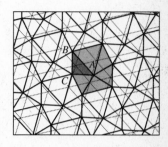

Exercise 7

Testing Your Knowledge (Chapter 5)

1-T, 2-T, 3-F, 4-F, 5-F, 6-T, 7-T, 8-T, 9-T, 10-F

Section 6.2

1. Subdivide triangle by drawing cevian, then use $k_1 = k_2 = k$.   **3.** Lines perpendicular to same line are parallel, hence $\diamondsuit ABCD$ is a parallelogram, with $AB = CD$ (depends on Property Z, which can be proven from Property U, which, in turn, is equivalent to Euclid's Fifth Postulate).   **5.** $PA \approx 517{,}381$ miles   **7.** Let A, B, and C be noncollinear points and let ℓ and m be the perpendicular bisectors of $\overline{AB}$ and $\overline{BC}$. Then ℓ and m meet at some point O (Problem 6), and O is equidistant from A, B, and C.   **9.** If the parallel to ray $\overrightarrow{BA}$ which passes through P does not meet $\overrightarrow{BC}$, Euclidean Parallel Postulate is denied. Hence, let point of intersection be Q. Choose R such that B-Q-R; line $\overrightarrow{RP}$ will then intersect both sides of $\angle ABC$.   **11.** Suppose $AD < BC$ in $\square ABCD$; on $\overline{BC}$ construct $\overline{BE} \cong \overline{AD} \to \diamondsuit ABED$ is Saccheri Quadrilateral $\to m\angle ADE = m\angle BED > 90$. $\to\leftarrow$ (Exterior Angle Inequality) $\therefore AD \geq BC$. Similarly, $BC \geq AD \to BC = AD$; $AB = CD$ follows.   **13.** Show quadrilateral is Saccheri Quadrilateral.   **15.** If $\square ABCD$ has $m\angle A = m\angle B = m\angle C = 90$ and $BC = AD$, it is a Saccheri Quadrilateral with congruent summit angles, $m\angle D = m\angle C = 90$.   **17.** Let $\diamondsuit ABCD$ be any Lambert Quadrilateral (right angles at A, B, and C), Problem 16 $\to \square BEGF$ exists, with B-A-E and B-C-F; let $\overleftrightarrow{CD}$ meet $\overline{EG}$ at H, and apply Problem 14 to $\diamondsuit BEHC$, then to $\diamondsuit ABCD$.   **19.** Double segment $\overline{AP}$ to P_1, let Q_1 be foot of perpendicular from P_1 on line $\overleftrightarrow{AQ}$, and let M be midpoint of $\overline{AQ_1}$. By a corollary of Theorem 3, §3.7, $PQ \leq PM \leq \frac{1}{2}P_1Q_1$ or $P_1Q_1 \geq 2PQ$. Repeat process, doubling AP_1 to P_2, with Q_2 = foot of perpendicular from P_2 on line $\overleftrightarrow{AP}$. $P_2Q_2 \geq 2P_1Q_1 \geq 4PQ$; nth step yields $P_nQ_n \geq 2^nPQ$. $\therefore x \to \infty$.

Section 6.3

3. (a) $x = 0$ ($y > 0$) (b) $x^2 + y^2 = 4$ ($y > 0$) (c) $x^2 + y^2 - 7x = 4$ ($y > 0$)   7. (a) Diameter of h-line $\overleftrightarrow{AB}$ has endpoints $M(5\sqrt{41} - 24, 0)$, $N(-5\sqrt{41} - 24, 0) \to c = AB^* = \left| \ln AM/AN \cdot BN/BM \right| = \frac{1}{2} \left| \ln 5(410 - 62\sqrt{41})/5(410 + 62\sqrt{41}) \cdot 25(82 + 10\sqrt{41})/25(82 - 10\sqrt{41}) \right|$
$\frac{1}{2} \left| \ln (25 - 3\sqrt{41})/(25 + 3\sqrt{41}) \right| = \frac{1}{2} \left| \ln (25 - 3\sqrt{41})/(25 + 3\sqrt{41}) \right| = \frac{1}{2} \ln (25 + 3\sqrt{41})^2/(25^2 - 9 \cdot 41) \equiv \ln d$, where $d =$

$(25 + 3\sqrt{41})/16$. From Problem 6, $a = b = \ln 2$ and $c = \ln d \neq \sqrt{a^2 + b^2}$, so Pythagorean Theorem fails. (b) $\cosh c = \cosh (\ln d) = \frac{1}{2}(e^{\ln d} + e^{-\ln d}) = \frac{1}{2}(d + 1/d) = \frac{25}{16}$; $\cosh a \cosh b = \cosh^2 a = \frac{25}{16} = \cosh c$.

Section 6.4

1. 15   3. $\delta = 20$, $\delta_1 = 2$, $\delta_2 = \delta_3 = \delta_4 = 6$ 5. Let $\triangle ABC$ be given, with base $\overline{BC}$ chosen such that angles at B and C are both acute; foot D of perpendicular from A on $\overleftrightarrow{BC}$ lies between B and C by Exterior Angle Inequality. By Theorem 1, $m\angle B + m\angle BAD < 90$ and $m\angle DAC + m\angle C < 90 \to m\angle B + m\angle BAD + m\angle DAC + m\angle C < 180$, or $m\angle B + m\angle BAC + m\angle C < 180$.   7. $\delta = 14$; because a Saccheri Quadrilateral may be divided into two congruent Lambert Quadrilaterals having equal areas   9. (a) 149 (b) $m\angle OAB = m\angle ABO = 74.5$; $m\angle AOB = 30$ (defect of each triangle equals 1)   11. 103, 36.5, 36.5 (exact)   13. Show each angle of one pentagon congruent to each angle of other, and central triangles congruent.   15. (a) No; rectangles do not exist in hyperbolic geometry. (b) Let $\triangle AOB$ be right triangle in Quadrant I $\to$ Area (quadrilateral) $= k(360 - 8m\angle OAB)$, which converges to 360k.

Section 6.5

1. Since $\triangle AB\Omega$ is isosceles, $\angle A \cong \angle B$; $MA = MB \to \triangle AM\Omega \cong \triangle BM\Omega$ (BA criterion) $\to \angle AMR \cong \angle RMB \to \overline{MR} \perp \overline{AB}$.   3. In $\triangle AB\Omega$, if $M =$ midpoint of $\overline{AB} \to$ asymptotic parallel $\overrightarrow{MR}$ (perpendicular to AB by Problem 1) coincides with perpendicular bisector of base $\overline{AB}$.   5. Choose $S \in \ell$ on P-side of $\overline{AB} \to$ either (1) $DS \leq$ Interior $\angle ADR$, or (2) $\overrightarrow{DS} \subseteq$ Interior $\angle RDB$. In (1), by transitivity/symmetry of asymptotic parallelism, $\overrightarrow{DR} \parallel_a \overrightarrow{AP} \to \overrightarrow{DS}$ meets $\overrightarrow{AP}$; in (2), $\overrightarrow{DR} \parallel_a \overrightarrow{BQ}$ (by construction) $\to \overrightarrow{DS}$ meets $\overrightarrow{BQ}$.   7. In Figure 6.35, Theorem 4 $\to m\angle 1 + m\angle 3 < 180$; Linear Pair Axiom $\to m\angle 2 + m\angle 3 = 180 \to m\angle 2 + m\angle 3 > m\angle 1 + m\angle 3 \to m\angle 2 > m\angle 1$.   9. Draw angle bisectors $\overrightarrow{AS}$, $\overrightarrow{BS}$, meeting at S (proof needed); drop perpendiculars $\overline{SD}$, $\overline{SE}$ to $\overrightarrow{AP}$, $\overrightarrow{BQ}$, then determine bisector $\overrightarrow{SR}$ of $\angle DSE$. Line $\overleftrightarrow{SR}$ meets $\overline{AB}$ at desired point C, and $\overrightarrow{CR} \parallel_a \overrightarrow{AP}$, $\overrightarrow{BQ}$.

Section 6.6

1. Since ray $\overrightarrow{AP}$ cannot meet ray $\overrightarrow{CR}$ (why?), it remains to prove if $\overrightarrow{AT} \subseteq$ Interior $\angle CAP$, then $\overrightarrow{AT}$ meets $\overrightarrow{CR}$; but $\overrightarrow{AT}$ meets $\overrightarrow{BQ}$ at some point S; assume A-S-T. T lies on C-side of $\overrightarrow{BS}$ and on S-side of $\overleftrightarrow{BC} = \overleftrightarrow{AC} \to T \in$ Interior $\angle CBS \to \overrightarrow{BT}$ meets $\overrightarrow{CR}$ at some point W, with B-T-W. By Postulate of Pasch for Absolute Geometry (applied to $\triangle BCW$), $\overrightarrow{AT}$ meets $\overrightarrow{CR}$.   3. As $a \to 0$, $1/e^{ka} \to 1/e^0 = 1$ and $\tan \frac{1}{2}A \to 1$ or $A \to 90$. As $a \to \infty$, $1/e^{ka} \to 0$ and $\tan \frac{1}{2}A \to 0$ or $A \to 0$.   5. 0.8814   7. Since $\ell \parallel m$, either $\ell \parallel_a m$ or $\ell \parallel_d m$. If $\ell \parallel_a m$, by result of Problem 6, $PQ \to 0$ as $AP \to \infty \to \leftarrow$. $\therefore \ell \parallel_a m$.   9. First prove (1) for every point Q on perpendicular $\overline{PA}$ to line ℓ such that A-P-Q, there are two lines through Q parallel to ℓ, then (2) if Q is the midpoint of $\overline{AP}$, there are two parallels through Q, and finally (3) every point Q on line $\overleftrightarrow{AP}$ has two parallels to ℓ. If R and m, $R \notin m$, are anywhere in the plane, drop the perpendicular $\overline{RB}$ to line m, and construct Q on $\overline{PA}$ such that $QA = RB$. The two parallels to ℓ through Q can then be "mapped" to two lines through R parallel to m.   11. $\tan A = \text{sech } 4 \approx 0.0366 \to A \approx 2.10°$; A tends to zero as limit.   13. $\cosh c = \cosh 0.03 \cosh 0.04 \approx 1.0012505 \to c \approx 0.0500048 \approx 0.05$ 16. Show that $\cosh a \cosh b = \cot A \cot B$; $\therefore \cosh c = \cot 20 \cot 60 \to c \approx 1.0359$, $a \approx 0.4096$, $b \approx 1.0664$.   17. In right triangle $\triangle ABC$ with base $\overline{AC}$ on line ℓ and side $\overline{BC}$ fixed, as $CA = b \to \infty$, $\tan B = \tanh b/\sinh a \to 1/\sinh a$ and $B \to A$ (angle of parallelism). Hence $\cot A = \sinh a \to a = \ln (\cot A + \sqrt{\cot^2 A + 1}) = \ln (\cot A + \csc A) = \ln \cot \frac{1}{2}A = -\ln \tan \frac{1}{2}A$ or $e^{-a} = \tan \frac{1}{2}A$.   19. By SASAA, $\diamond BLL'B' \cong \diamond CLL'C'$, $\diamond CMM'C' \cong \diamond AMM'A' \to BB' = CC' = AA'$. Then $\diamond AA'B'B$ is Saccheri Quadrilateral with ℓ as common perpendicular bisector of bases. As ℓ and $\overleftrightarrow{LL'}$ have m as a common perpendicular, $\ell \parallel_d \overleftrightarrow{LL'}$.

Section 6.7

1. Angle Sum Theorem   3. (a) Intersectors, (b) divergent parallels, (c) asymptotic parallels.   7. Lambert Quadrilateral with acute angle at A.   9. (a) Family of circles tangent to diameter at one of its endpoints.   11. (a) h-lines passing through A, h-ray $\overrightarrow{AU}$
13. $y - x = \ln (AQ, MN) - \ln (AP, MN) = \ln AM/AN \cdot QN/QM + \ln AN/AM \cdot PM/PN = \ln AM/AN \cdot QN/QM \cdot AN/AM \cdot PM/PN = \ln PM/PN \cdot QN/QM = \ln (PQ, MN) = PQ^*$.   17. If $BC < AD$, construct $\overline{AE} \cong \overline{BC}$ on $\overline{AD}$, draw $\overline{CE}$, and obtain a contradiction.

Section 6.8

1. 3.4   3. 120   5. By SAS, $\triangle ADE \cong \triangle PQR \to \varepsilon_1 = \varepsilon_4 = \varepsilon(\triangle ABC) \equiv \varepsilon$. Hence $\varepsilon = \varepsilon_1 + \varepsilon_2 + \varepsilon_3 > \varepsilon_1 = \varepsilon_4 = \varepsilon$. $\to \leftarrow$ 9. 180

11. Assuming any two lines intersect, let given lines ℓ, m meet at A, A'; locate the midway points L, M on ℓ, m, respectively. $(AL = LA'$, $AM = MA')$, and join L, M. Then line $\overleftrightarrow{LM}$ is the common perpendicular.   **13.** If $\varepsilon = \varepsilon(\triangle ABC)$, since excess is always positive and angle sum of any right triangle < 360, $0 < \varepsilon < 180 \rightarrow 0 < m\angle A + m\angle B + m\angle C - 180 < 180 \rightarrow 90 < m\angle A + m\angle B < 270$.   **15.** $c = \frac{1}{2}\pi$, $A = 90$, $B = 60$. This triangle is isosceles, birectangular with B at north pole and $\overline{AC}$ on equator; angle at B ($60°$) opens to $\frac{1}{3}$ of a hemisphere, so side opposite (on the equator) opens to $\frac{1}{3}$ of semicircle, or $b = \pi/3$.   **17.** $\cos c = (\cos 0.03)(\cos 0.04) \approx 0.9987505 \rightarrow c \approx 0.049995197 \approx 0.05$.   **19.** Since B and C lie on A-side of line $\overleftrightarrow{DE}$ (because $AB < \frac{1}{2}\alpha$, $AC < \frac{1}{2}\alpha \rightarrow A$-$B$-$D$ and A-C-E) then by convexity of half-planes, $M \in \overline{BC} \subseteq \{A\text{-side of } \overleftrightarrow{DE}\}$.   **21.** $AM = MC$, $AN = \frac{1}{2}AB = \frac{1}{2}BC = CL$, and $\angle A \cong \angle C$. By SAS, $\triangle AMN \cong \triangle CML$. Similarly, $\triangle AMN \cong \triangle BLN$. Thus, $\triangle MLN$ is equilateral/equiangular. Since $\varepsilon(\triangle ABC) > \varepsilon(\triangle LMN)$, then $3\theta - 180 > 3\phi - 180$ or $\theta > \phi$. With $t = m\angle NMA = m\angle LMC$, then $\phi + 2t = 180$ and $\varepsilon(\triangle AMN) = \theta + 2t - 180 = \theta + (180 - \phi) - 180 = \theta - \phi$.   **23.** Let ℓ, m be any two lines. On line ℓ, locate points A, B such that $AB = \alpha$, and let $C \in m$. If $AC = \alpha$, then $C = B$ and we are done. Otherwise, A-C-B. Locate D on $\overrightarrow{AC}$ and E on ℓ so that A-C-D-B and A-E-$B \rightarrow A$, E, D noncollinear (unless C already lies on line $\overleftrightarrow{AE}$) $\rightarrow$ by Postulate of Pasch applied to $\triangle ADE$, and, if necessary, to $\triangle DEB$, m meets either $\overleftrightarrow{AE}$, or $\overline{EB}$.

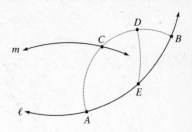

Exercise 23

Testing Your Knowledge (Chapter 6)

1-T, 2-F, 3-T, 4-T, 5-F, 6-T, 7-F, 8-F, 9-T, 10-T

APPENDIX C: SYMBOLS, AXIOMS, THEOREMS

SYMBOLS USED IN GEOMETRY

SYMBOL	DESCRIPTION	LOCATION (PAGE)
$\overleftrightarrow{AB}$	Line determined by A and B	90
AB	Distance from A to B	97
$A\text{-}B\text{-}C$	B lies between A and C	98
$A\text{-}B\text{-}C\text{-}D$	Betweenness relation for A, B, C, and D	99
$\overline{AB}$	Segment joining A and B	100
$\overrightarrow{AB}$	Ray through B with origin A	100
$\angle ABC$	Angle with sides $\overrightarrow{BA}$ and $\overrightarrow{BC}$	100
$m\angle ABC$	Measure of $\angle ABC$	113
$H(A, \ell)$	Half-plane, the A-side of line ℓ	107
$\overrightarrow{AB}\text{-}\overrightarrow{AC}\text{-}\overrightarrow{AD}$	Betweenness for rays	114
$\ell \perp m$	Line ℓ is perpendicular to line m	125
$\ell \parallel m$	Line ℓ is parallel to line m	208
$\triangle ABC$	Triangle with vertices A, B, and C	132
$\diamondsuit ABCD$	Quadrilateral with vertices A, B, C, and D	182
$\square ABCD$	Parallelogram $ABCD$	217
$\square ABCD$	Rectangle $ABCD$, square $ABCD$	182
$\overline{AB} \cong \overline{XY}$	$\overline{AB}$ is congruent to $\overline{XY}$	134
$\angle ABC \cong \angle XYZ$	$\angle ABC$ is congruent to $\angle XYZ$	134
$\triangle ABC \cong \triangle XYZ$	$\triangle ABC$ is congruent to $\triangle XYZ$	136
$\diamondsuit ABCD \cong \diamondsuit XYZW$	$\diamondsuit ABCD$ is congruent to $\diamondsuit XYZW$	184
$\triangle ABC \sim \triangle XYZ$	$\triangle ABC$ is similar to $\triangle XYZ$	226

AXIOMS FOR ABSOLUTE GEOMETRY

UNDEFINED TERMS: *Point, line, plane,* and *space*

AXIOM I-1 Each two distinct points determine a line.

AXIOM I-2 Three noncollinear points determine a plane.

AXIOM I-3 If two points lie in a plane, then any line determined by those two points lies in that plane.

AXIOM I-4 If two planes meet, their intersection is a line.

AXIOM I-5 Space consists of at least four noncoplanar points, and contains three non-collinear points. Each plane contains at least three noncollinear points, and each line contains at least two distinct points.

AXIOM D-1 (Existence) Each pair of points (A, B) is associated with a unique real number $AB \geq 0$, called the *distance* from A to B.

AXIOM D-2 (Positive Definiteness) For all points A and B, $AB > 0$ unless $A = B$.

AXIOM D-3 (Symmetry) For all points A and B, $AB = BA$.

AXIOM D-4 Given any three points A, B, and C on line ℓ such that A-B-C, $\overrightarrow{BA} \cup \overrightarrow{BC} = \ell$.

AXIOM D-5 (Ruler Postulate) The points of each line ℓ may be assigned to the real numbers x, $-\infty < x < \infty$, called *coordinates,* in such a manner that

(1) each point on ℓ is assigned to a unique coordinate,

(2) each coordinate is assigned to a unique point on ℓ,

(3) any point may be assigned the zero coordinate and any other point a positive coordinate,

(4) for any two points A and B of ℓ having coordinates a and b, $AB = |a - b|$.

AXIOM H-1 (Plane Separation Postulate) Let ℓ be any line lying in any plane P. The set of all points in P not on ℓ consists of the union of two subsets H_1 and H_2 of P such that:

(1) H_1 and H_2 are convex sets.

(2) H_1 and H_2 have no points in common.

(3) If $A \in H_1$ and $B \in H_2$, line ℓ intersects the segment $\overline{AB}$.

AXIOM A-1 (Existence of Angle-Measure) To every angle there corresponds a unique real number θ, $0 < \theta < 180$.

AXIOM A-2 (Angle Addition Postulate) If D lies in the interior of $\angle ABC$, then $m\angle ABC = m\angle ABD + m\angle DBC$, and conversely.

AXIOM A-3 (Protractor Postulate) The set of rays having a common origin O and lying on one side of line $\ell = \overleftrightarrow{OA}$, including ray $\overrightarrow{OA}$, may be assigned to the real numbers θ for which $0 \leq \theta < 180$, called *coordinates,* in such a manner that

(1) each ray is assigned to a unique coordinate θ,

(2) each coordinate is assigned to a unique ray,

(3) the coordinate of $\overrightarrow{OA}$ is 0,

(4) if rays $\overrightarrow{OP}$ and $\overrightarrow{OQ}$ have coordinates θ and ϕ, then $m\angle POQ = |\theta - \phi|$.

AXIOM A-4 (Linear Pair Axiom) A linear pair of angles is a supplementary pair.

AXIOM C-1 (SAS Postulate) If $\overline{AB} \cong \overline{XY}$, $\angle B \cong \angle Y$, and $\overline{BC} \cong \overline{YZ}$, then $\triangle ABC \cong \triangle XYZ$.

MAJOR THEOREMS (CHAPTERS 1–5)

1.5

THEOREM 2 *Steiner's Theorem.* As a point revolves about the circumcircle of a triangle, the corresponding Simson Line with respect to that point generates the family of tangents to a deltoid, whose center is the Nine-Point Center and whose inner tangent circle is the Nine-Point Circle of the triangle, and whose cusps are located on the three lines through the Nine-Point Center which are perpendicular to the sides of the Morley Triangle.

2.3

THEOREM 1 If R is a rectangle with base of length b units and height of length h units, then Area $R = bh$.

2.4

THEOREM 1   If $C \in \overrightarrow{AB}$, $D \in \overrightarrow{AB}$, and $C \neq D$, then $\overleftrightarrow{CD} = \overleftrightarrow{AB}$.

2.5

THEOREM 1   If A-B-C, then C-B-A, and neither A-C-B nor B-A-C.

THEOREM 3   (a) $\overline{AB} = \overline{BA}$ (b) $\overline{AB} \subseteq \overrightarrow{AB}$ (c) $\overrightarrow{AB} \subseteq \overleftrightarrow{AB}$.

THEOREM 4 $\overrightarrow{AB} \cap \overrightarrow{BA} = \overline{AB}$.

2.6

LEMMA   Suppose that B lies on line ℓ, A lies in one of the half-planes H_1 determined by ℓ, and that A-B-C holds. Then point C will lie in the opposite half-plane, H_2.

THEOREM 1   If one point of a segment or ray lies in a half-plane H_1 determined by some line ℓ, and the endpoint of the segment or ray itself lies on ℓ, then the entire open segment or open ray lies in H_1.

COROLLARY   Let B and F lie on opposite sides of a line ℓ and let A and G be any two points on ℓ. Then segment $\overline{GB}$ and ray $\overrightarrow{AF}$ have no points in common.

THEOREM 2   *(Postulate of Pasch)* Suppose A, B, and C are any three distinct noncollinear points in a plane, and ℓ is any line which also lies in that plane and passes through an interior point D of segment $\overline{AB}$ but not through A, B, nor C. Then ℓ meets either $\overline{AC}$ at some interior point E, or $\overline{BC}$ at some interior point F, the cases being mutually exclusive.

THEOREM 3   If A and C lie on the sides of $\angle B$, then, except for endpoints, segment $\overline{AC} \subseteq$ Interior $\angle B$. If $D \in$ Interior $\angle B$, then, except for B, ray $\overrightarrow{BD} \in$ Interior $\angle B$.

2.7

THEOREM 1   If $A[a]$, $B[b]$, and $C[c]$ are three collinear points (and $\overrightarrow{OA}[a]$, $\overrightarrow{OB}[b]$, $\overrightarrow{OC}[c]$ three concurrent rays) with their coordinates, then A-B-C ($\overrightarrow{OA}$-$\overrightarrow{OB}$-$\overrightarrow{OC}$) iff $a < b < c$ or $c < b < a$.

COROLLARY   Suppose that four distinct collinear points are given with their coordinates: $A[a]$, $B[b]$, $C[c]$, $D[d]$. If A-B-C and A-C-D, then A-B-C-D, and similarly for rays $\overrightarrow{OA}$, $\overrightarrow{OB}$, $\overrightarrow{OC}$, and $\overrightarrow{OD}$.

THEOREM 2   If $C \in \overrightarrow{AB}$ and $A \neq C$, then $\overrightarrow{AB} = \overrightarrow{AC}$.

THEOREM 3   *(Segment Construction Theorem)* If $\overline{AB}$ and $\overline{XY}$ are any two segments and $AB \neq XY$, then there is a unique point C on ray $\overrightarrow{AB}$ such that $AC = XY$, with A-C-B if $XY < AB$, or A-B-C if $XY > AB$.

THEOREM 5   *(Segment-Doubling Theorem)* There exists a unique point C on ray $\overrightarrow{AB}$ such that B is the midpoint of $\overline{AC}$.

THEOREM 3′   *(Angle Construction Theorem)* If $\angle ABC$ and $\angle XYZ$ are any two nondegenerate angles and $m\angle ABC \neq m\angle XYZ$, then there exists a unique ray $\overrightarrow{BD}$ on the C-side of $\overrightarrow{AB}$ such that $m\angle XYZ = m\angle ABD$, and either $\overrightarrow{BA}$-$\overrightarrow{BD}$-$\overrightarrow{BC}$ if $m\angle XYZ < m\angle ABC$, or $\overrightarrow{BA}$-$\overrightarrow{BC}$-$\overrightarrow{BD}$ if $m\angle XYZ > m\angle ABC$.

THEOREM 5′   *(Angle-Doubling Theorem)* Given any angle $\angle ABC$ having measure < 90, there exists a ray $\overrightarrow{BD}$ such that BC is the bisector of $\angle ABD$.

2.8

THEOREM 1   *(Crossbar Theorem)* If D is in the interior of $\angle BAC$, then ray $\overrightarrow{AD}$ meets segment $\overline{BC}$ at some interior point E.

THEOREM 2   Two angles which are supplementary (or complementary) to the same angle have equal measures.

THEOREM 3   *(Vertical Pair Theorem)* Vertical angles have equal measures.

COROLLARY A Two lines are perpendicular iff they contain the sides of a right angle.

THEOREM 5 *(Existence and Uniqueness of Perpendiculars)* Suppose that in some plane line m is given and an arbitrary point A on m is located. Then there exists a unique line ℓ that is perpendicular to m at A.

3.3

THEOREM 1 *(ASA Theorem)* If $\angle A \cong \angle D$, $\overline{AB} \cong \overline{DE}$, and $\angle B \cong \angle E$, then $\triangle ABC \cong \triangle DEF$.

THEOREM 2 *(Isosceles Triangle Theorem)* In $\triangle ABC$, $\overline{AB} \cong \overline{AC}$ iff $\angle B \cong \angle C$. Furthermore, in any isosceles triangle if line ℓ satisfies any two of the four symmetry properties *PB*, *BB*, *BV*, and *PV*, it satisfies all four, and is a line of symmetry for the triangle.

PERPENDICULAR BISECTOR THEOREM The set of all points equidistant from two fixed points A and B is the perpendicular bisector of $\overline{AB}$.

THEOREM 3 *(SSS Theorem)* If $\overline{AB} \cong \overline{DE}$, $\overline{BC} \cong \overline{EF}$, and $\overline{AC} \cong \overline{DF}$, then $\triangle ABC \cong \triangle DEF$.

THEOREM 4 Given line ℓ and any point A not on ℓ, there exists a unique line m passing through A perpendicular to ℓ.

3.4

THEOREM 1 *(Exterior Angle Inequality)* Given $\triangle ABC$ with B-C-D, $m\angle ACD > m\angle A$ and $m\angle ACD > m\angle B$.

COROLLARIES: (1) The sum of the measures of any two angles of a triangle is less than 180. (2) A triangle can have at most one right or obtuse angle. (3) The base angles of an isosceles triangle are acute.

THEOREM 2 *(Saccheri-Legendre Theorem)* The angle sum of any triangle is ≤ 180.

3.5

SCALENE INEQUALITY In $\triangle ABC$ with standard notation, $a > b$ iff $m\angle A > m\angle B$.

THEOREM 1 *(Triangle Inequality)* For any three distinct points A, B, and C, $AB + BC \geq AC$, with equality if and only if A-B-C.

COROLLARY *(Median Inequality)* Suppose that $\overline{AM}$ is the median to side $\overline{BC}$ of $\triangle ABC$. Then $AM < \frac{1}{2}(AB + AC)$.

THEOREM 2 *(SAS Inequality)* If in $\triangle ABC$ and $\triangle XYZ$ we have $AB = XY$, $AC = YZ$, but $m\angle A > m\angle X$, then $BC > YZ$.

3.6

THEOREM 1 *(AAS Theorem)* If $\angle A \cong \angle D$, $\angle B \cong \angle E$, and $\overline{BC} \cong \overline{EF}$, then $\triangle ABC \cong \triangle DEF$.

THEOREM 2 *(SsA Theorem)* If $AB \cong DE$, $BC \cong EF$, $\angle A \cong \angle D$, and $BC > AB$, then $\triangle ABC \cong \triangle DEF$.

COROLLARY A *(HA Theorem)* Two right triangles are congruent if they have congruent hypotenuses and a pair of congruent acute angles.

COROLLARY B *(HL Theorem)* Two right triangles are congruent if they have congruent hypotenuses and a pair of congruent legs.

3.7

THEOREM 1 *(SASAS Congruence)* Suppose that two convex quadrilaterals $\Diamond ABCD$ and $\Diamond XYZW$ have $\overline{AB} \cong \overline{XY}$, $\angle B \cong \angle Y$, $\overline{BC} \cong \overline{YZ}$, $\angle C \cong \angle Z$, and $\overline{CD} \cong \overline{ZW}$. Then $\Diamond ABCD \cong \Diamond XYZW$.

FURTHER CONGRUENCE CRITERIA: ASASA, SASAA, and SASSS.

THEOREM 2 The summit angles of a Saccheri Quadrilateral are congruent.

COROLLARIES:

(1) The diagonals of a Saccheri Quadrilateral are congruent.

(2) The line joining the midpoints of the base and summit of a Saccheri Quadrilateral is the perpendicular bisector of both the base and summit.

(3) If each of the summit angles of a Saccheri Quadrilateral is a right angle, the quadrilateral is a rectangle and the summit is congruent to the base.

(4) If the summit angles of a Saccheri Quadrilateral are acute, the summit has greater length than the base.

THEOREM 3   The angle sum of any triangle is one-half the measure of either summit angle of the associated Saccheri Quadrilateral, and the corresponding line joining midpoints of two of the sides of the triangle has length $1/2$ that of the base of the quadrilateral.

COROLLARY (5) The line joining the midpoints of two sides of a triangle has length less than or equal to one-half that of the third side.

3.8

THEOREM 1 *(Additivity of Arc Measure)* Suppose arcs $A_1 = \overset{\frown}{ADC}$ and $A_2 = \overset{\frown}{CEB}$ are any two arcs of circle O having just one point C in common, and such that their union, $A_1 \cup A_2 = \overset{\frown}{ACB}$, is also an arc. Then $m(A_1 \cup A_2) = mA_1 + mA_2$.

THEOREM 2 *(Tangent Theorem)* A line is tangent to a circle iff it is perpendicular to the radius at the point of contact.

COROLLARY   Two segments lying on tangents to a circle joining the points of contact to a common external point are congruent.

THEOREM 3 *(Secant Theorem)* If a line ℓ passes through an interior point A of a circle, it is a secant of the circle, intersecting the circle in precisely two points.

4.1

LEMMA *(Parallelism in Absolute Geometry)* If two lines in the same plane are cut by a transversal so that a pair of alternate interior angles are congruent, the lines are parallel.

THEOREM 1   If two parallel lines are cut by a transversal, then either pair of interior angles on the same side of the transversal are supplementary, and any pair of alternate interior angles are congruent.

COROLLARY A *(The Z Property)* If two lines in the same plane are cut by a transversal, the two lines are parallel iff a pair of alternate interior angles are congruent.

COROLLARY B *(The F Property)* If two lines in the same plane are cut by a transversal, the two lines are parallel iff a pair of corresponding angles are congruent.

COROLLARY C *(The U Property)* If two lines in the same plane are cut by a transversal, the two lines are parallel iff a pair of interior angles on the same side of the transversal are supplementary.

THEOREM 2   The sum of the measures of the angles of any triangle equals 180.

COROLLARY   The acute angles of a right triangle are complementary.

THEOREM 3 *(The Midpoint Connector Theorem)* The segment joining the midpoints of two sides of a triangle has length one-half that of the third side and is parallel to it.

COROLLARY   If a line bisects one side of a triangle and is parallel to the second, it also bisects the third side.

4.2

THEOREM 1   A diagonal of a parallelogram divides it into two congruent triangles.

COROLLARIES:

(1) A convex quadrilateral is a parallelogram iff opposite sides are congruent, or a pair of opposite sides are parallel and congruent.

(2) A parallelogram is a rectangle iff its diagonals are congruent; it is a square iff its diagonals are both perpendicular and congruent.

THEOREM 2   *(Midpoint-Connector Theorem for Trapezoids)* If a line segment bisects one leg of a trapezoid and is parallel to the base, then it is the median and its length is one-half the sum of the length of the two bases. Conversely, the median of a trapezoid is parallel to the bases.

LEMMA   If $\overleftrightarrow{EF}$ is not parallel to $\overleftrightarrow{BC}$, then $AE/AB \neq DF/DC$.

COROLLARY   If $AE/AB = DF/DC$, then $\overleftrightarrow{EF} \parallel \overleftrightarrow{BC}$.

COROLLARY   *(The Side-Splitting Theorem)* If a line parallel to the base $\overline{BC}$ of $\triangle ABC$ cuts the other two sides $\overline{AB}$ and $\overline{AC}$ at E and F, respectively, then $AE/AB = AF/FC$.

4.3

THEOREM 1   *(AA Similarity Criterion)* If under some correspondence two triangles have two pairs of corresponding angles congruent, the triangles are similar under that correspondence.

THEOREM 2   *(SAS Similarity Criterion)* If in $\triangle ABC$ and $\triangle XYZ$ we have $AB/XY = AC/XZ$ and $\angle A \cong \angle X$, then $\triangle ABC \sim \triangle XYZ$.

THEOREM 3   *(SSS Similarity Criterion)* If in $\triangle ABC$ and $\triangle XYZ$ we have $AB/XY = BC/YZ = AC/XZ$, then $\angle A \cong \angle X$, $\angle B \cong \angle Y$, $\angle C \cong \angle Z$ and $\triangle ABC \sim \triangle XYZ$.

PYTHAGOREAN THEOREM   If $\triangle ABC$ has a right angle at C and $a = BC$, $b = AC$, $c = AB$, then $a^2 + b^2 = c^2$.

4.4

THEOREM 2   The angle sum of a convex n-gon is $180(n - 2)$.

COROLLARY A   Each interior angle of a regular polygon has the measure $\phi = 180(n - 2)/n$

4.5

THEOREM 1   *(Inscribed Angle Theorem)* The measure of an inscribed angle of a circle equals one-half that of its intercepted arc.

COROLLARY   Any angle inscribed in a semicircle is a right angle.

THEOREM 2   *(Two-Chord Theorem)* When two chords of a circle intersect, the product of the lengths of the segments formed on one chord equals that on the other.

THEOREM 3   *(Secant-Tangent Theorem)* If a secant $\overrightarrow{PA}$ and tangent $\overrightarrow{PC}$ meet a circle at the respective points A, B, and C (point of contact), then $PC^2 = PA \cdot PB$.

COROLLARY   *(Two-Secant Theorem)* If two secants $\overrightarrow{PA}$ and $\overrightarrow{PC}$ of a circle meet the circle at A, B, and C, D, respectively, then $PA \cdot PB = PC \cdot PD$.

4.6

LAW OF COSINES FOR VECTORS $uv = \|u\| \, \|v\| \, \cos \theta$ $(\theta = m\angle(u, v))$

4.7

LEMMA A   *(Analytic Criterion for Orthogonality).* The circles $x^2 + y^2 + a_1 x + b_1 y + c_1 = 0$ and $x^2 + y^2 + a_2 x + b_2 y + c_2 = 0$ are orthogonal iff $a_1 a_2 + b_1 b_2 = 2c_1 + 2c_2$.

COSINE OF ANGLE BETWEEN TWO CIRCLES

$$\cos \theta = \frac{a_1 a_2 + b_1 b_2 - 2c_1 - 2c_2}{\sqrt{a_1^2 + b_1^2 - 4c_1} \, \sqrt{a_2^2 + b_2^2 - 4c_2}}$$

THEOREM 1   Given two orthogonal circles, one of them the unit circle C, there exist two one-parameter families of circles, $\mathcal{F}$ and $\mathcal{G}$ containing the two given circles as particular members (with $C \in \mathcal{G}$) such that each circle in $\mathcal{F}$ is orthogonal to each circle in $\mathcal{G}$.

THEOREM 2   Let D be a circle with center D and radius r, and suppose two points A and B are located on a ray from the center D such that the distances DA and DB satisfy the relation $DA \cdot DB = r^2$. Then every circle C' passing through A and B is orthogonal to the given circle D.

PROPERTIES OF CIRCULAR INVERSION:

(1) A circle orthogonal to the circle of inversion C is invariant.

(2) A circle maps to either another circle or a line. A circle maps to a line iff the given circle passes through O, the center of inversion, and the line is perpendicular to the line passing through O and the center of the given circle.

(3) A circular inversion is conformal (Theorem 3).

(4) A circular inversion preserves the generalized cross-ratio $(AB, CD) \equiv (AC/AD)(BD/BC)$ of any four distinct points A, B, C, and D in the plane.

4.8

THEOREM 1   *(Ceva's Theorem)* The cevians AD, BE, CF of $\triangle ABC$ are concurrent iff $AF/FB \cdot BD/DC \cdot CE/EA = 1$.

THEOREM 2   *(Menelaus' Theoerm).* If points D, E, and F lie on the sides of $\triangle ABC$ opposite A, B, and C, respectively, then D, E, and F are collinear iff $AF/FB \cdot BD/DC \cdot CE/EA = -1$.

5.3

THEOREM 1   Given any two congruent triangles $\triangle ABC$ and $\triangle PQR$, there exists a unique isometry which maps one triangle onto another.

THEOREM 2   *(Fundamental Theorem on Isometries)* Every isometry in the plane is the product of at most three reflections, exactly two if the isometry is direct and not the identity.

5.4

THEOREM 2   Given any two similar triangles $\triangle ABC$ and $\triangle A'B'C'$, there exists a unique similitude which maps the first triangle onto the second.

COROLLARY   Any similitude is the product of a dilation and at most three reflections.

5.7

THEOREM 6   Suppose that triangles $\triangle ABC$ and $\triangle XYZ$ satisfy the SAS Hypothesis under the correspondence $ABC \leftrightarrow XYZ$. then $\triangle ABC \cong \triangle XYZ$. (Proved from Axiom M.)

6.4

THEOREM 1   The sum of the measures of the angles of any right triangle is less than 180.

COROLLARY   The angle sum of any triangle is less than 180.

THEOREM 2   *(AAA Congruence Criterion for Hyperbolic Geometry)* If two triangles have the three angles of one congruent, respectively, to the three angles of the other, the triangles are congruent.

6.5

THEOREM 1   *(Symmetry Law for Asymptotic Parallelism)* In any asymptotic triangle $\triangle AB\Omega$, if $\overrightarrow{AC} \parallel_a \overrightarrow{BD}$, then $\overrightarrow{BD} \parallel_a \overrightarrow{AD}$, and each leg is asymptotically parallel to the other.

6.6

THEOREM 1   *(Transitive Law for Asymptotic Parallelism)* Let A, B, and C be any three collinear points. In terms of the preceding notation, if $\overrightarrow{AP} \parallel_a \overrightarrow{BQ}$ and $\overrightarrow{BD} \parallel_a \overrightarrow{CR}$, then $\overrightarrow{AP} \parallel_a \overrightarrow{CR}$.

THEOREM 3   *(Hilbert's Theorem on Parallels)* Two lines are divergently parallel iff they possess a common perpendicular.

6.8

THEOREM 3   The angle sum of any triangle exceeds 180.

INDEX